METHODS IN MOLECULAR BIOLOGY

For further volumes:
http://www.springer.com/series/7651

Plant Long Non-Coding RNAs

Methods and Protocols

Edited by

Julia A. Chekanova

Guangxi Key Laboratory of Sugarcane Biology, Guangxi University, Nanning, Guangxi, China

Hsiao-Lin V. Wang

Guangxi Key Laboratory of Sugarcane Biology, Guangxi University, Nanning, Guangxi, China

Department of Biology, Emory University, Atlanta, Georgia, USA

Humana Press

Editors
Julia A. Chekanova
Guangxi Key Laboratory
of Sugarcane Biology
Guangxi University
Nanning, Guangxi, China

Hsiao-Lin V. Wang
Guangxi Key Laboratory
of Sugarcane Biology
Guangxi University
Nanning, Guangxi, China

Department of Biology
Emory University
Atlanta, Georgia, USA

ISSN 1064-3745 ISSN 1940-6029 (electronic)
Methods in Molecular Biology
ISBN 978-1-4939-9044-3 ISBN 978-1-4939-9045-0 (eBook)
https://doi.org/10.1007/978-1-4939-9045-0

Library of Congress Control Number: 2019930835

This Humana Press imprint is published by the registered company Springer Science+Business Media, LLC, part of Springer Nature.
The registered company address is: 233 Spring Street, New York, NY 10013, U.S.A.

Dedication

We dedicate this book to the memory of Dmitry Belostotsky, whose pioneering discovery of pervasive transcription in plants opened so many important avenues of discovery, particularly on the importance of noncoding RNAs in nearly every aspect of plant genetics and molecular biology.

Preface

The surprising discovery of pervasive transcription in eukaryotic genomes created a race to understand the functions of all the RNAs produced from genomic "dark matter." The decade that passed since then coincided with the meteoric rise of modern genomics techniques and the discovery of thousands of noncoding RNAs (ncRNAs). This dramatically changed our conception of the functional regions of the genome and began to unveil the wide variety of molecular mechanisms by which ncRNAs act, as well as the systems governing the diverse processes that ncRNAs influence in eukaryotic cells, both in the nucleus and the cytoplasm. Thus, long noncoding RNAs (lncRNAs) have emerged as a globally important topic for gene regulation, development, and environmental responses, and the study of lncRNAs has launched myriad, diverse lines of research.

Our emerging understanding of lncRNAs has only begun to encompass the types and origins of lncRNAs and map the remarkable variety of their mechanisms of action. At the molecular level, lncRNAs regulate gene expression by many complex mechanisms that act at multiple levels. They function in *cis* or in *trans*, via sequence complementarity or RNA structure, and function alone or with proteins. Transcription of lncRNAs can also affect gene regulation; therefore, even lncRNAs that are promptly degraded by the cellular machinery can affect transcription or nuclear architecture. Indeed, recent discoveries have unveiled the important roles lncRNAs play in plants, such as regulation of flowering, the responses to biotic and abiotic stresses, gene silencing, root organogenesis, seedling photomorphogenesis, reproduction, and others. However, although we have explored the roles of many lncRNAs, efforts to elucidate the functions of most lncRNAs remain preliminary.

To support these efforts, this volume of *Methods in Molecular Biology* focuses on approaches to study lncRNAs, including methods for finding lncRNAs, determining their localization, and analyzing their functions. We also collected readily usable approaches to understand the role of lncRNAs in plants. For those who are interested in researching the lncRNAs already identified in various plant species, this book includes chapters on how to use various databases that have nicely cataloged various lncRNAs in different plant species. We also provide resources for studying lncRNAs, including such subjects as determining their subcellular localization, protein interactions, structures, and RNA modifications. We further give substantial attention to quantitative bioinformatic approaches for lncRNA analysis.

In Chapter 1, we detail the topics covered in the book and provide a curated overview of high-throughput methodologies that can be applied to study lncRNAs. The chapter aims to lead the reader to information on the technologies that we could not include in this book or that have yet to be used to study lncRNAs in plants. Thus far, research in this field has only touched the surface of this new area of ncRNA biology. As technology and methodologies evolve and more genomes of non-model plants are sequenced, we will continue to gain a deeper understanding of the functional consequences of pervasive transcription and the resulting ncRNAs in different plant species, and this will doubtless enable us to explore currently uncharted aspects of plant biology.

Julia A. Chekanova
Hsiao-Lin V. Wang

Contents

Dedication . . . . *v*
Preface . . . . *vii*
Contributors . . . . *xiii*

1 An Overview of Methodologies in Studying lncRNAs in the High-Throughput Era: When Acronyms ATTACK! 1
Hsiao-Lin V. Wang and Julia A. Chekanova

PART I ROLE OF NCRNAS IN REGULATION OF GENE EXPRESSION AND DEVELOPMENT IN PLANTS

2 Analysis of siRNA Precursors Generated by RNA Polymerase IV and RNA-Dependent RNA Polymerase 2 in Arabidopsis 33
Todd Blevins, Ram Podicheti, and Craig S. Pikaard

3 Identification of Long Noncoding RNAs in the Developing Endosperm of Maize 49
Eundeok Kim, Yuqing Xiong, Byung-Ho Kang, and Sibum Sung

4 RNA Isolation and Analysis of LncRNAs from Gametophytes of Maize 67
Linqian Han, Lin Li, Gary J. Muehlbauer, John E. Fowler, and Matthew M. S. Evans

PART II STUDYING THE TISSUE AND CELL-TYPE SPECIFIC LNCRNAS

5 Improved Method of RNA Isolation from Laser Capture Microdissection (LCM)-Derived Plant Tissues 89
Vibhav Gautam, Archita Singh, Sharmila Singh, Swati Verma, and Ananda K. Sarkar

6 Medium-Throughput RNA In Situ Hybridization of Serial Sections from Paraffin-Embedded Tissue Microarrays 99
Edith Francoz, Philippe Ranocha, Christophe Dunand, and Vincent Burlat

7 Purification and Functional Analysis of Plant Long Noncoding RNAs (lncRNA) 131
Trung Do, Zhipeng Qu, and Iain Searle

PART III STUDYING LNCRNAS ASSOCIATED WITH RESPONSE TO ABIOTIC AND BIOTIC STRESSES IN PLANTS

8 The Involvement of Long Noncoding RNAs in Response to Plant Stress 151
Akihiro Matsui and Motoaki Seki

9 Subcellular Localization and Functions of Plant lncRNAs in Drought and Salt Stress Tolerance 173
Tao Qin and Liming Xiong

10 Discovery, Identification, and Functional Characterization of Plant Long Intergenic Noncoding RNAs After Virus Infection 187
Ruimin Gao, Peng Liu, Nadia Irwanto, De Rong Loh and Sek-Man Wong

PART IV IDENTIFICATION AND FUNCTIONAL ANALYSIS OF LNCRNAS

11 Bioinformatics Approaches to Studying Plant Long Noncoding RNAs (lncRNAs): Identification and Functional Interpretation of lncRNAs from RNA-Seq Data Sets 197
Hai-Xi Sun and Nam-Hai Chua

12 Identification of Novel lincRNA and Co-Expression Network Analysis Using RNA-Sequencing Data in Plants 207
Song Qi, Shamima Akter, and Song Li

13 An Easy-to-Follow Pipeline for Long Noncoding RNA Identification: A Case Study in Diploid Strawberry *Fragaria vesca* 223
Chunying Kang and Zhongchi Liu

14 Reference-Based Identification of Long Noncoding RNAs in Plants with Strand-Specific RNA-Sequencing Data 245
Xiao Lin, Meng Ni, Zhixia Xiao, Ting-Fung Chan, and Hon-Ming Lam

15 NAMS: Noncoding Assessment of long RNAs in Magnoliophyta Species 257
Gaurav Sablok, Kun Sun, and Hao Sun

16 De Novo Plant Transcriptome Assembly and Annotation Using Illumina RNA-Seq Reads 265
Stephanie C. Kerr, Federico Gaiti, and Milos Tanurdzic

PART V PURIFICATION OF LNCRNAS AND RIBONUCLEOPROTEIN COMPLEXES, RNA SECONDARY STRUCTURE, AND RNA MODIFICATIONS

17 Identification of Long Noncoding RNA-Protein Interactions Through In Vitro RNA Pull-Down Assay with Plant Nuclear Extracts 279
Jun Sung Seo and Nam-Hai Chua

18 Analysis of Interaction Between Long Noncoding RNAs and Protein by RNA Immunoprecipitation in *Arabidopsis* 289
Jun Sung Seo and Nam-Hai Chua

19 Trimolecular Fluorescence Complementation (TriFC) Assay for Visualization of RNA-Protein Interaction in Plants 297
Jun Sung Seo and Nam-Hai Chua

20 In Vivo Genome-Wide RNA Structure Probing with Structure-seq 305
Laura E. Ritchey, Zhao Su, Sarah M. Assmann, and Philip C. Bevilacqua

21 Using Protein Interaction Profile Sequencing (PIP-seq) to Identify RNA Secondary Structure and RNA–Protein Interaction Sites of Long Noncoding RNAs in Plants 343
Marianne C. Kramer and Brian D. Gregory

22 Computationally Characterizing Protein-Bound Long Noncoding RNAs and Their Secondary Structure Using Protein Interaction Profile Sequencing (PIP-Seq) in Plants 363
Mengge Shan, Zachary D. Anderson, and Brian D. Gregory

23 Stalking Structure in Plant Long Noncoding RNAs 381
Karissa Y. Sanbonmatsu

24 Transcriptome-Wide Mapping 5-Methylcytosine by m^5C RNA Immunoprecipitation Followed by Deep Sequencing in Plant 389
Xiaofeng Gu and Zhe Liang

PART VI DATABASES OF PLANT LNCRNAS AND HOW TO USE THEM

25 A Walkthrough to the Use of GreeNC: The Plant lncRNA Database 397
Andreu Paytuvi-Gallart, Walter Sanseverino, and Riccardo Aiese Cigliano

26 CANTATAdb 2.0: Expanding the Collection of Plant Long Noncoding RNAs..... 415
Michał Wojciech Szcześniak, Oleksii Bryzghalov, Joanna Ciomborowska-Basheer, and Izabela Makałowska

27 Experimentally Validated Plant lncRNAs in EVLncRNAs Database 431
Bailing Zhou, Huiying Zhao, Jiafeng Yu, Chengang Guo, Xianghua Dou, Feng Song, Guodong Hu, Zanxia Cao, Yuanxu Qu, Yuedong Yang, Yaoqi Zhou, and Jihua Wang

PART VII LONG-RANGE CHROMATIN INTERACTIONS AND CHROMOSOME CONFORMATION CAPTURE

28 In Situ Hi-C for Plants: An Improved Method to Detect Long-Range Chromatin Interactions..... 441
Sudharsan Padmarasu, Axel Himmelbach, Martin Mascher, and Nils Stein

Index *473*

Contributors

RICCARDO AIESE CIGLIANO • *Sequentia Biotech SL, Carrer Comte d'Urgell 240, Barcelona, Spain*

SHAMIMA AKTER • *Department of Crop & Soil Environmental Sciences, Virginia Polytechnic Institute and State University, Blacksburg, VA, USA*

ZACHARY D. ANDERSON • *Department of Biology, University of Pennsylvania, Philadelphia, PA, USA*

SARAH M. ASSMANN • *Center for RNA Molecular Biology, Pennsylvania State University, University Park, PA, USA; Department of Biology, Pennsylvania State University, University Park, PA, USA*

PHILIP C. BEVILACQUA • *Department of Chemistry, Pennsylvania State University, University Park, PA, USA; Center for RNA Molecular Biology, Pennsylvania State University, University Park, PA, USA; Department of Biochemistry and Molecular Biology, Pennsylvania State University, University Park, PA, USA*

TODD BLEVINS • *Howard Hughes Medical Institute, Indiana University, Bloomington, IN, USA; Department of Biology, Indiana University, Bloomington, IN, USA; Department of Molecular and Cellular Biochemistry, Indiana University, Bloomington, IN, USA; Institut de Biologie Moléculaire des Plantes, CNRS UPR2357, Université de Strasbourg, Strasbourg, France*

OLEKSII BRYZGHALOV • *Laboratory of Integrative Genomics, Institute of Anthropology, Adam Mickiewicz University, Poznan, Poland*

VINCENT BURLAT • *Laboratoire de Recherche en Sciences Végétales, Université de Toulouse, CNRS, UPS, Castanet Tolosan, France*

ZANXIA CAO • *Shandong Provincial Key Laboratory of Biophysics, Institute of Biophysics, Dezhou University, Dezhou, China; College of Physics and Electronic Information, Dezhou University, Dezhou, China*

TING-FUNG CHAN • *School of Life Sciences and Center for Soybean Research of the Partner State Key Laboratory of Agrobiotechnology, The Chinese University of Hong Kong, Shatin, Hong Kong, China*

JULIA A. CHEKANOVA • *Guangxi Key Laboratory of Sugarcane Biology, Guangxi University, Nanning, Guangxi, China*

NAM-HAI CHUA • *Laboratory of Plant Molecular Biology, Rockefeller University, New York, NY, USA; TEMASEK Life Sciences Laboratory, National University of Singapore, Singapore, Singapore*

JOANNA CIOMBOROWSKA-BASHEER • *Laboratory of Integrative Genomics, Institute of Anthropology, Adam Mickiewicz University, Poznan, Poland*

TRUNG DO • *Department of Molecular and Biomedical Sciences, School of Biological Sciences, The University of Adelaide, Adelaide, SA, Australia*

XIANGHUA DOU • *Shandong Provincial Key Laboratory of Biophysics, Institute of Biophysics, Dezhou University, Dezhou, China; College of Physics and Electronic Information, Dezhou University, Dezhou, China*

CHRISTOPHE DUNAND • *Laboratoire de Recherche en Sciences Végétales, Université de Toulouse, CNRS, UPS, Castanet Tolosan, France*

Matthew M. S. Evans • *Department of Plant Biology, Carnegie Institution for Science, Stanford, CA, USA*

John E. Fowler • *Department of Botany and Plant Pathology, Oregon State University, Corvallis, OR, USA*

Edith Francoz • *Laboratoire de Recherche en Sciences Végétales, Université de Toulouse, CNRS, UPS, Castanet Tolosan, France*

Federico Gaiti • *New York Genome Center and Department of Medicine, Weill Cornell Medicine, New York, NY, USA*

Ruimin Gao • *Department of Biological Sciences, National University of Singapore, Singapore, Singapore*

Vibhav Gautam • *National Institute of Plant Genome Research (NIPGR), New Delhi, India*

Brian D. Gregory • *Department of Biology, University of Pennsylvania, Philadelphia, PA, USA; Genomics and Computational Biology Graduate Group, University of Pennsylvania, Philadelphia, PA, USA*

Xiaofeng Gu • *Biotechnology Research Institute, Chinese Academy of Agricultural Sciences, Beijing, China*

Chengang Guo • *Shandong Provincial Key Laboratory of Biophysics, Institute of Biophysics, Dezhou University, Dezhou, China; College of Physics and Electronic Information, Dezhou University, Dezhou, China*

Linqian Han • *College of Plant Sciences and Technology, Huazhong Agricultural University, Wuhan, China*

Axel Himmelbach • *Leibniz Institute of Plant Genetics and Crop Plant Research (IPK), Seeland, Germany*

Guodong Hu • *Shandong Provincial Key Laboratory of Biophysics, Institute of Biophysics, Dezhou University, Dezhou, China; College of Physics and Electronic Information, Dezhou University, Dezhou, China*

Nadia Irwanto • *NUS High School of Mathematics and Science, Singapore, Singapore*

Byung-Ho Kang • *Microbiology and Cell Science, University of Florida, Gainesville, FL, USA; School of Life Sciences, State Key Laboratory for Agrobiotechnology, The Chinese University of Hong Kong, Shatin, Hong Kong, China*

Chunying Kang • *Key Laboratory of Horticultural Plant Biology (Ministry of Education), College of Horticulture and Forestry Sciences, Huazhong Agricultural University, Wuhan, China*

Stephanie C. Kerr • *School of Biological Sciences, The University of Queensland, St Lucia, QLD, Australia*

Eundeok Kim • *Department of Molecular Biosciences and Institute for Cellular and Molecular Biology, The University of Texas at Austin, Austin, TX, USA; Department of Biology, University of Washington, Seattle, WA, USA*

Marianne C. Kramer • *Department of Biology, University of Pennsylvania, Philadelphia, PA, USA; Cell and Molecular Biology Graduate Group, University of Pennsylvania, Philadelphia, PA, USA*

Hon-Ming Lam • *School of Life Sciences and Center for Soybean Research of the Partner State Key Laboratory of Agrobiotechnology, The Chinese University of Hong Kong, Shatin, Hong Kong, China*

Lin Li • *College of Plant Sciences and Technology, Huazhong Agricultural University, Wuhan, China*

SONG LI • *Department of Crop & Soil Environmental Sciences, Virginia Polytechnic Institute and State University, Blacksburg, VA, USA*
ZHE LIANG • *Department of Biological Sciences, National University of Singapore, Singapore, Singapore*
XIAO LIN • *School of Life Sciences and Center for Soybean Research of the Partner State Key Laboratory of Agrobiotechnology, The Chinese University of Hong Kong, Shatin, Hong Kong, China*
PENG LIU • *Department of Biological Sciences, National University of Singapore, Singapore, Singapore*
ZHONGCHI LIU • *Key Laboratory of Horticultural Plant Biology (Ministry of Education), College of Horticulture and Forestry Sciences, Huazhong Agricultural University, Wuhan, China; Department of Cell Biology and Molecular Genetics, University of Maryland, College Park, MD, USA*
DE RONG LOH • *NUS High School of Mathematics and Science, Singapore, Singapore*
IZABELA MAKAŁOWSKA • *Laboratory of Integrative Genomics, Institute of Anthropology, Adam Mickiewicz University, Poznan, Poland*
MARTIN MASCHER • *Leibniz Institute of Plant Genetics and Crop Plant Research (IPK), Seeland, Germany*
AKIHIRO MATSUI • *Plant Genomic Network Research Team, RIKEN Center for Sustainable Resource Science, Yokohama, Kanagawa, Japan; Plant Epigenome Regulation Laboratory, RIKEN Cluster for Pioneering Research, Wako, Saitama, Japan*
GARY J. MUEHLBAUER • *Department of Agronomy and Plant Genetics, University of Minnesota, Saint Paul, MN, USA; Department of Plant and Microbial Biology, University of Minnesota, Saint Paul, MN, USA*
MENG NI • *School of Life Sciences and Center for Soybean Research of the Partner State Key Laboratory of Agrobiotechnology, The Chinese University of Hong Kong, Shatin, Hong Kong, China*
SUDHARSAN PADMARASU • *Leibniz Institute of Plant Genetics and Crop Plant Research (IPK), Seeland, Germany*
ANDREU PAYTUVI-GALLART • *Sequentia Biotech SL, Carrer Comte d'Urgell 240, Barcelona, Spain*
CRAIG S. PIKAARD • *Howard Hughes Medical Institute, Indiana University, Bloomington, IN, USA; Department of Biology, Indiana University, Bloomington, IN, USA; Department of Molecular and Cellular Biochemistry, Indiana University, Bloomington, IN, USA*
RAM PODICHETI • *Center for Genomics and Bioinformatics, Indiana University, Bloomington, IN, USA; School of Informatics, Computing and Engineering, Indiana University, Bloomington, IN, USA*
SONG QI • *Ph.D. Program in Genetics, Bioinformatics and Computational Biology, Virginia Polytechnic Institute and State University, Blacksburg, VA, USA; Department of Crop & Soil Environmental Sciences, Virginia Polytechnic Institute and State University, Blacksburg, VA, USA*
TAO QIN • *Texas A&M AgriLife Research Center, Dallas, TX, USA*
ZHIPENG QU • *Department of Molecular and Biomedical Sciences, School of Biological Sciences, The University of Adelaide, Adelaide, SA, Australia*
YUANXU QU • *Department of Surgery Beijing Tiantan Hospital, Capital Medical University, Beijing, China*

PHILIPPE RANOCHA • *Laboratoire de Recherche en Sciences Végétales, Université de Toulouse, CNRS, UPS, Castanet Tolosan, France*

LAURA E. RITCHEY • *Department of Chemistry, Pennsylvania State University, University Park, PA, USA; Center for RNA Molecular Biology, Pennsylvania State University, University Park, PA, USA; Department of Chemistry, University of Pittsburgh at Johnstown, Johnstown, PA, USA*

GAURAV SABLOK • *Department of Biodiversity and Molecular Ecology, Research and Innovation Centre, Trento, Italy*

KARISSA Y. SANBONMATSU • *Theoretical Biology and Biophysics Group, Theoretical Division, Los Alamos National Laboratory, Los Alamos, NM, USA*

WALTER SANSEVERINO • *Sequentia Biotech SL, Carrer Comte d'Urgell 240, Barcelona, Spain*

ANANDA K. SARKAR • *National Institute of Plant Genome Research (NIPGR), New Delhi, India*

IAIN SEARLE • *Department of Molecular and Biomedical Sciences, School of Biological Sciences, The University of Adelaide, Adelaide, SA, Australia*

MOTOAKI SEKI • *Plant Genomic Network Research Team, RIKEN Center for Sustainable Resource Science, Yokohama, Kanagawa, Japan; Plant Epigenome Regulation Laboratory, RIKEN Cluster for Pioneering Research, Wako, Saitama, Japan; Kihara Institute for Biological Research, Yokohama City University, Yokohama, Kanagawa, Japan; Core Research for Evolutional Science and Technology, Japan Science and Technology, Kawaguchi, Saitama, Japan*

JUN SUNG SEO • *Laboratory of Plant Molecular Biology, Rockefeller University, New York, NY, USA; TEMASEK Life Sciences Laboratory, Singapore, Singapore*

MENGGE SHAN • *Department of Biology, University of Pennsylvania, Philadelphia, PA, USA; Genomics and Computational Biology Graduate Group, University of Pennsylvania, Philadelphia, PA, USA*

ARCHITA SINGH • *National Institute of Plant Genome Research (NIPGR), New Delhi, India*

SHARMILA SINGH • *National Institute of Plant Genome Research (NIPGR), New Delhi, India*

FENG SONG • *Shandong Provincial Key Laboratory of Biophysics, Institute of Biophysics, Dezhou University, Dezhou, China; College of Physics and Electronic Information, Dezhou University, Dezhou, China*

NILS STEIN • *Leibniz Institute of Plant Genetics and Crop Plant Research (IPK), Seeland, Germany*

ZHAO SU • *Department of Biology, Pennsylvania State University, University Park, PA, USA*

HAI-XI SUN • *Laboratory of Plant Molecular Biology, Rockefeller University, New York, NY, USA*

KUN SUN • *Department of Chemical Pathology, The Chinese University of Hong Kong, Shatin, Hong Kong, China; Li Ka Shing Institute of Health Sciences, The Chinese University of Hong Kong, Shatin, Hong Kong, China*

HAO SUN • *Department of Chemical Pathology, The Chinese University of Hong Kong, Shatin, Hong Kong, China; Li Ka Shing Institute of Health Sciences, The Chinese University of Hong Kong, Shatin, Hong Kong, China*

SIBUM SUNG • *Department of Molecular Biosciences and Institute for Cellular and Molecular Biology, The University of Texas at Austin, Austin, TX, USA; International Scholar, Kyung-Hee University, Suwon, South Korea*

MICHAŁ WOJCIECH SZCZEŚNIAK • *Laboratory of Integrative Genomics, Institute of Anthropology, Adam Mickiewicz University, Poznan, Poland*

MILOS TANURDZIC • *School of Biological Sciences, The University of Queensland, St Lucia, QLD, Australia*
SWATI VERMA • *National Institute of Plant Genome Research (NIPGR), New Delhi, India*
HSIAO-LIN V. WANG • *Guangxi Key Laboratory of Sugarcane Biology, Guangxi University, Nanning, Guangxi, China*
Present address: Department of Biology, Emory University, Atlanta, GA, USA
JIHUA WANG • *Shandong Provincial Key Laboratory of Biophysics, Institute of Biophysics, Dezhou University, Dezhou, China; College of Physics and Electronic Information, Dezhou University, Dezhou, China*
SEK-MAN WONG • *Department of Biological Sciences, National University of Singapore, Singapore, Singapore; Temasek Life Sciences Laboratory, Singapore, Singapore; National University of Singapore Suzhou Research Institute, Suzhou, Jiangsu, China*
ZHIXIA XIAO • *School of Life Sciences and Center for Soybean Research of the Partner State Key Laboratory of Agrobiotechnology, The Chinese University of Hong Kong, Shatin, Hong Kong, China*
YUQING XIONG • *Microbiology and Cell Science, University of Florida, Gainesville, FL, USA*
LIMING XIONG • *Texas A&M AgriLife Research Center, Dallas, TX, USA; Department of Biology, Hong Kong Baptist University, Kowloon Tong, Hong Kong*
YUEDONG YANG • *School of Data and Computer Science, Sun Yat-sen University, Guangzhou, China*
JIAFENG YU • *Shandong Provincial Key Laboratory of Biophysics, Institute of Biophysics, Dezhou University, Dezhou, China; College of Physics and Electronic Information, Dezhou University, Dezhou, China*
HUIYING ZHAO • *Sun Yat-sen Memorial Hospital, Sun Yat-sen University, Guangzhou, China*
BAILING ZHOU • *Shandong Provincial Key Laboratory of Biophysics, Institute of Biophysics, Dezhou University, Dezhou, China; College of Physics and Electronic Information, Dezhou University, Dezhou, China*
YAOQI ZHOU • *Shandong Provincial Key Laboratory of Biophysics, Institute of Biophysics, Dezhou University, Dezhou, China; Institute for Glycomics and School of Information and Communication Technology, Griffith University, Gold Coast, QLD, Australia*

Chapter 1

An Overview of Methodologies in Studying lncRNAs in the High-Throughput Era: When Acronyms ATTACK!

Hsiao-Lin V. Wang and Julia A. Chekanova

Abstract

The discovery of pervasive transcription in eukaryotic genomes provided one of many surprising (and perhaps most surprising) findings of the genomic era and led to the uncovering of a large number of previously unstudied transcriptional events. This pervasive transcription leads to the production of large numbers of noncoding RNAs (ncRNAs) and thus opened the window to study these diverse, abundant transcripts of unclear relevance and unknown function. Since that discovery, recent advances in high-throughput sequencing technologies have identified a large collection of ncRNAs, from microRNAs to long noncoding RNAs (lncRNAs). Subsequent discoveries have shown that many lncRNAs play important roles in various eukaryotic processes; these discoveries have profoundly altered our understanding of the regulation of eukaryotic gene expression. Although the identification of ncRNAs has become a standard experimental approach, the functional characterization of these diverse ncRNAs remains a major challenge. In this chapter, we highlight recent progress in the methods to identify lncRNAs and the techniques to study the molecular function of these lncRNAs and the application of these techniques to the study of plant lncRNAs.

Key words High-throughput methods, RNA methods, Noncoding RNAs, lncRNAs, Plant lncRNAs, RNA secondary structures, RNA interactions

1 Introduction

Recent studies using high-throughput technologies have identified increasing numbers of lncRNAs in various eukaryotic transcriptomes. Functional studies have shown that some of these lncRNAs have diverse and important functions [1–3] in gene silencing and imprinting, transcription, mRNA splicing, translation, trafficking of nuclear factors, genome rearrangements, and regulation of chromatin modifications. In plants, lncRNAs are involved in the regulation of flowering, root development, plant immunity, responses to biotic and abiotic stresses, and many other important biological processes [1, 2].

Julia A. Chekanova and Hsiao-Lin V. Wang (eds.), *Plant Long Non-Coding RNAs: Methods and Protocols*,
Methods in Molecular Biology, vol. 1933, https://doi.org/10.1007/978-1-4939-9045-0_1,

The detection of lncRNA transcripts has become easier, particularly due to the recent development and improvement of high-throughput sequencing technologies, but the functions of most lncRNA remain largely unknown. Therefore, how to decipher the functions of lncRNAs has become an important topic in genome research. In this chapter, we provide an overview of methods for the identification and functional characterization of lncRNAs and focus on how these techniques could be adapted to study plant lncRNAs. Following this Introduction (Subheading 1), this chapter has two parts: one focuses on the identification of lncRNAs (Subheading 2), and the other focuses on analysis of the biological and molecular functions of lncRNAs (Subheading 3). In Subheading 2, we focus on methods to identify different populations of lncRNAs, revolving around different adaptations of RNA-seq methodology. We include some tag-based methods that were developed before the next-generation sequencing (NGS) era and present them through a historical lens. The details of how RNA-seq and high-throughput sequencing can be applied in a wide range of applications, including lncRNAs, have been extensively summarized by Wang and Snyder [4] and Reuter et al. [5]. As the functions of most lncRNAs largely remain to be elucidated, in Subheading 3 we present several selected methods to address various aspects of lncRNA functions in a high-throughput manner. The functional aspects addressed include tissue or cell type-specific analysis and examination of RNA-protein interactions, RNA-DNA/chromatin interactions, RNA-RNA interactions, RNA secondary structures, as well as RNA modifications. In addition to providing an overview of selected methods that are available to study lncRNAs, we detail the topics covered in this volume of *Methods in Molecular Biology, Plant Long Non-Coding RNAs: Methods and Protocols*. Another aim of this chapter is to give the reader links to information on specific technologies that were beyond the scope of this book or that have not yet been used in plants.

With the massive amount of data generated in each high-throughput sequencing experiment, data analysis has become a crucially important subject. Therefore, several chapters in this book provide step-by-step protocols for analyzing the large-scale sequencing data produced in the high-throughput experiments. However, a general summary of data analysis is a massive topic that requires separate, dedicated reviews and is thus outside of the scope of this chapter; therefore, we direct the reader to additional reviews and resources.

RNA biology is a vast, fast-moving field with myriad methods to study RNAs that are being improved and spawning variants as fast as discoveries can be disseminated to the research community; therefore this chapter only scratches the surface of the methodologies available to study lncRNAs.

2 Identification of lncRNAs Using High-Throughput Methodologies

The identification of lncRNAs is the first step in elucidating the role of lncRNAs in plants, and in recent years, most studies on plant ncRNAs have focused on identification of plant lncRNAs [1, 2]. Many of these lncRNAs are curated in various lncRNA databases, which we summarize in Table 1. The easy-to-follow protocols for how to use the three plant lncRNA databases, GreeNC [9], CANTATAdb 2.0 [11], and EVLncRNAs [13], are described in Part VI of this book, Chapters 25–27, respectively. lncRNAs are largely tissue specific and typically have a relatively low expression level; therefore, choosing the appropriate experimental techniques to identify and study lncRNAs is extremely important. In addition to the identification of novel lncRNAs, the methods described below can be used to examine the expression levels of known lncRNAs.

Before the high-throughput era, RNAs were traditionally detected using Northern blotting analysis, nuclease protection assays (NPA), in situ hybridization, reverse transcription-polymerase chain reaction (RT-PCR), etc. Although most of these methods are not used in the initial steps of identifying lncRNAs genome-wide, they are often employed to validate the expression of lncRNAs and to examine them in the context of specific molecular functions. In this section of the chapter, we first provide an overview of selected high-throughput methods for identifying lncRNAs; these are summarized in Table 2 and described in Subheadings 2.1–2.6. We include widely used high-throughput sequencing techniques (Subheadings 2.3–2.6), hybridization-based approaches, and tag-based methods, which were developed before the high-throughput era (Subheadings 2.1 and 2.2, respectively).

2.1 Hybridization-Based Approaches

Before the emergence of high-throughput sequencing technologies, hybridization-based methods, including custom-designed and high-density oligo microarrays and genomic tiling microarrays, have been developed to analyze the transcriptome quantitatively [15–20]. In these approaches, cDNAs produced from a population of RNAs are hybridized to microarrays of tiled oligonucleotides that cover the non-repetitive sequences of the target genome at a very high resolution. Since the cDNAs and tiled oligonucleotides are labeled with different fluorophores, the relative abundance of RNAs can be inferred from the differences in fluorescent signal produced upon hybridization. For example, in *Arabidopsis thaliana*, the Affymetrix ATH 1.0F arrays and 100 ATH 1.0R arrays have been used to determine the transcriptional activity of the *Arabidopsis* genome and identify ncRNAs [20–25, 70].

Although hybridization-based approaches are relatively inexpensive and high-throughput, they have several limitations [4]. These limitations include reliance on the coverage and density

Table 1
A selected list of plant lncRNA databases

Name	Descriptions/features	Link	Refs
The Arabidopsis Information Resource (TAIR)	TAIR and its latest version (TAIR10) offer a comprehensive database of the *Arabidopsis thaliana* genome. In addition to gene structures and transcriptome data for coding and nonprotein-coding loci, it offers many analytical tools	https://www.arabidopsis.org/	[6]
Araport11	Similar to TAIR, Araport11 curates comprehensive genomic information of *Arabidopsis thaliana* using the ecotype Col-0 version 11 genome	https://www.araport.org/	[7]
Plant long noncoding RNA database (PLncDB)	This is a plant-specific database that has >13,000 lncRNAs. The organ-specific expression and the differential expression in RNA-directed DNA methylation mutants of the curated lncRNAs are provided. A genome browser allows the user to examine the association between lincRNAs and epigenetic markers	http://chualab.rockefeller.edu/gbrowse2/homepage.html	[8]
Green Non-coding Database (GreeNC)[a]	Including lncRNAs from 37 plant species and algae, this database curates a total of >120,000 annotated lncRNAs and offers the coding potential and folding energy for each curated lncRNA	http://greenc.sciencedesigners.com/wiki/Main_Page	[9]
NONCODE v4.0	This database mainly curated noncoding RNAs from metazoans. Although *Arabidopsis* is the only plant species considered in this database, it has >500,000 lncRNAs from all 17 species and > 3500 *Arabidopsis* lncRNAs	http://www.noncode.org/index.php	[10]
CANTATAdb[a]	This database has >45,000 plant lncRNAs from 10 plant species. Each lncRNA is also evaluated based on potential roles in splicing regulation and miRNA modulations, as well as their tissue-specific expressions and coding potential	http://cantata.amu.edu.pl/	[11]
Plant ncRNA database (PNRD)	Includes lncRNAs from 150 plant species, this database curated a total of >25,000 ncRNAs of 11 types. In addition, this database offers several analytical tools and includes a customized genome browser, a coding potential calculator, and miRNA predictor	http://structuralbiology.cau.edu.cn/PNRD/	[12]
EVLncRNAs[a]	This database has lncRNAs that are validated by experiments and also integrated information from other databases. The database currently has >1500 lncRNAs	http://biophy.dzu.edu.cn/EVLncRNAs/	[13]

(continued)

Table 1 (continued)

Name	Descriptions/features	Link	Refs
	from >75 species, including animals, plants, and microbes		
Plant Natural Antisense Transcripts DataBase (PlantNATsDB)	This database has the natural antisense transcripts (NATs) of 70 plant species. Other information is housed in the database, including associated gene information, small RNA expression, and GO annotation	http://bis.zju.edu.cn/pnatdb/	[14]

[a]The descriptions and protocols for how to use them are included in the same book

of probes, sufficient knowledge of genome sequence and gene annotations, high background noise due to cross-hybridization, etc. Many of these limitations have made microarray analysis unsuitable for non-model plant species. However, despite these limitations, microarrays with probes representing already identified lncRNAs are now widely used to detect lncRNA expression with high sensitivity in many organisms, including plants. For example, Liu et al. used a custom microarray with 60-mer oligonucleotide probes for *Arabidopsis thaliana* long intergenic ncRNAs (lincRNAs; ATH lincRNAv1 array) to verify the expression of identified lincRNAs and to facilitate detection of lncRNA in different tissues, in response to biotic stresses, and in various mutants [70].

2.2 Sequence Tag-Based Approaches

Other large-scale methodologies for quantitatively analyzing expression of RNAs involve the production of very short sequence tags from the cDNAs derived from a given RNA sample. These short sequence tags are then sequenced using various platforms. The abundance of individual sequence tags corresponding to specific transcripts determines the relative abundance of each transcript. Unlike microarray probes, which must be preselected from known sequences, sequence tags are discovered by random sequencing; therefore, this approach allows researchers to find novel RNA sequences. For example, expressed sequence tags (ESTs) are a collection of short subsequences derived from pools of cDNAs. ESTs can be used to examine gene expression [71], but EST-based approaches are low throughput, costly, and nonquantitative.

Other tag-based methods have overcome these limitations [72]. These new methods include serial analysis of gene expression (SAGE) [26], cap analysis of gene expression (CAGE) [36, 37, 73], and massively parallel signature sequencing (MPSS) [31, 74, 75], which are described in Subheadings 2.2.1–2.2.3. These high-

Table 2
Selected high-throughput methods for identifying lncRNAs

Method	Purpose	Reference
Tiling microarray	Transcripts and transcriptome analysis	[15, 16]; plant references [17–20]; plant lncRNAs references [20–25]
Serial analysis of gene expression (SAGE)	Transcripts and transcriptome analysis	[26]; plant references [27–30]
Massively parallel signature sequencing (MPSS)	Transcripts and transcriptome analysis	[31]; plant references [32–35])
Cap analysis of gene expression (CAGE, CAGE-seq)	Identify transcripts with 5′ caps	[36, 37]; plant references [38–40]; protocol [41, 42]
RNA-seq	Transcripts and transcriptome analysis	First report [43] and comprehensive review by Wang et al. [4]; in this book: RNA-seq protocols (Part IV, Chapters 11–16), biotic and abiotic stress-related protocol (Part III, Chapters 8–10)
Parallel analysis of RNA ends (PARE)/genome-wide mapping of uncapped and cleaved transcripts (GMUCT)/degradome-seq	Identify transcripts that are being degraded and/or microRNA targets	PARE: [44], protocol [45]; GMUCT: [46], protocol [47]; degradome sequencing [48], protocol [49]
Transcript isoform sequencing (TIF-seq)	High-throughput identification of transcript isoforms with 5′ caps and poly(A) tails	[50]; protocol [51]
Global run-on sequencing (GRO-seq)	Identify the binding sites of transcriptionally active RNA Pol II and nascent RNAs	[52]; plant references [53, 54]; protocol [55, 56]
Precision nuclear run-on sequencing (PRO-seq)	Identify the binding sites of transcriptionally active RNA Pol II and nascent RNAs	[57]; protocol [58]
Native elongating transcript sequencing (NET-seq)	Identify the binding sites of elongating RNA Pol II and the associated nascent RNAs	[59–61]; protocols [62–64]
BRIC-Seq/BrU-Seq/BrUChase-Seq	Identify nascent RNAs and RNA half-lives, and measure RNA decay	BRIC-seq:[65], protocol [66, 67]; BrU-seq/BrUChase-seq: [68], protocol [69]

throughput tag-based approaches provide precise transcript levels. However, many of the short tags do not map uniquely to the reference genome. Moreover, these methods analyze only a small segment of each transcript and cannot distinguish transcript isoforms, which limits their use in studying the dynamic structures of many transcripts.

2.2.1 Serial Analysis of Gene Expression (SAGE)

SAGE was one of the first tag-based methods for high-throughput analysis of transcriptomes [26]. SAGE uses short sequence tags of cDNAs made from all the polyadenylated RNAs in a given sample. Each RNA is first converted into biotinylated cDNAs, which are captured on streptavidin beads. A few rounds of restriction enzyme digestions, ligation, and PCR result in a collection of short sequence tags representing each of the RNAs in the sample. The tag length must allow the tags to be mapped to the genes that they represent in the reference genome. Although, in theory, a short sequence tag of 9–10 nucleotides could be enough to identify individual transcripts, there is still the possibility that multiple genes could have the same tags. In practice, SAGE generally uses tags of 14–20 bp; the superSAGE variant uses tags of about 26 bp.

After the digested cDNAs are released from the beads, the tags are concatenated so that they can be cloned and sequenced in large groups. Counting the occurrences of each tag in the sequence data will give relative RNA expression levels. Because the SAGE technique maps the tags to a reference genome to identify genes, it works best in organisms that have a complete genome sequence. SAGE and superSAGE have been used in different plant species, including *Arabidopsis*, wheat, and chickpea, to analyze and detect existing transcripts and novel ncRNAs [27–30]. However, SAGE has been largely replaced by NGS technologies, which can examine more transcripts in greater depth. In addition, NGS methods generally skip the concatenation of tags, which SAGE uses to improve yields in Sanger sequencing.

2.2.2 Massively Parallel Signature Sequencing (MPSS)

Another sequence tag-based expression technique, massively parallel signature sequencing (MPSS), was developed to quantitatively analyze gene expression [31]. MPSS involves the acquisition of 17–20-nt tags (signatures) from cDNAs cloned on beads, using an unconventional, massively parallel sequencing method. MPSS uses a unique cloning strategy where every mRNA (and the corresponding cDNA) in a sample is represented by a single microbead; these microbeads are analyzed in a flow cell setup in an array format containing thousands of beads. The bases of mRNAs are systematically removed after the sequencer reads the mRNA bases by hybridization to a labeled coder. This produces a collection of 17–20 bp signature tags representing each of the mRNAs in the sample. Work in multiple plants, including *Arabidopsis*, has used MPSS for analysis of the transcriptome [32–35], but MPSS has also been largely replaced by NGS.

2.2.3 Cap Analysis of Gene Expression (CAGE)

In contrast to other tag-based sequencing methodologies, like SAGE and MPSS, which largely depend on 3′ end of the RNA transcript, cap analysis of gene expression (CAGE or 5′-SAGE) is designed to capture the expression of 5′-capped RNAs quantitatively by using sequence tags from the 5′ ends of cDNAs [36, 37, 76]. The original CAGE method used the biotinylated CAP-trapper method [76], in which the cap structure of capped and polyadenylated RNAs was chemically biotinylated. CAGE has been also used to analyze full-length cDNAs in *Arabidopsis* [38]. One of the advantages of CAGE is that it allows effective detection of the transcriptional activity around the promoter regions and RNA polymerase II-driven transcription start sites. However, the major limitation of CAGE is that non-capped RNAs are not detected.

In addition to the original tag-based CAGE, the same methodology now can be coupled with high-throughput sequencing (CAGE-seq) to examine the 5′-capped RNAs in a high-throughput manner not limiting to mRNAs [41, 42]. Several plant studies have used the high-throughput sequencing-based CAGE or nanoCAGE to analyze transcriptional activity around transcription start sites [39, 40]. Additionally, paired-end analysis of transcription start sites (PEAT) is another approach that has been developed to capture 5′-capped RNAs [77]. Morton et al. successfully analyzed the transcriptional activity around transcription start sites using PEAT in wild-type Columbia-0 *Arabidopsis thaliana* whole root tissues [78].

2.3 RNA Sequencing (RNA-seq)

RNA-seq is currently widely used for the detection of RNA expression and for the discovery of novel lncRNAs [4]. In addition, RNA-seq can be used to find alternatively spliced mRNAs and splice junctions [79], as well as different isoforms. For RNA-seq, transcripts are reverse-transcribed into a pool of cDNAs that are cloned into a library for sequencing. The first such libraries were reverse-transcribed with oligo (dT) primers, thus capturing mostly polyadenylated RNAs and only a few non-polyadenylated RNAs, particularly rRNA. Oligo (dT) priming also excludes non-polyadenylated transcripts and many transcripts from the degradome. Therefore, most RNA-seq libraries are now reverse-transcribed with random primers, using a pool of RNAs that has been depleted of rRNA.

RNA-seq typically produces 30–400 bp reads, depending on the platform and methods used. The high-throughput sequencing platforms include Illumina, ABI's SOLiD, Life Technologies/ThermoFisher/Ion Torrent, Oxford Nanopore Technologies, and Pacific Biosciences, as well as the recently retired Roche 454 sequencing platform. These various high-throughput sequencing platforms are discussed in detail by Reuter et al. [5]. Most platforms fragment RNA molecules (which generates a population of short sequences) and use short read-based technique, but Pacific

Biosciences's SMRT sequencing and Oxford Nanopore sequencing use single-molecule-based sequencing technology, with an average read length of >14 kb and individual reads as long as 60 kb for Pacific Biosciences's SMRT sequencing and a medium read length of 6 kb and maximum of >60 kb for Oxford Nanopore sequencing. Although both of these single-molecule-based sequencing technologies can provide longer reads compared to other sequencing platforms, their high error rates are one of the most cumbersome technical problems.

A large number of studies have conducted RNA-seq to identify and categorize lncRNAs in plants and metazoans. Several of the chapters in this book provide detailed protocols for analyzing lncRNAs using RNA-seq followed by extensive bioinformatic and functional analysis (Chapters 11–16, all chapters in Part IV of the book: identification and functional analysis of lncRNAs). Specifically, Chapter 2 from the Pikaard lab provides a comprehensive protocol for how to analyze the ncRNAs that are produced by RNA polymerase IV; these lncRNAs serve as precursors for small interfering RNAs (siRNAs) in the RNA-directed DNA methylation pathway. Additionally, Chapters 3 and 4 provide protocols for how to use RNA-seq to identify differentially expressed lncRNAs during development. Chapters 9 and 10 provide protocols for how to use RNA-seq to identify lncRNAs produced in response to biotic and abiotic stresses, including drought and salt tolerances (*see* Chapter 9) and virus infection (*see* Chapter 10). Additionally, Chapter 8, by Matsui and Seki, provides a comprehensive review on the subject of lncRNAs and stress responses in plants.

2.4 Parallel Analysis of RNA Ends (PARE)/Genome-Wide Mapping of Uncapped and Cleaved Transcripts (GMUCT)/Degradome-Seq

Like all RNAs, lncRNAs are eventually degraded; moreover, some lncRNAs serve as targets for miRNAs, and degradation of some lncRNAs yields miRNAs [80]. The intense interest in the RNA interference pathway (RNAi) in the plant field has led to the development of techniques to examine transcripts that are in the process of being degraded or could be miRNA targets and precursors. These techniques include parallel analysis of RNA ends (PARE) [44] (protocol, [45]), degradome sequencing [48] (protocol, [49]), and genome-wide mapping of uncapped and cleaved transcripts (GMUCT) [46] (protocol, [47]). These three nearly identical techniques generate equivalent data and were developed using *Arabidopsis thaliana* and used in plants. These approaches all target RNA degradation products that have been uncapped at their 5′ end; these RNAs are ligated to an RNA adapter that allows them to be converted to cDNAs, which are then amplified by PCR and sequenced.

2.5 Transcript Isoform Sequencing (TIF-seq)

The techniques described above examine the 3′ or 5′ ends of transcripts. By contrast, TIF-seq examines both ends of transcripts [50, 51], thus enabling genome-wide assessment of transcripts based on the precise positions of their 5′ and 3′ ends. TIF-seq was originally designed to study transcriptional heterogeneity and unique transcript isoforms in *Saccharomyces cerevisiae* and relies on the usage of oligo-capping, which identifies the 5′ cap structure to allow for ligation of an oligo tag at the 5′ end. After oligo-capping, the resulting RNA molecules containing both 5′ cap structure and poly(A) tail undergo reverse transcription, generating full-length cDNAs with barcodes at both 5′ and 3′ ends. The barcoded cDNAs undergo intramolecular circularization, which allows sequencing of the junction of the 5′ and 3′ ends of the transcript. In contrast to approaches that target only the 3′ or 5′ ends of transcripts, pinpointing the 3′ and 5′ ends of transcripts by TIF-seq allows the researcher to distinguish full-length and truncated transcripts, transcription through multiple open reading frames (bicistronic messages), and transcripts that originate from different start sites or terminate at different end sites. However, TIF-seq has not yet been used in plants nor in eukaryotes other than *S. cerevisiae*.

2.6 Nascent RNA Sequencing

In many cases, it is important to detect nascent transcription and nascent transcripts to capture the RNAs that are in the process of being transcribed. However, nascent transcripts can be unstable and difficult to distinguish from degraded or complete transcripts. The abundance of RNA polymerase II (Pol II) is often utilized to determine the level of nascent transcription at particular genomic locus. For example, chromatin immunoprecipitation microarray (ChIP-ChIP) or ChIP sequencing (ChIP-seq) methods are typically used to immunoprecipitate Pol II and associated chromatin. However, IP of Pol II collects paused Pol II and active Pol II-RNA complexes [81]. On the other hand, simply sequencing total RNA by RNA-seq or CAGE-seq detects the pool of steady-state RNAs and is also inefficient for detecting unstable nascent RNAs.

Several methods were designed to capture nascent RNAs that are associated with Pol II; rather than immunoprecipitation, these methods rely on "run-on" extension of nascent transcripts. In nuclear run-on experiments, cells are treated to halt transcription in vivo; reinitiation of transcription in isolated nuclei supplied with labeled RNA precursors (often 5′-bromo-uridine, BrU) labels only the nascent RNAs. These nuclear run-on assays include generic run-on assays and global run-on sequencing assay (GRO-seq) [52, 82], precision nuclear run-on sequencing (PRO-seq) [57], and native elongating transcript sequencing (NET-seq) [59–61]. Additional methods, like BRIC-Seq/BrU-Seq/BrU-Chase-Seq, were also developed to capture nascent transcripts [65, 68]. Although each method was designed with the similar goal of capturing actively transcribed RNA Pol II transcripts and nascent RNAs, they differ in technical details and have specific limitations, as described in the subsections below. Moreover, only GRO-seq has been used in plants [53, 54].

2.6.1 Global Run-On Sequencing (GRO-Seq)

Nuclear run-on assays and global run-on sequencing (GRO-seq) were developed to capture nascent RNAs and to measure RNA half-life [52]. GRO-seq reveals Pol II-engaged transcripts genome-wide, with high resolution and specific information on the orientation and exact 5′ end of the transcript. In GRO-seq, nuclear run-on assays use BrU as the label and release paused Pol II with sarkosyl, to label only transcripts from engaged polymerases. The BrU-labeled transcripts are purified with anti-Br-UTP antibodies and deep sequenced. This very sensitive and specific method gives high-throughput, genome-wide data on nascent transcripts. However, purification of nuclei, reinitiation of transcription under non-physiological conditions, and precipitation of labeled RNAs have proven difficult. GRO-seq also has a limited resolution of 30–50 bases due to the necessity to allow polymerase to run on and incorporate labeled BrU into RNAs.

Recently, GRO-seq and 5′GRO-seq (also called GRO-cap, see description below), which use a 7-methylguanylate (m7G) cap, have been used to capture the characteristics of the nascent transcriptome in *Arabidopsis thaliana* seedlings [53] and in maize [54]. Moreover, protocols for GRO-seq have been described in multiple publications (e.g., *see* [55, 56]), and this technique will likely see a wider application in the plant studies in the future.

2.6.2 Precision Nuclear Run-On Sequencing (PRO-Seq)

Similar to GRO-seq, precision nuclear run-on sequencing (PRO-seq) was developed to examine Pol II that is actively engaged in transcription at high resolution and on a genome-wide scale [57] (protocol, [58]). However, in contrast to GRO-seq, PRO-seq can reveal the mapping and distribution of Pol II pausing at single-base resolution. Similar to traditional Sanger sequencing, PRO-seq uses chain-terminating ribonucleotide triphosphate analogs labeled with biotin (biotin-NTPs, either all four, or one with additional unlabeled NTPs) for run-on assays. The nascent RNAs can be purified using the biotin label and used for high-throughput sequencing.

PRO-seq can be modified to capture 5′ capped RNAs, a method termed PRO-cap [57]. In PRO-cap, uncapped RNAs are first removed, leaving only the pool of capped RNAs. The cap of each RNA is then modified to allow the ligation of adapters to the 5′ end. Therefore, PRO-cap allows identification of the transcription start sites at the RNA synthesis level. PRO-seq and PRO-cap can be coupled to compare the differences in the Pol II initiation and pause sites. However, PRO-seq has not been used in plants yet. As a nuclear run-on-based methodology, PRO-seq has the same technical difficulties as GRO-seq. Moreover, PRO-seq only identifies Pol II complexes that are competent to elongate nascent transcripts; it cannot map complexes that are backtracking or arrested.

2.6.3 Native Elongating Transcript Sequencing (NET-Seq)

NET-seq or mammalian NET-seq (mNET-seq) detects nascent, actively transcribed Pol II RNAs, through the capture of 3′ RNAs [59–61] (protocol, [62–64]). In this method, the affinity-tagged Pol II elongation complex is immunoprecipitated, and then coprecipitated RNA is extracted and reverse-transcribed into cDNA. Deep sequencing of the cDNAs produces 3′-end sequences of nascent RNA, providing nucleotide resolution mapping of transcripts. This immunoprecipitation-based method captures elongating complexes and complexes that are backtracked or arrested, an advantage, depending on the goal of the experiment, compared to GRO-seq and PRO-seq. However, immunoprecipitation requires that Pol II complexes be solubilized, which can be challenging in metazoan cells where they are typically insoluble and strongly associated with chromatin under native conditions. NET-seq and mNET-seq have been used in *Saccharomyces cerevisiae* and HeLa cells, but the NET-seq protocol has not been used in plants yet.

2.6.4 BRIC-Seq/BrU-Seq/BrUChase-Seq

In addition to examining nascent RNAs, other methods can measure the half-lives of mRNAs or lncRNAs, which can inform analysis of their physiological functions and regulation. In organisms with established cell cultures, endogenous transcripts can be pulse-labeled by adding BrU to the culture media. In different variants of the classic pulse-chase method, label can be added for different times and removed from the media; labeled RNA can be immunoprecipitated and sequenced. For example, to establish the half-lives of RNAs or lncRNAs, in 5′-BrU immunoprecipitation chase-deep sequencing analysis (BRIC-seq) [65] (protocol, [66, 67]), total RNAs containing BrU-labeled RNAs (BrU-RNAs) are isolated at sequential time intervals after removal of BrU from the culture medium. BrU-RNAs are then recovered by immunopurification, which is followed by RT-qPCR or deep sequencing.

BrU labeling and sequencing (BrU-seq) and BrU pulse-chase sequencing (BrUChase-Seq) also involve BrU pulse-labeling that is chased with uridine, giving pools of RNA of different ages [68] (protocol, [69]). Following immunocapture, the BrU-labeled RNA is deep sequenced. However, none of the BRIC-Seq/BrU-Seq/BrUChase-Seq protocols have been used in plants yet.

3 Analyzing the Biological and Molecular Functions of lncRNAs

After the identification of lncRNAs using high-throughput approaches, one next step would be determining if these lncRNAs have biological functions, followed by identification of these functions. However, these experimental approaches have proven challenging, particularly for high-throughput studies, because of the diverse functions of lncRNAs, their potential tissue and stage specificity, and the varied mechanisms by which lncRNAs achieve these functions. For example, as one approach, overexpression or knockdown of the target lncRNAs can be used to study the functions of lncRNAs; however, such approaches are difficult to conduct in a high-throughput fashion in multicellular organisms.

Despite these challenges, new methods are emerging that can examine lncRNA function in a high-throughput manner. For example, in addition to modulating gene expression, lncRNAs have been implicated in genome architecture. The local interaction and looping events of the genomic regions where ncRNAs originated from can be captured by chromosome conformation capture followed by massively parallel sequencing [83, 84]. In this book, Chapter 28 by Padmarasu et al. describes an improved method to detect long-range chromatin interactions using in situ Hi-C for plants. Below, we present several selected methods that can be used to analyze the functions of lncRNAs in a high-throughput manner; these techniques are summarized in Table 3 (Subheadings 3.1–3.6).

3.1 Tissue or Cell Type-Specific Analysis

In multicellular organisms, specialized cell types each have a specific phenotype, function, and transcriptional program. However, our knowledge of how cells implement these programs during differentiation, particularly the effects of lncRNAs, remains limited, in part because purifying individual cell types for transcriptional and epigenomic profiling remains challenging. However, ongoing research has developed multiple methods to study lncRNAs in specific plant cell types by purifying individual cell types for analysis. These methods include laser microdissection (LM or laser capture microdissection, LCM) of fixed tissue sections [85], fluorescence-activated cell sorting (FACS) of fluorescently labeled cells or nuclei [89], and isolation of nuclei tagged in specific cell types (INTACT) using affinity-based isolation [92]. All three of these methods are commonly used in different plant species, and additional information is provided below (Subheadings 3.1.1–3.1.3). Three chapters in the Part II of this book (Chapters 5–7) also provide different protocols in studying the tissue and cell type-specific lncRNAs.

In addition to these methods, cryosectioning and cryostat sectioning can be used to study specific cell types [143, 144]; this is often less invasive compared to other techniques. Cryosectioning is also the first step of LM (or LCM) for sample preparation and can be coupled with other cell type-specific or tissue-specific techniques to isolate and study plant lncRNAs [145, 146]. Chapter 3 in this book by Kim et al. describes a protocol that uses cryostat sectioning to isolate distinct tissue types in the developing endosperm in maize followed by transcriptome and epigenome analysis to identify lncRNAs. Other methods like ex vivo differentiation from progenitor cells and the use of cultured cell lines are commonly used in metazoan studies; however, cultured cell lines are not commonly used in plant studies.

Table 3
Selected methods for analyzing the functions of lncRNAs

Method	Purpose	Reference
Laser microdissection (LM)	Tissue or cell-type specific analysis	[85–88]; protocol can be found in Chapter 5 of this book
Fluorescence-activated cell sorting (FACS)	Tissue or cell-type specific analysis	[89–91]
Isolation of nuclei tagged in specific cell types (INTACT)	Tissue or cell-type specific analysis	[92]; protocol can be found in Chapter 7 of this book and [93, 94]
Cryosectioning	Tissue or cell-type specific analysis	Protocol can be found in Chapter 3 of this book
In situ hybridization (ISH) and fluorescence in situ hybridization (FISH)	Visualization of RNA and DNA; tissue or cell-type specific analysis	[95]; protocol can be found in Chapter 6 of this book and [96]
RNA immunoprecipitation followed by microarray or sequencing (RIP-Chip or RIP-seq)	High-throughput identification of RNA-protein interactions	[97–99]; protocol can be found in Chapters 17 and 18 of this book and [100, 101]
m^5C RNA immunoprecipitation followed by sequencing (m^5C-RIP-seq)	High-throughput identification of RNA modifications	[102]; protocol can be found in Chapter 24 of this book
High-throughput sequencing cross-linking immunoprecipitation (HITS-CLIP, CLIP-seq, eCLIP, irCLIP)	High-throughput identification of RNA-protein interactions	[103–107]; protocol [108, 109]
Photoactivatable ribonucleotide-enhanced cross-linking and immunoprecipitation (PAR-CLIP)	High-throughput identification of RNA-protein interactions	[110, 111]; protocol [112, 113]
Chromatin isolation by RNA purification sequencing (ChIRP, ChIRP-Seq)	Genome-wide identification of RNA-DNA/chromatin interactions	[114]; ChIRP-MS [115]; protocol [116, 117]
RNA antisense purification (RAP)	Genome-wide identification of RNA-DNA/chromatin interactions	[118]; protocol can be found on the Guttman lab's website at http://www.lncrna-test.caltech.edu/protocols.php and [119]
Capture hybridization analysis of RNA targets (CHART, CHART-seq)	Genome-wide identification of RNA-DNA interactions	[120, 121]; protocol [122]

(continued)

Table 3
(continued)

Method	Purpose	Reference
RNA antisense purification followed by RNA sequencing (RAP-RNA)	High-throughput mapping of RNA-RNA interactions	[123]; protocol can be found on the Guttman lab's website at http://www.lncrna-test.caltech.edu/protocols.php
Cross-linking, ligation, and sequencing of hybrids (CLASH)	High-throughput mapping of RNA-RNA interactions	[124, 125]; protocol [126, 127]
Selective 2′-hydroxyl acylation by primer extension sequencing (SHAPE-seq)	In vitro high-throughput profiling of RNA secondary structure	[128]; protocol [129, 130]
Structure-seq/Structure-seq2	In vivo high-throughput profiling of RNA secondary structure; method developed using *Arabidopsis thaliana*	[131–134]; protocol can be found in Chapter 20 of this book and [135]
Protein interaction profile sequencing (PIP-seq)	High-throughput profiling of RNA-protein interactions and RNA secondary structure; method developed using HeLa cells and *Arabidopsis thaliana*	[136, 137]; protocol can be found in Chapters 21 and 22 of this book and [138]
Parallel analysis of RNA structure (PARS)	High-throughput profiling of RNA secondary structure	[139]; protocol [140]
Fragmentation sequencing (frag-seq)	High-throughput profiling of RNA secondary structure	[141]; protocol [142]

3.1.1 Laser Microdissection (LM)

Laser microdissection (LM; laser-captured microdissection, LCM; laser-assisted microdissection, LMD or LAM) uses a laser beam and direct microscopic visualization to isolate specific cells from heterogeneous tissues [85–87]. When coupled with high-throughput sequencing or microarray analysis, LM allows genome-wide analysis of gene expression in specific cell types. The detailed methodology and technological requirements of LM are comprehensively discussed in the review by Bevilacqua and Ducos [147]. LM has been used in separation of specific cell types in *Arabidopsis* [148], maize [149, 150], and other plants. The recent advances and applications of LM in the context of plant biology and transcriptome studies were comprehensively reviewed by Gautam and Sarkar [88]. Chapter 5 in this book by Gautam et al. describes a protocol that adapts the LM to obtain high-quality RNA of low abundance from specific tissues, followed by RT-PCR or stem-loop RT-PCR.

3.1.2 Fluorescence-Activated Cell Sorting (FACS)

Fluorescence-activated cell sorting (FACS) to separate cells or nuclei into different populations is based on the green fluorescent protein (GFP) labeling of specific cells, which are then separated from unlabeled cells using flow cytometry [89, 90]. FACS is followed by RNA extraction from each subpopulation of cells and high-throughput sequencing or microarray analysis. Based on the use of enhancer trap lines or promoter-GFP fusions that express GFP in specific tissues, these techniques have been widely used in plants. This has allowed deep analysis of RNA expression and the transcriptome in distinct cell types, cells in different developmental stages, and cells in response to biotic or abiotic stresses (reviewed by Carter et al. [91]). For example, recently cell type expression analyses in *Arabidopsis* roots were used to characterize intergenic lncRNAs [151].

3.1.3 Isolation of Nuclei Tagged in Specific Cell Types (INTACT)

LM and FACS require specialized equipment and the manipulation of whole cells; by contrast, the isolation of transgenically tagged nuclei in specific cell types (INTACT) uses affinity-based methods to isolate tagged nuclei from total nuclei. INTACT does not require the dissociation and manipulation of whole cells [92] (protocol, [93, 94]). The INTACT method was initially developed to study cell types in the *Arabidopsis thaliana* root epidermis with high yield and efficiency. Although it has its advantages, in order to successfully obtain transgenically tagged nuclei, it requires a promoter or enhancer trap line that is expressed in the specific cell type to be examined. Chapter 7 in this book by Do et al. describes a protocol that adapts the INTACT methodology to isolate specific cell types with tagged nuclei, followed by RNA-seq and bioinformatic analysis to identify nuclear lncRNAs in *Arabidopsis.*

3.2 In Situ Hybridization (ISH) and Fluorescence In Situ Hybridization (FISH)

FISH can be used to visualize the subcellular localization of lncRNAs and possibly provide information on their potential functions [95]. DNA and RNA can be visualized in situ using DNA FISH and RNA FISH, respectively, and multiplex FISH can simultaneously assay multiple targets within the same specimen. FISH techniques have been used for decades; however, emerging work has brought FISH into the genomics era. For example, the fluorescent in situ RNA sequencing (FISSEQ) amplifies cDNAs in cells and tissues [96]; compared with RNA-FISH, FISSEQ gives a higher resolution and can identify more targets. FISSEQ produces fewer reads than regular RNA-seq, but could provide cell-specific spatial information on lncRNAs. Chapter 7 in this book by Francoz et al. describes a protocol that integrates ISH and transcriptomics resulting in a medium-throughput RNA in situ hybridization methodology.

3.3 RNA-Protein Interactions

Many lncRNAs function in complexes with proteins, but very few of the proteins that interact with lncRNAs have been identified. Identification of the protein interactors of lncRNAs will shed substantial light on the mechanisms of lncRNA function. Below, in Subheadings 3.3.1–3.3.3, we describe selected methods to analyze RNA-protein interactions. Also, three chapters in this book, from the Chua lab, provide three distinct and comprehensive protocols for identification and analysis of lncRNAs and protein interactions (*see* Chapters 17–19 in Part V of this book). Additional techniques that are not presented here are comprehensively summarized in the review by Ferre et al. [152].

3.3.1 RNA Immunoprecipitation Followed by Sequencing (RIP-Seq)

The versatile technique of RNA immunoprecipitation (RIP)-seq can examine multiple aspects of RNA-protein interactions, from either the RNA or the protein side. If the interacting protein is known, then antibodies against the target (or the affinity-tagged targets) can be used for RIP of RNAs that interact with the protein of interest [153]. RIP can be coupled with microarray (RIP-Chip) or high-throughput sequencing (RIP-seq) to identify the RNAs that interact with proteins genome-wide [97, 100]. Additionally, RIP can be used to identify the binding sites for specific proteins. RIP and RIP-seq have been widely used in metazoans and plants. For example, in *Arabidopsis*, RIP-seq was used to identify the transcriptome-wide RNA targets of SR34, a serine/arginine-rich (SR)-like RNA-binding protein that functions in constitutive and alternative splicing [98]. RIP was also used to identify Argonaute (AGO)-associated smRNAs (RIP smRNA-seq) [99] or RNAs (AGO RIP-seq) [101] in the RNA interference (RNAi) pathway.

RIP can also be used to identify regions in the RNA molecule that interact with proteins. Indeed, the first studies used RIP to find proteins that interact with the lncRNA Xist, which functions in X chromosome inactivation in mammals [154]. Usually, chemical agents are used to cross-link RNAs and proteins, but this can introduce artifacts. RIP can also be conducted without cross-linking, thus reducing the potential generation of artifacts. Moreover, various nuclease treatments can provide additional information on protein-nucleic acid interactions. For example, RNase H digests the RNA in RNA-DNA hybrids, while DNase I digests DNA, and the combination of the treatments with these nucleases can help distinguish the indirect binding of protein of interest to neighboring DNA from direct binding between protein of interest and RNAs. Different nuclease treatments can also distinguish protein-RNA interactions that involve single-stranded or stem-loop RNAs.

The modifications of RNAs play an important role in regulating the functions of RNA molecules, and RIP-seq can be adopted to map RNA modifications genome-wide, such as mapping 5-methylcytosine (m^5C) or *N6*-methyladenosine (m^5A) of RNAs (m^5C-

RIP-seq, m^5A-seq, respectively) [102, 155, 156]. Chapter 24 in this book by Liang and Gu describes a protocol for m^5C-RIP-seq to map m^5C RNA modifications in plants genome-wide. There are additional techniques for identifying RNA modification sites of both mRNAs and lncRNAs transcriptome-wide, such as coupling RNA bisulfite conversion with sequencing (bsRNA-seq) [157, 158] and 5-azacytidine-mediated RNA immunoprecipitation (Aza-IP) [159].

3.3.2 High-Throughput Sequencing Cross-Linking Immunoprecipitation (HITS-CLIP or CLIP-seq)

CLIP (cross-linking immunoprecipitation) examines RNA-protein interactions by UV cross-linking cells before immunoprecipitation [160] (protocol, [161]). CLIP-seq, also known as HITS-CLIP, is a method for genome-wide mapping of RNA-protein binding sites, by CLIP, followed by high-throughput sequencing of the RNA [103, 104]. In contrast to chromatin immunoprecipitation sequencing (ChIP-seq), which uses formaldehyde cross-linking, in CLIP-Seq UV cross-linking covalently links the RNA and protein. The pools of cross-linked and immunoprecipitated RNA molecules are first fragmented with RNase followed by proteinase digestion and purification. One of the main advantages of this method is that it identifies the essential protein-binding sites on the RNA molecule. However, the UV light can cause mutations and CLIP-Seq does not give full-length sequence of the immunoprecipitated RNA. These disadvantages can be particularly problematic for systems that lack collections of full-length, annotated lncRNA sequences and for lncRNAs that are present at very low levels.

CLIP-Seq/HITS-CLIP has been widely used in mammalian studies, including identifying genome-wide interactions of RNAs and the neuron-specific splicing factor Nova in mouse brains [103]; however, no study has used CLIP-Seq in plants so far. Several other methods have been developed to improve the efficiency of CLIP-Seq, including enhanced CLIP (eCLIP) and infrared-CLIP [105–107]. Protocols for CLIP-seq are described in these references [108, 109].

3.3.3 Photoactivatable Ribonucleotide-Enhanced Cross-Linking and Immunoprecipitation (PAR-CLIP)

PAR-CLIP, a variant of CLIP-Seq, has better cross-linking efficiency and resolution, as well as a higher signal-to-noise ration compared with other methods [110, 112]. PAR-CLIP uses the ribonucleoside analogs, 4-thiouridine (4SU) and 6-thioguanosine (6SG). Photoactivation of 4SU and 6SG by UV light produces strong cross-links and specific mutations of the nucleic acid sequence: 4SU produces T to C changes, and 6SG produces G to A. PAR-CLIP can therefore be used to identify the binding sites of RNA-binding proteins. Moreover, PAR-CLIP can be used to identify miRNA targets [113]. PAR-CLIP has been widely implemented in metazoan studies [111]; however, no study has used PAR-CLIP in plants so far. Protocols for PAR-CLIP are described in these references [112, 113].

3.4 RNA-DNA/Chromatin Interactions

lncRNAs can physically associate with chromatin, indirectly through an RNA-protein interaction [162] or directly through RNA-DNA hybridization in a triple helix [163, 164]. DNA-RNA FISH can show this association only at low resolution; new technologies can examine these lncRNA-DNA interactions at higher resolution. In Subheadings 3.4.1–3.4.3, we describe three recently developed methodologies for mapping the interactions of RNAs with chromatin in a high-throughput manner; however, none of these methods have been used in plants so far. These three techniques are very similar and detect the lncRNA by probing, differing only in some specifics.

3.4.1 Chromatin Isolation by RNA Purification Sequencing (ChIRP or ChIRP-Seq)

The techniques described above in Subheading 3.3 examine RNA-protein interactions to identify the RNAs that bind to a known protein. ChIRP can be used to identify the proteins and chromatin regions that are bound by a known RNA [114]. After cross-linking and sonication of chromatin, ChIRP uses tiled biotinylated oligonucleotides (20-mers) to affinity purify a known lncRNA in complex with its associated chromatin and proteins. The DNA genomic regions associated with the RNA of interest can be identified by sequencing (ChIRP-Seq), and the RNA can be quantified by qPCR. Chu et al. used ChIRP to identify DNA regions associated with the lncRNA HOTAIR. In addition to identifying associated DNA regions, RNA-associated proteins can be purified from ChIRP reactions and examined by mass spectrometry (ChIRP-MS) [115]. For example, ChIRP-MS identified 81 proteins associated with the Xist lncRNA, which plays key roles in X chromosome silencing in mammals. A comprehensive protocol, with video of ChIRP-seq, can be found in this article [116].

3.4.2 RNA Antisense Purification (RAP)

RNA antisense purification (RAP) or RNA antisense purification followed by DNA sequencing (RAP-DNA) can be used to find the chromatin regions that associate with a specific RNA [118]. In contrast to ChIRP, which uses the tiled biotinylated 20-nt oligonucleotides, RAP uses 120-nt-long antisense RNA probes, thus improving target lncRNA binding and increasing the signal-to-noise ratio. Like ChIRP, RAP uses tiled overlapping probes that cover target transcripts without considering whether the regions are accessible for hybridization. DNase I digestion then produces genomic DNA fragments of <300 bp; these fragments are then sequenced. RAP was used to find the exact X chromosome binding sites of Xist in mouse ES cells [118]. A detailed protocol for RAP-DNA can be found on the Guttman lab's website (http://www.lncrna-test.caltech.edu/protocols.php) and [119].

3.4.3 Capture Hybridization Analysis of RNA Targets (CHART)

Like ChIRP and RAP, CHART involves cross-linking and purification of RNA-DNA complexes [120] (protocol, [122]). However, CHART uses capture oligonucleotides (C-oligos) that carry a tag such as biotin for affinity purification. CHART probe design involves the detection of open binding sites by hybridization of oligonucleotides and RNase H mapping; then this information is used to design biotinylated 24-nt-long C-oligos that target the open sites. Similar to other methods, CHART can recover the chromatin and proteins associated with the target lncRNA. CHART has been used to study the function of roX2 lncRNA in *Drosophila* (lncRNA involved in dosage compensation) [120] and Xist lncRNA [121].

3.5 RNA-RNA Interactions

Interactions between RNAs involve direct base-pairing interactions (including miRNA-mRNA and mRNA-lncRNA sense-antisense interactions) and indirect interactions mediated by protein intermediates. Some lncRNAs associate with RNA-processing proteins, indicating that these lncRNAs may target other RNAs via proteins [165, 166]. Below, in Subheadings 3.5.1 and 3.5.2, we present two recently developed methods for mapping RNA-RNA interactions for a RNA molecule of interest genome-wide. However, none of these methods have been used in plants so far.

3.5.1 RNA Antisense Purification Followed by RNA Sequencing (RAP-RNA)

RNA antisense purification (RAP) and variants thereof detect RNA-RNA interactions. For example, to identify the intermolecular contacts of lncRNA-RNA interactions, RAP-RNA, a method based on RAP, was developed to systematically map RNA-RNA interactions [123]. RAP-RNA cross-links RNAs in vivo and then uses antisense oligonucleotide to purify the RNA, followed by high-throughput RNA sequencing. Cross-linking reagents that differ in their specificity for proteins and nucleic acids can give additional information on the interactions. Additional variants, RAP-RNA$^{[AMT]}$, RAP-RNA$^{[FA]}$, and RAP-RNA$^{[FA\text{-}DSG]}$, can identify direct and indirect RNA-RNA interactions [123]. Each variant uses a different cross-linking agent, 4′-aminomethyltrioxsalen (AMT), formaldehyde (FA), and both FA and disuccinimidyl glutarate (DSG), respectively. These methodologies were applied to investigate two ncRNAs implicated in RNA processing: U1 small nuclear RNA, a component of the spliceosome, and Malat1, a large lncRNA that localizes to nuclear speckles [123]. This study revealed that U1 hybridizes to the 5′ splice sites of RNAs and Malat1 indirectly binds pre-mRNAs by interacting with proteins. A detailed protocol for RAP-RNA can be found on the Guttman lab's website (http://www.lncrna-test.caltech.edu/protocols.php).

3.5.2 Cross-Linking, Ligation, and Sequencing of Hybrids (CLASH)

CLASH uses UV cross-linking to capture direct RNA-RNA hybridization in RNA-protein complexes [124] (protocol, [126, 127]). UV light cross-links interacting nucleic acids and interacting

RNA-protein complexes, and UV cross-linking offers advantages compared to chemical cross-linking. For examples, chemical cross-linking can also cross-link between interacting protein-protein complexes, making it hard to differentiate direct and indirect (i.e., protein-mediated) interactions. UV cross-linked RNA-protein complexes are affinity-purified for the protein of interest. RNA-RNA hybrids are ligated together to generate chimeric RNAs and isolated, thus giving high-throughput data on RNA-RNA interactions. Despite the advantages of high-throughput data, ligation of the RNAs remains a challenging step. Work in yeast used CLASH to find novel snoRNA-rRNA interactions [124], and work in humans found miRNA-mRNA interactions in complexes with Argonaute [125]. This technique has a potential to be used to capture lncRNA-RNA interactions.

3.6 RNA Secondary Structures

Functional RNAs act via sequence complementarity with RNAs or DNA and via their structures, which involve intricate base pairing with multiple stems and loops. These RNA structures can give RNAs the ability to catalyze reactions, scaffold macromolecular complexes, and bind ligands. RNAs can thus affect epigenetic regulation; mRNA splicing, stability, and translation; and signal transduction. Therefore, understanding the structure of lncRNAs provides key information for understanding their functions [167–169].

Computational prediction of RNA secondary structure works well for smaller RNAs, but for lncRNAs, the number of possible structures increases exponentially. For experimental methods, recent studies have developed genome-wide techniques to examine RNA secondary structures [169]. Different methodologies have been extensively summarized in the review articles by Wan et al. [169], Vandivier et al. [167], and Bevilacqua et al. [168]. Below, in Subheadings 3.6.1–3.6.4, we present five high-throughput methods that have been (or are being) used to solve RNA secondary structures in plants or have the potential to be adapted in the plant fields. Some of these methods are also provided in Part V of this book in Chapters 20–22. Recently, the secondary structure of COOLAIR, an antisense RNAs involved in the regulation of *Flowering Locus C* (*FLC*) and vernalization, was determined using chemical probing across the Brassicaceae [170]; the methodology is also described in Chapter 23 of this book.

3.6.1 Selective 2′-Hydroxyl Acylation by Primer Extension Sequencing (SHAPE-Seq)

Selective 2′-hydroxyl acylation by primer extension (SHAPE) provides large-scale data on RNA secondary structure and could be used to determine the secondary structure of lncRNAs [169, 171]. SHAPE uses the electrophile N-methylisatoic anhydride (NMIA), which can attack the 2′OHs of bases in flexible regions of the RNA, forming 2-O adducts, which can be detected because they terminate reverse transcription. SHAPE reverse transcription followed by capillary sequencing has been used to examine the structures of the 16S rRNA and the 9-kb HIV RNA genome [169, 172]. Recently, SHAPE has also been coupled to high-throughput sequencing (SHAPE-seq) to provide the secondary and tertiary structural information of seven RNAs in vitro [128] (protocol, [129, 130]).

3.6.2 Structure-Seq and Structure-Seq2

RNA structures, which are influenced by the cellular environment, often differ in vitro and in vivo [173]. Structure-seq2 and its predecessor, Structure-seq, are high-throughput methods that provide an efficient way to study RNA structures in vivo [131, 132]. They do this by using DMS (dimethyl suberimidate) to react with RNA in vivo, similar to DMS-seq [174] and Mod-seq [175], and in contrast to SHAPE. The Galaxy environment (https://usegalaxy.org/) provides open access to the user-friendly Structure-seq computational pipeline [133]. Structure-seq2 offers improvements compared to its predecessor and can be used in additional plant species [132]. A comprehensive computational pipeline, StructureFold2, accompanies this new and improved method [134].

Comprehensive protocol for Structure-seq is described in this article [135]. Also, Chapter 20 in this book, by Ritchey et al. (the authors are the developers of this method), describes two versions of the Structure-seq2 protocol and provides detailed step-by-step instructions on how to produce and analyze Structure-seq2 data. This protocol has been used in plants and can be easily applied to other organisms.

3.6.3 Protein Interaction Profile Sequencing (PIP-Seq)

To simultaneously examine protein-RNA binding sites and RNA secondary structure globally and in an unbiased manner, protein interaction profile sequencing (PIP-seq) uses RNase-based footprinting assays for RNA-associated proteins, which are cross-linked on the RNA [136, 137]. RNase digestion then removes the regions of the RNA that are not bound by the protein, leaving the bound regions as the protein footprint. After reversal of the cross-linking, the bound regions are recovered by ligation to linkers, PCR amplification, and cloning as strand-specific libraries for sequencing. To identify RNase-resistant regions, which would appear as protein-bound regions, control reactions are conducted in which the proteins are denatured and partially digested before RNase treatment. Moreover, to examine RNA structure, PIP-seq uses RNases specific for double- and single-stranded RNA; comparison of the results reveals the secondary structure of the RNAs. Comprehensive protocols for PIP-seq can be found in the following articles [138, 176]. Also, two chapters in this book, Chapters 21 and 22, from the Gregory lab (the developer of this method) provide detailed step-by-step instructions on how to perform PIP-seq experiments and analyze the data in the context of plant transcriptomes.

3.6.4 Parallel Analysis of RNA Structure (PARS) and Fragmentation Sequencing (Frag-Seq)

RNA structures can be determined in vitro by parallel analysis of RNA structure (PARS) and fragmentation sequencing (Frag-seq) [139, 141]. PARS uses RNase V1 and S1 nucleases which are specific to double- or single-stranded regions of RNAs, respectively, and compares the results [139]. PARS uses genome-wide RNA structure probing and was validated on the known structure of the HOTAIR lncRNA. Frag-seq uses RNase P1, which is specific to single-stranded nucleic acids, to sequence RNA [141]. PARS and Frag-seq both use nucleases with specific activities on different structures, but differ in how they map RNA structure and thus provide complementary information. Structure prediction programs can be used to examine Frag-seq and PARS data to improve the prediction of RNA secondary structure. However, PARS and Frag-seq have not been used to examine RNA secondary structure in plants. Comprehensive protocols can be found here [140] for PARS and here for Fraq-seq [142].

4 Conclusions and Future Outlook

lncRNAs function at multiple levels in gene regulatory networks, where they modulate complex biological processes in eukaryotes. A large amount of effort has been spent studying lncRNAs in the metazoan field, effort that has translated into a better understanding of lncRNA functions in animals. However, unlike protein-coding genes, functional RNA coding genes (i.e., tRNAs, rRNAs, etc.), and microRNAs, lncRNAs show little conservation between species, making it difficult to identify functional lncRNAs and use the findings from one species to inform the study of another. This is why it is so important to study plant lncRNAs; however, very few plant lncRNAs have been examined. It will be crucial to identify factors controlling the expression and biogenesis of plants lncRNAs, as well as the protein and RNA components functioning with these lncRNAs. Together, this information will contribute to the much-needed understanding of plant lncRNA functions. Furthermore, these studies should be extended to a wide range of plant species.

Perturbation of lncRNA expression, in situ hybridization, structural characterization, and genomics are among the major tools available to dissect the molecular mechanisms of lncRNA functions. Revealing the regulatory roles of lncRNAs may require improvements in various techniques, as well as adoption of technologies from other fields to plants. These technologies include in vivo imaging of RNAs through single-molecule techniques, identifying the binding partners of lncRNAs by high-throughput methods, as well as using single-cell methods to decipher the heterogeneity of the transcriptome in a cellular population. We hope that the continuing interest in the biology of lncRNAs will bring new insights and discoveries to the functional mechanisms of lncRNAs in plants.

References

1. Chekanova JA (2015) Long non-coding RNAs and their functions in plants. Curr Opin Plant Biol 27:207–216
2. Wang H-LV, Chekanova JA (2017) Long Noncoding RNAs in Plants. Adv Exp Med Biol 1008:133–154
3. Quinn JJ, Chang HY (2016) Unique features of long non-coding RNA biogenesis and function. Nat Rev Genet 17:47–62
4. Wang Z, Gerstein M, Snyder M (2009) RNA-Seq: a revolutionary tool for transcriptomics. Nat Rev Genet 10:57–63
5. Reuter JA, Spacek DV, Snyder MP (2015) High-throughput sequencing technologies. Mol Cell 58:586–597
6. Lamesch P, Berardini TZ, Li D et al (2012) The Arabidopsis Information Resource (TAIR): improved gene annotation and new tools. Nucleic Acids Res 40:D1202–D1210
7. Cheng C-Y, Krishnakumar V, Chan AP et al (2017) Araport11: a complete reannotation of the Arabidopsis thaliana reference genome. Plant J 89:789–804
8. Jin J, Liu J, Wang H et al (2013) PLncDB: plant long non-coding RNA database. Bioinformatics 29:1068–1071
9. Paytuví Gallart A, Hermoso Pulido A, Anzar Martínez de Lagrán I et al (2015) GREENC: a Wiki-based database of plant lncRNAs. Nucleic Acids Res 44(Database issue): D1161–D1166. https://doi.org/10.1093/nar/gkv1215
10. Zhao Y, Li H, Fang S et al (2016) NONCODE 2016: an informative and valuable data source of long non-coding RNAs. Nucleic Acids Res 44:D203–D208
11. Szcześniak MW, Rosikiewicz W, Makałowska I (2016) CANTATAdb: a collection of plant long non-coding RNAs. Plant Cell Physiol 57:e8–e8
12. Yi X, Zhang Z, Ling Y et al (2015) PNRD: a plant non-coding RNA database. Nucleic Acids Res 43:D982–D989
13. Zhou B, Zhao H, Yu J et al (2018) EVLncRNAs: a manually curated database for long non-coding RNAs validated by low-throughput experiments. Nucleic Acids Res 46(D1):D100–D105
14. Chen D, Yuan C, Zhang J et al (2012) PlantNATsDB: a comprehensive database of plant natural antisense transcripts. Nucleic Acids Res 40:D1187–D1193
15. Kapranov P, Cawley SE, Drenkow J et al (2002) Large-scale transcriptional activity in chromosomes 21 and 22. Science 296:916–919
16. Shoemaker DD, Schadt EE, Armour CD et al (2001) Experimental annotation of the human genome using microarray technology. Nature 409:922–927
17. Yazaki J, Gregory BD, Ecker JR (2007) Mapping the genome landscape using tiling array technology. Curr Opin Plant Biol 10:534–542
18. Gregory BD, Yazaki J, Ecker JR (2008) Utilizing tiling microarrays for whole-genome analysis in plants. Plant J 53:636–644
19. Yamada K, Lim J, Dale JM et al (2003) Empirical analysis of transcriptional activity in the Arabidopsis genome. Science 302:842–846
20. Chekanova JA, Gregory BD, Reverdatto SV et al (2007) Genome-wide high-resolution mapping of exosome substrates reveals hidden features in the Arabidopsis transcriptome. Cell 131:1340–1353
21. MacIntosh GC, Wilkerson C, Green PJ (2001) Identification and analysis of Arabidopsis expressed sequence tags characteristic of non-coding RNAs. Plant Physiol 127:765–776
22. Marker C, Zemann A, Terhörst T et al (2002) Experimental RNomics: identification of 140 candidates for small non-messenger RNAs in the plant Arabidopsis thaliana. Curr Biol 12:2002–2013
23. Rymarquis LA, Kastenmayer JP, Hüttenhofer AG et al (2008) Diamonds in the rough: mRNA-like non-coding RNAs. Trends Plant Sci 13:329–334
24. Song D, Yang Y, Yu B et al (2009) Computational prediction of novel non-coding RNAs in Arabidopsis thaliana. BMC Bioinformatics 10(Suppl 1):S36
25. Jouannet V, Crespi M (2011) Long nonprotein-coding RNAs in plants. Prog Mol Subcell Biol 51:179–200
26. Velculescu VE, Zhang L, Vogelstein B et al (1995) Serial analysis of gene expression. Science 270:484–487
27. Robinson SJ, Cram DJ, Lewis CT et al (2004) Maximizing the efficacy of SAGE analysis identifies novel transcripts in Arabidopsis. Plant Physiol 136:3223–3233
28. Robinson SJ, Parkin IAP (2008) Differential SAGE analysis in Arabidopsis uncovers increased transcriptome complexity in response to low temperature. BMC Genomics 9:434
29. Poole RL, Barker GLA, Werner K et al (2008) Analysis of wheat SAGE tags reveals evidence

for widespread antisense transcription. BMC Genomics 9:475
30. Molina C, Rotter B, Horres R et al (2008) SuperSAGE: the drought stress-responsive transcriptome of chickpea roots. BMC Genomics 9:553
31. Brenner S, Johnson M, Bridgham J et al (2000) Gene expression analysis by massively parallel signature sequencing (MPSS) on microbead arrays. Nat Biotechnol 18:630–634
32. Meyers BC, Vu TH, Tej SS et al (2004) Analysis of the transcriptional complexity of Arabidopsis thaliana by massively parallel signature sequencing. Nat Biotechnol 22:1006–1011
33. Meyers BC, Tej SS, Vu TH et al (2004) The use of MPSS for whole-genome transcriptional analysis in Arabidopsis. Genome Res 14:1641–1653
34. Meyers BC, Lee DK, Vu TH et al (2004) Arabidopsis MPSS. An online resource for quantitative expression analysis. Plant Physiol 135:801–813
35. Quattro CD, Enrico Pè M, Bertolini E (2017) Long noncoding RNAs in the model species Brachypodium distachyon. Sci Rep 7:11252
36. Kodzius R, Kojima M, Nishiyori H et al (2006) CAGE: cap analysis of gene expression. Nat Methods 3:211–222
37. Shiraki T, Kondo S, Katayama S et al (2003) Cap analysis gene expression for high-throughput analysis of transcriptional starting point and identification of promoter usage. Proc Natl Acad Sci U S A 100:15776–15781
38. Seki M, Carninci P, Nishiyama Y et al (1998) High-efficiency cloning of Arabidopsis full-length cDNA by biotinylated CAP trapper. Plant J 15:707–720
39. Mejia-Guerra MK, Li W, Galeano NF et al (2015) Core promoter plasticity between maize tissues and genotypes contrasts with predominance of sharp transcription initiation sites. Plant Cell 27:3309–3320
40. Cumbie JS, Ivanchenko MG, Megraw M (2015) NanoCAGE-XL and CapFilter: an approach to genome wide identification of high confidence transcription start sites. BMC Genomics 16:597
41. Takahashi H, Lassmann T, Murata M et al (2012) 5′ end-centered expression profiling using cap-analysis gene expression and next-generation sequencing. Nat Protoc 7:542–561
42. Takahashi H, Kato S, Murata M et al (2012) CAGE (cap analysis of gene expression): a protocol for the detection of promoter and transcriptional networks. Methods Mol Biol 786:181–200
43. Nagalakshmi U, Wang Z, Waern K et al (2008) The transcriptional landscape of the yeast genome defined by RNA sequencing. Science 320:1344–1349
44. German MA, Pillay M, Jeong D-H et al (2008) Global identification of microRNA-target RNA pairs by parallel analysis of RNA ends. Nat Biotechnol 26:941–946
45. Zhai J, Arikit S, Simon SA et al (2013) Rapid construction of parallel analysis of RNA end (PARE) libraries for Illumina sequencing. Methods 67(1):84–90
46. Gregory BD, O'Malley RC, Lister R et al (2008) A link between RNA metabolism and silencing affecting Arabidopsis development. Dev Cell 14:854–866
47. Willmann MR, Berkowitz ND, Gregory BD (2014) Improved genome-wide mapping of uncapped and cleaved transcripts in eukaryotes--GMUCT 2.0. Methods 67:64–73
48. Addo-Quaye C, Eshoo TW, Bartel DP et al (2008) Endogenous siRNA and miRNA targets identified by sequencing of the Arabidopsis degradome. Curr Biol 18:758–762
49. Addo-Quaye C, Miller W, Axtell MJ (2009) CleaveLand: a pipeline for using degradome data to find cleaved small RNA targets. Bioinformatics 25:130–131
50. Pelechano V, Wei W, Steinmetz LM (2013) Extensive transcriptional heterogeneity revealed by isoform profiling. Nature 497:127–131
51. Pelechano V, Wei W, Jakob P et al (2014) Genome-wide identification of transcript start and end sites by transcript isoform sequencing. Nat Protoc 9:1740–1759
52. Core LJ, Waterfall JJ, Lis JT (2008) Nascent RNA sequencing reveals widespread pausing and divergent initiation at human promoters. Science 322:1845–1848
53. Hetzel J, Duttke SH, Benner C et al (2016) Nascent RNA sequencing reveals distinct features in plant transcription. Proc Natl Acad Sci 113:12316–12321
54. Erhard KF, Talbot J-ERB, Deans NC et al (2015) Nascent transcription affected by RNA polymerase IV in Zea mays. Genetics 199:1107–1125
55. Gardini A (2017) Global Run-On Sequencing (GRO-Seq). Methods Mol Biol 1468:111–120
56. Lopes R, Agami R, Korkmaz G (2017) GRO-seq, A tool for identification of transcripts regulating gene expression. Methods Mol Biol 1543:45–55
57. Kwak H, Fuda NJ, Core LJ et al (2013) Precise maps of RNA polymerase reveal how

promoters direct initiation and pausing. Science 339:950–953
58. Mahat DB, Kwak H, Booth GT et al (2016) Base-pair-resolution genome-wide mapping of active RNA polymerases using precision nuclear run-on (PRO-seq). Nat Protoc 11:1455–1476
59. Churchman LS, Weissman JS (2011) Nascent transcript sequencing visualizes transcription at nucleotide resolution. Nature 469:368–373
60. Larson MH, Mooney RA, Peters JM et al (2014) A pause sequence enriched at translation start sites drives transcription dynamics in vivo. Science 344:1042–1047
61. Nojima T, Gomes T, Grosso ARF et al (2015) Mammalian NET-seq reveals genome-wide nascent transcription coupled to RNA processing. Cell 161:526–540
62. Nojima T, Gomes T, Carmo-Fonseca M et al (2016) Mammalian NET-seq analysis defines nascent RNA profiles and associated RNA processing genome-wide. Nat Protoc 11:413–428
63. Churchman LS, Weissman JS (2012) Native elongating transcript sequencing (NET-seq). Curr Protoc Mol Biol edited by Frederick M. Ausubel ... [et al.] Chapter 4:Unit 4.14.1–Unit 4.1417
64. Mayer A, Churchman LS (2016) Genome-wide profiling of RNA polymerase transcription at nucleotide resolution in human cells with native elongating transcript sequencing. Nat Protoc 11:813–833
65. Tani H, Mizutani R, Salam KA et al (2012) Genome-wide determination of RNA stability reveals hundreds of short-lived noncoding transcripts in mammals. Genome Res 22:947–956
66. Yamada T, Imamachi N, Onoguchi-Mizutani R et al (2018) 5′-Bromouridine IP Chase (BRIC)-Seq to determine RNA half-lives. Methods Mol Biol 1720:1–13
67. Imamachi N, Tani H, Mizutani R et al (2014) BRIC-seq: a genome-wide approach for determining RNA stability in mammalian cells. Methods 67:55–63
68. Paulsen MT, Veloso A, Prasad J et al (2013) Coordinated regulation of synthesis and stability of RNA during the acute TNF-induced proinflammatory response. Proc Natl Acad Sci 110:2240–2245
69. Paulsen MT, Veloso A, Prasad J et al (2014) Use of Bru-Seq and BruChase-Seq for genome-wide assessment of the synthesis and stability of RNA. Methods 67(1):45–54
70. Liu J, Jung C, Xu J et al (2012) Genome-wide analysis uncovers regulation of long intergenic noncoding RNAs in Arabidopsis. Plant Cell 24:4333–4345
71. Adams MD, Kerlavage AR, Fleischmann RD et al (1995) Initial assessment of human gene diversity and expression patterns based upon 83 million nucleotides of cDNA sequence. Nature 377:3–174
72. Harbers M, Carninci P (2005) Tag-based approaches for transcriptome research and genome annotation. Nat Methods 2:495–502
73. Nakamura M, Carninci P (2004) [Cap analysis gene expression: CAGE], Tanpakushitsu kakusan koso. Protein, nucleic acid, enzyme 49, 2688–2693
74. Peiffer JA, Kaushik S, Sakai H et al (2008) A spatial dissection of the Arabidopsis floral transcriptome by MPSS. BMC Plant Biol 8:43
75. Reinartz J, Bruyns E, Lin J-Z et al (2002) Massively parallel signature sequencing (MPSS) as a tool for in-depth quantitative gene expression profiling in all organisms. Brief Funct Genomic Proteomic 1:95–104
76. Carninci P, Kvam C, Kitamura A et al (1996) High-efficiency full-length cDNA cloning by biotinylated CAP trapper. Genomics 37:327–336
77. Ni T, Corcoran DL, Rach EA et al (2010) A paired-end sequencing strategy to map the complex landscape of transcription initiation. Nat Methods 7:521–527
78. Morton T, Petricka J, Corcoran DL et al (2014) Paired-end analysis of transcription start sites in Arabidopsis reveals plant-specific promoter signatures. Plant Cell 26:2746–2760
79. Trapnell C, Pachter L, Salzberg SL (2009) TopHat: discovering splice junctions with RNA-Seq. Bioinformatics 25:1105–1111
80. Wang H-LV, Chekanova JA (2016) Small RNAs: essential regulators of gene expression and defenses against environmental stresses in plants. Wiley Interdiscip Rev RNA 7:356–381
81. Nechaev S, Adelman K (2011) Pol II waiting in the starting gates: regulating the transition from transcription initiation into productive elongation. Biochim Biophys Acta 1809:34–45
82. Min IM, Waterfall JJ, Core LJ et al (2011) Regulating RNA polymerase pausing and transcription elongation in embryonic stem cells. Genes Dev 25:742–754
83. Lieberman-Aiden E, van Berkum NL, Williams L et al (2009) Comprehensive mapping of long-range interactions reveals folding

principles of the human genome. Science 326:289–293

84. Dekker J, Rippe K, Dekker M et al (2002) Capturing chromosome conformation. Science 295:1306–1311
85. Emmert-Buck MR, Bonner RF, Smith PD et al (1996) Laser capture microdissection. Science 274:998–1001
86. Kerk NM, Ceserani T, Tausta SL et al (2003) Laser capture microdissection of cells from plant tissues. Plant Physiol 132:27–35
87. Ohtsu K, Takahashi H, Schnable PS et al (2007) Cell type-specific gene expression profiling in plants by using a combination of laser microdissection and high-throughput technologies. Plant Cell Physiol 48:3–7
88. Gautam V, Sarkar AK (2014) Laser assisted microdissection, an efficient technique to understand tissue specific gene expression patterns and functional genomics in plants. Mol Biotechnol 57:299–308
89. Birnbaum K, Jung JW, Wang JY et al (2005) Cell type-specific expression profiling in plants via cell sorting of protoplasts from fluorescent reporter lines. Nat Methods 2:615–619
90. Birnbaum K, Shasha DE, Wang JY et al (2003) A gene expression map of the Arabidopsis root. Science 302:1956–1960
91. Carter AD, Bonyadi R, Gifford ML (2013) The use of fluorescence-activated cell sorting in studying plant development and environmental responses. Int J Dev Biol 57:545–552
92. Deal RB, Henikoff S (2010) A simple method for gene expression and chromatin profiling of individual cell types within a tissue. Dev Cell 18:1030–1040
93. Deal RB, Henikoff S (2011) The INTACT method for cell type-specific gene expression and chromatin profiling in Arabidopsis thaliana. Nat Protoc 6:56–68
94. Wang D, Deal RB (2015) Epigenome profiling of specific plant cell types using a streamlined INTACT protocol and ChIP-seq. Methods Mol Biol 1284:3–25
95. Speicher MR, Carter NP (2005) The new cytogenetics: blurring the boundaries with molecular biology. Nat Rev Genet 6:782–792
96. Lee JH, Daugharthy ER, Scheiman J et al (2015) Fluorescent in situ sequencing (FISSEQ) of RNA for gene expression profiling in intact cells and tissues. Nat Protoc 10:442–458
97. Sephton CF, Cenik C, Kucukural A et al (2011) Identification of neuronal RNA targets of TDP-43-containing ribonucleoprotein complexes. J Biol Chem 286:1204–1215
98. Xing D, Wang Y, Hamilton M et al (2015) Transcriptome-wide identification of RNA targets of Arabidopsis SERINE/ARGININE-RICH45 uncovers the unexpected roles of this RNA binding protein in RNA processing. Plant Cell 27:3294–3308
99. Mi S, Cai T, Hu Y et al (2008) Sorting of small RNAs into Arabidopsis argonaute complexes is directed by the 5′ terminal nucleotide. Cell 133:116–127
100. Keene JD, Komisarow JM, Friedersdorf MB (2006) RIP-Chip: the isolation and identification of mRNAs, microRNAs and protein components of ribonucleoprotein complexes from cell extracts. Nat Protoc 1:302–307
101. Carbonell A (2017) Immunoprecipitation and high-throughput sequencing of ARGONAUTE-bound target RNAs from plants. Methods Mol Biol 1640:93–112
102. Cui X, Liang Z, Shen L et al (2017) 5-methylcytosine RNA methylation in Arabidopsis Thaliana. Mol Plant 10:1387–1399
103. Licatalosi DD, Mele A, Fak JJ et al (2008) HITS-CLIP yields genome-wide insights into brain alternative RNA processing. Nature 456:464–469
104. Wang T, Xiao G, Chu Y et al (2015) Design and bioinformatics analysis of genome-wide CLIP experiments. Nucleic Acids Res 43:5263–5274
105. Haque N, Hogg JR (2016) Easier, better, faster, stronger: improved methods for RNA-protein interaction studies. Mol Cell 62:650–651
106. Van Nostrand EL, Pratt GA, Shishkin AA et al (2016) Robust transcriptome-wide discovery of RNA-binding protein binding sites with enhanced CLIP (eCLIP). Nat Methods 13:508–514
107. Zarnegar BJ, Flynn RA, Shen Y et al (2016) irCLIP platform for efficient characterization of protein-RNA interactions. Nat Methods 13:489–492
108. Moore MJ, Zhang C, Gantman EC et al (2014) Mapping Argonaute and conventional RNA-binding protein interactions with RNA at single-nucleotide resolution using HITS-CLIP and CIMS analysis. Nat Protoc 9:263–293
109. König J, Zarnack K, Rot G et al (2011) iCLIP--transcriptome-wide mapping of protein-RNA interactions with individual nucleotide resolution. J Vis Exp 50:e2638
110. Hafner M, Landthaler M, Burger L et al (2010) Transcriptome-wide identification of RNA-binding protein and microRNA target sites by PAR-CLIP. Cell 141:129–141

111. Ascano M, Hafner M, Cekan P et al (2012) Identification of RNA-protein interaction networks using PAR-CLIP. Wiley Interdiscip Rev RNA 3:159–177
112. Garzia A, Meyer C, Morozov P et al (2017) Optimization of PAR-CLIP for transcriptome-wide identification of binding sites of RNA-binding proteins. Methods 118–119:24–40
113. Hafner M, Lianoglou S, Tuschl T et al (2012) Genome-wide identification of miRNA targets by PAR-CLIP. Methods 58:94–105
114. Chu C, Qu K, Zhong FL et al (2011) Genomic maps of long noncoding RNA occupancy reveal principles of RNA-chromatin interactions. Mol Cell 44:667–678
115. Chu C, Zhang QC, da Rocha ST et al (2015) Systematic discovery of Xist RNA binding proteins. Cell 161:404–416
116. Chu C, Quinn J, Chang HY (2012) Chromatin isolation by RNA purification (ChIRP). J Vis Exp 61:e3912
117. Chu C, Chang HY (2016) Understanding RNA-Chromatin Interactions Using Chromatin Isolation by RNA Purification (ChIRP). Methods Mol Biol 1480:115–123
118. Engreitz JM, Pandya-Jones A, McDonel P et al (2013) The Xist lncRNA exploits three-dimensional genome architecture to spread across the X chromosome. Science 341:1237973
119. Engreitz J, Lander ES, Guttman M (2015) RNA antisense purification (RAP) for mapping RNA interactions with chromatin. Methods Mol Biol 1262:183–197
120. Simon MD, Wang CI, Kharchenko PV et al (2011) The genomic binding sites of a noncoding RNA. Proc Natl Acad Sci 108:20497–20502
121. Simon MD, Pinter SF, Fang R et al (2013) High-resolution Xist binding maps reveal two-step spreading during X-chromosome inactivation. Nature 504:465–469
122. Simon MD (2013) Capture hybridization analysis of RNA targets (CHART). Curr Protoc Mol Biol edited by Frederick M. Ausubel ... [et al.] Chapter 21:Unit 21.25
123. Engreitz JM, Sirokman K, McDonel P et al (2014) RNA-RNA interactions enable specific targeting of noncoding RNAs to nascent Pre-mRNAs and chromatin sites. Cell 159:188–199
124. Kudla G, Granneman S, Hahn D et al (2011) Cross-linking, ligation, and sequencing of hybrids reveals RNA-RNA interactions in yeast. Proc Natl Acad Sci 108:10010–10015
125. Helwak A, Kudla G, Dudnakova T et al (2013) Mapping the human miRNA interactome by CLASH reveals frequent noncanonical binding. Cell 153:654–665
126. Helwak A, Tollervey D (2014) Mapping the miRNA interactome by cross-linking ligation and sequencing of hybrids (CLASH). Nat Protoc 9:711–728
127. Helwak A, Tollervey D (2016) Identification of miRNA-target RNA interactions using CLASH. Methods Mol Biol 1358:229–251
128. Lucks JB, Mortimer SA, Trapnell C et al (2011) Multiplexed RNA structure characterization with selective 2′-hydroxyl acylation analyzed by primer extension sequencing (SHAPE-Seq). Proc Natl Acad Sci U S A 108:11063–11068
129. Watters KE, Yu AM, Strobel EJ et al (2016) Characterizing RNA structures in vitro and in vivo with selective 2′-hydroxyl acylation analyzed by primer extension sequencing (SHAPE-Seq). Methods 103:34–48
130. Mortimer SA, Trapnell C, Aviran S et al (2012) SHAPE-seq: high-throughput RNA structure analysis. Curr Protoc Chem Biol 4:275–297
131. Ding Y, Tang Y, Kwok CK et al (2014) In vivo genome-wide profiling of RNA secondary structure reveals novel regulatory features. Nature 505:696–700
132. Ritchey LE, Su Z, Tang Y et al (2017) Structure-seq2: sensitive and accurate genome-wide profiling of RNA structure in vivo. Nucleic Acids Res 45:e135–e135
133. Tang Y, Bouvier E, Kwok CK et al (2015) StructureFold: genome-wide RNA secondary structure mapping and reconstruction in vivo. Bioinformatics 31:2668–2675
134. Tack DC, Tang Y, Ritchey LE et al (2018) StructureFold2: bringing chemical probing data into the computational fold of RNA structural analysis. Methods 143:12–15
135. Ding Y, Kwok CK, Tang Y et al (2015) Genome-wide profiling of in vivo RNA structure at single-nucleotide resolution using structure-seq. Nat Protoc 10:1050–1066
136. Silverman IM, Li F, Alexander A et al (2014) RNase-mediated protein footprint sequencing reveals protein-binding sites throughout the human transcriptome. Genome Biol 15: R3
137. Gosai SJ, Foley SW, Wang D et al (2015) Global analysis of the RNA-protein interaction and RNA secondary structure landscapes of the Arabidopsis nucleus. Mol Cell 57:376–388

138. Foley SW, Gregory BD (2016) Protein Interaction Profile Sequencing (PIP-seq). Curr Protoc Mol Biol. / edited by Frederick M. Ausubel ... [et al.] 116:27.5.1–27.5.15
139. Kertesz M, Wan Y, Mazor E et al (2010) Genome-wide measurement of RNA secondary structure in yeast. Nature 467:103–107
140. Wan Y, Qu K, Ouyang Z et al (2013) Genome-wide mapping of RNA structure using nuclease digestion and high-throughput sequencing. Nat Protoc 8:849–869
141. Underwood JG, Uzilov AV, Katzman S et al (2010) FragSeq: transcriptome-wide RNA structure probing using high-throughput sequencing. Nat Methods 7:995–1001
142. Uzilov AV, Underwood JG (2016) High-throughput nuclease probing of RNA structures using FragSeq. Methods Mol Biol 1490:105–134
143. Lazof DB, GOLDSMITH JKG, RUFTY TW et al (2011) The preparation of cryosections from plant tissue: an alternative method appropriate for secondary ion mass spectrometry studies of nutrient tracers and trace metals. J Microsc 176:99–109
144. Kim E-D, Xiong Y, Pyo Y et al (2017) Spatiotemporal analysis of coding and long noncoding transcripts during maize endosperm development. Sci Rep 7:3838
145. Barcala M, Fenoll C, Escobar C (2012) Laser microdissection of cells and isolation of high-quality RNA after cryosectioning. Methods Mol Biol 883:87–95
146. Blokhina O, Valerio C, Sokołowska K et al (2016) Laser capture microdissection protocol for xylem tissues of woody plants. Front Plant Sci 7:1965
147. Bevilacqua C, Ducos B (2018) Laser microdissection: a powerful tool for genomics at cell level. Mol Aspects Med 59:5–27
148. Gautam V, Singh A, Singh S et al (2016) An efficient LCM-based method for tissue specific expression analysis of genes and miRNAs. Sci Rep 6:21577
149. Nakazono M, Qiu F, Borsuk LA et al (2003) Laser-capture microdissection, a tool for the global analysis of gene expression in specific plant cell types: identification of genes expressed differentially in epidermal cells or vascular tissues of maize. Plant Cell 15:583–596
150. Ohtsu K, Smith MB, Emrich SJ et al (2007) Global gene expression analysis of the shoot apical meristem of maize (Zea mays L.). Plant J 52:391–404
151. Li S, Yamada M, Han X et al (2016) High-resolution expression map of the Arabidopsis root reveals alternative splicing and lincRNA regulation. Dev Cell 39:508–522
152. Ferrè F, Colantoni A, Helmer-Citterich M (2016) Revealing protein-lncRNA interaction. Brief Bioinform 17:106–116
153. Wheeler EC, Van Nostrand EL, Yeo GW (2018) Advances and challenges in the detection of transcriptome-wide protein-RNA interactions. Wiley Interdiscip Rev RNA 9: e1436
154. Augui S, Nora EP, Heard E (2011) Regulation of X-chromosome inactivation by the X-inactivation centre. Nat Rev Genet 12:429–442
155. Dominissini D, Moshitch-Moshkovitz S, Salmon-Divon M et al (2013) Transcriptome-wide mapping of N6-methyladenosine by m6A-seq based on immunocapturing and massively parallel sequencing. Nat Protoc 8:176–189
156. Dominissini D, Moshitch-Moshkovitz S, Schwartz S et al (2012) Topology of the human and mouse m6A RNA methylomes revealed by m6A-seq. Nature 485:201–206
157. Squires JE, Patel HR, Nousch M et al (2012) Widespread occurrence of 5-methylcytosine in human coding and non-coding RNA. Nucleic Acids Res 40:5023–5033
158. David R, Burgess A, Parker B et al (2017) Transcriptome-wide mapping of RNA 5-methylcytosine in Arabidopsis mRNAs and noncoding RNAs. Plant Cell 29:445–460
159. Khoddami V, Cairns BR (2013) Identification of direct targets and modified bases of RNA cytosine methyltransferases. Nat Biotechnol 31:458–464
160. Ule J, Jensen KB, Ruggiu M et al (2003) CLIP identifies Nova-regulated RNA networks in the brain. Science 302:1212–1215
161. Ule J, Jensen K, Mele A et al (2005) CLIP: a method for identifying protein-RNA interaction sites in living cells. Methods 37:376–386
162. Jeon Y, Lee JT (2011) YY1 tethers Xist RNA to the inactive X nucleation center. Cell 146:119–133
163. Martianov I, Ramadass A, Serra Barros A et al (2007) Repression of the human dihydrofolate reductase gene by a non-coding interfering transcript. Nature 445:666–670
164. Schmitz K-M, Mayer C, Postepska A et al (2010) Interaction of noncoding RNA with the rDNA promoter mediates recruitment of DNMT3b and silencing of rRNA genes. Genes Dev 24:2264–2269

165. Tollervey JR, Curk T, Rogelj B et al (2011) Characterizing the RNA targets and position-dependent splicing regulation by TDP-43. Nat Neurosci 14:452–458
166. Wang G, Chen H-W, Oktay Y et al (2010) PNPASE regulates RNA import into mitochondria. Cell 142:456–467
167. Vandivier LE, Anderson SJ, Foley SW et al (2016) The conservation and function of RNA secondary structure in plants. Annu Rev Plant Biol 67:463–488
168. Bevilacqua PC, Ritchey LE, Su Z et al (2016) Genome-wide analysis of RNA secondary structure. Annu Rev Genet 50:235–266
169. Wan Y, Kertesz M, Spitale RC et al (2011) Understanding the transcriptome through RNA structure. Nat Rev Genet 12:641–655
170. Hawkes EJ, Hennelly SP, Novikova IV et al (2016) COOLAIR antisense RNAs form evolutionarily conserved elaborate secondary structures. Cell Rep 16:3087–3096
171. Merino EJ, Wilkinson KA, Coughlan JL et al (2005) RNA structure analysis at single nucleotide resolution by selective 2′-hydroxyl acylation and primer extension (SHAPE). J Am Chem Soc 127:4223–4231
172. Watts JM, Dang KK, Gorelick RJ et al (2009) Architecture and secondary structure of an entire HIV-1 RNA genome. Nature 460:711–716
173. Zaug AJ, Cech TR (1995) Analysis of the structure of Tetrahymena nuclear RNAs in vivo: telomerase RNA, the self-splicing rRNA intron, and U2 snRNA. RNA 1:363–374
174. Rouskin S, Zubradt M, Washietl S et al (2014) Genome-wide probing of RNA structure reveals active unfolding of mRNA structures in vivo. Nature 505:701–705
175. Talkish J, May G, Lin Y et al (2014) Mod-seq: high-throughput sequencing for chemical probing of RNA structure. RNA 20:713–720
176. Silverman IM, Gregory BD (2015) Transcriptome-wide ribonuclease-mediated protein footprinting to identify RNA-protein interaction sites. Methods 72:76–85

Part I

Role of ncRNAs in Regulation of Gene Expression and Development in Plants

Chapter 2

Analysis of siRNA Precursors Generated by RNA Polymerase IV and RNA-Dependent RNA Polymerase 2 in Arabidopsis

Todd Blevins, Ram Podicheti, and Craig S. Pikaard

Abstract

Noncoding RNAs perform diverse regulatory functions in living cells. In plants, two RNA polymerase II-related enzymes, RNA polymerases IV and V (Pol IV and V), specialize in the synthesis of noncoding RNAs that silence a subset of transposable elements and genes via RNA-directed DNA methylation (RdDM). In this process, Pol IV partners with RNA-dependent RNA polymerase 2 (RDR2) to produce double-stranded RNAs that are then cut by an RNase III enzyme, Dicer-like 3 (DCL3), into 24 nt small interfering RNAs (siRNAs). The siRNAs are loaded into an Argonaute family protein, primarily AGO4, and guide the complex to complementary DNA target sequences where RdDM and repressive chromatin modifications ensue. The dependence of 24 nt siRNA biogenesis on Pol IV and RDR2 has been known for more than a decade, but the elusive pre-siRNA transcripts synthesized by Pol IV and RDR2 have only recently been identified. This chapter describes the approaches that enabled our identification of Pol IV/ RDR2-dependent RNAs (P4R2 RNAs) in *Arabidopsis thaliana*. These included the use of a triple Dicer mutant (*dcl2 dcl3 dcl4*) to cause P4R2 RNAs to accumulate, genome-wide identification and mapping of P4R2 RNAs using a modified Illumina small RNA-Seq protocol, and multiple bioinformatic pipelines for data analysis and displaying results.

Key words Nuclear multisubunit RNA polymerase IV (Pol IV), RNA-dependent RNA polymerase 2 (RDR2), RNA-directed DNA methylation (RdDM), siRNAs, P4R2 RNAs

1 Introduction

Diverse noncoding RNAs mediate chromatin modification, RNA silencing, and genome surveillance in eukaryotic cells. An evolutionarily conserved mode of RNA silencing is to use 20–25 nucleotide (nt) small interfering RNAs (siRNAs) to guide Argonaute (AGO) effector proteins to complementary target sequences [1, 2]. In terrestrial plants, duplication and subfunctionalization of genes involved in this mode of silencing have spawned multiple microRNA and siRNA biogenesis pathways with distinct silencing repertoires [3–5]. Specialized AGO proteins, loaded with appropriate small RNA guides, can cleave mRNAs to regulate gene

Julia A. Chekanova and Hsiao-Lin V. Wang (eds.), *Plant Long Non-Coding RNAs: Methods and Protocols*, Methods in Molecular Biology, vol. 1933, https://doi.org/10.1007/978-1-4939-9045-0_2,

expression [6], inhibit mRNA translation [7], or silence transposable elements (TEs) via RNA-directed DNA methylation (RdDM) [8]. In most eukaryotes, the RNAs involved in RNA silencing are initially synthesized by RNA polymerase II (Pol II), but in the case of RNA-directed DNA methylation in plants, two specialized forms of Pol II, Pol IV and Pol V, play leading roles.

RNA polymerase IV (Pol IV) is thought to initiate the canonical RdDM pathway [9, 10]. This 12-subunit enzyme appears to be recruited to target sites, primarily transposable elements (TEs) and tandem repeats [11, 12], by chromatin marks that include DNA methylation (at CG motifs) and posttranslational modifications of the histone proteins that wrap the DNA. Pol IV is thought to synthesize primary transcripts that serve as templates for second-strand synthesis by RNA-dependent RNA polymerase 2 (RDR2), which physically associates with Pol IV [13]. Resulting double-stranded RNAs (dsRNAs) are then processed into 24 nt siRNA duplexes by the RNase III enzyme Dicer-like 3 (DCL3) (Fig. 1a). The siRNAs are 2′-*O*-methylated by HEN1 and loaded into AGO4, or a related Argonaute family protein, and are deduced to guide AGO4-siRNA complexes to their chromosomal targets via base pairing with longer "scaffold" transcripts synthesized by Pol V [14], although direct AGO4-siRNA pairing to DNA in the Pol V transcription bubble has also been posited [15]. AGO4-siRNA complexes bound to Pol V nascent transcripts are then thought to engage in protein-protein interactions that recruit the de novo cytosine methyltransferase, DRM2 [16], leading to DNA methylation in all sequence contexts (CG, CHG, and CHH, where H is a nucleotide other than G) and a transcriptionally silenced state (Fig. 1b).

In this chapter we present methods for characterizing the noncoding RNAs that are precursors of the 24 nt siRNAs that guide RdDM (Fig. 2a). Initial evidence for the existence of these precursors came from small RNA blots (northern blots), which revealed that the loss of 24 nt siRNAs in a *dcl3* mutant was accompanied by the detection of larger RNAs that, like the siRNAs, were dependent on both Pol IV and RDR2; this observation suggested that these so-called P4R2 RNAs might be the undiced precursors of the siRNAs (Fig. 2b) [17–20]. In support of this hypothesis, addition of affinity-purified DCL3 to RNA isolated from a *dcl2 dcl3 dcl4* triple mutant, in which P4R2 RNAs accumulate to their highest levels, resulted in the cutting of the P4R2 RNAs into 24 nt products. This led us to gel purify and sequence the P4R2 RNAs, allowing their chromosomal origins, length profiles, and sequence characteristics to be identified [18]. Whereas small RNA profiles in wild-type (WT) plants reveal a prominent peak at 24 nt, *dcl2/3/4* triple mutants lack this peak and instead show accumulation of P4R2 RNAs of ~26–45 nt, consistent with northern blot results (Fig. 2b).

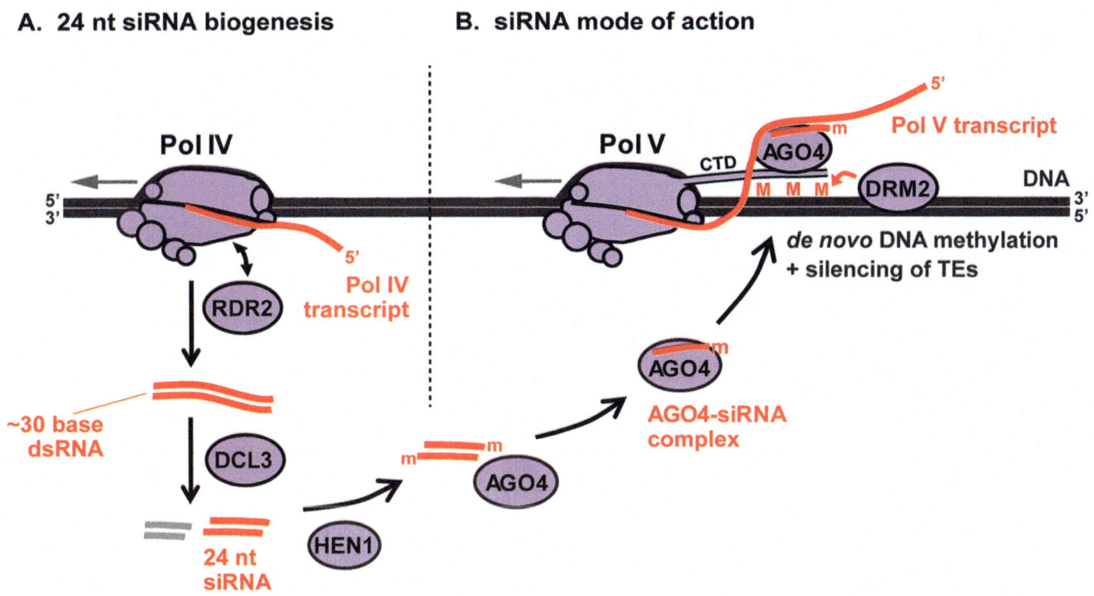

Fig. 1 Simplified model of RNA-directed DNA methylation (RdDM) in plants. (**a**) To initiate small RNA biogenesis, nuclear RNA polymerase IV (Pol IV) transcribes chromosomal DNA into noncoding RNA transcripts. RNA-dependent RNA polymerase 2 (RDR2) physically associates with Pol IV and is presumed to use Pol IV transcripts as templates to synthesize double-stranded RNAs (dsRNAs), which are then processed by Dicer-like 3 (DCL3) into 24 nt siRNA duplexes. The siRNA duplexes are 2′-*O*-methylated by HEN1 (lowercase "m"), and then an individual siRNA strand becomes associated with Argonaute 4 (AGO4). (**b**) 24 nt siRNA-AGO4 complexes find sites of action by binding to Pol V transcripts and by interacting with the C-terminal domain (CTD) of Pol V's largest subunit. These events facilitate recruitment of the DNA methyltransferase DRM2, which de novo methylates cytosines (uppercase "M"), resulting in chromatin-level silencing of transposable elements (TEs) and other RdDM targets

To study the Pol IV and RDR2-dependent RNAs (P4R2 RNAs) genome-wide, we modified an existing analysis pipeline for small RNA-Seq, which is the highly parallel sequencing of cDNA derived from low-molecular-weight RNAs, such as endogenous microRNAs and siRNAs, or virus-derived siRNAs from plant or insect antiviral defense pathways [21–24]. As in forerunner protocols [25, 26], Illumina's TruSeq small RNA library preparation protocol captures RNA fragments that have a 5′-monophosphate and 3′-OH at their termini. Purifying ~15–90 nt RNAs by polyacrylamide gel electrophoresis, we successfully captured and sequenced P4R2 RNAs via the Illumina protocol and then mapped the sequence reads to the *Arabidopsis* genome (browser shot, Fig. 2c) [18]. We found that P4R2 RNAs average ~30 nt in length, suggesting that they are only long enough to be diced once, releasing a solitary siRNA, a conclusion also reached in a contemporaneous study by another group [18, 27].

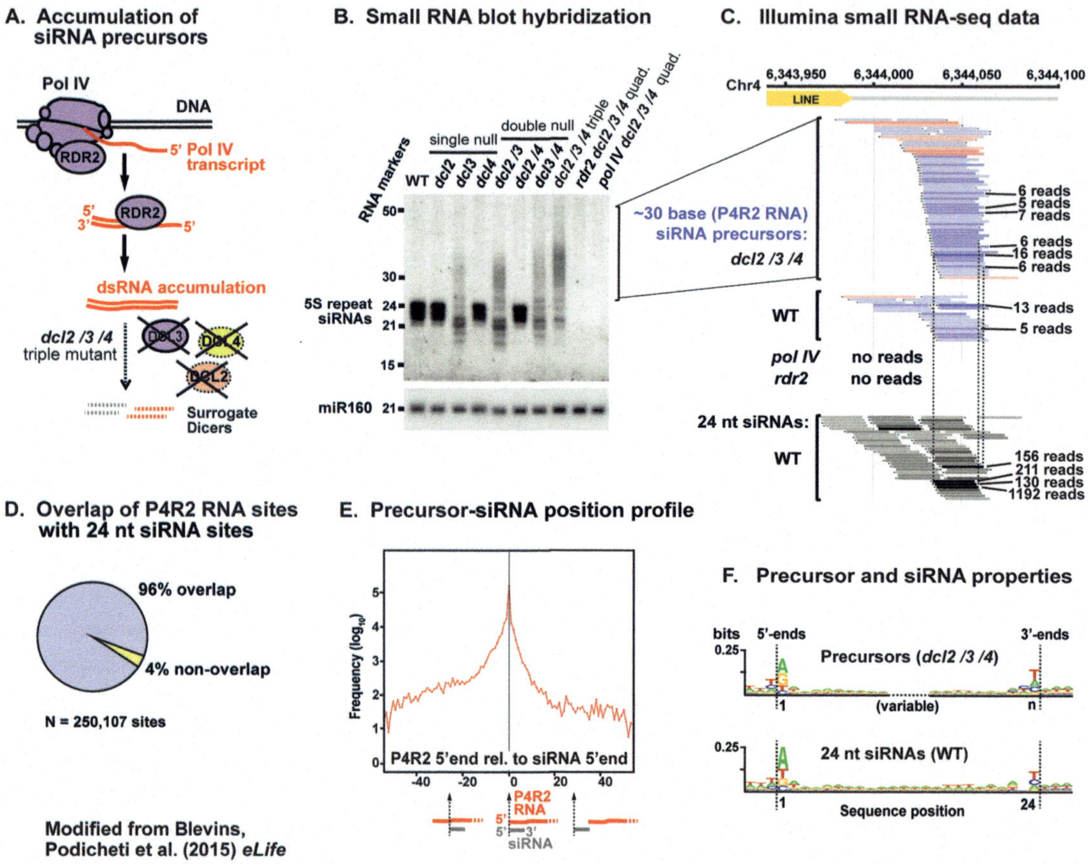

Fig. 2 Pol IV and RDR2-dependent RNAs analyzed by blot hybridization and sequencing. (**a**) Accumulation of Pol IV and RDR2-dependent RNAs (P4R2 RNAs) in plants deficient for DCL3 and two alternative Dicers, DCL2 and DCL4. In *dcl2 dcl3 dcl4* (*dcl2/3/4*) triple mutant plants, dsRNAs synthesized by RDR2 remain undiced, allowing P4R2 RNAs to be detected on blots and isolated for sequencing. (**b**) Small RNA blot analysis of 24 nt siRNAs and their precursors. The blot was first hybridized to a probe matching the intergenic region of 5S rRNA gene repeats (siR1003) and then stripped and re-probed to detect a microRNA, miR160, which serves as a loading control. (**c**) JBrowse view of an siRNA locus flanking a LINE retrotransposon, at which P4R2 RNAs accumulate in the *dcl2/3/4* triple mutant; these P4R2 RNAs are also detected at low levels in the wild type (WT), but not in *pol IV* or *rdr2* mutants. P4R2 RNA clusters in *dcl2/3/4* tend to align with abundant 24 nt siRNAs in the WT. Vertical dotted lines provide alignments between abundant siRNAs and P4R2 RNAs. P4R2 RNAs longer than 40 nt are shown in shades of pink, whereas RNAs of 25–39 nt are shown in shades of blue. (**d**) Correspondence between P4R2 RNA and siRNA loci. P4R2 RNAs were mapped to the *Arabidopsis* reference genome (TAIR10), and the frequency at which 24 nt siRNAs overlap these P4R2 genomic positions was calculated. To be considered for this analysis, specific siRNAs had to be represented by at least five reads in WT plants. (**e**) P4R2 RNA and 24 nt siRNA spatial relationships. The plot depicts the frequency distribution of P4R2 RNA 5′-end positions relative to siRNA 5′-ends. At position zero on the x-axis, P4R2 RNAs and siRNAs share the same 5′-terminus. Negative values indicate how far (in nt) the 5′-end of the P4R2 RNA is located upstream of an siRNA start position. (**f**) Sequence logos for chromosomal DNA sequences corresponding to P4R2 RNAs observed in *dcl2/3/4* triple mutants (top panel) or 24 nt siRNAs identified in WT plants (bottom panel). The logos include three nucleotides upstream and three nucleotides downstream of the chromosomal DNA sequences that match the RNAs. Each unique RNA sequence is represented only once in the input data

2 Materials

The experiments described in this protocol were performed using RNA extracted from *Arabidopsis thaliana* plants [18]. The work was facilitated, in particular, by the *A. thaliana dcl2/3/4* triple mutant line. However, the basic approach should permit P4R2 RNA detection and sequencing from any plant species in which the DCL responsible for dicing Pol IV and RDR2-dependent dsRNAs can be mutated or significantly knocked down (e.g., DCL3 in *Physcomitrella patens* [28] or DCL3a in *Oryza sativa* [29]).

2.1 Plant Materials

1. *A. thaliana* (Col-0 ecotype) mutants *rdr2-1* and *nrpd1-3 (pol IV)* are used as controls that cause the loss of P4R2 RNAs and 24 nt siRNAs in comparison to wild-type (WT) Col-0 samples. These T-DNA insertion mutant lines are described in Xie et al. (2004) and Onodera et al. (2005), respectively [3, 30].
2. Single DCL-deficient T-DNA insertion mutants *dcl2-5, dcl3-1, and dcl4-2*; double mutants *dcl2-5 dcl3-1*, *dcl2-5 dcl4-2*, and *dcl3-1 dcl4-2*; and triple mutant *dcl2-5 dcl3-1 dcl4-2* (*dcl2/3/4*) plant lines were described in Blevins et al. [31].
3. Quadruple mutants *rdr2-1 dcl2-5 dcl3-1 dcl4-2* (*rdr2 dcl2/3/4*) and *nrpd1-3 dcl2-5 dcl3-1 dcl4-2* (*pol IV dcl2/3/4*) [18] are used as negative controls for the enhanced P4R2 RNA accumulation observed in the *dcl2/3/4* triple mutant.
4. For the experiments described below, about 4 g of seedling tissue for each genotype was harvested 2 weeks post-germination and frozen in liquid nitrogen (*see* **Note 1**).

2.2 RNA Analyses

1. TRIzol reagent (Invitrogen, Carlsbad, CA, USA).
2. RNeasy Mini Spin columns and kit (Qiagen, Venlo, Netherlands).
3. NanoDrop spectrophotometer (NanoDrop Technologies, Wilmington, DE, USA).
4. Vertical polyacrylamide gel electrophoresis (PAGE) apparatus, with matching glass gel plates (~14 cm × 16 cm), spacers (1 mm thick), and gel comb (1 mm thick with 20 teeth).
5. TBE buffer stock solution (5×): 450 mM Tris-borate, 10 mM EDTA, pH 8.0. Store at room temperature.
6. Acrylamide/urea stock solution for 16% PAGE: 40 mL of 40% acrylamide/bis-acrylamide (19:1) (Sigma-Aldrich, St. Louis, MO, USA), 42 g urea and 20 mL of 5× TBE are combined in a 250 mL flask with bidistilled water to a 100 mL final volume, mixed until urea dissolves and then filtered through a Millipore Express Plus (0.22 μm) presterilized filter (Millipore, Billerica, MA, USA).

Table 1
Synthetic RNA oligonucleotide size standards for Urea-PAGE

Oligo name	Sequence
GFP-15 nt-RNA	5′-GUAAACGGCCACAAG-3′
GFP-21 nt-RNA	5′-GUAAACGGCCACAAGUUCAGC-3′
GFP-24 nt-RNA	5′-GUAAACGGCCACAAGUUCAGCGUG-3′
GFP-30 nt-RNA	5′-GUAAACGGCCACAAGUUCAGCGUGUCCGGC-3′
GFP-40 nt-RNA	5′-GUAAACGGCCACAAGUUCAGCGUGUCCGGCGAGGGCGAGG-3′

7. N, N, N′, N′-Tetramethylethylenediamine (TEMED), ~99% (Sigma-Aldrich). Store at 4 °C.
8. Ammonium persulfate (USB Corp., Cleveland, OH, USA); 10% (w/v) in bidistilled water.
9. Size standards for PAGE: synthetic RNA oligonucleotides (oligos) of 15, 21, 24, 30, and 40 nt lengths (Integrated DNA Technologies, Coralville, IA, USA) matching an exogenous sequence (Table 1). These synthetic oligos are combined with the "Low Range ssRNA Ladder" (New England Biolabs, Ipswich, MA, USA), which includes 50, 80, and 150 nt RNAs. For a less expensive alternative, *see* **Note 2**.
10. RNA loading buffer: 95% (v/v) formamide, 0.025% (w/v) bromophenol blue, 0.025% (w/v) xylene cyanol FF, 5 mM EDTA, 0.025% (w/v) SDS, pH 8.5. Store in 500 μL aliquots at −20 °C.
11. Ethidium bromide (10 mg/mL stock), diluted 10,000× in 1× TBE for staining.
12. Hybond-N+ positively charged nylon membrane (GE Healthcare, Chicago, IL, USA).
13. Immersed electroblotting apparatus (e.g., Trans-Blot Cell system; Bio-Rad, Hercules, CA, USA).
14. An ultraviolet cross-linking device (e.g., Stratalinker apparatus; Stratagene, La Jolla, CA, USA).
15. PerfectHyb Plus hybridization buffer (Sigma-Aldrich). Store at room temperature.
16. A hybridization (carrousel) oven and fitted tubes with a 20–30 mL buffer/wash capacity.
17. ^{32}P-end-labeled DNA oligonucleotides for blot hybridizations: the probe for "5S repeat siRNAs" (siR1003 sense) was 5′-ATG CCT ATG TTG GCC TCA CGG TCT-3′ and for "miR160" was 5′-TGG CAT ACA GGG AGC CAG GCA-3′ (Integrated

DNA Technologies). Radioactive probe preparation is described in Blevins et al. [32].

18. Post-hybridization washing solution (1×): 2× SSC, 0.5% SDS. Resuspend SDS at 65 °C for 20 min if it precipitates during storage at room temperature.
19. Phosphorimager screen with a cassette and a Typhoon Variable Mode Imager (GE Healthcare).
20. Illumina TruSeq small RNA kit for 24 reactions RS-200-0012 (Illumina, San Diego, CA, USA), SuperScript II (Invitrogen), and T4 RNA Ligase 2, truncated (New England Biolabs). For a less expensive alternative, *see* **Note 3**.

2.3 Bioinformatics Resources

1. Small RNA-Seq data generated using this protocol [18] are publicly available at the NCBI Sequence Read Archive (Table 2).
2. The reference sequence for the *A. thaliana* Col-0 ecotype genome (FASTA format) and corresponding TAIR10 functional annotation files (GFF3 format)—used to map, filter, and analyze P4R2 RNA sequencing data—can be obtained at https://www.araport.org/downloads (*see* **Note 4**).
3. Bioinformatics analyses described below were conducted on an x86_64 Intel Xeon server running Ubuntu (version 14.04 LTS) with 16 cores and 128 Gb RAM. The following software is recommended: Trimmomatic (version ≥0.33); bowtie (version ≥0.12.8); SAMtools (version ≥1.3); BEDtools (version ≥2.20.0); Perl (version ≥5.18); FEATnotator (version ≥1.2.2); WebLogo (version ≥2.8.2); and R for generating plots.

3 Methods

3.1 RNA Blot Hybridization Analysis of P4R2 RNAs

1. For each sample/replicate, about 1–2 g of tissue from 2-week-old plants is ground to a fine powder in liquid nitrogen with a mortar and pestle. This frozen powder is extracted using TRIzol (Invitrogen) to obtain 200–400 μg total RNA (caution: *see* **Note 5**).
2. Size-fractionate ~90 μg of total RNA using RNeasy Mini Columns (Qiagen) as described in Blevins et al. [32], except resuspend the low-molecular-weight (LMW) RNA fraction in 15 μL RNase-free water. Quantify this LMW RNA on a NanoDrop spectrophotometer (yield is typically 8–10 μg). Transfer equal-mass aliquots of each sample to new microfuge tubes, and bring each volume to 12 μL by adding sufficient RNase-free water.
3. Prepare the RNA size standard mixture in a microfuge tube by combining RNA oligos as follows: 3 μL of 15 nt oligo, 1 μL of

Table 2
Small RNA-Seq data from Blevins, Podicheti et al. (2015)

Dataset name	Replicate	SRA accession
WT Col-0	Rep1	SRR2075815
dcl2 dcl3 dcl4	Rep1	SRR2075816
nrpd1-3	Rep1	SRR2505369
rdr2-1	Rep1	SRR2505433
WT Col-0	Rep2	SRR2075819
dcl2 dcl3 dcl4	Rep2	SRR2075821
nrpd1-3	Rep2	SRR2505408
rdr2-1	Rep2	SRR2505479

Raw FASTQ-formatted datasets can be downloaded from the following address: https://www.ncbi.nlm.nih.gov//sra/?term=SRP059814

21 nt oligo, 1 μL of 24 nt oligo, 0.5 μL of 30 nt oligo, and 0.5 μL of 40 nt oligo (each from 100 μM stock solutions). Then add 1.5 μL of "Low Range ssRNA Ladder" (New England Biolabs). Bring the total volume to 12 μL by adding 4.5 μL RNase-free water (*see* **Note 2**).

4. Prepare a denaturing 16% polyacrylamide (Urea-PAGE) gel for RNA size separation. In brief, transfer 25 mL of acrylamide/urea stock solution to a 50 mL falcon tube. Add 25 μL TEMED and 250 μL APS (10%), mix quickly by inversion, and then pipette the mixture between the glass plates in the gel assembly. Insert the gel comb, taking care to avoid bubble formation. Wait 20 min for polymerization, then set up the PAGE apparatus, add 1× TBE to both buffer reservoirs, remove the comb, and pre-run for 20 min at a voltage sufficient for ~12 W, or typically 450–550 V (Caution: *see* **Note 6**).

5. Add 12 μL RNA loading buffer to each LMW RNA sample and to the size standard mixture. Heat these tubes for 3 min at 95 °C to minimize RNA secondary structure, and then immediately return them to ice for polyacrylamide gel loading.

6. Disconnect the power from the pre-running gel, and thoroughly rinse out the sample slots with 1× TBE. Load the denatured RNA samples and size standard mixture into adjacent gel slots, and then run the gel for 1–2 h, so as to maintain ~12 W. Bromophenol blue in the loading buffer will run off into the lower buffer reservoir. Electrophoresis may be complete when the xylene cyanol FF marker has migrated to the midpoint of the gel.

7. Slide the gel, supported underneath by a glass plate, into ethidium bromide solution (prepare fresh: 30 μL of 10 mg/mL ethidium bromide in 300 mL 1× TBE) and stain for 20 min on an orbital shaker. Image the gel on an ultraviolet (UV) trans-illuminator (Caution: *see* **Note 7**).
8. Depending on gel dimensions, cut a ~12 × 14 cm rectangle of Hybond-N+ nylon membrane (GE Healthcare) and two equivalent rectangles of Whatman paper (GE Healthcare). Label the membrane's edge in pencil to indicate the side to which RNA will be blotted. Immerse the transfer sponges, membrane, and paper in 1× TBE. Assemble the electroblotting stack with the gel holder cassette's *black* half underneath: (1) sponge, (2) paper, (3) gel, (4) membrane, (5) paper, and (6) sponge. Remove air bubbles between each layer by rolling a serological pipette across the stack as it is assembled. Close the cassette, and place it in the electroblotting tank with the *black* half facing the *negative* electrode. Fill the tank with 1× TBE, connect the electrode lid to the power supply, and transfer for 2 h at ~10 V.
9. Expose the membrane surface to an energy of 140 mJ in a UV cross-linker, then slide the membrane into a hybridization tube with the RNA-transfer side facing the interior. Add 10 mL PerfectHyb Plus buffer and pre-hybridize for at least 4 h at 35 °C.
10. Prepare a ^{32}P-end-labeled DNA oligonucleotide probe (e.g., siR1003 probe for 5S rRNA intergenic repeats, as described in [18, 32]), and then add 20 μL of the concentrated probe to the hybridization tube, taking care that the probe mixes well with the hybridization buffer. Hybridize for at least 8 h at 35 °C, and then wash the membrane three times, 20 min each, with 2× SSC, 0.5% SDS at 40 °C (Caution: *see* **Note 8**).
11. Seal the washed membrane in a transparent plastic bag (or in kitchen plastic wrap), and then tape the membrane into a cassette with an erased phosphorimager screen. Remove the screen after 24 h, and scan it on the Typhoon Variable Mode Imager to check for specific RNA hybridization signals. A longer exposure of 3–5 days will be necessary to detect P4R2 RNA signals in their entirety (Fig. 2b). Exposure to X-ray film for conventional autoradiography can be done as an alternative to phosphorimaging.

3.2 Sequencing of P4R2 RNAs and Related siRNAs

1. Load at least 5 μg total RNA onto a 16% denaturing polyacrylamide gel (Urea-PAGE, as previously, but without prior Qiagen column-based size fractionation). Take care to space RNA samples apart from one another by at least 3–4 empty gel lanes. After the gel run is complete, stain the gel for 20 min in ethidium bromide (as previously). Excise ~15–90 nt RNAs

with a razor on a UV trans-illuminator, noting the 15 and 80 nt size markers to estimate the correct size range (*see* **Note 9**).

2. To elute the RNAs, cut the gel slices into 1 mm cubes, transfer to a 15 mL falcon tube, add 3 mL 0.3 M NaCl, and rotate the tubes overnight at 4 °C. Pipette the solution to a new tube. Add 2 mL fresh 0.3 M NaCl to the gel fragments and rotate at 4 °C for another 4 h. Combine the two elutions. Centrifuge briefly to pellet any residual gel fragments, and then transfer the solution to a new tube.
3. Precipitate the size-selected RNA by adding 1/20 volume 4 M sodium acetate (pH 5.2), 10 μL glycogen (20 mg/mL), and 2 volumes ethanol, and then store overnight at −20 °C. Centrifuge at 4500 × *g* for 45 min (4 °C). Wash the pellet twice with cold 75% ethanol, transfer the pellet to a low-bind microfuge tube, dry the pellet briefly, and then resuspend in 20 μL bidistilled water.
4. Ligate standard Illumina 3′- and 5′-adapters (RA3 and RA5) to the purified ~15–90 nt RNAs, and perform reverse transcription, PCR amplification, quality control, and normalization steps following the TruSeq small RNA library procedure in the supplier's manual (Illumina, *see* **Note 3**). For the experiments in [18], two replicates of RNA samples from four different *A. thaliana* genotypes were barcoded for sequencing on a single HiSeq 2500 instrument by Fasteris SA (http://www.fasteris.com/), following the manufacturer's recommendations (TruSeq SBS Kit version 3, Illumina).

3.3 Mapping of the Small RNA-Seq Reads

1. Raw FASTQ-formatted reads were trimmed using Trimmomatic 0.33 (trimming options: ILLUMINACLIP:adapters.fa:2:20:7 MINLEN:15) [33]. Reads lacking an identifiable 3′-adapter sequence (RA3: 5′-TGGAATTCTCGGGTGC…3′) and reads shorter than 15 nt after trimming were excluded. No base quality trimming was applied, because it could alter the size class to which an RNA insert belongs. However, reads containing any bases with quality scores less than Q25 were excluded. The clean FASTQ inserts ranged from 15 to 94 nt in length after removal of the 3′-adapter sequence.
2. Before small RNA-Seq read alignment, a custom reference genome was prepared by combining the *A. thaliana* TAIR10 assembly—five nuclear chromosomes (Chr1 to Chr5), chloroplast genome (ChrC), and mitochondrial genome (ChrM)—with the ~10 kb sequence of a 45S rRNA gene repeat (45S_rRNA) [34]. This custom genome was indexed using bowtie-build, and then the trimmed FASTQ inserts were mapped against it using bowtie (version 0.12.8, command line options: -v 1 --best --strata -k 50) [35], allowing up to

one mismatch per alignment and reporting up to 50 alignments per read ID.

3. The resulting Sequence Alignment/Map (SAM) files were converted to Binary Alignment/Map (BAM) format [36] and cleaned of ribosomal RNA gene repeat and organelle genome alignments as follows: first, a list of aligned read IDs (BAM file, first column) was prepared including all reads for which the reference sequence name (BAM file, third column) was 45S_rRNA, ChrC or ChrM. Second, the BAM file was re-scanned, comparing values in the first column against this list. Any row in the BAM file with a read ID matching an entry in the list was excluded.
4. To profile RNA size-class abundances throughout the genome, occurrences of individual 15–94 nt inserts were tabulated at each genomic position. For each sample library, reads of the same size class with coinciding 5′-ends on the same strand were merged into counts. These counts were normalized by dividing by the total count of mapped reads for the library and then multiplying by the maximum total read count among all the libraries. Based on TAIR10 functional annotations, reads mapped to the genomic positions of transfer RNA (tRNA) genes, small nuclear RNA (snRNAs) genes, and small nucleolar RNA (snoRNAs) genes were excluded from further analysis.
5. Loci in wild-type (WT) samples represented by at least five 24 nt reads were designated as siRNA positions, and those loci in *dcl2/3/4* triple mutant samples with at least 5 reads mapping of any size class between 26 and 45 nt were designated as putative P4R2 RNA positions.

3.4 Data Visualization and Analysis

1. The filtered BAM files containing P4R2 RNA and siRNA alignments were streamlined for JBrowse visualization (*see* Fig. 2c) [37] by collapsing any reoccurring alignments to a single entry and storing the alignment's frequency instead as a custom "XC" tag. Using the "hooks/modify" parameter in JBrowse configuration (http://gmod.org/wiki/JBrowse_FAQ), alignments for P4R2 RNAs are depicted in shades of pink or blue, while those for siRNAs are shown in shades of gray to black, with color intensity reflecting the alignment frequency stored in the "XC" tag. This minimizes rendering time in web browsers and makes the display more intuitive at loci with many clustered P4R2 RNA and siRNA reads.
2. The spatial relationships of P4R2 RNA 5′- and 3′-ends relative to the corresponding ends of 24 nt siRNAs were computed using FEATnotator (version 1.2.2) [38]. P4R2 RNA 5′- and 3′-end positions, respectively, were saved to separate tab-delimited text files composed of three columns:

chromosome, 5′- or 3′-end position, and genomic strand. Then, the genomic alignment coordinates for 5′- and 3′-end positions of each siRNA were saved in General Feature Format (GFF) [39].

3. Each table of P4R2 RNA alignment positions (three-column text file) was annotated with corresponding siRNA ends (GFF file) by providing the two files as input to FEATnotator. For each P4R2 RNA 5′- or 3′-end, FEATnotator reports if there is an siRNA end at the same position (i.e., a position difference of 0). If not, then the software calculates the distance from the P4R2 RNA end to the nearest mapped 24 nt siRNA end. These P4R2 RNA-siRNA relative alignment positions (x-axis) and their corresponding frequencies on a $\log_{10}$ scale (y-axis) were plotted using the R plot function (Fig. 2e).

4. The consensus sequences for loci corresponding to P4R2 RNAs and siRNAs were obtained using WebLogo [40]. P4R2 RNA sequence logos were generated by collating the 5-’ and 3′-ends, respectively, from all TAIR10 genomic source loci of 26–45 nt RNAs supported by at least five identical reads in the *dcl2/3/4* sample. Ungapped lists of these termini, along with three flanking bases, yielded two files:

   File 1

   P4R2 locus 1: 3 upstream bases +10 bases of the 5′-end

   P4R2 locus 2: 3 upstream bases +10 bases of the 5′-end

   . . .

   File 2

   P4R2 locus 1: 10 bases of the 3′-end +3 downstream bases

   P4R2 locus 2: 10 bases of the 3′-end +3 downstream bases

   . . .

   These files were then supplied as input to WebLogo (version 2.8.2, command line options: -k 1 -n -Y -b -c -a -s -2) (Fig. 2f, top panel).

5. For the 24 nt siRNA consensus sequence (derived from WT Col-0 data), the set of unique sequences including three bases upstream and three bases downstream (30 nt total) was extracted from the TAIR10 reference. These siRNA alignment positions with upstream and downstream context were supplied as input to WebLogo using the same command line options as in **step 4** (Fig. 2f, bottom panel).

4 Notes

1. To obtain better yields of low-molecular-weight RNA (including P4R2 RNAs and siRNAs), inflorescences from 6-week-old plants can be collected instead of whole 2-week-old seedlings.
2. A less expensive alternative for PAGE size standards: combine the "NEB microRNA Marker," which includes 17, 21, and 25 nt oligonucleotides, with the "NEB Low Range ssRNA Ladder" (both from New England Biolabs), which includes 50, 80, 150, 300, 500, and 1000 nt RNAs.
3. Constructing libraries using the Illumina TruSeq small RNA kit (24 reactions, RS-200-0012) requires the additional purchase of two other enzymes: SuperScript II reverse transcriptase (Invitrogen) and T4 RNA Ligase 2, truncated mutant (New England Biolabs). By contrast, the NEBNext Multiplex Small RNA library Prep Set for Illumina (24 reactions, NEB #E7300S) includes the latter reagents and is itself less expensive, making the total cost about half that of using Illumina-branded reagents.
4. Reference sequences for numerous other plant genomes can be easily downloaded using the Phytozome [41] portal hosted at the Joint Genome Institute of the US Department of Energy: https://genome.jgi.doe.gov/portal/pages/tree-of-life.jsf
5. TRIzol is toxic and corrosive: always wear gloves and work in a fume hood while pipetting TRIzol and during subsequent manipulation of solutions containing phenol traces.
6. Acrylamide monomer is a neurotoxin and potential carcinogen. Wear gloves and clean any acrylamide spillage thoroughly.
7. Ethidium bromide is an irritant, a mutagen, and a suspected carcinogen. Wear nitrile gloves for handling this chemical.
8. Radiation protection measures must be taken for probe preparation, hybridization, detection, and washing steps. Wear gloves and a lab coat and monitor the work area for contamination.
9. NGS services sometimes offer to replace labor-intensive PAGE purification by bead-based enrichment of low-molecular-weight RNA (18–200 nt) prior to Illumina TruSeq Small RNA library preparation. We compared both protocols side by side from the same total RNA samples and determined that bead-based size selection resulted in ~fourfold less recovery of 20–25 nt small RNAs while maintaining efficient recovery of P4R2 RNAs. Agarose gel-based size selection of library DNA was performed simultaneously for both protocols at the indexing step, so we hypothesize that the problem with bead-based RNA size selection in this particular application is the

poor binding and/or recovery of 20–25 nt small RNAs (including microRNAs and siRNAs) on the beads themselves.

Acknowledgments

The authors thank Doug Rusch and Haixu Tang for helping guide the informatics analyses, as well as Ross Cocklin for generating the *rdr2 dcl2/3/4* quadruple mutant. C.S.P. is an Investigator at the Howard Hughes Medical Institute (HHMI) and former Plant Investigator of the Gordon and Betty Moore Foundation (GBMF). This work was supported by the National Institutes of Health (NIH) Grant GM077590, GBMF Grant GBMF3036 (to C.S.P.), and Investigator support funding of C.S.P. from HHMI. T.B. was supported, in part, by an NIH Ruth L. Kirschstein National Research Service Award. T.B. is currently supported by the LabEx consortium ANR-10-LABX-0036_NETRNA ("Investissements d'Avenir") and by the French Agence Nationale de la Recherche (ANR) Grant ANR-17-CE20-0004-01.

References

1. Wilson RC, Doudna JA (2013) Molecular mechanisms of RNA interference. Annu Rev Biophys 42:217–239. https://doi.org/10.1146/annurev-biophys-083012-130404
2. Vaucheret H (2008) Plant ARGONAUTES. Trends Plant Sci 13(7):350–358. https://doi.org/10.1016/j.tplants.2008.04.007
3. Xie Z, Johansen LK, Gustafson AM, Kasschau KD, Lellis AD, Zilberman D, Jacobsen SE, Carrington JC (2004) Genetic and functional diversification of small RNA pathways in plants. PLoS Biol 2(5):E104. https://doi.org/10.1371/journal.pbio.0020104
4. Yoshikawa M, Peragine A, Park MY, Poethig RS (2005) A pathway for the biogenesis of trans-acting siRNAs in Arabidopsis. Genes Dev 19(18):2164–2175. https://doi.org/10.1101/gad.1352605
5. Vazquez F, Vaucheret H, Rajagopalan R, Lepers C, Gasciolli V, Mallory AC, Hilbert JL, Bartel DP, Crete P (2004) Endogenous trans-acting siRNAs regulate the accumulation of Arabidopsis mRNAs. Mol Cell 16(1):69–79
6. Llave C, Xie Z, Kasschau KD, Carrington JC (2002) Cleavage of Scarecrow-like mRNA targets directed by a class of Arabidopsis miRNA. Science 297(5589):2053–2056. https://doi.org/10.1126/science.1076311
7. Brodersen P, Sakvarelidze-Achard L, Bruun-Rasmussen M, Dunoyer P, Yamamoto YY, Sieburth L, Voinnet O (2008) Widespread translational inhibition by plant miRNAs and siRNAs. Science 320(5880):1185–1190. https://doi.org/10.1126/science.1159151
8. Matzke MA, Kanno T, Matzke AJ (2015) RNA-Directed DNA methylation: the evolution of a complex epigenetic pathway in flowering plants. Annu Rev Plant Biol 66:243–267. https://doi.org/10.1146/annurev-arplant-043014-114633
9. Ream TS, Haag JR, Wierzbicki AT, Nicora CD, Norbeck AD, Zhu JK, Hagen G, Guilfoyle TJ, Pasa-Tolic L, Pikaard CS (2009) Subunit compositions of the RNA-silencing enzymes Pol IV and Pol V reveal their origins as specialized forms of RNA polymerase II. Mol Cell 33 (2):192–203. https://doi.org/10.1016/j.molcel.2008.12.015
10. Wendte JM, Pikaard CS (2017) The RNAs of RNA-directed DNA methylation. Biochim Biophys Acta 1860(1):140–148. https://doi.org/10.1016/j.bbagrm.2016.08.004
11. Law JA, Du J, Hale CJ, Feng S, Krajewski K, Palanca AM, Strahl BD, Patel DJ, Jacobsen SE (2013) Polymerase IV occupancy at RNA-directed DNA methylation sites requires SHH1. Nature 498(7454):385–389. https://doi.org/10.1038/nature12178

12. Blevins T, Pontvianne F, Cocklin R, Podicheti R, Chandrasekhara C, Yerneni S, Braun C, Lee B, Rusch D, Mockaitis K, Tang H, Pikaard CS (2014) A two-step process for epigenetic inheritance in Arabidopsis. Mol Cell 54(1):30–42. https://doi.org/10.1016/j.molcel.2014.02.019
13. Haag JR, Ream TS, Marasco M, Nicora CD, Norbeck AD, Pasa-Tolic L, Pikaard CS (2012) In vitro transcription activities of Pol IV, Pol V, and RDR2 reveal coupling of Pol IV and RDR2 for dsRNA synthesis in plant RNA silencing. Mol Cell 48(5):811–818. https://doi.org/10.1016/j.molcel.2012.09.027
14. Wierzbicki AT, Ream TS, Haag JR, Pikaard CS (2009) RNA polymerase V transcription guides ARGONAUTE4 to chromatin. Nat Genet 41 (5):630–634. https://doi.org/10.1038/ng.365
15. Lahmy S, Pontier D, Bies-Etheve N, Laudie M, Feng S, Jobet E, Hale CJ, Cooke R, Hakimi MA, Angelov D, Jacobsen SE, Lagrange T (2016) Evidence for ARGONAUTE4-DNA interactions in RNA-directed DNA methylation in plants. Genes Dev 30(23):2565–2570. https://doi.org/10.1101/gad.289553.116
16. Zhong X, Du J, Hale CJ, Gallego-Bartolome J, Feng S, Vashisht AA, Chory J, Wohlschlegel JA, Patel DJ, Jacobsen SE (2014) Molecular mechanism of action of plant DRM de novo DNA methyltransferases. Cell 157 (5):1050–1060. https://doi.org/10.1016/j.cell.2014.03.056
17. Blevins T, Pontes O, Pikaard CS, Meins F Jr (2009) Heterochromatic siRNAs and DDM1 independently silence aberrant 5S rDNA transcripts in Arabidopsis. PLoS One 4(6):e5932. https://doi.org/10.1371/journal.pone.0005932
18. Blevins T, Podicheti R, Mishra V, Marasco M, Wang J, Rusch D, Tang H, Pikaard CS (2015) Identification of Pol IV and RDR2-dependent precursors of 24 nt siRNAs guiding de novo DNA methylation in Arabidopsis. elife 4: e09591. https://doi.org/10.7554/eLife.09591
19. Pontes O, Li CF, Nunes PC, Haag J, Ream T, Vitins A, Jacobsen SE, Pikaard CS (2006) The Arabidopsis chromatin-modifying nuclear siRNA pathway involves a nucleolar RNA processing center. Cell 126(1):79–92
20. Henderson IR, Zhang X, Lu C, Johnson L, Meyers BC, Green PJ, Jacobsen SE (2006) Dissecting Arabidopsis thaliana DICER function in small RNA processing, gene silencing and DNA methylation patterning. Nat Genet 38(6):721–725 Epub 2006 May 2014
21. Blevins T, Rajeswaran R, Aregger M, Borah BK, Schepetilnikov M, Baerlocher L, Farinelli L, Meins F Jr, Hohn T, Pooggin MM (2011) Massive production of small RNAs from a non-coding region of Cauliflower mosaic virus in plant defense and viral counter-defense. Nucleic Acids Res 39 (12):5003–5014. https://doi.org/10.1093/nar/gkr119
22. Axtell MJ (2013) Classification and comparison of small RNAs from plants. Annu Rev Plant Biol 64:137–159. https://doi.org/10.1146/annurev-arplant-050312-120043
23. Rajagopalan R, Vaucheret H, Trejo J, Bartel DP (2006) A diverse and evolutionarily fluid set of microRNAs in Arabidopsis thaliana. Genes Dev 20(24):3407–3425
24. Kasschau KD, Fahlgren N, Chapman EJ, Sullivan CM, Cumbie JS, Givan SA, Carrington JC (2007) Genome-wide profiling and analysis of Arabidopsis siRNAs. PLoS Biol 5(3):e57
25. Elbashir SM, Lendeckel W, Tuschl T (2001) RNA interference is mediated by 21- and 22-nucleotide RNAs. Genes Dev 15 (2):188–200
26. Lu C, Meyers BC, Green PJ (2007) Construction of small RNA cDNA libraries for deep sequencing. Methods 43(2):110–117. https://doi.org/10.1016/j.ymeth.2007.05.002
27. Zhai J, Bischof S, Wang H, Feng S, Lee TF, Teng C, Chen X, Park SY, Liu L, Gallego-Bartolome J, Liu W, Henderson IR, Meyers BC, Ausin I, Jacobsen SE (2015) A one precursor one siRNA model for Pol IV-dependent siRNA biogenesis. Cell 163(2):445–455. https://doi.org/10.1016/j.cell.2015.09.032
28. Coruh C, Cho SH, Shahid S, Liu Q, Wierzbicki A, Axtell MJ (2015) Comprehensive annotation of Physcomitrella patens small RNA loci reveals that the heterochromatic short interfering RNA pathway is largely conserved in land plants. Plant Cell 27 (8):2148–2162. https://doi.org/10.1105/tpc.15.00228
29. Fei Q, Xia R, Meyers BC (2013) Phased, secondary, small interfering RNAs in posttranscriptional regulatory networks. Plant Cell 25 (7):2400–2415. https://doi.org/10.1105/tpc.113.114652
30. Onodera Y, Haag JR, Ream T, Costa Nunes P, Pontes O, Pikaard CS (2005) Plant nuclear RNA polymerase IV mediates siRNA and DNA methylation-dependent heterochromatin formation. Cell 120(5):613–622. https://doi.org/10.1016/j.cell.2005.02.007

31. Blevins T, Rajeswaran R, Shivaprasad PV, Beknazariants D, Si-Ammour A, Park HS, Vazquez F, Robertson D, Meins F Jr, Hohn T, Pooggin MM (2006) Four plant Dicers mediate viral small RNA biogenesis and DNA virus induced silencing. Nucleic Acids Res 34(21):6233–6246. https://doi.org/10.1093/nar/gkl886
32. Blevins T (2017) Northern blotting techniques for small RNAs. Methods Mol Biol 1456:141–162. https://doi.org/10.1007/978-1-4899-7708-3_12
33. Bolger AM, Lohse M, Usadel B (2014) Trimmomatic: a flexible trimmer for Illumina sequence data. Bioinformatics 30 (15):2114–2120. https://doi.org/10.1093/bioinformatics/btu170
34. Chandrasekhara C, Mohannath G, Blevins T, Pontvianne F, Pikaard CS (2016) Chromosome-specific NOR inactivation explains selective rRNA gene silencing and dosage control in Arabidopsis. Genes Dev 30 (2):177–190. https://doi.org/10.1101/gad.273755.115
35. Langmead B, Trapnell C, Pop M, Salzberg SL (2009) Ultrafast and memory-efficient alignment of short DNA sequences to the human genome. Genome Biol 10(3):R25. https://doi.org/10.1186/gb-2009-10-3-r25
36. Li H, Handsaker B, Wysoker A, Fennell T, Ruan J, Homer N, Marth G, Abecasis G, Durbin R (2009) The sequence alignment/map format and SAMtools. Bioinformatics 25 (16):2078–2079. https://doi.org/10.1093/bioinformatics/btp352
37. Skinner ME, Uzilov AV, Stein LD, Mungall CJ, Holmes IH (2009) JBrowse: a next-generation genome browser. Genome Res 19 (9):1630–1638. https://doi.org/10.1101/gr.094607.109
38. Podicheti R, Mockaitis K (2015) FEATnotator: a tool for integrated annotation of sequence features and variation, facilitating interpretation in genomics experiments. Methods 79–80:11–17. https://doi.org/10.1016/j.ymeth.2015.04.028
39. Stein L (2013) Generic feature format version 3. https://github.com/The-Sequence-Ontology/Specifications/blob/master/gff3.md
40. Crooks GE, Hon G, Chandonia JM, Brenner SE (2004) WebLogo: a sequence logo generator. Genome Res 14(6):1188–1190. https://doi.org/10.1101/gr.849004
41. Goodstein DM, Shu S, Howson R, Neupane R, Hayes RD, Fazo J, Mitros T, Dirks W, Hellsten U, Putnam N, Rokhsar DS (2012) Phytozome: a comparative platform for green plant genomics. Nucleic Acids Res 40(Database issue):D1178–D1186. https://doi.org/10.1093/nar/gkr944

Chapter 3

Identification of Long Noncoding RNAs in the Developing Endosperm of Maize

Eundeok Kim, Yuqing Xiong, Byung-Ho Kang, and Sibum Sung

Abstract

Maize endosperm consists of three distinct types of tissues, including the starchy endosperm (SE), the basal endosperm transfer cell layer (BETL), and the aleurone cell layer (AL). Compartmentalization of these tissues during endosperm differentiation makes the endosperm development an excellent model to study changes in gene expression during development. By utilizing cryo-dissection of developing endosperm, morphologically distinct samples can be obtained for transcriptome and epigenome analysis. Here, we describe methods for the isolation of tissues from developing maize endosperm and for the transcriptome analysis to identify novel long noncoding RNAs. The transcriptome data can be further analyzed to illustrate spatiotemporal changes in both coding and noncoding transcripts during the endosperm development.

Key words Endosperm, Long noncoding RNAs, Transcriptome analysis, Cryo-dissection, Maize, In situ hybridization

1 Introduction

High-throughput RNA sequencing (RNA-Seq) has identified rare transcripts, including long noncoding RNAs (lncRNAs). Eukaryotic transcriptomes include a large number of long noncoding RNAs, some of which plays roles in various developmental programs [1]. A number of recent studies have shown that lncRNAs are essential *cis* and/or *trans* regulators of gene activity that can function as scaffolds for chromatin-modifying complexes and precursors for regulatory RNAs [2–5]. Although several lncRNAs have been known for decades, it has only been appreciated recently with the advent of high-throughput RNA sequencing (RNA-Seq) [6–8]. Despite growing numbers of identified lncRNAs in

Electronic supplementary material: The online version of this chapter (https://doi.org/10.1007/978-1-4939-9045-0_3) contains supplementary material, which is available to authorized users.

Julia A. Chekanova and Hsiao-Lin V. Wang (eds.), *Plant Long Non-Coding RNAs: Methods and Protocols*,
Methods in Molecular Biology, vol. 1933, https://doi.org/10.1007/978-1-4939-9045-0_3,

eukaryotes, functions and underlying mechanisms of lncRNA action are still poorly understood. As the first step to decipher the roles of lncRNAs, the establishment of robust bioinformatics analysis of transcriptome data is essential to correctly annotate lncRNAs. In addition, identification of differentially expressed lncRNAs during development would help identify potential regulatory lncRNAs.

Development of maize endosperm includes compartmentalization of differentiated tissues, making it feasible to micro-dissect endosperm to enrich tissue samples that are at different developmental stages of each specific cell types. Laser capture microdissection (LCM) [9] or cryo-dissection [10] can be used to isolate specific cell types. Cryo-dissection method is less invasive compared to the LCM, although it is rather labor-intensive. Despite of the large number of transcriptome data obtained from maize endosperm [11–16], spatiotemporal transcriptome data analysis using developing endosperm has been poorly reported. Here, we described methods of transcriptome analysis using cell-type-specific tissues at different developmental stages, by which spatiotemporally expressed lncRNAs can be identified. We also described methods to validate identified lncRNAs by reverse transcription followed by quantitative PCR (RT-PCR) and by in situ hybridization in maize endosperm.

2 Materials

2.1 RNA Extraction from Three Cell Types of the Maize Endosperm

2.1.1 Equipment

1. Cryostat (Leica Microsystem, CM1950).
2. Dissecting stereomicroscope (Meiji Techno, EMZ-200).
3. Staining jars.
4. Scalpels.
5. NanoDrop ND-1000 Spectrophotometer (Thermo Fisher Scientific).

2.1.2 Solutions and Chemicals

1. Liquid nitrogen.
2. Tissue-Tek optimum cutting temperature-embedding compound (VWR).
3. HistoGene LCM Frozen Section Staining Kit (Applied Biosystems).
4. PEN membrane glass slides (Applied Biosystems).
5. PicoPure RNA Isolation Kit (Applied Biosystems).
6. RapidOut DNA Removal Kit (Thermo Fisher Scientific).
7. RiboMinus Plant Kit for RNA-Seq (Invitrogen).
8. Glycogen (Sigma).

9. Sodium acetate.
10. 100% ethanol.
11. Xylene.

2.2 Preparation of cDNA Libraries for Illumina Sequencing

2.2.1 Equipment

1. Agilent Bioanalyzer.

2.2.2 Solutions and Chemicals

1. Invitrogen Dynabeads®.
2. Oligo(dT)$_{25}$ (Invitrogen).
3. Actinomycin D (Sigma).
4. AMPure XP (Agencourt).
5. RNAClean XP (Agencourt).
6. RNasin Plus RNase Inhibitor (Promega).
7. PicoGreen dsDNA Assay Kit (Invitrogen).
8. dNTP (10 mM each of dATP, dTTP, dGTP, dCTP).
9. Superscript III Reverse Transcriptase (Invitrogen).
10. DTT.
11. DNA Polymerase I (NEB).
12. End-Repair Mix (Enzymatics).
13. T4 DNA Ligase (NEB).
14. Klenow 3′-5′ exo- (NEB).
15. RNase H (NEB).
16. End-Repair Mix (NEB).
17. Phusion® High-Fidelity DNA Polymerase (NEB).
18. $MgCl_2$.
19. DMSO.
20. 100% ethanol.
21. Nuclease-free water.

2.2.3 Buffers

1. Binding buffer: 20 mM Tris–HCl, pH 7.5, 1.0 M LiCl, 2 mM EDTA.
2. Washing buffer: 10 mM Tris–HCl, pH 7.5, 0.15 M LiCl, 1 mM EDTA.

2.3 Equipment Used to Identification of Long Noncoding RNAs

2.3.1 Hardware

1. 64-bit computer with 1 TB hard disk and 16 GB of memory.

2.3.2 Software

1. Unix/Linux platform (e.g., Mac OS or Ubuntu).
2. Python 2.7.12 with NumPy, SciPy, and Pylab package.
3. Bedtools.
4. Cufflinks.
5. Biopython.

2.4 Quantitative RT-PCR

2.4.1 Equipment

1. Pellet pestle or equivalent.
2. Liquid nitrogen.
3. Thermal cycler.
4. Real-time PCR system.
5. DNase- and RNase-free microfuge tubes, 1.5 mL.
6. DNase- and RNase-free PCR tubes, 0.2 mL.

2.4.2 Solutions and Chemicals

1. Nuclease-free water.
2. Random primer.
3. SYBR Green/ROX qPCR master mix (Thermo Fisher Scientific).
4. M-MLV Reverse Transcriptase (Promega).
5. RNase H (Promega).
6. DNase I (Invitrogen).
7. Dithiothreitol (DTT).
8. Plant RNA Extraction Kit (Qiagen).
9. PCR Purification Kit (Qiagen).

2.5 In Situ Hybridization

2.5.1 Equipment

1. NanoDrop ND-1000 Spectrophotometer (Thermo Fisher Scientific).

2.5.2 Solutions and Chemicals

1. AccuPrime^{TM} *Pfx* DNA Polymerase (Invitrogen).
2. QIAquick PCR Purification Kit.
3. Zero Blunt^{R} TOPO^{R} PCR Cloning Kit (Invitrogen).
4. M13-F (or M13-R) primers.
5. DIG RNA Labeling Mix (Sigma-Aldrich).
6. Protector RNase Inhibitor (Sigma-Aldrich).
7. SP6 RNA Polymerase (Sigma-Aldrich).

8. T7 RNA Polymerase (or SP6 RNA Polymerase, Sigma-Aldrich).
9. Qiagen RNeasy MinElute Cleanup Kit (Qiagen).
10. Nitro blue tetrazolium chloride (Sigma-Aldrich).
11. BCIP (Sigma-Aldrich).
12. Acetic acid.
13. Sodium acetate.
14. 100% ethanol.
15. Diethyl pyrocarbonate (DEPC).
16. Acetone.
17. Formamide.
18. Polyvinyl alcohol.

2.5.3 Buffers

1. Sodium carbonate buffer: 200 mM sodium carbonate, 200.
2. Maleic acid buffer: 100 mM maleic acid, 150 mM NaCl, pH 7.5
3. Developing solution: 100 mM Tris–HCl pH 9.5, 150 mM NaCl, 50 mM $MgCl_2$.
4. 20× SSC buffer: 3 M NaCl, 0.3 M sodium citrate, pH 7.0.
5. 20× SSPE buffer: 0.02M EDTA, 2.98 M NaCl, 0.2 M phosphate buffer (pH 7.4).
6. Washing buffer: 50% formamide, 2× SSC.
7. NTE buffer: 500 mM NaCl, 10 mM Tris–HCl pH 7.5, 1 mM EDTA
8. 10× Phosphate buffer saline: 1.27 M NaCl, 27 mM KCl, 100 mM Na_2HPO_4, 18 mM KH_2PO_4.
9. Detection Buffer 1: 100 mM Tris–HCl pH 7.5, 150 mM NaCl.
10. Detection Buffer 2: 100 mM Tris–HCl pH 9.5, 100 mM NaCl, 50 mM $MgCl_2$.

3 Methods

3.1 RNA Extraction from Three Cell Types in the Maize Endosperm

3.1.1 Cryosectioning and Isolating Total RNA Samples from Maize Endosperm

1. Collect maize kernels at the field and freeze them immediately with liquid nitrogen.
2. Mount frozen kernels on cryostat stubs, and prepare 30-μm sections in cryostat (−22 °C).
3. Collect sections on PEN membrane glass slides. After thawing the frozen sections, quickly stain the sections with HistoGene LCM Frozen Section Staining Solution.

4. Dehydrate the stained sections in ethanol (75, 90, and 100%) and finally in 100% xylene.
5. After drying 5 min in the fume hood, dissect out starchy endosperm, basal endosperm cell layer, and aleurone cell layer with scalpel. The PEN membrane facilitates separation of tissue specimens from the glass slide.
6. Isolate total RNA samples from dissected tissue specimens with a PicoPure RNA Isolation Kit according to the manufacturer's direction, and remove DNA from the total RNA samples with a RapidOut DNA Removal Kit. RNA samples should be more than 1 μg for Illumina library preparation.

3.1.2 Removal of Ribosomal RNA

1. To a sterile, RNase-free 1.5 mL microcentrifuge tube, add the following:

Total RNA (1–5 μg):	4.8 μL
Spike-in control RNA (50 pg/μL):	0.2 μL
RiboMinus™ Probe (15 pmol/μL):	2.5 μL
Hybridization Buffer:	22.5 μL

2. Incubate the tube at 70–75 °C for 5 min to denature RNA.
3. Allow the sample to cool to 37 °C slowly over a period of 30 min by placing the tube in a 37 °C water bath. It is important to allow slow cooling for sequence-specific hybridization.
4. After the 37 °C incubation step of the hybridized sample (above), briefly centrifuge the tube to collect the sample to the bottom of the tube.
5. Transfer the sample to the RiboMinus magnetic beads in the kit. Mix by pipetting up and down or low-speed vortexing.
6. Incubate the tube at 37 °C for 15 min. During incubation, gently mix the contents occasionally. Briefly centrifuge the tube to collect the sample to the bottom.
7. Place the tube on a magnetic separator for 1 min to pellet the rRNA-probe complex. Collect the supernatant containing RiboMinus RNA.
8. Repeat **steps 2–4** to further remove rRNA-probe complex.
9. Transfer the RiboMinus RNA sample into a clean, nuclease-free 1.5 mL microcentrifuge tube, and add the following components to precipitate RNA.

   1 μL glycogen (20 mg/mL).

   1/10th sample (eluted RNA) volume of 3M sodium acetate.

   2.5× sample volumes of 100% ethanol.

10. Centrifuge the tube for 30 min at the highest speed at 4 °C. Carefully discard the supernatant without disturbing the pellet.
11. Wash the pellet with 70% cold ethanol twice.
12. Air-dry the pellet for ~5 min. Resuspend the RNA pellet in 10 μL DEPC-treated water.

3.2 Preparation of cDNA Libraries for Illumina Sequencing

3.2.1 RNA Isolation and Fragmentation

1. Add 10 μL of binding buffer to RNA samples from Subheading 3.1, and heat up to 65 °C to disrupt secondary structures. After 2 min at 65 °C, place on ice immediately.
2. To the 20 μL of RNA samples, add 10 μL Dynabeads. Mix the beads and solution thoroughly, and anneal by rotating continuously on a mixer for 5 min at room temperature.
3. Place the tube on the magnet for 1–2 min and carefully remove all the supernatants.
4. Remove the tube from the magnet and add 20 μL washing buffer. Mix by pipetting carefully a couple of times. Use the magnet to pull the beads to the side of the tube. Carefully remove all the supernatants.
5. Repeat the washing step as described in **step 4**.
6. Remove the tube from the magnet, and resuspend the beads using 10 μL 2× SuperScript III buffer with 10 mM DTT.
7. Incubate at 94 °C for 10 min to fragment the mRNA. The incubation time, however, must be optimized for your samples.
8. Briefly spin the tube and place it on the magnet. Transfer the supernatant quickly to a new 0.5 mL microcentrifuge tube.

3.2.2 First-Strand cDNA Synthesis

1. Assemble the following RT reaction:

Fragmented mRNA	10 μL
Random hexamer	0.5 μL
Rnasin Plus	0.5 μL

2. Heat at 50 °C for 1 min. Immediately place on ice. Add 9 μL of the following master mix to the fragmented mRNA

H_2O	6.88 μL
Actinomycin D (1 μg/μL)	0.12 μL
DTT (100 mM)	1 μL
dNTP (25 mM)	0.5 μL
SuperScript III	0.5 μL

3. Perform the RT reaction (25 °C for 10 min and 50 °C for 50 min).

4. Add 36 μL of RNAClean XP and incubate the mixture on ice for 15 min. (The solution is viscous; mix thoroughly by pipetting up and down.)
5. Collect RNA/cDNA hybrid molecules with solid-phase reversible immobilization (SPRI) beads, and wash the beads twice with 75% ethanol gently.
6. Air-dry the beads for 5 min. Elute the RNA/cDNA hybrid with 10 μL distilled water.

3.2.3 Second-Strand Synthesis and End Repair

1. Prepare the second-strand reaction master mix on ice as follows:

NEBuffer 2	1.5 μL
dNTP mix (10 mM each)	1 μL
RNase H (5 U/μL)	0.2 μL
DNA Polymerase I (10 U/μL)	1 μL
H_2O	1.3 μL

2. Add 5 μL the master mix to each 10 μL of RNA/cDNA hybrid samples. Incubate at 16 °C for 2.5 h.
3. Purify double-stranded DNA (dsDNA) using 1.8 volumes of AMPure XP beads. Elute with 10 μL of H_2O.
4. Prepare an appropriate amount of the end-repair master mix on ice as follows:

End-Repair Buffer (10×)	1.5 μL
dNTP mix (1 mM)	3 μL
End-Repair Mix	0.5 μL

5. Add 5 μL of the master mix to each 10 μL of dsDNA. Incubate at 20 °C for 30 min.
6. Purify using 1.8 volumes of AMPure XP beads. Elute with 10 μL of H_2O.

3.2.4 A-Tailing and Y-Shape Adapter Ligation

1. Prepare A-tailing master mix on ice:

NEBuffer 2	1.5 μL
dATP (10 mM)	0.5 μL
H_2O	2.5 μL
Klenow 3′-5′ exo-	0.5 μL

2. Add 5 μL of the master mix to each sample and incubate at 37 °C for 30 min.
3. Purify using 1.8 volumes of AMPure XP beads. Elute with 8 μL of H_2O.
4. Add 1 μL of the desired barcode adapter (5 μM) to each sample.
5. Prepare the master mix on ice as follows:

Rapid ligation buffer (2×)	8.5 μL
T4 DNA ligase (600 U/μL)	0.5 μL

6. Add 9 μL of the master mix to each sample and incubate at room temperature for 15 min.
7. Purify DNA using one volume (18 μL) of AMPure XP beads. Elute with 10 μL of TE buffer.
8. Purify the library again using the precisely same amount of AMPure XP beads (10 μL) and elute with 10 μL of TE buffer.

3.2.5 PCR Enrichment

1. Set up PCR reactions as follows:

dsDNA sample	5 μL
Primer A (10 μM)	0.2 μL
Primer B (10 μM)	0.2 μL
Phusion HF Buffer (5×)	4 μL
dNTP (10 mM)	0.2 μL
H_2O	10 μL
Phusion II	0.5 μL

2. Perform initial denaturation at 94 °C for 2 min, followed by 10–12 cycles of amplification (98 °C for 10 s, 65 °C for 30 s, 72 °C for 30 s). (This is a critical step.) To avoid overamplification, the PCR cycle should be optimized for each cell/tissue type. For many plant RNA samples, use 14–15 cycles for 0.5–1 μg total RNA, 10–12 cycles for 1–5 μg RNA, and 8–10 cycles if >5 μg of RNA has been used.
3. If needed, remove adapter dimers (~100 bp) from PCR reactions with SPRI cleanup, or isolate 200–500 bp PCR products using Qiagen QIAquick Gel Extraction Kit.
4. Purify using one volume of AMPure XP beads. Elute with 20 μL of TE buffer.

3.3 Identification of Long Noncoding RNAs

3.3.1 Identify Long Noncoding Transcripts

1. Generate FASTA file using BED file which includes the coordinate information of unannotated transcript from an intergenic region by using bedtools (http://bedtools.readthedocs.io/en/latest/index.html)

```
$ bedtools getfatsta [OPTIONS] -fi < input FASTA file> -
bed <BED/GFF/VCF>
```

2. Identify coding regions within the transcript sequence generated by de novo transcript assembly by using coding potential program (TransDecoder tool). TransDecoder tool will identify coding sequences that have open reading frame and homology to known proteins through blast or Pfam searches.

```
$ TransDecoder.LongOrfs -t <input FASTA file>

$ TransDecoder.Predict -t <input FASTA file>
```

3. Remove the transcripts that are identified from **step 2**.
4. Remove putative precursors of all small RNAs using the python script from [17]—also see Supplementary File 1 for the python script.
5. Remove transposable element (TE) transcripts using a modified python script (Supplementary File 2, python script 2) (*see* **Note 1**).

3.3.2 Cluster Analysis by Using the Self-Organizing Map (SOM) Analysis Tool

1. Transform the expression data by shifting FPKM values so that the minimum value be zero and normalize the maximum values to 10 (*see* **Note 2**).
2. Upload the processed value to https://genepattern.broadinstitute.org to use SOM module embedded in the GenePattern.
3. Choose SOM module and perform clustering analysis with 4,200,000 repetitions of the process (*see* **Note 3**).

3.3.3 Identification of Cis-regulatory Elements

1. Extract the genomic sequences from 1kb upstream and downstream of long noncoding RNA transcripts from each cluster from Subheading 3.3.2.
2. Upload the extracted sequences to http://meme-suite.org/ to identify de novo motifs in those regions [18].
3. Only those with *q*-values lower than 0.01 from **step 2** were taken for further analysis to identify motifs with the significant similarity between coding RNA and lncRNA clusters by using TOMTOM program [19] (*see* **Note 4**).

3.4 Validate Expression Patterns of lncRNAs by Quantitative Real-Time PCR

3.4.1 Total RNA Isolation and First-Strand cDNA Synthesis for Quantitative RT-PCR

1. Grind approximately 150 cryoselected samples (from Subheading 3.2) to a fine powder in liquid nitrogen using pellet pestle (*see* **Note 5**).
2. Isolate total RNA using RNA Isolation Kit according to the manufacturer's instructions (*see* **Note 6**).
3. Remove genomic DNA using DNase I according to the manufacturer's instructions.
4. Purify DNase I treated total RNA using RNA Purification Kit according to the manufacturer's instructions.
5. Synthesize first-strand cDNA with 1ug of total RNA from step 4 by using random primer according to the manufacturer's instructions using thermal cycler.
6. Purify the reaction from step 5 using PCR Purification Kit according to the manufacturer's instructions, and add nuclease-free water to make a total 100 μL volume.

3.4.2 Measure the Relative Expression of lncRNA by Quantitative Real-Time PCR

1. Design the primer set targeting the identified lncRNA transcripts using a Primer3 program including *ZmTXN* as an internal reference (*see* **Note 7**).
2. Prepare the qPCR amplification mix in a 0.2 mL tube on ice.

PCR amplification reaction mix	
Component	**Volume (μL)**
2× SYBR master mix	7.5
Water	4.5
10 μM primer F	0.5
10 μM primer R	0.5
Purified cDNA	2
Total	15

3. Mix well and perform PCR cycling as described Table X using a real-time machine.

Cycle number	Temperature (°C)	Time
1	95	10 min
40 cycles	95 60 72	15 s 30 s 30 s
1 cycle	4	Hold

4. Calculate the relative transcript levels of lncRNAs and analyze the data according to manufacturer's protocol.

3.5 In Situ Hybridization

3.5.1 In Vitro Transcription and Probe Shortening

1. Set up in vitro transcription reactions on ice as follows.

DNA template	5 μL (~0.5 μg)
H_2O	8 μL
10× NTP mix	2 μL
10× transcription buffer	2 μL
SP6 (or T7) Polymerase	2 μL
RNase inhibitor	1 μL

2. Mix gently, centrifuge briefly, and incubate at 37 °C, 2 h.
3. Check 1 μL reaction on 1% agarose gel.
4. Purify the RNA using the Qiagen RNeasy MinElute Cleanup Kit. Elute with 50 μL of RNase-free water.
5. Add 50 μL 200 mM carbonate buffer pH 10.2 and incubate at 60 °C for calculated times to prepare particular sizes of probe (Time = [size (kb) of the probe – 0.15]/[0.11 × size (kb) of the probe × 0.15]).
6. Transfer tubes to ice and terminate probe hydrolysis reaction by adding 10 μL 10% acetic acid and 12 μL 3 M NaOAc. (Gas bubbles will appear.)
7. Add 312 μL ethanol and incubate at −20 °C for 60 min.
8. Spin down at 4 °C for 10 min and wash the pellet with 100% ethanol.
9. Air-dry the pellet about 10 min and resuspend in 50 μL DEPC water. This solution can be stored at −20 °C.

3.5.2 Hybridization

1. Prepare hybridization buffer by mixing the following (for 12 slides):

100 μL	3 M NaCl, 0.1 M Tris–HCl pH 6.8, 0.1 M $NaPO_4$ buffer, 50 mM EDTA
400 μL	Deionized formamide
200 μL	50% Dextran sulfate
10 μL	100 mg/mL tRNA
20 μL	50× Denhardt's solution
70 μL	H_2O

2. Add 16 μL probe mix to 64 μL hybridization buffer and leave in 60 °C water bath.
3. Spread 80 μL of mix evenly on the slides. Cover with a clean coverslip.

4. Incubate overnight at 50 °C in an airtight, moist box lined with tissue soaked in 50% formamide + 2× SSC (*see* **Note 8**). Seal the box with tape and cover with foil.

3.5.3 Washing

For washing, incubate in washing buffer, NTE buffer, and PBS according to the following order and conditions.

1.	Washing buffer	10–15 min	50 °C
2.	Washing buffer	30–60 min	50 °C
3.	Washing buffer	30–60 min	50 °C
4.	NTE	5 min	37 °C with slowly shaking in incubator
5.	NTE	5 min	37 °C with slowly shaking in incubator
6.	RNase A	30 min	37 °C with slowly shaking in incubator
7.	NTE	5 min	Room temperature shaking
8.	NTE	5 min	Room temperature shaking
9.	Washing buffer	30–60 min	50 °C
10.	PBS	5 min	Room temperature shaking

3.5.4 Detection

Detection can be performed under the following order. Each incubation can be done in a small tray containing 100 mL of buffer at RT with gentle shaking. For color reaction, add 1.5 μL of NBT and BCIP per mL of 10% polyvinyl alcohol prepared in Detection Buffer 2.

1.	Detection Buffer 1	5 min
2.	Detection Buffer 1 with 0.5% blocking reagent	60 min
3.	Detection Buffer 1 with 1% BSA, 0.3% Triton	30–60 min
4.	Anti-DIG-AP	60 min
	(diluted 1:3K in Buffer 1 with 1% BSA, 0.3% Triton)	
5.	Detection Buffer 1 with 0.3% Triton	10–20 min
6.	Detection Buffer 1 with 0.3% Triton	10–20 min
7.	Detection Buffer 1 with 0.3% Triton	10–20 min
8.	Detection Buffer 1 with 0.3% Triton	10–20 min
9.	Detection Buffer 1	5 min
10.	Detection Buffer 2	5 min
11.	Color reaction	1–3 days
	(under darkness for overnight up to 3 days at room temperature)	

4 Notes

1. To execute this script, you need to create the blastable database of transposable elements from FASTA file including transposable element sequence information (http://maizetedb.org/~maize/). Use makeblastdb command to create blastable database from FASTA file from http://maizetedb.org/~maize/TE_12-Feb-2015_15-35.fa. And all databases should be the same directory.

   (e.g., `$ makeblastdb -in TE.fasta -dbtype nucl`)

2. In many cases, normalizing the values improves the clustering since all FPKM values are weighted equally by shifting and normalizing. This ensures that individual expression values are from the shape of their expression patterns, not from their absolute levels.

3. The parameter below was used for SOM analysis as follow. For coding gene clusters, the generated clusters were tested by GO enrichment analysis to adjust a parameter to get a better fit to the researched model.

Parameter	
output stub	output name
cluster range	cluster range you predict
seed range	42
iterations	4,200,000
cluster by	rows
som rows	0
som cols	0
initialization	Random_Datapoints
neighborhood	Bubble
alpha initial	0.1
alpha final	0.005
sigma initial	5.0
sigma final	0.5

Table 1
Bioinformatics tools and databases

Resource	URL
Primer3	http://frodo.wi.mit.edu/primer3/
NCBI Blast	http://blast.ncbi.nlm.nih.gov
MEME	http://meme-suite.org/
Maize Genetics and Genomics Database	https://www.maizegdb.org/
Maize transposable element (TE) database	http://maizetedb.org/~maize/
TransDecoder program	https://github.com/TransDecoder/TransDecoder/wiki
Biopython	http://biopython.org/
Blast software and database	https://blast.ncbi.nlm.nih.gov/Blast.cgi?CMD=Web&PAGE_TYPE=BlastDocs&DOC_TYPE=Download
GenePattern	https://genepattern.broadinstitute.org/gp/pages/login.jsf
Bedtools	http://bedtools.readthedocs.io/en/latest/index.html

4. TOMTOM program allows you to identify shared or unique *cis*-regulatory elements between each cluster.
5. Grind tissue using pellet pestle in e-tube because the amount of tissue is about 20–30 μL. If you grind the tissue using mortar and pestle, you will lose the most of sample.
6. Ensure that all work surfaces, pipettes, and reagents needed for isolation and preparation of first-strand cDNA synthesis are free of RNase contamination by wiping down the work surfaces and pipettes with a nucleic acid decontamination detergent before starting work.
7. *ZmTXN* was used as an internal reference gene because it shows the least level of variation among different tissues. And all primer pairs need to be tested for the specificity and efficiency by visualizing bands on 3% agarose gel and blast primer sequences back to maize genome.
8. 50% formamide + 2× SSC can be replaced with 1× SSPE (Table 1).

Acknowledgments

B.H. Kang and S. Sung are supported by USDA NIFA Award AFRI grant (2011-67013-30119). B.H. Kang is also supported by the grants from the Research Grants Council (RGC) of Hong Kong

(GRF14126116, C4011-14R, and AoE/M-05/12), Cooperative Research Program for Agriculture Science and Technology Development (Project No. 10953092018), and Rural Development Administration, Republic of Korea, and S. Sung is also supported by NIH (GM100108) and NSF (1656764).

References

1. Fatica A, Bozzoni I (2014) Long non-coding RNAs: new players in cell differentiation and development. Nat Rev Genet 15(1):7–21. https://doi.org/10.1038/nrg3606
2. Kim DH, Sung S (2017) Vernalization-triggered intragenic chromatin loop formation by long noncoding RNAs. Dev Cell 40 (3):302–312 e304. https://doi.org/10.1016/j.devcel.2016.12.021
3. Heo JB, Sung S (2011) Vernalization-mediated epigenetic silencing by a long intronic noncoding RNA. Science 331 (6013):76–79. https://doi.org/10.1126/science.1197349
4. Quinn JJ, Chang HY (2016) Unique features of long non-coding RNA biogenesis and function. Nat Rev Genet 17(1):47–62. https://doi.org/10.1038/nrg.2015.10
5. Chu C, Zhang QC, da Rocha ST, Flynn RA, Bharadwaj M, Calabrese JM, Magnuson T, Heard E, Chang HY (2015) Systematic discovery of Xist RNA binding proteins. Cell 161 (2):404–416. https://doi.org/10.1016/j.cell.2015.03.025
6. Derrien T, Johnson R, Bussotti G, Tanzer A, Djebali S, Tilgner H, Guernec G, Martin D, Merkel A, Knowles DG, Lagarde J, Veeravalli L, Ruan X, Ruan Y, Lassmann T, Carninci P, Brown JB, Lipovich L, Gonzalez JM, Thomas M, Davis CA, Shiekhattar R, Gingeras TR, Hubbard TJ, Notredame C, Harrow J, Guigo R (2012) The GENCODE v7 catalog of human long noncoding RNAs: analysis of their gene structure, evolution, and expression. Genome Res 22(9):1775–1789. https://doi.org/10.1101/gr.132159.111
7. Rutenberg-Schoenberg M, Sexton AN, Simon MD (2016) The properties of long noncoding RNAs that regulate chromatin. Annu Rev Genomics Hum Genet 17:69–94. https://doi.org/10.1146/annurev-genom-090314-024939
8. Goff LA, Rinn JL (2015) Linking RNA biology to lncRNAs. Genome Res 25 (10):1456–1465. https://doi.org/10.1101/gr.191122.115
9. Zhang S, Thakare D, Yadegari R (2018) Laser-capture microdissection of maize kernel compartments for RNA-Seq-based expression analysis. Methods Mol Biol 1676:153–163. https://doi.org/10.1007/978-1-4939-7315-6_9
10. Kim ED, Xiong Y, Pyo Y, Kim DH, Kang BH, Sung S (2017) Spatio-temporal analysis of coding and long noncoding transcripts during maize endosperm development. Sci Rep 7 (1):3838. https://doi.org/10.1038/s41598-017-03878-4
11. Kiyota E, Pena IA, Arruda P (2015) The saccharopine pathway in seed development and stress response of maize. Plant Cell Environ. https://doi.org/10.1111/pce.12563
12. Zhang Z, Yang J, Wu Y (2015) Transcriptional regulation of zein gene expression in maize through the additive and synergistic action of opaque2, prolamine-box binding factor, and O_2 heterodimerizing proteins. Plant Cell 27 (4):1162–1172. https://doi.org/10.1105/tpc.15.00035
13. Zhan J, Thakare D, Ma C, Lloyd A, Nixon NM, Arakaki AM, Burnett WJ, Logan KO, Wang D, Wang X, Drews GN, Yadegari R (2015) RNA sequencing of laser-capture microdissected compartments of the maize kernel identifies regulatory modules associated with endosperm cell differentiation. Plant Cell 27(3):513–531. https://doi.org/10.1105/tpc.114.135657
14. Zhang W, Yan H, Chen W, Liu J, Jiang C, Jiang H, Zhu S, Cheng B (2014) Genome-wide identification and characterization of maize expansion genes expressed in endosperm. Mol Genet Genomics 289 (6):1061–1074. https://doi.org/10.1007/s00438-014-0867-8
15. Xiong Y, Mei W, Kim ED, Mukherjee K, Hassanein H, Barbazuk WB, Sung S, Kolaczkowski B, Kang BH (2014) Adaptive expansion of the maize maternally expressed gene (Meg) family involves changes in expression patterns and protein secondary structures of its members. BMC Plant Biol 14:204. https://doi.org/10.1186/s12870-014-0204-8
16. Xiong YQ, Li QB, Kang BH, Chourey PS (2011) Discovery of genes expressed in basal

endosperm transfer cells in maize using 454 transcriptome sequencing. Plant Molecular Biology Reporter 29(4):835–847. https://doi.org/10.1007/s11105-011-0291-8

17. Boerner S, McGinnis KM (2012) Computational identification and functional predictions of long noncoding RNA in Zea mays. PLoS One 7(8):e43047. https://doi.org/10.1371/journal.pone.0043047 PONE-D-12-12079 [pii].

18. Bailey TL, Elkan C (1994) Fitting a mixture model by expectation maximization to discover motifs in biopolymers. Proc Int Conf Intell Syst Mol Biol 2:28–36

19. Gupta M (2007) Generalized hierarchical markov models for the discovery of length-constrained sequence features from genome tiling arrays. Biometrics 63(3):797–805. https://doi.org/10.1111/j.1541-0420.2007.00760.x

Chapter 4

RNA Isolation and Analysis of LncRNAs from Gametophytes of Maize

Linqian Han, Lin Li, Gary J. Muehlbauer, John E. Fowler, and Matthew M. S. Evans

Abstract

The explosion of RNA-Seq data has enabled the identification of expressed genes without relying on gene models with biases toward open reading frames, allowing the identification of many more long noncoding RNAs (lncRNAs) in eukaryotes. Various tissue enrichment strategies and deep sequencing have also enabled the identification of an extensive list of genes expressed in maize gametophytes, tissues that are intractable to both traditional genetic and gene expression analyses. However, the function of very few genes from the lncRNA and gametophyte sets (or from their intersection) has been tested. Methods for isolating and identifying lncRNAs from gametophyte samples of maize are described here. This method is transferable to any maize gametophyte mutant enabling the development of gene networks involving both protein-coding genes and lncRNAs. Additionally, these methods can be adapted to apply to other grass model systems to test for evolutionary conservation of lncRNA expression patterns.

Key words LncRNAs, Maize, Gametophyte, Pollen, Embryo sac

1 Introduction

Long noncoding RNAs (lncRNAs) arbitrarily refer to RNAs that are 200 bp or longer and are not likely to encode a protein. With the advent of next-generation sequencing, numerous studies have identified lncRNAs, and recent studies are beginning to show their roles in gene expression regulation, such as X chromosome silencing, genomic imprinting, chromatin modification, transcriptional activation, transcriptional interference, and other cellular developmental processes (e.g., [1–3]).

In humans, Iyer et al. [4] applied ab initio assembly methodology to 7,256 RNA-Seq libraries and identified tens of thousands of expressed gene loci, of which over 68% were classified as lncRNAs. Of these lncRNAs in humans, 79% were previously not annotated [4]. In Drosophila, Wen et al. [5] identified 128 testis-specific

Julia A. Chekanova and Hsiao-Lin V. Wang (eds.), *Plant Long Non-Coding RNAs: Methods and Protocols*, Methods in Molecular Biology, vol. 1933, https://doi.org/10.1007/978-1-4939-9045-0_4,

lncRNAs and created knockout mutations for 105. Interestingly, 31% exhibited a partial or complete loss of male fertility [5]. In plants, vernalization is influenced by two types of lncRNAs termed COOLAIR and COLDAIR. These two lncRNAs are involved in flowering by inhibiting expression of FLC when plants sense cold at different stages [6, 7]. These examples indicate that lncRNA-derived sequences are not solely transcriptional noise but function in a wide range of biological processes. Here, we describe a detailed procedure to isolate and sequence lncRNAs from maize gametophyte tissues and detect novel lncRNAs in the maize genome.

2 Materials

1. Ears of the desired stage of *Zea mays* plants of the desired genotype (previously performed for the wild-type B73 reference inbred line).
2. Tassels of mature *Zea mays* plants in the middle of the pollen shedding period.
3. Forceps (fine tip), dissecting needles, Pasteur pipet, petri dish, mortar and pestle, and microfuge tubes.
4. Cell wall enzyme digesting mix: 0.75% pectinase, 0.25% pectolyase, 0.5% cellulase, 0.5% hemicellulase buffered in 0.55 M Mannitol pH 5.0 and 0.53 M Mannitol adjusted to pH 5.0 with 1 M monobasic potassium phosphate and 0.1 M potassium hydroxide [8–10].
5. Dissecting microscope with bright-field illumination.
6. TRIzol (Invitrogen).
7. Tungsten carbide beads and a MixerMill300 (Qiagen).
8. High-molecular-weight (at least 15,000 MW) polyethylene glycol (HMW PEG).
9. Pollen germination media (PGM): 10% sucrose, 0.0005% H_3BO_3, 10 mM $CaCl_2$, 0.05 mM KH_2PO_4, and 6% PEG 4000 ([11] and erratum).
10. Microcentrifuge.
11. SMARTer PCR cDNA Synthesis Kit with SMARTScribe Reverse Transcriptase (Clontech Laboratories, Inc.), Advantage 2 PCR kit (Clontech Laboratories, Inc.), and SMART PCR cDNA Synthesis Kit with SMART MMLV Reverse Transcriptase (Clontech Laboratories, Inc.).
12. Software: FastQC (http://www.bioinformatics.babraham.ac.uk/projects/fastqc),Trimmomatic [12], bowtie2 [13], TopHat [14], SAMtools [15], Cufflinks [16], LncRNA_Finder [17], RSEM [18], and IGV [19].

3 Methods

3.1 Sample Collection

3.1.1 Female Gametophyte Collection and RNA Isolation

Embryo sacs of different stages can be collected depending upon the stage of the florets (flowers) from which they are collected. Each maize floret has a single ovule with a single embryo sac. Silk length is an accurate proxy for embryo sac stage [20]. Florets with a silk length of 10 cm or more can be considered mature. For pre-cellularization embryo sacs, flowers with silks less than 1 cm in length would need to be used. To prevent pollen contamination and fertilization of the embryo sacs, ears need to be covered with ear shoot bags prior to silk emergence. At the time of silk emergence, most of the florets have mature embryo sacs, except for a few at the tip of the ear depending upon the inbred line used. For younger florets, ears need to be used before silk emergence, typically when the ears are less than 5 cm in length. Because the florets at the base of the ear are initiated first, these are older than those at the tip, and consequently there is a gradient of floret and embryo sac stages in these younger ears, as can be seen by the variation in silk length (Fig. 1). To collect embryo sacs at the same stage in these ears, it is therefore necessary to collect them from florets at the same position from the base of the ears (plus or minus a few florets). These florets will have silks of the same length. To avoid any abnormalities in development or stage that may occur at the tip or base of the ear, collect embryo sacs from the middle third of each ear. For mature ears, all of the embryo sacs in this region will be at the same stage.

Once the husk leaves are removed from the ear, the florets will start to dry out, so keep the time to a minimum. First, remove the glumes, lemma, and palea surrounding the ovary (*see* **Note 1**), and then the silk and ovary wall need to be removed to access the ovule (Fig. 2). Using one tine of a pair of fine forceps, cut an arc halfway or more around the base of the silk on the side of the floret toward the base of the ear. Insert the side of the forceps into this slit, and lift the silk like a cap away from the floret to expose the ovule. Then using the forceps again, make an incision at the base of the ovule by sliding the edge of the forceps from the ear base side of the ovule to the ear tip side going deeper toward the cob as you make the cut. The embryo sac is located on the apical-basal midline of the ovule on the ear tip side and in the mature ovule is located tight against the cob side (*see* **Note 2**). It is therefore necessary to make sure the cut used to remove the ovule from the ear is deep enough on the ear tip side to ensure that the embryo sac comes off with the ovule and is not left behind on the ear. Each ovule is immediately placed in a petri dish containing a cell wall enzyme digesting mix. Ovules are left in the cell wall enzyme mix for at least 1 h and up to 4 h at 37 °C.

Fig. 1 Maize ear with mature florets at the base and middle of the ear and immature florets with shorter silks at the tip of the ear. Scale bar = 10 cm

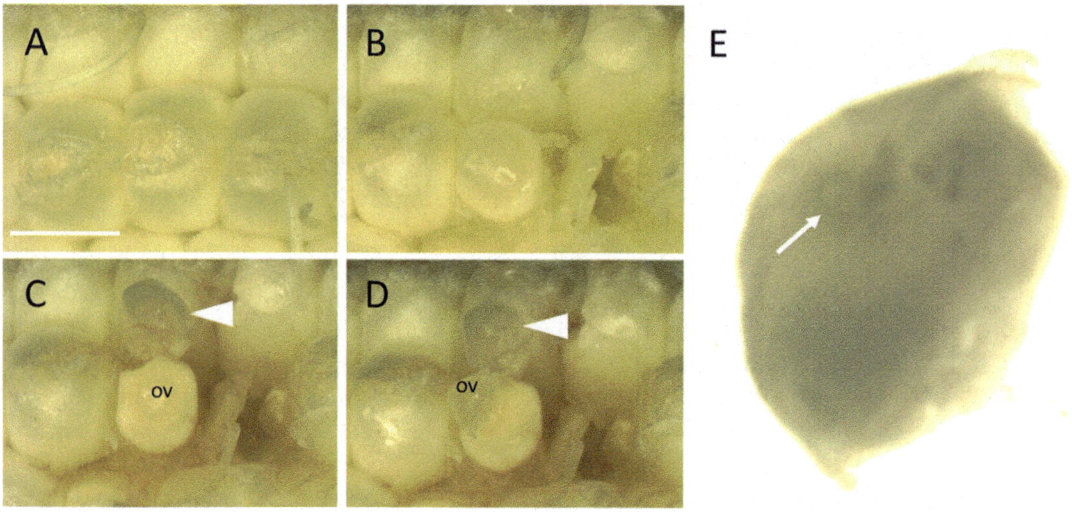

Fig. 2 Ovule dissection. (**a**) Undissected florets from the middle portion of a mature ear. (**b**) A floret with the glumes removed. The floret immediately basal to the floret being dissected has been removed to facilitate dissection. (**c**) A floret with the ovary wall/base of the silk (arrowhead) cut off from the base of the floret to expose the ovule(ov). (**d**) The ovule(ov) has been cut off from the base of the floret for transfer to the cell wall digestion buffer. (**e**) An isolated ovule under bright-field illumination with the outline of the embryo sac (arrow) inside the ovule visible for dissection. Scale bar = 2 mm in (**a–d**) and 0.25 mm in (**e**)

It is typically only possible to successfully dissect an embryo sac out of 25% or fewer of the ovules that have been collected from the ear, so collect many more than are needed for the experiment. Successful RNA-Seq has been performed with as few as 15 partially purified embryo sacs after amplification of cDNA prior to RNA-Seq. With current technologies this is sure to be even lower, although keeping the amount of amplification before sequencing to a minimum reduces the probability of introducing artifacts into the data.

After enzymatic digestion embryo sacs are removed from the surrounding nucellus using fine tungsten dissecting needles, either manually or with a micromanipulator on a dissecting microscope equipped with bright-field illumination. Stand the ovule on one edge to find the micropyle that is on the side toward the tip of the ear (*see* **Note 3**). The outline of the embryo sac and the micropyle can be seen inside the ovule. Use the dissecting needles to remove the embryo sac and some of the surrounding nucellus from the rest of the ovule. Once it is closer to the surface, the embryo sac can be distinguished from the rest of the ovule because the extremely large vacuole of the central cell appears as a clear space in the ovule and because of the optically dense cluster of antipodal cells. However, if the general location of the embryo sac is not identified at the start of dissection, this becomes less useful after the dissecting needles have been inserted into the ovule creating holes in the tissue. Use the dissecting needles to remove as much of surrounding nucellar cells as possible without rupturing the central cell (*see* **Note 4**). Collect the embryo sac with a Pasteur pipet, and deposit it into a 1.5 mL microfuge tube containing 500 μL 0.55 M Mannitol pH 5.0 (*see* **Note 5**). Deposit the main mass of the ovule from which the embryo sac was removed into a separate 1.5 mL microfuge tube containing 500 μL 0.55 M Mannitol pH 5.0 as a control sample if desired.

After collecting embryo sacs (and ovules, if desired), spin the tissue to the bottom of the tube in a microfuge at 1000 × *g* for 1 min. Remove the excess buffer with a pipet, and add 400 μL of TRIzol (Invitrogen) to each tube. Vortex to mix and suspend the tissue and transfer to a 1.2 mL tube with one tungsten carbide bead and shake at high speed for 3 min on a MixerMill300 (Qiagen). After disruption in TRIzol, samples can be flash frozen in liquid nitrogen and stored at −80 °C. RNA was extracted according to manufacturer's specifications to isolate total RNA using glycogen as a carrier for precipitation because of the small amount of RNA per sample. Twenty partially purified embryo sacs yield approximately 100 ng of total RNA.

3.1.2 Male Gametophyte Collection and RNA Isolation

Fresh mature pollen can be collected in the lab from tassels grown in the field or in the greenhouse. Identify tassels that have exerted a large number of anthers on the current day but still have a large number of closed florets. Three tassels that are shedding well provide more than 100 mg of pollen if pooled, enough for a double-size RNA prep, resulting in 5–20 μg of total RNA. Cut the stem several nodes below the tassel, place in water, and then cut off the bottom of the stem (at least 2 cm) under water. Transport to the lab, "clean" off the old anthers, and bag each tassel to reduce lab mess. The next morning at 10–11 a.m., remove the bag and again clean off the old anthers, and then re-bag to collect freshly shed pollen. Leave the bag on the tassel for 45 min–2 h, and then collect the pollen, pool on weigh paper, and weigh the sample.

Approximately 10 mg of pollen can be used to test for pollen tube germination, if desired. To test germination, vigorously mix ~10 mg pollen in 1 mL liquid pollen germination media (PGM), and then plate in a small (15 mm) glass petri dish, removing excess media to leave a thin film. Cover and keep humid at room temperature. Pollen tubes should be visible within 15 min, with near full germination by 30 min after mixing in PGM. The germination rate at 30 min should be at least 65%, with rates up to 80–90% achievable. Be careful to distinguish between growing pollen tubes and pollen grain "rupture," in which the cytoplasm and contents of the pollen grain are rapid extruded from the pollen grain pore. Because ruptured contents can sometimes be distributed as a continuous line, an untrained eye can mistake them for pollen tubes. In our hands, the percentage of pollen grains and tubes rupturing increases significantly after 30–45 min in PGM.

The standard TRIzol RNA isolation procedure (Invitrogen) is followed, with two modifications, due to the presence of high concentrations of polysaccharides in the pollen grain. First, high-molecular-weight (at least 15,000 MW) polyethylene glycol (HMW PEG) is added to the TRIzol at 2% (20 mg/mL), prior to the extraction (based on [21]). Second, a High Salt Precipitation Solution (0.8 M sodium citrate/1.2 M NaCl, nuclease-free) is added to the first RNA precipitation step (as described in the original TRIzol protocol).

Extraction of the pollen grains uses mortar and pestle but, in contrast to most plant tissue extractions, is not done at freezing temperatures (which makes the grains recalcitrant to breaking). For 100 mg pollen, first coat mortar and pestle with 900 μL of TRIzol/HMW PEG. Add the pollen and grind quickly and thoroughly, adding another 900 μL TRIzol/HMW PEG during grinding. When completely ground, remove as much as possible into two 2 mL microfuge tubes (~1400 μL), and then wash the mortar and pestle with 800 μL TRIzol/HMW PEG, adding the wash to the microfuge tubes, up to ~1 mL in each tube. Alternate between frequent vortexing and incubation of the two tubes at room temperature (RT) for 5 min. (50 mg pollen samples can be extracted with half the amount of TRIzol/HMW PEG, if pollen is limiting.) Samples can now be frozen at −80 °C for several months, if desired.

Following the TRIzol protocol, briefly, pollen/TRIzol samples are spun at 12,000 × *g* for 10 min at 2–8°C, and then the soluble portion is moved into a new 2 mL tube, vortexed and allowed to return to RT (2–5 min). Add chloroform to each tube (200 μL/1 mL TRIzol), vortex, and incubate at RT (2–3 min). Centrifuge at

12,000 × *g* for 15 min at 2–8 °C to separate phases; remove aqueous phase (500–600 μL) into a new 2 mL tube. For the High Salt Precipitation of RNA, add 0.25 mL isopropanol, mix, and then add 0.25 mL High Salt Precipitation Solution; vortex and incubate at RT (5–10 min). Centrifuge at 12,000 × *g* for 10 min at 2–8 °C, discard the supernatant, wash with 1 mL 75% ethanol, and then centrifuge at 7500 × *g* for 5 min at 2–8 °C. Discard the supernatant, air-dry the pellet at RT (~10 min), and then resuspend in 100 μL nuclease-free water (incubating at 55–60 °C for 10 min, if necessary). Assessing total RNA quality via $A_{260/280}$ at this point often shows contamination; therefore, the RNA is usually further purified using a column prep method (e.g., RNeasy, Qiagen) or via direct selection of polyA-RNA (e.g., with biotinylated polyT oligonucleotides and streptavidin [22]).

3.1.3 cDNA Synthesis and RNA-Seq

For embryo sacs, cDNA was generated from 50 ng of total RNA. First-strand cDNA was synthesized using the SMARTer PCR cDNA Synthesis Kit with SMARTScribe Reverse Transcriptase (Clontech Laboratories, Inc.) for embryo sacs and ovules. The second strand was synthesized with the Advantage 2 PCR kit (Clontech Laboratories, Inc.). After second-strand synthesis, cDNAs from the embryo sac samples were amplified using the Advantage 2 PCR kit (for 26 cycles) (Clontech Laboratories, Inc.) to produce sufficient cDNA for generating Illumina libraries.

In an earlier iteration of the Illumina RNA-Seq methodology for pollen [9], cDNA libraries were generated from 0.5 to 20 μg total RNA via the following two steps. First-strand cDNA was synthesized using the SMART PCR cDNA Synthesis Kit with SMART MMLV Reverse Transcriptase (Clontech Laboratories, Inc.). After second-strand synthesis, cDNAs from pollen were amplified using the Advantage 2 PCR kit (for 15 to 17 cycles) (Clontech Laboratories, Inc.) to produce sufficient cDNA for generating Illumina libraries. More recent library preparation protocols provide alternatives to these steps (e.g., Illumina TruSeq or BrAD-seq [22]).

RNA-Seq data from many platforms can be analyzed in the lncRNA pipeline below. For the analysis of gametophytic tissues here, 80-mer paired-end reads were produced using Illumina next-generation sequencing. The cDNA libraries were prepared for Illumina sequencing using a nebulizer for fragmentation and the Illumina paired-end sequencing preparation kit per the manufacturer's protocol.

3.2 LncRNA Identification and Analysis from RNA-Seq Data

Recently, transcriptome data has been accumulated through expressed sequence tag (EST) sequencing and full-length cDNA sequencing, tilling microarray, and mRNA sequencing (RNA-Seq) and has been utilized to identify thousands of lncRNAs in diverse eukaryotic genomes [23]. Ample transcriptomic datasets provide

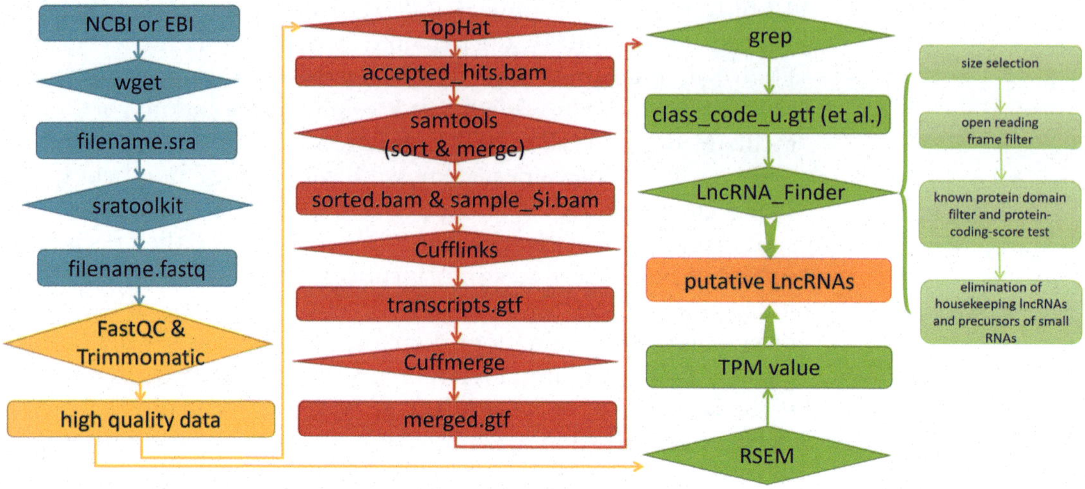

Fig. 3 Flowchart of a bioinformatic pipeline for identifying lncRNAs in plants. Four major steps, including data collection (in blue), data quality control (in yellow), transcriptome assembly (in red), and lncRNA filtering (in green), are employed for the identification of lncRNAs in plants

the opportunity to dissect all possible transcriptomic isoforms, which enables isolation of lncRNAs. In spite of the wide existence and functional importance of lncRNAs, standard procedures to detect and annotate lncRNAs from existing databases are lacking. Here, we describe a detailed procedure to detect novel lncRNAs in the maize genome. This procedure can be adopted for any species that have available transcriptome data and a reference genome. Briefly, four major steps are employed including data collection, data quality control, transcriptome assembly, and lncRNA filtering (Fig. 3).

3.2.1 Transcriptome Data Collection

1. Data Downloading

Transcriptome data can also be downloaded from the National Center for Biotechnology Information (NCBI; https://www.ncbi.nlm.nih.gov/) and the European Bioinformatics Institute (EBI; http://www.ebi.ac.uk/) public databases. Here, we show an example of obtaining RNA-Seq data from NCBI. First, go to the NCBI website at https://www.ncbi.nlm.nih.gov/, and then click and open the link "Sequence Read Archive" (SRA) at the bottom of the page. Sequencing data is stored as "runs" in the SRA. Second, input query information in the bar "search," such as "B73 rna seq." According to some basic descriptive information in the "title," click the "Accession" link on the left to enter a page to view the corresponding research project abstract, biosample information, experiment information, etc. If the data are appropriate for your study, click "Runs," to get the accession number with the prefix of

"SRR, ERR, or DRR." The data can be downloaded through these accession numbers. Third, after obtaining the annotations of the transcriptome datasets and the accession numbers of these data, the raw sequence data can be downloaded. For example, open the home page of NCBI, choose the link "Download," enter "FTP" at ftp://ftp.ncbi.nlm.nih.gov/sra/sra-instant/reads/ByRun/sra/, and provide the appropriate accession number to start downloading.

Data can be downloaded to a PC and then transferred to the Linux environment for analysis. Alternatively, data can be downloaded directly to the Linux environment using the command "wget." The collection of other data such as the reference genome or annotation file is obtained in a similar manner. Ensembl Plants (http://plants.ensembl.org/index.html) is a bioinformatic database which hosts plant genomic sequence and annotation information. Open the Ensembl Plants home page, choose the appropriate species, and then click the link "Download DNA sequence (FASTA)" or gene annotation gff file. To download multiple datasets in parallel, employing recurrent scripts will be helpful. An example of downloading the "SRR531202" and "SRR531203" data is shown here. First, save all the URLs into a txt file (here is "SRR.txt"), and write a shell script (wget.sh) to download a batch of data as follows:

```
for i in $(cat SRR.txt)
do
wget ftp://ftp.ncbi.nlm.nih.gov/sra/sra-instant/reads/By-
Run/sra/$i;
done
```

Run "sh wget.sh" and the data will be downloaded into a directory. This process is usually time-consuming; therefore, the "nohup" command can be employed to run the download in the background. In this manner, the process will not be interrupted if the Linux terminal is closed. The download progress can be monitored in the instant "nohup.out" file.

The "jobs -l" command can be used to view the running program in the current terminal or the "ps -ef|grep wget" command to view the running program when the terminal is reopened. When finished, the output information can be obtained from the "nohup.out" file.

Paired-end sequencing datasets (ERR361362.sra and ERR361363.sra) can be downloaded in the same way for subsequent analyses. Downloading small RNA, protein-coding, and housekeeping RNA sequence data is similar and requires entering the proper website and obtaining the appropriate download link.

2. Data Format Conversion

The original downloaded files with a suffix of ".sra" are compressed binary files, which need to be uncompressed and converted into fastq format for subsequent analyses. The sratoolkit software (http://www.ncbi.nlm.nih.gov/Traces/sra/?view=toolkit_doc) is employed to convert the data to the fastq format. The Portable Batch System (PBS) is applied to submit command lines to the computing node of the Linux server. First, save all data files (without suffix) into a txt file (SRRid.txt), and then write a script file (fastq.pbs), and use the "qsub" command in the PBS system to run this script for single-end sequencing data. The converting method of paired-end sequencing datasets is similar but with slightly different "sratoolkit" parameters. After running the converting script, fastq files can be obtained that are named as the original sra files but with a suffix fastq.

The script of fastq.pbs for single-end data is as follows:

```
#!/bin/bash
#PBS -N fastq.pbs
#PBS -l nodes=1:ppn=1
#PBS -l walltime=7200:00:00
#PBS -q batch
cd /public/home/lqhan/lncRNA_protocol/single_end/
for i in $(cat SRRid.txt)
do
/public/home/lqhan/software/sratoolkit.2.6.2-centos_linux64/
bin/fastq-dump $i.sra -O./
done
```

The script of fastq.pbs for paired-end data is as follows:

```
#!/bin/bash
#PBS -N fastq.pbs
#PBS -l nodes=1:ppn=1
#PBS -l walltime=7200:00:00
#PBS -q batch
cd /public/home/lqhan/lncRNA_protocol/paired_end/
for i in $(cat SRRid.txt)
do
/public/home/lqhan/software/sratoolkit.2.6.2-centos_linux64/
bin/fastq-dump --split-3 $i.sra -O ./
done
```

For paired-end original sequencing datasets that are transferred from the sra to the fastq format, two files are generated including forward and reverse chains named *_1.fastq and *_2.fastq.

```
@SRR531202.1 AMELIA:220:D154TACXX:8:1101:1242:2141 length=100
AGAAGCCATACTCCAGGTCAATTGTGGGTGCTTCCTTGGAAGTAATCCAGAAGAAGAGAGCTGAGAAGCCAGAGGTCCG
CGATGCTGCTAGAGAAGCTGC
+SRR531202.1 AMELIA:220:D154TACXX:8:1101:1242:2141 length=100
@@<DDFADDFDDFFHHB+ACHFGE@ACG@FGHED?<FGI<<D9BGGDBBF9FG*?C;88@FFC:;@@DGH>>EEBCCCB
:;'8=?AC3>>C3((9<?AA9
@SRR531202.2 AMELIA:220:D154TACXX:8:1101:1392:2122 length=100
CCTGGGCTCACTGCAAAAGTGTTCCGTACATATAATGCTTCCATAACATTGGACGGAATCTTGCATGAAGAAACAGAAG
ATGGAACTGTTTTAGAAAAAA
+SRR531202.2 AMELIA:220:D154TACXX:8:1101:1392:2122 length=100
@CCFFFFFHHHHHJJJJJHHFHJIJHBHGHIJJJJJJJJJJJJJEIJIJJJIJJIIJJJJJFIIJIJJDFGFFFFFF
ECEEEEDDDDCCDCCCCC?BB
```

Fig. 4 The first four lines of "SRR531202.fastq" file

The "less" command in the Linux terminal can be used to view the fastq file. Four lines correspond to one read: the first line begins with "@" and shows the name of the read, followed by the descriptive information; the second line is the sequence; the third line begins with "+" and the same information as the first line; and the fourth line shows the quality scores of each nucleotide in the read (Fig. 4).

3.2.2 Quality Control

Different sources of transcriptome data have different sequence qualities. Therefore, quality control is a prerequisite before further analyses. First, the FastQC software (http://www.bioinformatics.babraham.ac.uk/projects/fastqc) can be used to evaluate raw data quality, which generates a user-friendly report. The command lines are as follows:

```
FastQC/fastqc -o ./ -f fastq filename.fastq
```

It generates "html" and "zip" files; the quality report can be viewed in the "html" file in a web browser. Sequence scores in the "per base sequence quality" report less than 20 are poor quality (Fig. 5).

Use the "unzip filename" command to uncompress the "fastqc.zip" package, and obtain six files: "fastqc_data.txt, fastqc.fo, fastqc_report.html, Icons, Images, and summary.txt." Use the "less" command to view relevant files, such as fastqc_data.txt. The method of quality checking paired-end sequence data is similar; just separate forward and reverse files as two files.

In the early stage of next-generation sequencing, fastq files were encoded in a phred 33 code table, which has been replaced by phred 64. The encoding system is important for quality control and following analyses. It is strongly recommended to convert the old code into the new version. You can use a perl script named fastq_phred.pl to determine the encoding system (http://blog.

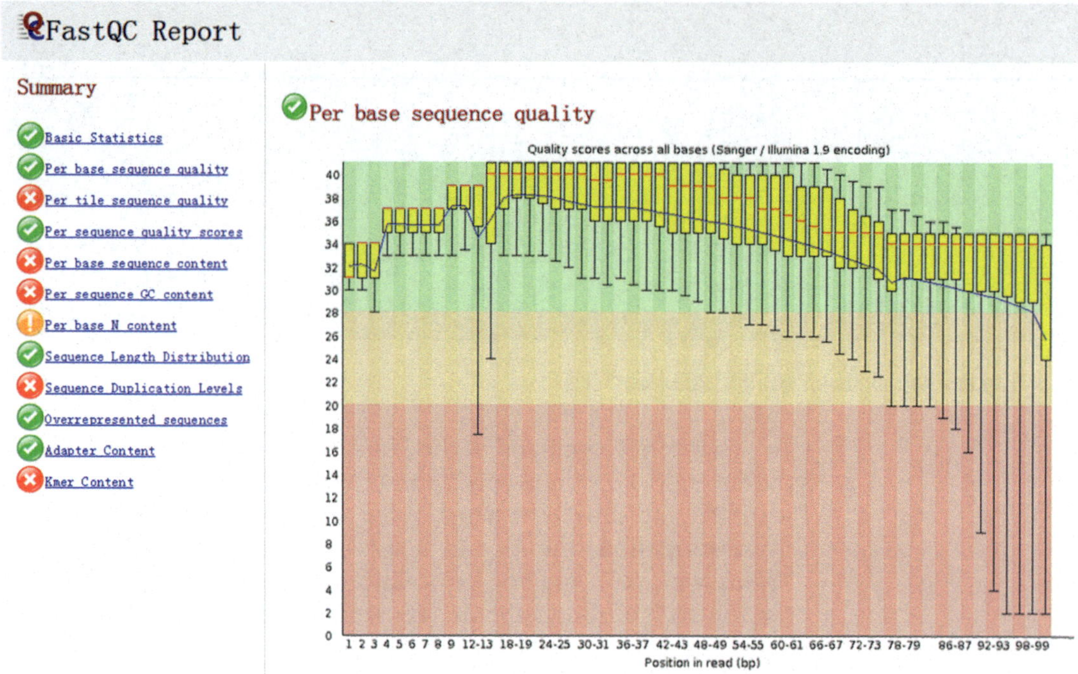

Fig. 5 "Summary" and "per base sequence quality" in FastQC report of "SRR531202_fastqc.html" file

sciencenet.cn/home.php?mod=attachment&filename=fastq_phred.pl&id=57063).

Trimmomatic [12] is a read trimming tool based on the quality of reads derived from Illumina sequencing platforms (http://www.illumina.com/). A recurrent script (trimmomatic.pbs) is used to deal with multiple raw fastq files. The analysis of single-end sequencing is slightly different from paired-end data.

The codes of trimmomatic.pbs for single-end data are as follows:

```
#!/bin/bash
#PBS -N trimmomatic.pbs
#PBS -l nodes=1:ppn=5
#PBS -l walltime=7200:00:00
#PBS -q batch
cd /public/home/lqhan/lncRNA_protocol/single_end/
for i in $(cat SRRid.txt)
do
java -jar /public/home/lqhan/software/Trimmomatic-0.33/trim-
momatic-0.33.jar SE -threads 5 -phred33 $i.fastq $i.fastq_t
ILLUMINACLIP:/public/home/lqhan/software/Trimmomatic-0.33/
adapters/TruSeq3-SE.fa:1:30:10 MINLEN:36 SLIDINGWINDOW:3:20;
done
```

The codes of trimmomatic.pbs for paired-end data are as follows:

```
#!/bin/bash
#PBS -N trimmomatic.pbs
#PBS -l nodes=1:ppn=10
#PBS -l walltime=7200:00:00
#PBS -q batch
cd /public/home/lqhan/lncRNA_protocol/paired_end/
for i in $(cat SRRid.txt)
do
java -jar /public/home/lqhan/software/Trimmomatic-0.33/trim-
momatic-0.33.jar PE -threads 10 -phred33 $i\_1.fastq $i\_2.
fastq $i-paired_1.fastq $i-unpaired_1.fastq $i-paired_2.fastq
$i-unpaired_2.fastq ILLUMINACLIP:/public/home/lqhan/software/
Trimmomatic-0.33/adapters/TruSeq3-PE.fa:1:30:10 MINLEN:36
SLIDINGWINDOW:3:20;
done
```

FastQC software can be used again to evaluate the quality of fastq files after being processed by Trimmomatic (previously mentioned). The new fastq files show higher quality of sequencing reads than the original fastq files (Fig. 6).

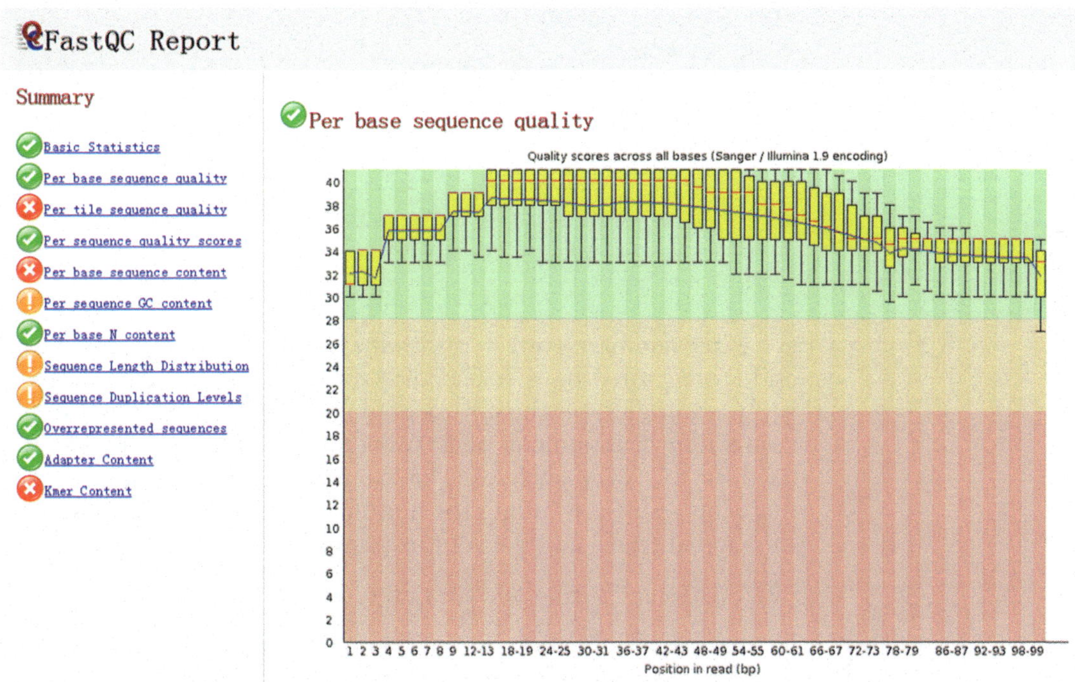

Fig. 6 Higher "per base sequence quality" of "SRR531202.fastq_t_fastqc.html" processed by "Trimmomatic"

Comparison of pre-and post-processed fastq files by Trimmomatic shows that low-quality reads have been effectively removed. The quality score of reads after processing by Trimmomatic is significantly higher than 20, which ensures high-quality transcriptome data for further analyses.

3.2.3 Transcriptome Assembly

1. Read Mapping

To map transcriptome data against the reference genome, download the reference genome from the Ensembl Plants database, and build a reference genome index for mapping. In maize, the reference genome is derived from the widely used B73 inbred line [24]. The software bowtie2 [13] is employed to build a reference genome index using command "bowtie2-build," which generates several index files as the reference genome for mapping. Then, the transcriptome read mapping software TopHat [14] can be used to accurately and efficiently map RNA-Seq reads to the reference genome.

Before running the TopHat program, a directory is created and named as the accession number for the RNA-Seq data to store corresponding files generated by TopHat. The commands of running TopHat for single-end and paired-end data are as follows:

```
#!/bin/bash
#PBS -N tophat.pbs
#PBS -l nodes=1:ppn=15
#PBS -l walltime=7200:00:00
#PBS -q batch
cd /public/home/lqhan/lncRNA_protocol/
for i in $(cat single_end/SRRid.txt)
do
/public/home/lqhan/software/tophat-2.0.13.Linux_x86_64/tophat
-G Zea_mays.AGPv4.32.gff3 --max-multihits 1 -p 15 -o single_end/
$i/ Zea_mays.AGPv4.dna_sm.toplevel single_end/$i.fastq_t;
done
```

After running TopHat successfully, output files are generated and stored into the specified directory. TopHat output contains seven result files, of which "accepted_hits.bam" is a binary bam file containing all alignments against the reference genome and will be used for following analyses. The "align_summary.txt" summarizes mapping results, including the number of input reads, the number of mapped reads, and the read mapping rate.

2. BAM File Sorting and Merging

To remove redundancy, output bam files from different biological samples can be merged. Before merging, SAMtools [15] can be used to sort all bam files generated by TopHat. Different sorted bam files can be merged by SAMtools merge, which output the merged bam file sample_1.bam.

The code for sorting is as follows:

```
samtools-0.1.13/samtools sort -m 5000000000 $i/accepted_hits.
bam $i/sorted;
```

The code for merging is as follows:

```
samtools-0.1.13/samtools merge sample_1.bam SRR531202/sorted.
bam SRR531203/sorted.bam
```

3. Transcriptome Assembly

Several assembly tools have been devised to assemble transcriptomes, such as Trinity and Cufflinks. Cufflinks [16], which is a counterpart of TopHat, is shown here for a transcriptome assembly. Before running Cufflinks, an output directory (sample_1) is created according to the name of the sample. Parameters "-g" for the reference genome-annotation-guided assembly and "--max-frag-multihits 1" for the uniquely mapping reads are used (as follows):

```
#!/bin/bash
#PBS -N cufflinks.pbs
#PBS -l nodes=1:ppn=15
#PBS -l walltime=7200:00:00
#PBS -q batch
cd /public/home/lqhan/lncRNA_protocol/
for ((i=1;i<2;i++))
do
/public/home/lqhan/software/cufflinks-2.2.1.Linux_x86_64/
cufflinks -p 15 -g Zea_mays.AGPv4.32.gff3 --max-frag-
multihits 1 -L sample_$i -o single_end/sample_$i/ single_-
end/sample_$i.bam
done
```

Cufflinks can generate four assembly files: genes.fpkm_tracking, isoforms.fpkm_tracking, skipped.gtf, and transcripts.gtf. The "less -S" command is used to view the annotated file transcripts.gtf, showing different transcript assemblies. The method for paired-end read assembly is similar to that of single-end read data.

Next, the "transcripts.gtf" files are merged together using cuffmerge for a genome-wide landscape of the transcriptome. This process can integrate all the transcriptome assemblies from different sources, such as previous genome annotations, EST information, and newly generated transcripts by mRNA-Seq. For each transcript, a class_code is generated to show the relationships

Table 1
The meaning of each letter in class_code [16]

=	Complete match of intron chain
j	Potentially novel isoform (fragment): at least one splice junction is shared with a reference transcript
i	A transfrag falling entirely within a reference intron
o	Generic exonic overlap with a reference transcript
p	Possible polymerase run-on fragment (within 2Kbases of a reference transcript)
u	Unknown, intergenic transcript
s	An intron of the transfrag overlaps a reference intron on the opposite strand (likely due to read mapping errors)
x	Exonic overlap with reference on the opposite strand
e	Single-exon transfrag overlapping a reference exon and at least 10 bp of a reference intron, indicating a possible pre-mRNA fragment
c	Contained
r	Repeat. Currently determined by looking at the soft-masked reference sequence and applied to transcripts where at least 50% of the bases are lower case
.	(.tracking file only, indicates multiple classifications)

among different transcriptome sources. Different class_code represent different relationships (Table 1).

The code of cuffmerge is as follows ("mergelist.txt" contains all of the transcripts.gtf files with right path):

```
cufflinks-2.2.1.Linux_x86_64/cuffmerge -p 15 -g Zea_mays.
AGPv4.32.gff3 mergelist.txt
```

To reduce the computational load, assembled transcripts are removed from the class_code "=," meaning that they completely match with previously annotated transcripts. The command "grep" in Linux is used to group the merged.gtf file according to class_code. Here is an example of class_code "u," meaning novel transcripts, which are more likely to be lncRNAs than that with class_code "=."

3.2.4 LncRNA Identification and Expression-Level Quantification

1. Identification of lncRNAs by LncRNA_Finder

Usually putative lncRNAs are defined as transcripts that are >=200 bp and have no or weak protein-coding ability. The LncRNA_Finder.pl is used to identify putative LncRNAs. LncRNA_Finder is a Perl script which enables the discovery of lncRNAs using native sequence fasta files. This script essentially uses the results from external alignment programs and performs lncRNA filtering with a set of custom parameters. When using

LncRNA_Finder.pl, all of the input files are fasta formatted. So, gffread in Cufflinks is used to obtain the fasta-formatted transcript sequences according to the class_code_u.gtf. The command "gffread" is:

```
    gffread -w calss_code_u.fa -g /path/to/referencegenome.fa
class_code_u.gtf
```

To rule out housekeeping, protein-encoding, and small RNA precursors, transcript sequences from calss_code_u.fa were aligned against these datasets. All of these data can be obtained directly except small RNA datasets. You can get raw small RNA-Seq data from NCBI, followed by a series of conversion and quality control parameters using softwares such as sratools and Trimmomatic to get distinct high-quality small RNA reads with length ranging from 18 to 30.

Next use LncRNA_Finder to identify putative lncRNAs. During this process, size selection, open reading frame filter, known protein domain filter, protein-coding-score test, and elimination of housekeeping lncRNAs and precursors of small RNAs will be carried out in a row by LncRNA_Finder [17]. A pbs script can be created to run LncRNA_Finder as follows:

```
    LncRNA_Finder.pl -i class_code_u.fa -p uniprot_sprot.fasta
-k housekeeping.fa -s smallRNA_uniq.fa -o lncrna/lncrna -t 5
```

After running LncRNA_Finder successfully, one fasta file containing lncRNAs will be created.

2. Quantification of lncRNA Expression Levels

Based on the newly generated merged.gtf file derived from Cufflinks, the software RSEM [18] can be used to remap, normalize RNA-Seq, and calculate "TPM" values. The procedure of using RSEM software is as follows: (1) build transcript reference index files using the command of "rsem-prepare-reference" in RSEM; (2) use "rsem-calculate-expression" in RSEM with parameters "-p 8 –bowtie2 –estimate-rspd –append-names –output-genome-bam" to calculate TPM value for each transcript; and (3) use the "grep" command in Linux to obtain the expression level (TPM) of lncRNAs across different tissues according to the output from LncRNA_Finder.

3. Visualization of lncRNAs by IGV

IGV software provides the ability to visualize transcripts across the whole genome. We input sample_1.bam.tdf processed by "tools" in IGV software, class_code_u.gtf newly merged annotated file, and annotation.gtf from the original reference genome annotation file. Then, a putative LncRNA locus such as

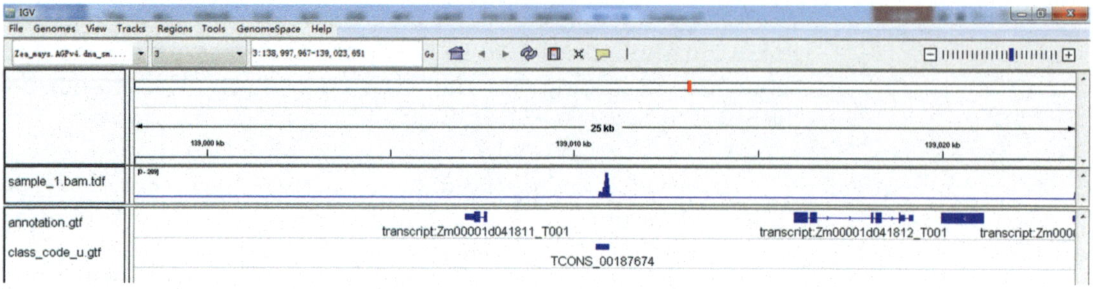

Fig. 7 IGV display showing the distribution of a putative lncRNA locus "TCONS_00187674"

"TCONS_00187674" can be closely examined. This locus is only annotated by the newly synthesized annotation file and is located in the middle of two previously annotated genes (Fig. 7).

4 Notes

1. Working on florets in the same row makes access and removal of glumes, lemma, and palea easier.
2. When removing the ovule from the cob, be sure to cut as close as possible to the cob at the base of the ovule so that the embryo sac is not left behind.
3. In preparation for dissecting the embryo sacs from the ovules, remove enough of the cell wall digesting mix so that the ovules are resting on the bottom of the petri dish. This stabilizes them so that they remain steady while removing the nucellar tissue from around the embryo sac.
4. Before collecting the embryo sac in the Pasteur pipet, remove the rest of the ovule tissue from its vicinity so that it is not inadvertently collected with the embryo sac.
5. When collecting the isolated embryo sac with a Pasteur pipet, apply gently positive pressure so that the media is not sucked into the pipet tip as soon as it touches the surface. This precaution prevents the embryo sac from being suddenly pushed out of the field of view or pulling unwanted tissue into the pipet.

Acknowledgments

We thank Z. Vejlupkova and R. Cole for their contributions to the development of the RNA isolation and library preparation methodology. This work was supported by National Science Foundation Plant Genome Research Program Awards, DBI-0701731 and DBI-1340050, to Matthew Evans and by Huazhong Agricultural University Scientific & Technological Self-innovation Foundation to Lin Li and Linqian Han (Program No. 2662016PY096).

References

1. Batista PJ, Chang HY (2013) Long noncoding RNAs: cellular address codes in development and disease. Cell 152(6):1298–1307. https://doi.org/10.1016/j.cell.2013.02.012
2. Rinn JL, Chang HY (2012) Genome regulation by long noncoding RNAs. Ann Rev Biochem 81:145–166. https://doi.org/10.1146/annurev-biochem-051410-092902
3. Su Y, Zhang C, Wei Q (2014) Advances of long noncoding RNA. Acta Botanica Boreali-Occidentalia Sinica 11:31
4. Iyer MK, Niknafs YS, Malik R, Singhal U, Sahu A, Hosono Y, Barrette TR, Prensner JR, Evans JR, Zhao S, Poliakov A, Cao X, Dhanasekaran SM, Wu YM, Robinson DR, Beer DG, Feng FY, Iyer HK, Chinnaiyan AM (2015) The landscape of long noncoding RNAs in the human transcriptome. Nat Genet 47 (3):199–208. https://doi.org/10.1038/ng.3192
5. Wen K, Yang L, Xiong T, Di C, Ma D, Wu M, Xue Z, Zhang X, Long L, Zhang W, Zhang J, Bi X, Dai J, Zhang Q, Lu ZJ, Gao G (2016) Critical roles of long noncoding RNAs in Drosophila spermatogenesis. Genome Res 26 (9):1233–1244. https://doi.org/10.1101/gr.199547.115
6. Heo JB, Sung S (2011) Vernalization-mediated epigenetic silencing by a long intronic noncoding RNA. Science 331 (6013):76–79. https://doi.org/10.1126/science.1197349
7. Chekanova JA (2015) Long non-coding RNAs and their functions in plants. Curr Opin Plant Biol 27:207–216. https://doi.org/10.1016/j.pbi.2015.08.003
8. Chettoor AM, Givan SA, Cole RA, Coker CT, Unger-Wallace E, Vejlupkova Z, Vollbrecht E, Fowler JE, Evans M (2014) Discovery of novel transcripts and gametophytic functions via RNA-seq analysis of maize gametophytic transcriptomes. Genome Biol 15(7):414. https://doi.org/10.1186/s13059-014-0414-2
9. Kranz E, Bautor J, Lorz H (1991) In vitro ferilization of single, isolated gametes of maize mediated by electrofusion. Sex Plant Reprod 4:12–16
10. Yang H, Kaur N, Kiriakopolos S, McCormick S (2006) EST generation and analyses towards identifying female gametophyte-specific genes in Zea mays L. Planta 224(5):1004–1014
11. Schreiber DN, Dresselhaus T (2003) In vitro pollen germination and transient transformation of Zea mays and other plant species. Plant Mol Biol Rep 21:31–41
12. Bolger AM, Lohse M, Usadel B (2014) Trimmomatic: a flexible trimmer for Illumina sequence data. Bioinformatics 30 (15):2114–2120. https://doi.org/10.1093/bioinformatics/btu170
13. Langmead B, Salzberg SL (2012) Fast gapped-read alignment with Bowtie 2. Nat Methods 9 (4):357–359. https://doi.org/10.1038/nmeth.1923
14. Trapnell C, Pachter L, Salzberg SL (2009) TopHat: discovering splice junctions with RNA-Seq. Bioinformatics 25(9):1105–1111. https://doi.org/10.1093/bioinformatics/btp120
15. Li H, Handsaker B, Wysoker A, Fennell T, Ruan J, Homer N, Marth G, Abecasis G, Durbin R, Genome Project Data Processing S (2009) The sequence alignment/map format and SAMtools. Bioinformatics 25 (16):2078–2079. https://doi.org/10.1093/bioinformatics/btp352
16. Mortazavi A, Williams BA, McCue K, Schaeffer L, Wold B (2008) Mapping and quantifying mammalian transcriptomes by RNA-Seq. Nat Methods 5(7):621–628. https://doi.org/10.1038/nmeth.1226
17. Li L, Eichten SR, Shimizu R, Petsch K, Yeh CT, Wu W, Chettoor AM, Givan SA, Cole RA, Fowler JE, Evans MMS, Scanlon MJ, Yu J, Schnable PS, Timmermans MC, Springer NM, Muehlbauer GJ (2014) Genome-wide discovery and characterization of maize long non-coding RNAs. Genome Biol 15(2):R40. https://doi.org/10.1186/gb-2014-15-2-r40
18. Li B, Dewey CN (2011) RSEM: accurate transcript quantification from RNA-Seq data with or without a reference genome. BMC Bioinformatics 12:323. https://doi.org/10.1186/1471-2105-12-323
19. Robinson J, Thorvaldsdottir H, Winckler W, Guttman M, Lander ES, Getz G, & Mesirov JP (2011) Integrative genomics viewer. Nature Biotechnology, 29(1):24–26
20. Huang BQ, Sheridan WF (1994) Female gametophyte development in maize: microtubular organization and embryo sac polarity. Plant Cell 6(6):845–861
21. Tattersall EAR, Ergul A, AlKayal F, DeLuc L, Cushman JC, Cramer GR (2005) A comparison of methods for isolating high-quality RNA from leaves of grapevine. Am J Enol Vitic 56 (4):400–406
22. Townsley BT, Covington MF, Ichihashi Y, Zumstein K, Sinha NR (2015) BrAD-seq: breath Adapter Directional sequencing: a

streamlined, ultra-simple and fast library preparation protocol for strand specific mRNA library construction. Front Plant Sci 6:366. https://doi.org/10.3389/fpls.2015.00366

23. Ulitsky I (2016) Evolution to the rescue: using comparative genomics to understand long non-coding RNAs. Nat Rev Genetics 17 (10):601–614. https://doi.org/10.1038/nrg.2016.85

24. Schnable PS, Ware D, Fulton RS, Stein JC, Wei F, Pasternak S, Liang C, Zhang J, Fulton L, Graves TA, Minx P, Reily AD, Courtney L, Kruchowski SS, Tomlinson C, Strong C, Delehaunty K, Fronick C, Courtney B, Rock SM, Belter E, Du F, Kim K, Abbott RM, Cotton M, Levy A, Marchetto P, Ochoa K, Jackson SM, Gillam B, Chen W, Yan L, Higginbotham J, Cardenas M, Waligorski J, Applebaum E, Phelps L, Falcone J, Kanchi K, Thane T, Scimone A, Thane N, Henke J, Wang T, Ruppert J, Shah N, Rotter K, Hodges J, Ingenthron E, Cordes M, Kohlberg S, Sgro J, Delgado B, Mead K, Chinwalla A, Leonard S, Crouse K, Collura K, Kudrna D, Currie J, He R, Angelova A, Rajasekar S, Mueller T, Lomeli R, Scara G, Ko A, Delaney K, Wissotski M, Lopez G, Campos D, Braidotti M, Ashley E, Golser W, Kim H, Lee S, Lin J, Dujmic Z, Kim W, Talag J, Zuccolo A, Fan C, Sebastian A, Kramer M, Spiegel L, Nascimento L, Zutavern T, Miller B, Ambroise C, Muller S, Spooner W, Narechania A, Ren L, Wei S, Kumari S, Faga B, Levy MJ, McMahan L, Van Buren P, Vaughn MW, Ying K, Yeh CT, Emrich SJ, Jia Y, Kalyanaraman A, Hsia AP, Barbazuk WB, Baucom RS, Brutnell TP, Carpita NC, Chaparro C, Chia JM, Deragon JM, Estill JC, Fu Y, Jeddeloh JA, Han Y, Lee H, Li P, Lisch DR, Liu S, Liu Z, Nagel DH, McCann MC, SanMiguel P, Myers AM, Nettleton D, Nguyen J, Penning BW, Ponnala L, Schneider KL, Schwartz DC, Sharma A, Soderlund C, Springer NM, Sun Q, Wang H, Waterman M, Westerman R, Wolfgruber TK, Yang L, Yu Y, Zhang L, Zhou S, Zhu Q, Bennetzen JL, Dawe RK, Jiang J, Jiang N, Presting GG, Wessler SR, Aluru S, Martienssen RA, Clifton SW, McCombie WR, Wing RA, Wilson RK (2009) The B73 maize genome: complexity, diversity, and dynamics. Science 326(5956):1112–1115

Part II

Studying the Tissue and Cell-Type Specific lncRNAs

Chapter 5

Improved Method of RNA Isolation from Laser Capture Microdissection (LCM)-Derived Plant Tissues

Vibhav Gautam, Archita Singh, Sharmila Singh, Swati Verma, and Ananda K. Sarkar

Abstract

Laser capture microdissection (LCM) is a tool to isolate desired and/or less accessible cells or tissues from a heterogeneous population. In the current method, we describe an efficient and cost-effective method to obtain both high-quality mRNA and miRNAs in sufficient quantity from LCM-derived plant tissues. The quality of the isolated RNA can be assessed using Bioanalyzer. Using modified stem-loop RT-PCR, we confirmed the presence of 21–24 nucleotide (nt) long mature miRNAs. This modified LCM-based method has been found to be suitable for the tissue-specific expression analysis of both genes and small RNAs (miRNAs).

Key words LCM, Laser microdissection, RNA isolation, Gene expression, miRNA, Root meristem

1 Introduction

Isolation of good-quality RNA is a prerequisite for gene expression studies in both plants and animals. Fluorescence-activated cell sorting (FACS) and LCM are the two important techniques frequently used to study tissue- or cell-specific gene expression patterns. In FACS, RNA is isolated from sorted cells, labeled with a fluorescent marker, such as green fluorescent protein (GFP), which is further used for downstream applications [1–4]. The use of FACS is limited by the availability of desired cell-specific molecular marker, accessibility to tissue, and vulnerability of isolated plant protoplasts to damage. Development of LCM-based cell or tissue excision approach helped to overcome these limitations [5, 6]. LCM offers flexibility in observing a specific population of cells under microscope, mark them on screen, and microdissect and collect them. RNA isolated from collected cells/tissues can be used for many downstream applications [5, 7]. LCM coupled with quantitative RT-PCR (qRT-PCR), microarray, or next-generation sequencing

Julia A. Chekanova and Hsiao-Lin V. Wang (eds.), *Plant Long Non-Coding RNAs: Methods and Protocols*, Methods in Molecular Biology, vol. 1933, https://doi.org/10.1007/978-1-4939-9045-0_5,

(NGS), has been recently used for studying cell- or tissue-specific gene expression patterns [5, 7, 8].

LCM of plant tissue requires fixation, embedding (in wax), and thin sectioning. RNA isolated from LCM-excised tissue generally suffers from poor integrity, low quantity, and less abundance of 21–24 nt mature small RNAs. The use of expensive kits, which are often used for isolation of RNA from wax or paraplast-embedded tissue, generally yields low quantity of total RNA. The quality and quantity of LCM-derived RNA are also affected by tissue handling, tissue fixation procedures, and post-LCM RNA isolation method. To overcome aforesaid difficulties, we have tried to significantly improve the existing protocols for tissue fixation, sectioning, tissue handling, and RNA isolation [7–9]. A flowchart showing steps for tissue fixation, LCM, and RNA isolation has been shown (*see* Fig. 1). With the present modified protocol, we were able to isolate high-quality RNA from paraplast-embedded tissue using LCM (*see* Fig. 2a, b). Quality assessment of RNA samples from *Zea mays* (maize) shoot apical meristem (SAM), isolated using present method, showed good RNA integrity number (RIN) value (greater than 7), indicating high RNA integrity (*see* Fig. 2c, d). Both quantity and quality of RNA were found to be good, without the use of expensive kits. Total RNA isolated from LCM-captured SAM of *Zea mays* contained mature miRNAs. We analyzed the expression of selected genes and mature miRNAs in the SAM using RT-PCR and modified stem-loop RT-PCR methods, respectively (*see* Fig. 3a, b). Our results confirm that this LCM-based method is suitable for isolation and expression analysis of tissue-specific mRNAs and miRNAs [10, 11]. The method presented in the current chapter is originally described in Gautam et al. [11].

2 Materials

2.1 Reagents

1. *Zea mays* shoot apex of 7-day-old seedlings containing SAM were used as tissue material.
2. Diethyl pyrocarbonate (DEPC) (Sigma-Aldrich, Cat. No. D5758).
3. Acetone (Fisher Scientific, Cat. No. A18-4).
4. Xylene (Fisher Scientific, Cat. No. X5-4).
5. Paraplast (Sigma-Aldrich, Cat. No. P3558).
6. TRI Reagent (Sigma-Aldrich, Cat. No. T9424).
7. 100% Ethanol (Merck, Cat. No. 1009830511).
8. 100% Isopropanol (Fisher Scientific, Cat. No. 43566).
9. Chloroform (Ranbaxy, Cat. No. C0200).
10. Agarose (LONZA, Cat. No. 50004L).

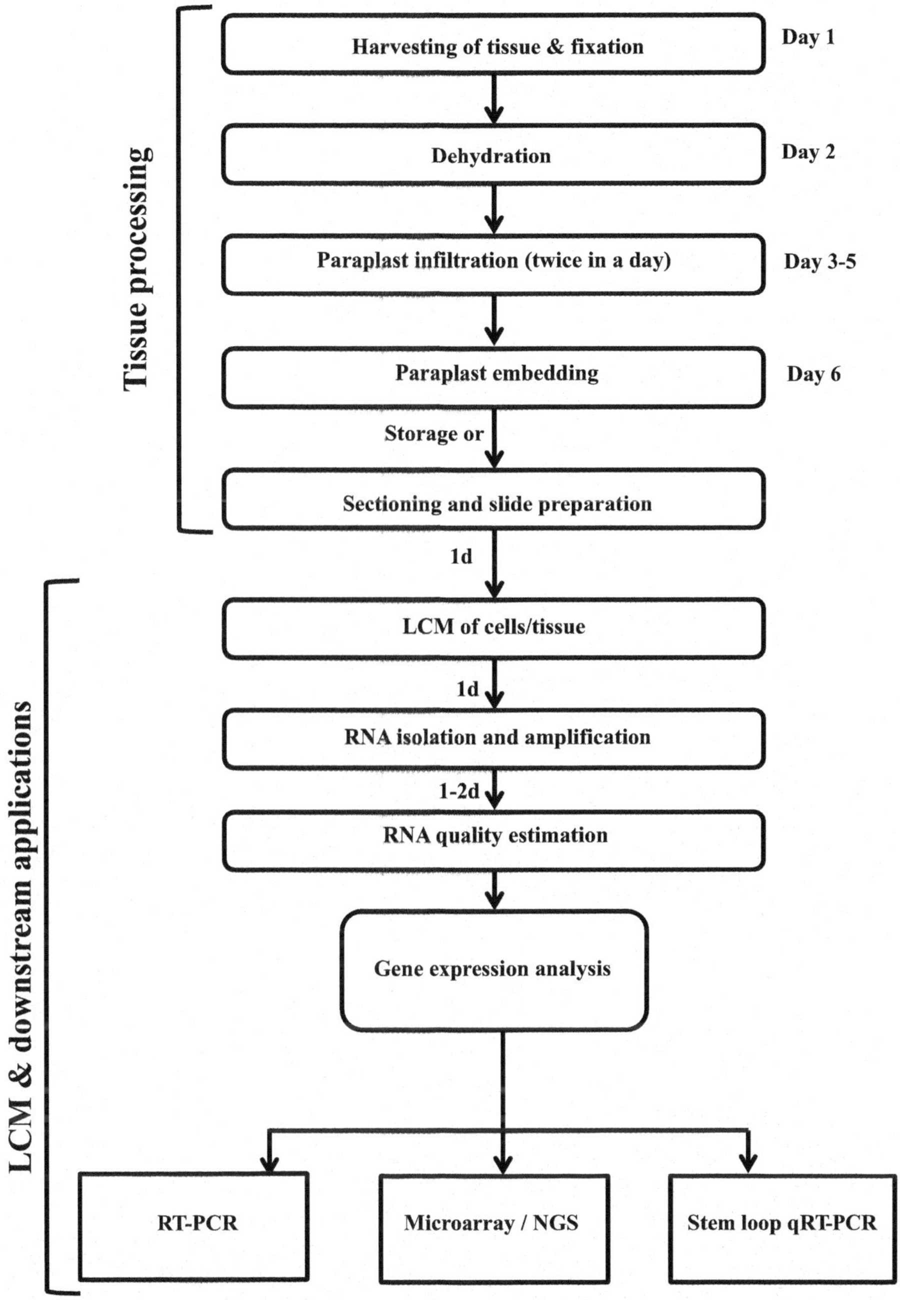

Fig. 1 Schematic outline of the LCM-based method of RNA isolation. The method is divided into two parts: tissue processing and LCM and its downstream applications. The flowchart highlights the approximate time required in each step. "d" denotes day. This flowchart is reproduced from our previous publication (Gautam et al., Sci Rep. 2016)

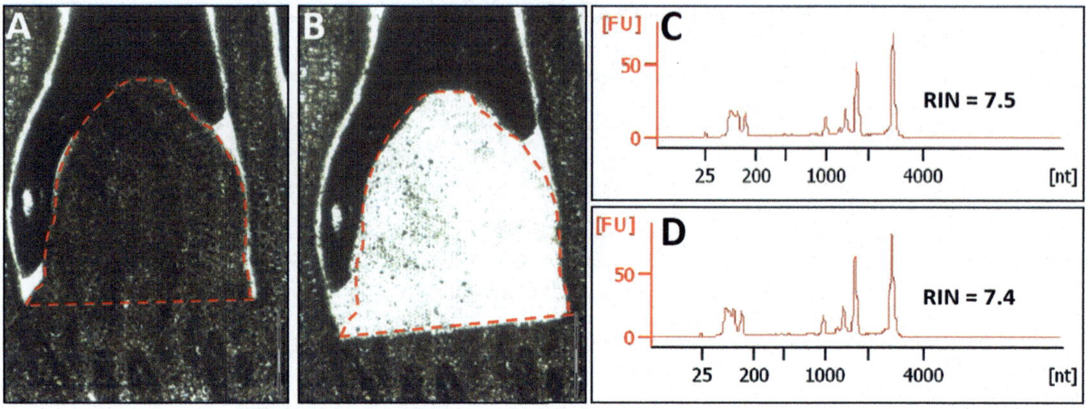

Fig. 2 LCM of 7-day-old seedling of *Zea mays* SAM. (**a**) Marked *Zea mays* SAM before and (**b**) after LCM, and (**c**, **d**) Bioanalyzer-based analysis of LCM-derived RNA (two replicates)

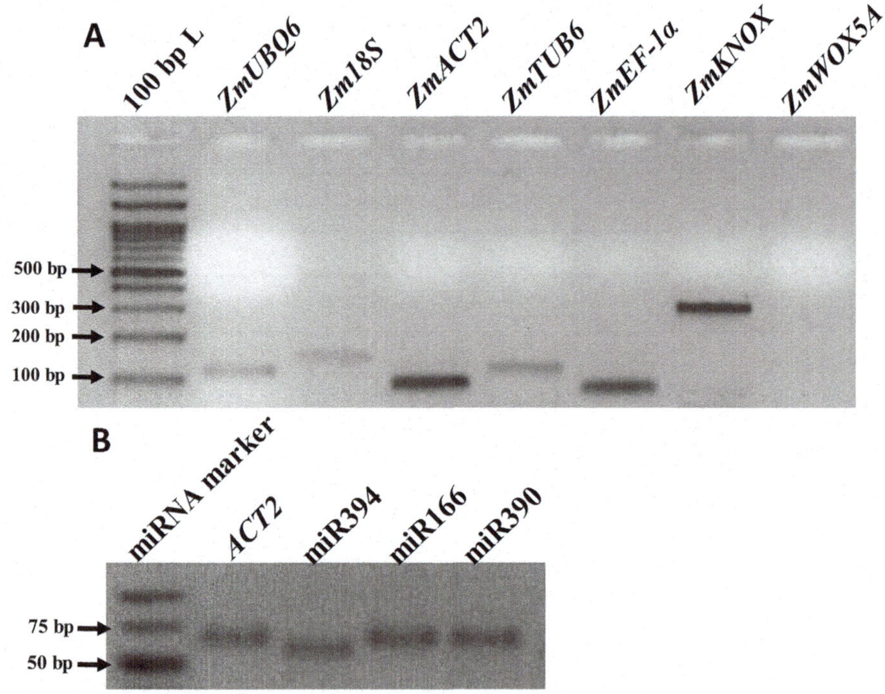

Fig. 3 LCM-based RNA isolation and expression analysis of housekeeping genes and miRNAs. (**a**) qRT-PCR showing the expression of housekeeping genes like *ZmUbiquitin6* (*ZmUBQ6*), *Zm18S*, *ZmActin2* (*ZmACT2*), *ZmTubulin6* (*ZmTUB6*), *ZmElongation Factor-1 alpha* (*ZmEF-1*α), and positive control *ZmKnotted 1-like homeobox* (*ZmKNOX*) (known to express in SAM), and negative control *ZmWuschel-related homeobox 5* (*ZmWOX5A*) (known to express in RAM). (**b**) Stem-loop RT-PCR showing the expression of *ACT2*, and miRNAs like miR394, miR166, and miR390

11. Histoclear (Sigma-Aldrich, Cat. No. H2779).
12. RiboAmp HS plus RNA amplification kit (Arcturus, Applied Biosystems, Cat. No KIT0525).
13. Mineral oil (Amresco, Cat. No. J217).
14. SuperScript III reverse transcriptase (RT) enzyme (Invitrogen, Cat. No. 18080-051).
15. RNase-free water (Sigma-Aldrich, Cat. No. W4502).
16. RNaseOUT spray (GE Biosciences, Cat. No. 786-71).
17. 3B DNA polymerase (BlackBio Biotech, Cat. No. 3B009).
18. DNase I (Thermo Fisher Scientific).

2.2 Equipments

1. LCM Microscope (Carl Zeiss, Jena, Germany).
2. Rotary microtome (Leica RM2265, Germany).
3. Bioanalyzer (Agilent Technologies).
4. Agilent RNA 6000 Nano Chips (Agilent Technologies).
5. High-temperature oven (Scientific Systems, India).
6. Metal hot plate (Scientific Systems, India).
7. Slide warmer (Medite OTS400).
8. Tissue floating water bath (Medite TFB55).
9. Peel-A-Way embedding mold (Sigma-Aldrich, Cat. No. E6032).
10. Disposable steel knife/blade (Leica Microsystems, Cat. No. 140358838925).
11. Painting brush (narrow).
12. Aluminum foil.
13. Forceps.
14. Steel needles.
15. RNase-free charged glass slides (HistoBond^{+}, Marienfeld Cat. No. 0810411).
16. Glass cover slips (HiMedia, Cat. No. CG115).
17. Slide rack (Tarsons).
18. Polypropylene conical tubes (50 mL).
19. Plastic microcentrifuge tubes (1.5 mL).
20. Plastic microcentrifuge tubes (0.5 mL).
21. Nitrile gloves.
22. Thermal cycler (Applied Biosystems).
23. NanoDrop 1000 (Thermo Fisher Scientific).

3 Methods

3.1 Fixation

An RNase-free practice should be followed in all the steps which include tissue collection, fixation, embedding, LCM, RNA isolation, and other downstream experiments.

1. Dissect the tissue (*Zea mays* shoot apex) using the RNase-free scalpel, and transfer to 2 mL RNase-free microfuge tubes containing ice-cold 100% acetone [7, 9, 12].
2. Immediately vacuum infiltrate the tissue kept in 100% acetone at 4 °C under 350 mmHg pressure for a minimum of 15 min (for better penetration of the fixative) or until all the tissue settles at the bottom completely (*see* **Note 1**).
3. Replace the acetone with fresh ice-cold acetone (100%), and keep it at room temperature (RT) for 1 h with mild agitation.
4. Perform the second replacement of acetone with fresh ice-cold 100% acetone; leave it overnight at 4 °C with gentle agitation.
5. Dehydrate the tissue by passing through a series of acetone/xylene gradients of the ratio 3:1, 1:1, and 1:3 for 1 h each, replace old solution with fresh xylene (100%), and incubate at RT for 1 h with mild agitation.
6. Add few paraplast chips into the vial and leave it overnight at RT with agitation.
7. Keep the tubes at 57 °C in the oven to melt the paraplast; on the next day, replace the old paraplast with the fresh molten paraplast at 57 °C, and repeat it twice a day for the next 3 days (*see* **Note 2**).

3.2 Paraplast Embedding

1. After performing the above mentioned six changes of paraplast at 57 °C, embed the shoot apex using Peel-A-Way molds or steel mold and embedding ring (*see* **Note 2**).
2. Dispense the contents of the vial in a mold placed on a hot plate at 57 °C, using RNase-free forceps, and arrange the siliques in the desired orientation so as to make small blocks.
3. Solidify the paraplast by placing the molds at RT and store at 4 °C. Molds can be stored in a sealed container at 4 °C for years.

3.3 Tissue Sectioning

1. Trim the tissue block into a desired shape, and place it in a desired orientation on the plastic embedding rings.
2. Fix the rings on the holding clamp of rotary microtome, and make 8–10 μm thin tissue sections.
3. Flatten the tissue sections for 3–5 min in water heated to 50–55 °C (below the melting temperature of paraplast),

transfer to HistoBond$^+$ charged slides, dry for 30 min on hot plate at 42 °C, and store at 4 °C until proceeding for LCM (*see* **Note 3**).

3.4 LCM

1. Dewax the tissue sections by dipping the slides twice in histo-clear for 2 min, and air-dry the slide at RT (*see* **Note 4**).
2. Use stereo microscope for an initial observation of the tissue. Without much delay, observe the slides under LCM microscope, identify SAM, and mark the SAM using on-screen PALM MicroBeam tool (*see* Fig. 2a, b).
3. Laser-dissect the tissues along the marking, and follow laser catapulting (*see* Fig. 2b). Collect the LCM-based catapulted tissues in RNase-free 0.5 mL tubes containing a drop of mineral oil, and store temporarily at –80 °C (*see* **Note 5**).

3.5 RNA Isolation and Amplification

1. Centrifuge the collection tubes containing LCM-derived tissues at 1844 × *g* for 1 min, add 150 μL of TRI Reagent, and centrifuge at 1844 × *g* for 2 min.
2. Add 100 μL of chloroform to the tube, vortex it for a few seconds, incubate at RT for 15 min, and centrifuge at 12,662 × *g* for 30 min.
3. Transfer the upper aqueous phase into a fresh 1.5 mL microfuge tube, add equal volume of isopropanol, mix properly, keep at −20 °C for 1 h, and centrifuge at 12,662 × *g* for 1 h.
4. Discard the supernatant without disturbing the pellet, and wash the pellet with 100 μL of 70% ethanol by carrying out centrifugation at 4150 × *g* for 15 min.
5. Air-dry the pellet and dissolve in 10 μL of nuclease-free water.
6. Assess the RNA concentration and RIN of the RNA samples using NanoDrop 1000 and Bioanalyzer nanochip, respectively (*see* Fig. 2c, d).
7. Perform two rounds of RNA amplification using RiboAmp HS Plus kit as per company's manual [12, 13].
8. Carry out precipitation of RNA using isopropanol (as described above).

3.6 RT-PCR

1. Treat approximately 1–2 μg of amplified RNA with DNase I as per company's manual, heat inactivate at 65 °C for 10 min, and proceed for first-strand cDNA synthesis using SuperScript III RT.
2. Use 2 μL of diluted (1:5) cDNA as template for a 20 μL PCR reaction using gene-specific primers. The PCR thermal reaction profile is as follows: 94 °C for 5′:1×; 94 °C for 15″, 60 °C for 40″, 72 °C for 1′:40×; 72 °C for 10′ and holding at 4 °C.
3. Run the PCR product on 1% agarose gel (*see* Fig. 3a).

3.7 Stem-Loop RT-PCR

Stem-loop RT-PCR is used to detect and amplify mature miRNAs from an RNA pool. A stem-loop reverse primer (SLP) is designed such that on ligation to 3′ end of the mature miRNA, it forms a hairpin and extends a 3′ overhang complementary to the miRNA. The miRNA-specific forward primer (FP) contains a 5′ adapter to match the Tm with reverse primer (RP). Along with the FP, a universal reverse primer (URP) is used for PCR amplification of the mature miRNA sequences (*see* Fig. 3b) [14, 15].

1. For reverse transcription reaction, prepare 11.5 μL of "reaction mix-A" containing 0.5 μL of 10 mM dNTPs, DNase I-treated RNA (10–80 ng), and RNase-free water heated for 5 min at 65 °C and cooled on ice for 2 min. Add reaction mix-A to 6.5 μL of "reaction mix-B" containing first-strand buffer (1×), 2 μL of 0.1 M DTT, 10 U of RNaseOUT, and 50 U of SuperScript III. Add 1 μL each of 1 μM SLP and 1 μM endogenous control RP.
2. Perform the RT reaction in a thermal cycler with the following thermal reaction profile: 16 °C for 30′. 1×; 30 °C for 10″, 42 °C for 10″, 50 °C for 1″: 60×; 85 °C for 5′: 1× followed by incubation at 4 °C [14].
3. A standard endpoint PCR can be used for RT-PCR-based expression analysis of miRNAs (thermal reaction program: 94 °C for 5′:1×; 94 °C for 30″, 60 °C for 40″, 72 °C for 1′:40×; 72 °C for 10′ and holding at 4 °C). The 20 μL of an endpoint PCR reaction includes miRNA-specific FP (0.25 μM), URP (0.25 μM), dNTPs (0.2 mM), 3B DNA polymerase (0.5 U), buffer, sterile water, and 1 μL of stem-loop RT product.
4. Run the PCR product on 5% agarose gel (*see* Fig. 3b).

4 Notes

1. During tissue fixation, if the sample does not settle at the bottom of the tube, which could be due to the improper vacuum infiltration, then turn off the vacuum, release the pressure slowly, tap the vial, and put it back for vacuum infiltration, check after 5 min, and repeat until the sample settles.
2. Perform the wax exchange swiftly without losing tissue, and maintain the oven temperature strictly at 57 °C.
3. Tissue sections should be flattened on water bath, which is set at not more than 55 °C, and the flattening duration should not be more than 5 min.

4. Make sure to dewax the tissue sections using fresh histoclear for 2 min (may be increased by 1 min), drain off the histoclear from slide, and air-dry at RT.
5. When collecting LCM catapulted tissue, focus on the tube cap, and observe carefully if the tissue got collected into the cap. While isolating RNA, carefully discard the supernatant without disturbing the pellet.

Acknowledgments

V.G., A.S., and S.S. thank the Council for Scientific and Industrial Research (CSIR), India, and the National Institute of Plant Genome Research (NIPGR), New Delhi, India, for funding and internal grants. V.G. and A.S. also acknowledge the Department of Biotechnology (DBT), India, for fellowship. S.V. thanks the Department of Science and Technology-Science and Engineering Research Board (DST-SERB), India, for National Post Doctoral Fellowship (N-PDF). A.K.S. thanks NIPGR and DBT (Project Grant No. BT/PR12766/BPA/118/63/2015), New Delhi, India, for fellowship and grants. The authors declare no conflict of interests.

References

1. Iyer-Pascuzzi AS, Benfey PN (2010) Fluorescence-activated cell sorting in plant developmental biology. Methods Mol Biol 655:313–319. https://doi.org/10.1007/978-1-60761-765-5_21
2. Herzenberg LA, Sweet RG (1976) Fluorescence-activated cell sorting. Sci Am 234(3):108–117
3. Birnbaum K, Jung JW, Wang JY, Lambert GM, Hirst JA, Galbraith DW, Benfey PN (2005) Cell type-specific expression profiling in plants via cell sorting of protoplasts from fluorescent reporter lines. Nat Methods 2(8):615–619. https://doi.org/10.1038/nmeth0805-615
4. Nawy T, Lee JY, Colinas J, Wang JY, Thongrod SC, Malamy JE, Birnbaum K, Benfey PN (2005) Transcriptional profile of the Arabidopsis root quiescent center. Plant Cell 17 (7):1908–1925. https://doi.org/10.1105/tpc.105.031724
5. Emmert-Buck MR, Bonner RF, Smith PD, Chuaqui RF, Zhuang Z, Goldstein SR, Weiss RA, Liotta LA (1996) Laser capture microdissection. Science 274(5289):998–1001
6. Domazet B, Maclennan GT, Lopez-Beltran A, Montironi R, Cheng L (2008) Laser capture microdissection in the genomic and proteomic era: targeting the genetic basis of cancer. Int J Clin Exp Pathol 1(6):475–488
7. Kerk NM, Ceserani T, Tausta SL, Sussex IM, Nelson TM (2003) Laser capture microdissection of cells from plant tissues. Plant Physiol 132(1):27–35. https://doi.org/10.1104/pp.102.018127
8. Ohtsu K, Smith MB, Emrich SJ, Borsuk LA, Zhou R, Chen T, Zhang X, Timmermans MC, Beck J, Buckner B, Janick-Buckner D, Nettleton D, Scanlon MJ, Schnable PS (2007) Global gene expression analysis of the shoot apical meristem of maize (Zea mays L.). Plant J 52(3):391–404. https://doi.org/10.1111/j.1365-313X.2007.03244.x
9. Brooks L 3rd, Strable J, Zhang X, Ohtsu K, Zhou R, Sarkar A, Hargreaves S, Elshire RJ, Eudy D, Pawlowska T, Ware D, Janick-Buckner D, Buckner B, Timmermans MC, Schnable PS, Nettleton D, Scanlon MJ (2009) Microdissection of shoot meristem functional domains. PLoS Genet 5(5): e1000476. https://doi.org/10.1371/journal.pgen.1000476
10. Gautam V, Sarkar AK (2015) Laser assisted microdissection, an efficient technique to understand tissue specific gene expression

patterns and functional genomics in plants. Mol Biotechnol 57(4):299–308. https://doi.org/10.1007/s12033-014-9824-3

11. Gautam V, Singh A, Singh S, Sarkar AK (2016) An efficient LCM-based method for tissue specific expression analysis of genes and miRNAs. Sci Rep 6:21577. https://doi.org/10.1038/srep21577
12. Scanlon MJ, Ohtsu K, Timmermans MC, Schnable PS (2009) Laser microdissection-mediated isolation and in vitro transcriptional amplification of plant RNA. Curr Protoc Mol Biol Chapter 25:Unit 25A.23. https://doi.org/10.1002/0471142727.mb25a03s87
13. Ohtsu K, Schnable PS (2007) T7-based RNA amplification for genotyping from maize shoot apical meristem. CSH Protoc 2007:pdb prot4785
14. Varkonyi-Gasic E, Wu R, Wood M, Walton EF, Hellens RP (2007) Protocol: a highly sensitive RT-PCR method for detection and quantification of microRNAs. Plant Methods 3:12. https://doi.org/10.1186/1746-4811-3-12
15. Benes V, Castoldi M (2010) Expression profiling of microRNA using real-time quantitative PCR, how to use it and what is available. Methods 50(4):244–249. https://doi.org/10.1016/j.ymeth.2010.01.026

Chapter 6

Medium-Throughput RNA In Situ Hybridization of Serial Sections from Paraffin-Embedded Tissue Microarrays

Edith Francoz, Philippe Ranocha, Christophe Dunand, and Vincent Burlat

Abstract

(m)RNA spatiotemporal pattern of distribution is of key importance to decipher gene function. In this post-genomic era, numerous transcriptomic studies are made publicly available, sometimes reaching a tissular resolution and even more rarely the cellular level. This "one tissue-numerous genes" information can be completed by the reverse "one gene-numerous tissues" picture through traditional RNA in situ hybridization (ISH). Here, we present a method including (1) principles of transcriptomic data mining to be performed prior and following ISH and (2) a detailed step-by-step medium-throughput ISH protocol performed on serial sections from tissue microarrays. In a recent work, we implemented this method for 39 selected genes studied by medium-throughput ISH complementing an existing tissue-specific transcriptomic dataset focused on the model plant *Arabidopsis* seed development kinetics (Francoz et al., Scientific Reports 6:24644, 2016). This full integration of ISH and transcriptomics demonstrated the complementarity of both techniques in terms of tissue/cell specificity, signal sensitivity, gene specificity, and spatiotemporal resolution.

Key words Medium-throughput RNA in situ hybridization, Tissue-specific transcriptomics, Plants, *Arabidopsis*, Seed development, Tissue microarray paraffin serial sections, Digoxigenin-labeled riboprobes, Slide scanner, Data integration

1 Introduction

The post-genomic era leads to the generation of increasing amount of publicly available large transcriptomic datasets. In December 2017, 4348 array- and sequence-based datasets covering 2,291,555 samples from all living kingdoms were deposited on the Gene Expression Omnibus (http://www.ncbi.nlm.nih.gov/geo/). For the sole *Arabidopsis* model species, 104 datasets covering 46,234 samples were deposited at the same date. Nowadays, this type of information is also available for non-model plant species through increasing RNA-seq experiments providing a transcriptome overview (e.g., http://medicinalplantgenomics.msu.edu/; https://sites.google.com/a/ualberta.ca/onekp/home). In some

Julia A. Chekanova and Hsiao-Lin V. Wang (eds.), *Plant Long Non-Coding RNAs: Methods and Protocols*,
Methods in Molecular Biology, vol. 1933, https://doi.org/10.1007/978-1-4939-9045-0_6,

instance, transcriptomics can reach cellular/tissular levels using accurate sampling methods such as fluorescence-activated cell sorting (FACS) of protoplast populations [1, 2], isolation of tagged nuclei in specific cell types [3], or laser-capture microdissection (LCM) of serial tissue sections [4–7]. These methods are fully complementary to microscopic analyses focused on selected genes. These include promoter-reporter genes or RNA in situ hybridization (ISH). ISH is a powerful approach often underestimated probably due to its perception as being difficult to perform, its relative sensitivity, and its moderate throughput. The most popular ISH protocols use whole mount permeabilized samples [8] or Paraplast-embedded sections [9, 10]. In the last example, we recently performed a fully integrated study establishing systematic comparison between (1) the whole transcriptomic data obtained for 42 LCM samples from *Arabidopsis* seed development kinetics and (2) the cell-specific ISH profile obtained for 39 selected genes on serial sections from paraffin-embedded tissue microarrays covering the same seed development kinetics [9]. From a technical point of view, this work demonstrated the complementarity of both datasets. ISH obviously provided a higher spatial resolution not reached by the LCM tissular samples. Tissue-specific transcriptomic high throughput was not challenged by ISH medium throughput. However, ISH was more sensitive than transcriptomic in specific tissues/developmental stages. Moreover, ISH allowed obtaining specific signals for genes with high sequence identity that were not discriminated by tissue-specific transcriptomics. Full integration of the results from both datasets finally allowed estimating, for each of the 42 LCM samples, thresholds of tissue-specific transcriptomic expression values for which a specific ISH signal could be reasonably detected for lists of hundreds to tens of thousands of genes [9].

This chapter provides a detailed step-by-step ISH protocol including digoxigenin-labeled riboprobe in vitro transcription, plant tissue microarray preparation, and the ISH protocol itself. It also summarizes the principle of transcriptomic data mining to be performed prior (candidate gene selection) and after the ISH protocol (full data integration) (Fig. 1). This method also provides the mean to reach medium throughput during the ISH protocol. It is not restricted to *Arabidopsis* seed development and is also suitable for other organs and for non-model plant species [11–14].

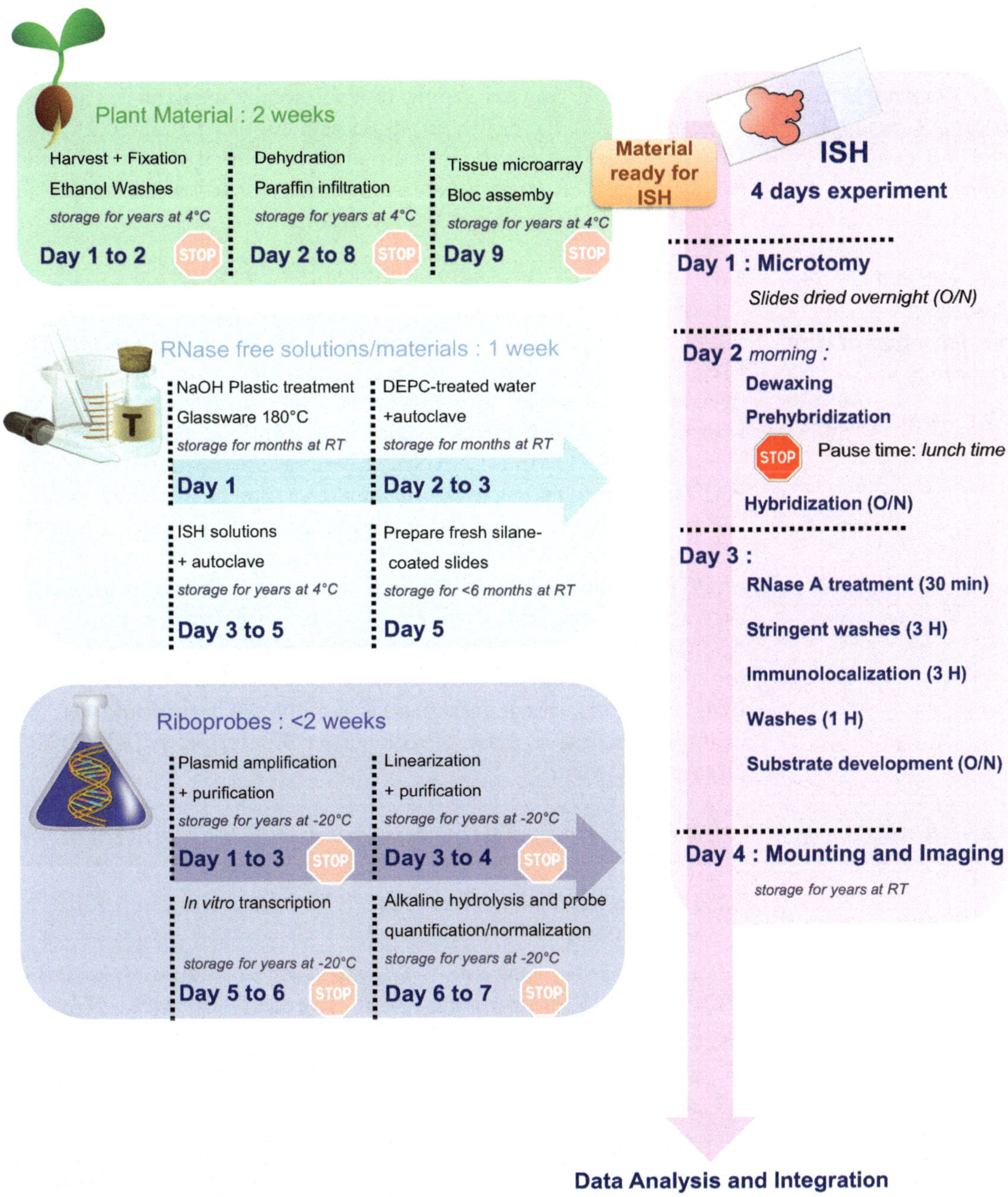

Fig. 1 Flow chart of the ISH protocol including timing and pause times

2 Materials

2.1 Transcriptomic Dataset Repositories and Data Mining Software

Raw published transcriptomic data is usually available either as a large Microsoft Excel file provided as a supplementary data (e.g., [4]) or as a GSE file to be downloaded from http://www.ncbi.nlm.nih.gov/geo/. Alternative sources of transcriptomic data and original RNA-seq files will not be detailed here.

2.2 Plasmid Resources and Riboprobe In Vitro Transcription

2.2.1 Plasmids Ordering or cDNA Cloning

For *Arabidopsis*, convenient full-length cDNA resources are "ready-to-use" pda clones ordered at the RIKEN BioResource Center (*http://www.brc.riken.jp/lab/epd/catalog/cdnaclone.html*; [15, 16]). These cDNA clones are provided within pBluescript-derived vector series allowing their direct utilization for in vitro transcription of riboprobes since the multiple cloning sites are framed by the T3 and T7 RNA polymerase promoters (*see* http://epd.brc.riken.jp/en/qa/rafl to view the four RAFL clone maps). When unavailable, the full-length cDNA to be studied should be obtained from mRNA after reverse transcription and amplification by PCR, including if possible the 5′ and 3′ untranslated regions that increase the specificity of the probes. Any vector containing orientated T3, T7, or SP6 RNA polymerase promoters surrounding a multicloning site with as large as possible choice in 5′ overhang or blunt restriction sites is suitable. These could be, for example, pGEM-T Easy (Promega), pBluescript, pCRII-TOPO (Invitrogen) vectors.

2.2.2 Molecular Biology Reagents and Kits

1. Bacterial strain: DH5α *Escherichia coli* (Z-Competent *E. coli* Transformation Kit; Zymo Research), stored at −80 °C.
2. Plasmid purification kit: GeneJET Plasmid Miniprep Kit (e.g., Fermentas).
3. Linearization enzymes: 5′ overhang or blunt restriction enzymes (e.g., Promega, Roche) stored at −20 °C (*see* **Note 1**).
4. In vitro transcription enzymes: T3, T7, or SP6 RNA polymerase and 5× RNA polymerase transcription buffer (Promega) stored at −20 °C (*see* **Note 2**).
5. RNA probe labeling kit: DIG RNA labeling mix (Roche) stored at −20 °C (*see* **Note 3**): 10 mM ATP, 10 mM CTP, 10 mM GTP, 6.5 mM UTP, 3.5 mM Dig-UTP.
6. Unlabeled NTP mix (Promega): 2.5 mM ATP, 2.5 mM CTP, 2.5 mM GTP, 2.5 mM UTP.
7. RNA transcription stop solution (4 °C dedicated RNase-free aliquot): 0.5 M EDTA pH 8.
8. Riboprobe precipitation solutions (stored at 4 °C, dedicated RNase-free aliquots): 3 M sodium acetate pH 5.2, 100% ethanol, 70% ethanol, DEPC-treated water [17].

9. Riboprobe alkaline hydrolysis solution: 100 mM sodium carbonate/Na_2CO_3 buffer pH 10.2 (stored as ready-to-use aliquots at −20 °C).
10. Riboprobe alkaline hydrolysis stop solution and precipitation solutions: 10% acetic acid, 3 M sodium acetate pH 5.2, 100% ethanol.

2.2.3 Sequence Analysis and Gel Image Analysis Software

1. Your favorite sequence analysis software, e.g., Serial Cloner (http://serialbasics.free.fr/Serial_Cloner.html)
2. Image J (https://imagej.nih.gov/ij/)

2.3 Materials Needed for Fixation, Dehydration, Paraffin Infiltration, and Embedding Steps

1. Plant culture: *Arabidopsis thaliana* (Wassilewskija, Ws, Columbia-0 or T-DNA insertion lines) were cultivated under 115 μmol/m^2/s continuous light, 22 °C, 75% relative humidity [4, 9].
2. Vacuum chamber.
3. 50 mL RNase-free tubes (e.g., Falcon or Corning).
4. Razor blades /!\ (*see* **Note 4**).
5. FAA: 10% formalin (37% formaldehyde solution, Sigma-Aldrich), 50% ethyl alcohol, 5% acetic acid, 35% DEPC-treated water /!\ (*see* **Note 5**).
6. Absolute ethanol.
7. *tert*-butanol /!\ (*see* **Note 5**).
8. Paraplast Plus.
9. 1% erythrosine ethanolic solution.
10. 60 °C oven.
11. Metallic embedding base molds (38 × 25 × 12 mm, Tissue-Tek).
12. White plastic embedding ring fitting on the embedding base molds.
13. Large slide warmer (e.g., 64 × 20 cm, LabScientific).

2.4 Medium-Throughput In Situ Hybridization Conditions

The respect of RNase-free conditions, whenever reasonably possible, is crucial for the success of in situ hybridization.

2.4.1 RNase-Free Conditions and Treatment of Material

1. Water means either DEPC-treated water [17] prepared in 5 L Schott bottle using caps that can be safely baked at 180 °C or fresh Milli-Q water contained in RNase-free bottles.
2. Solvents and chemicals were dedicated for ISH and stored separately.
3. Stock solutions and concentrated buffers were prepared in RNase-free conditions and stored either at −20 °C for small aliquots or at 4 °C for bottles.

4. All glassware (except microscopy slides) and metallic materials have to be sequentially rinsed with distilled water, wrapped in aluminum foil, baked at 180 °C for 8 h [17], and stored at room temperature in a dedicated cabinet for weeks to months.
5. Non-disposable plastic materials (e.g., tweezers, magnetic bars, polypropylene staining racks for 20 slides) were treated as follows: wash for 20 min in 0.2 N NaOH, rinse three times in DEPC-treated water, dry overnight at 37 °C in RNase-free glassware, and store at room temperature in a dedicated cabinet for weeks to months.
6. All disposable material purchased has to be RNase-free.

2.4.2 Bench Material and Large Material

1. Classical rotary microtome.
2. Tissue-Tek A35 disposable microtome blades /!\ (*see* **Note 4**).
3. At least three magnetic stirrers enabling stirring of up to 2 L solutions.
4. Two water baths (37 °C and 65 °C).
5. Large slide warmer (e.g., 64 × 20 cm, LabScientific). It can easily accommodate 40 slides used in one ISH experiment.
6. Hybridization oven (interior 400 × 330 × 400 mm) with accurate temperature setting (Binder BF53).
7. Light microscope with digital camera.
8. Nanozoomer HT slide scanner (Hamamatsu, TRI imaging platform facilities, Toulouse, France).

2.4.3 Specific Glassware, Plastic Material, and Other Materials

1. Classical ready-to-use 76 × 26 mm Starfrost microscopy slides (not RNase-free, use clean new boxes properly stored): Dispose with clean gloves or tweezers 100 slides (2 packs of 50) in 5 polypropylene staining racks for 20 slides. For slide coating, sequentially process each rack as follows (*see* **Note 6**): Dip for 5 min in 200 mL of fresh 2% (v/v) 3-aminopropyltriethoxysilane solution in acetone within a glass staining jar disposed under a fume hood, dry for 15 min under the hood on a clean Kimwipe, and sequentially wash in three glass staining jars each containing 200 mL of DEPC-treated water and dry (*see* **Note 7**).
2. Coverslips (25 × 60 mm).
3. Glass staining jars SPC (115 × 73 × 74 mm, Rogo-Sampaic). This material is relatively inexpensive and at least 20–30 jars are necessary.
4. Polypropylene staining racks for 20 slides (Kartell polypropylene staining racks 354), treated to be RNase-free. This material is inexpensive and stocks of at least 10–20 racks are convenient.

5. Sterile Thermo Scientific Nunc BioAssay square dishes ($245 \times 245 \times 25$ mm).
6. Schott bottles of various volumes (0.5–5 L) with red caps supporting heating at 180 °C.
7. Benchkote.
8. RNase-free plastic tweezers.
9. Fume hood.

2.4.4 Solvents and Chemicals

1. Acetone /!\ (*see* **Note 5**).
2. Xylene /!\ (*see* **Note 5**).
3. Absolute ethanol.
4. Dimethylformamide /!\ (*see* **Note 5**).
5. Sodium acetate.
6. Sodium carbonate (Na_2CO_3).
7. Sodium bicarbonate ($NaHCO_3$).
8. Formalin (37% formaldehyde solution, Sigma-Aldrich) /!\ (*see* **Note 5**).
9. Diethyl pyrocarbonate (DEPC) /!\ (*see* **Note 5**).
10. 3-(Aminopropyl)triethoxysilane /!\ (*see* **Note 5**)
11. Tris base.
12. Ethylenediaminetetraacetic acid (EDTA).
13. Sodium chloride (NaCl).
14. Proteinase K (Sigma-Aldrich).
15. Glycine.
16. Sodium phosphate monobasic (NaH_2PO_4).
17. Sodium phosphate dibasic (Na_2HPO_4).
18. Sodium citrate (citric acid, trisodium salt dehydrate).
19. RNase A (Sigma-Aldrich).
20. Deionized formamide (Sigma-Aldrich) /!\ (*see* **Note 5**).
21. Formamide (Sigma-Aldrich) /!\ (*see* **Note 5**).
22. Dextran sulfate (Sigma-Aldrich).
23. Yeast RNA (Roche).
24. BSA (Fraction V).
25. Anti-digoxigenin-AP Fab fragments (Roche).
26. Nitro blue tetrazolium (Kalys).
27. X-phosphate *p*-toluidine salt (BCIP p-toluidine salt; Kalys).
28. Eukitt quick-hardening mounting medium /!\ (*see* **Note 5**).

2.5 Buffers and Stock Solutions for In Situ Hybridization

Buffers are prepared in RNase-free liquid cylinders, with DEPC-treated water using a pH meter electrode cleaned with DEPC-treated water. We usually prepare 10× stock solutions stored at 4 °C for months/years in RNase-free Schott bottles, following autoclaving. Working 1× solutions are freshly made during ISH experiment. Smaller aliquots are stored in microtubes at −20 °C as specified.

1. 1 M Tris–HCl pH 8.0 (10× stock solution) for proteinase K: Weight 48.44 g of Tris, add DEPC-treated water (less than 400 mL), and adjust pH at 8.0 with approximately 10–20 mL of 37% HCl. /!\ (see Note 5). Adjust the volume up to 400 mL, autoclave, and store at 4 °C.
2. 0.5 M EDTA pH 8.0 (10× stock solution) for proteinase K: Weight 74.4 g of EDTA, add DEPC-treated water (less than 400 mL), and adjust pH at 8.0 with approximately 8–10 g of NaOH (pellets). Complete volume up to 400 mL, autoclave, and store at 4 °C.
3. 1× Tris-EDTA working solution for proteinase K: Extemporaneously dilute 10× Tris-HCl pH 8.0 and 10× EDTA stock solutions with DEPC-treated water (about 400 mL is necessary for one 40-slide ISH experiment) (*see* Subheading 3.4.5, **step 4**).
4. Phosphate buffer saline—10× PBS stock solution (main buffer for ISH): Mix 112 mL of 1 M Na_2HPO_4 (70 mM final concentration), 48 mL of 1 M NaH_2PO_4 (30 mM final concentration), 121.6 g of NaCl (1.3 M final concentration), and about 1.2 L DEPC-treated water, adjust pH to 7.0, and complete volume up to 1.6 L with DEPC-treated water. Autoclave and store at 4 °C.
5. 1× PBS working solution: Extemporaneously dilute 10× PBS stock solution with DEPC-treated water (about 2.6 L is necessary for one 40-slide ISH experiment) (*see* Subheading 3.4.2, **step 7**).
6. 0.2% (w/v) glycine in 1× PBS working solution: Mix 1.6 g of glycine and 80 mL of 10× PBS, complete volume up to 800 mL with DEPC-treated water, autoclave, and store at 4 °C. 200 mL of this 1× working solution is necessary for one 40-slide ISH experiment (*see* Subheading 3.4.2, **step 9**).
7. 10× salts (hybridization buffer stock solution): Mix 30 mL of 5 M NaCl (3.0 M final concentration), 5 mL of 1 M Tris–HCl pH 6.8 (0.1 M final concentration), 5 mL of 1 M sodium phosphate pH 6.8 (0.1 M final concentration), 5 mL of 0.5 M EDTA (50 mM final concentration), and up to 50 mL of DEPC-treated water, aliquot the 10× mix in microtubes, and store at −20 °C (*see* Subheading 3.4.3).

8. Yeast RNA (hybridization buffer stock solution): This should be deproteinated by phenol/chloroform extraction [17] and diluted to 10 mg/mL in DEPC-treated water. Aliquots are stored in microtubes at −20 °C (*see* Subheading 3.4.3).
9. 50% (w/v) dextran sulfate solution (hybridization buffer stock solution): Solubilization in DEPC-treated water is tedious due to the viscosity. Aliquots are stored in microtubes at −20 °C (*see* Subheading 3.4.3).
10. Deionized formamide (hybridization buffer stock solution): A new bottle is directly aliquoted in microtubes and stored at −20 °C (*see* Subheading 3.4.3).
11. TBS-EDTA (10 mM Tris–HCl pH 7.5, 500 mM NaCl, 1 mM EDTA) 1× working solution for RNase A: Mix 0.9688 g of Tris, 23.376 g of NaCl, and 1.6 mL of 0.5 M EDTA pH 8.0, add DEPC-treated water (less than 800 mL) to adjust pH at 7.5 with HCl, complete volume up to 800 mL with DEPC-treated water, autoclave, and store at 4 °C. 400 mL of this 1× working solution is necessary for one 40-slide ISH experiment (*see* Subheading 3.4.4).
12. RNase A: Aliquots of 10 mg are stored at −20 °C in microtubes (*see* **Note 8**).
13. 20× SSC (for stringent post-hybridization washing): Mix 140.24 g of NaCl and 70.56 g sodium citrate (tri Na citrate, $2H_2O$) in about 600 mL DEPC-treated water, adjust pH to 7.0 with HCl, and fill with DEPC-treated water up to 800 mL; autoclave and store at 4 °C. About 272 mL is necessary to prepare 2×, 1×, and 0.1× SSC, 1.5 L each for one 40-slide ISH experiment (*see* Subheading 3.4.4).
14. 10× TBS stock solution (1 M Tris–HCl pH 7.5, 1.5 M NaCl) for immunodetection: Weight 193.76 g of Tris and 140.256 g of NaCl, add about 1.2 L DEPC-treated water and adjust pH at 7.5 with about 160 mL 37% HCl (*see* **Note 5**), fill with DEPC-treated water up to 1.6 L, and autoclave and store at 4 °C.
15. 1× TTBS working solution (0.1 M Tris–HCl pH 7.5, 150 mM Nacl, 0.3% (w/v) Triton X100) for immunodetection: Extemporaneously mix 4.8 g of Triton X100 with 160 mL of 10× TBS stock solution, and fill with DEPC-treated water up to 1.6 L for one 40-slide ISH experiment (*see* Subheading 3.4.5).
16. 10× TBS stock solution (1 M Tris, 1 M NaCl, pH 9.5 for alkaline phosphatase reaction): Weight 96.88 g of Tris and 46.75 g of NaCl, add DEPC-treated water (less than 800 mL), adjust pH at 9.5, adjust volume to 800 mL, autoclave, and store at 4 °C.

17. 10× $MgCl_2$ stock solution (0.1 M $MgCl_2$): Weight 16.26 g $MgCl_2$ ($6H_2O$), fill with DEPC-treated water up to 800 mL, autoclave, and store at 4 °C.
18. 1× TBS-$MgCl_2$ working solution (0.1 M Tris, 100 mM NaCl, pH 9.5): Extemporaneously dilute 10× TBS and 10× $MgCl_2$ stock solutions with DEPC-treated water (about 900 mL is necessary for one 40-slide ISH experiment) (*see* Subheadings 3.4.5 and 3.4.6).
19. BCIP stock solution: Dissolve 1 g BCIP in DMF (20 mL) in a 50 mL Falcon tube to get a 50 mg/mL stock solution. Store at −20 °C (*see* Subheading 3.4.6).
20. NBT stock solution: Prepare 20 mL of 70% (v/v) DMF with fresh Milli-Q water. Dissolve 1 g NBT in 70% DMF (20 mL) in a 50 mL Falcon tube to get a 50 mg/mL stock solution. Store at −20 °C (*see* Subheading 3.4.6).

3 Methods

3.1 Transcriptomic Data Mining

This section does not intend to give an exhaustive step-by-step protocol but rather to provide tracks of possibilities for transcriptomic analysis. Please refer to the original publication to see how this approach was implemented in the context of a fully integrated study [9].

3.1.1 Download of Whole Data Series GSE File from http:/www.ncbi.nlm.nih.gov/geo/

1. Download the file of interest as "series matrix files."
2. Decompress the .txt.
3. Open the .txt file with Microsoft Excel and save it as a .xls or .xlsx file.

3.1.2 Editing of the Transcriptomic Dataset Using Microsoft Excel

This step intends to get a more user-friendly file allowing to better target the genes of interest to be studied by ISH.

1. Edit the Microsoft Excel file by clearly identifying the individual GSM references to individual samples (top of the column) (*see* **Note 9**).
2. Edit the Microsoft Excel file by attributing the correspondence between the individual chip ID reference for all genes available on the microarray (first column) and the individual locus reference (AGI number) (*see* **Note 10**).
3. Edit the Microsoft Excel file by organizing the samples in a logical order and by calculating the means of repeats.
4. Edit the Microsoft Excel file by, e.g., sorting the data by expression level (highest to lowest expression value) or sorting the data using a selected candidate gene to build a co-expression network (*see* **Note 11**).

3.2 Digoxigenin-Labeled Riboprobe In Vitro Transcription

3.2.1 Plasmid Amplification in Bacteria

1. Dissolve 100 ng of RIKEN plasmids (pda clones) in 100 μL Tris-EDTA (100 mM Tris–HCl pH 8.0, 50 mM EDTA pH 8.0), and add 1–50 μL of competent DH5α *Escherichia coli* on ice (*see* **Note 12**).
2. Directly spread this mix on 37 °C hot LB plate (pBluescript selection needs ampicillin, 50 μg/mL) for heat shock transformation, and cultivate overnight at 37 °C.
3. Isolate 1 colony for liquid LB culture (3 × 3 mL for each plasmid, pBluescript selection needs ampicillin, 50 μg/mL), and cultivate overnight under shaking (250 RPM) at 37 °C.
4. Concentrate the bacteria from the 3 × 3 mL, and perform a plasmid purification using GeneJET miniprep kits from Thermo Fisher Scientific, and strictly follow their instructions (*see* **Note 13**).
5. Measure the purified plasmid yield with *NanoDrop* Nucleic Acid Quantification (Thermo Fisher Scientific) or a similar spectrophotometer (*see* **Note 14**).
6. Check the cDNA sequence and orientation (*see* **Note 15**).

3.2.2 Plasmid Linearization

Select 5′ overhang or, if not available, blunt restriction enzymes that single cut the multicloning site in 5′ and 3′ end of the insert. Avoid the use of restriction enzymes producing a 3′ protruding end in order to prevent the initiation of transcription of opposite transcription products [18] (*see* **Note 16**).

1. Sequentially add in a 1.5 mL microtube for a final volume of 100 μL:
   (a) (87–*x*) μL of RNase-free water.
   (b) 10 μL of the buffer (10×) corresponding to the chosen restriction enzyme(s).
   (c) 1 μL of BSA (10 mg/mL).
   (d) *x* μL of purified plasmid containing DNA of interest (5–7 μg).
   (e) 2 μL of the restriction enzyme (to be added last).
2. Incubate for 3–4 h at 37 °C.
3. Control the linearization efficiency on a 5 μL aliquot by agarose gel electrophoresis (*see* **Note 17**).
4. When linearization is complete, inactivate the restriction enzyme for 15 min at 65 °C (*see* **Note 18**).
5. Purify the samples from remaining enzymes and salts (*see* **Note 19**).
6. Measure the linearized and purified DNA yield with *NanoDrop* Nucleic Acid Quantification or a similar spectrophotometer.
7. Store the samples at −20 °C for days or proceed directly to the in vitro transcription.

*3.2.3 Digoxigenin-Labeled Riboprobe In Vitro Transcription (See **Note 20**)*

1. Add and carefully mix in a RNase-free 1.5 mL tube:
   (a) 1 μg of linearized plasmid (*see* **Note 21**).
   (b) 2 μL of DIG RNA labeling mix or 8 μL of unlabeled NTP mix (for competitive inhibition in situ hybridization [9].
   (c) 8 μL of 5× RNA polymerase transcription buffer (Promega) (*see* **Note 22**).
   (d) 4 μL of 100 mM DTT (provided with RNA polymerases in Promega's kit).
   (e) 1 μL of RNasine (40 unit/μL).
   (f) 1.5 μL of T3, T7, or SP6 RNA polymerase (15–17 unit/μL).
   (g) Complete to 40 μL with DEPC-treated water.
2. Do a short spin and incubate for 2 h in a 37 °C oven using the pre-warmed rack.
3. Stop the reaction by sequentially adding 6 μL 3 M NaOAc pH 5.2 and 140 μL 100% ethanol (*see* **Note 23**).
4. Precipitate on ice for 1 h or up to overnight at −20 °C.
5. Centrifuge at 20,000 × *g* during 15 min.
6. Carefully remove the supernatant and gently wash the pellet with 1 mL of 70% ethanol.
7. Centrifuge at 20,000 × *g* during 15 min at room temperature.
8. Carefully remove the ethanol without disturbing the pellet, and evaporate the remaining droplets at 40–45 °C (*see* **Note 24**).
9. Resuspend the pellet by pipetting (30 s up to 1 min) and vortexing in 50 μL of DEPC-treated water (*see* **Note 25**).

3.2.4 Alkaline Hydrolysis of Digoxigenin-Labeled Riboprobes

The remaining 48 μL will be used for the alkaline hydrolysis. The aim is to obtain fragmented riboprobes to facilitate accessibility to the mRNA targets while retaining the specificity of the ISH. We routinely fragment the probes to an average size of about 400 bases. That is why the treatment duration depends on the transcript size (*see* **step 2**):

1. Add 50 μL of 100 mM sodium carbonate buffer pH 10.2.
2. Mix gently and incubate at 60 °C for 9 up to 20 min (*see* **Note 26**). For convenience, we sequentially start the treatment tube by tube (depending on the duration of the hydrolysis needed) and end the treatment at the same time for all tubes.
3. Stop the reaction by sequentially adding the following in that order:
   (a) 5 μL of 10% acetic acid; mix by pipetting.

(b) 11 μL of 3 M sodium acetate/NaOAc pH 5.2; mix by pipetting.

(c) 300 μL of 100% ethanol; mix by pipetting and spin.

4. Incubate for 15 min at room temperature.
5. Centrifuge at 20,000 × *g* during 15 min at room temperature.
6. Carefully discard the supernatant.
7. Gently wash the pellet with 0.5 mL 70% ethanol.
8. Centrifuge at 20,000 × *g* during 15 min at room temperature.
9. Gently discard the supernatant, without disturbing the "invisible pellet" (*see* **Note 27**).
10. Dry carefully the pellet in a 40–45 °C oven.
11. Resuspend the pellet by pipetting (30 s up to 1 min) in 50 μL of DEPC-treated water.
12. Store the tubes at −20 °C until use, or proceed directly to the quantification of hydrolyzed riboprobes.

3.2.5 Quantification of Hydrolyzed Riboprobes

This step consists in analyzing riboprobes on 1% (w/v) agarose gel electrophoresis in TAE 1× (Fig. 2a) in order (1) to assess the transcription and hydrolysis efficiency comparing the migration profile of non-hydrolyzed and hydrolyzed probes and (2) to estimate the hydrolyzed riboprobe concentrations by comparison with a quantitative RNA ladder (RiboRuler Low Range RNA Ladder, Fermentas).

1. Mix the probe samples (2 μL for non-hydrolyzed and 4 μL for hydrolyzed probes) and RNA ladder with H_2O and 1 μL RNA 10× dye to reach a final volume of 10 μL (*see* **Note 28**).
2. Heat the riboprobe aliquots for 10 min at 65 °C, cool on ice, spin, and load on a 1% (w/v) agarose gel electrophoresis in TAE 1×.
3. Use the "gel plot lanes" option of the ImageJ software [19] to quantify the BET colored riboprobe smear (*see* **Note 29**) (Fig. 2b).
4. Collect peak areas (Fig. 2c).
5. Analyze results using Microsoft Excel to compare standard (RNA ladder) vs respective peak areas of riboprobes and to calculate absolute concentration (ng/μL). For convenience, we normalize our probe stock at a concentration of 25 ng/kb/μL by dilution in DEPC-treated water (*see* **Note 30**).

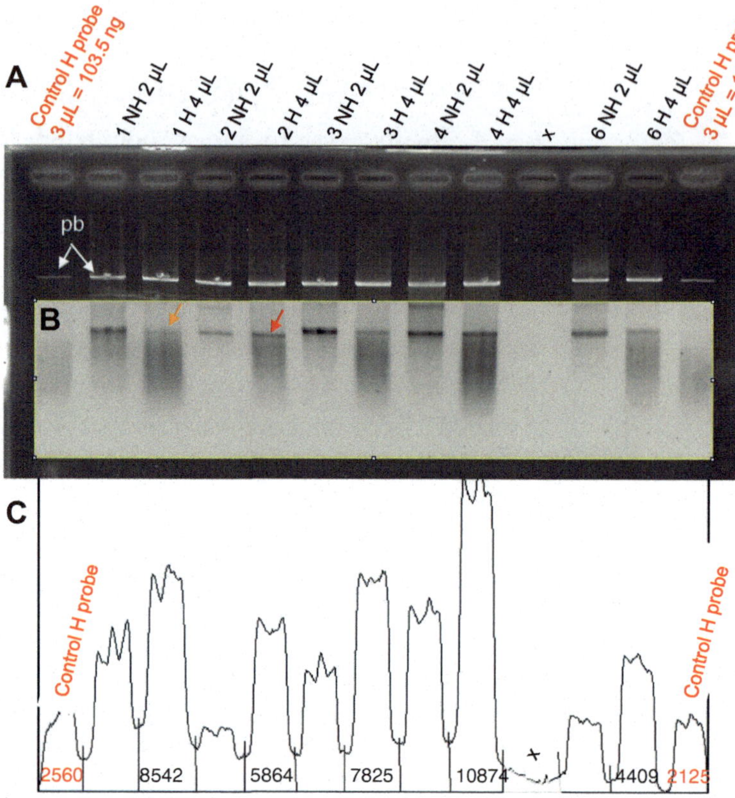

Fig. 2 Agarose gel of riboprobes before and after hydrolysis and quantification (Subheading 3.2.5). (**a**) 10 µL of denaturated samples containing 2 µL non-hydrolyzed (NH) or 4 µL hydrolyzed (H) riboprobes was loaded and runned on gel. 3 µL of hydrolyzed and denaturated control riboprobes (in red, 103.5 ng) is also loaded. (**b**) Ethidium bromide-colored riboprobe smears were selected, and colors were inverted using ImageJ software (with the "gel plot lanes" function) for quantification. (**c**) Resulting "gel plot lanes," with corresponding peak areas only for (H) hydrolyzed riboprobes (obtained with the "magic wand" tools in ImageJ). pb, remaining plasmid band; orange arrow, complete alkaline hydrolysis; red arrow, remaining non-hydrolyzed probes

3.3 Tissue Sampling, Fixation, Dehydration, Paraffin Infiltration, and Embedding in Tissue Microarrays

RNase-free conditions have to be followed for these steps (*see* Subheading 2.4.1). For our study [9], we harvested developing seeds from 39- to 45-day-old plants, following the culture conditions of [4]. Other types of samples/plant species can be used similarly, e.g., [11–14].

3.3.1 Fixation

1. Harvest the material (in our case dozens of whole siliques).
2. Immediately immerse in 20–30 mL of FAA solution in 50 mL tubes under a fume hood /!\ (*see* **Note 5**).

3. Perform five cycles of vacuum infiltration (1 min)/vacuum release in a vacuum chamber to infiltrate the samples.
4. Incubate in FAA for 6 to 16 h (overnight) at 4 °C.
5. Rinse four times in 50% ethanol in DEPC-treated water (10 min each) under a fume hood (*see* **Notes 5** /!\ and **31**).

3.3.2 Dehydration and Paraffin Infiltration

The infiltration protocol was adapted from [20]. Ethanol/DEPC-treated water/*tert*-butanol gradient series are carried out at room temperature with the following proportions (*see* **Notes 5** /!\ **and 32**). The steps with paraffin (Paraplast) are performed in a 62 °C oven.

1. (Day 1) 50/40/10 = 50 mL of DEPC-treated water + 40 mL 100% ethanol + 10 mL *tert*-butanol for 1 h.
2. 30/50/20 for 1 h.
3. 15/50/35 for 1 h.
4. 0/45/55 for 1 h (*see* **Note 33**).
5. 0/25/75 for 1 h.
6. Pure *tert*-butanol (overnight).
7. (Day 2) pure *tert*-butanol (renewed solution, 1 h).
8. Replace half of *tert*-butanol by pure Paraplast (no need to be precise), and place the samples in a 60 °C oven overnight (*see* **Note 34**).
9. (Day 3–5), replace by pure Paraplast (renew pure Paraplast, i.e., in the morning and/or in the evening) (*see* **Note 35**).

3.3.3 Embedding and Tissue Microarray Bloc Preparation

1. Pre-warm the tubes in a 60 °C oven the day before in case of storage at 4 °C.
2. Transfer the samples in paraffin within 120 × 120 × 17 mm square Petri dishes positioned on a slide warmer protected from paraffin with aluminum foils and set at 60–65 °C; this helps to select and manipulate samples when preparing your molds (Fig. 3a).
3. Warm the embedding metallic molds and manipulating metallic spatula and tweezers at 60–65 °C (Fig. 3a) (*see* **Note 36**).
4. Dispose and organize the samples in the molds filled with melted paraffin (*see* **Note 37**). When everything is in place, carefully cover the top of the mold with a white embedding ring with written reference/number.
5. Gently move the molds from the slide warmer to a cooled metallic plate or directly to the bench in order to secure the

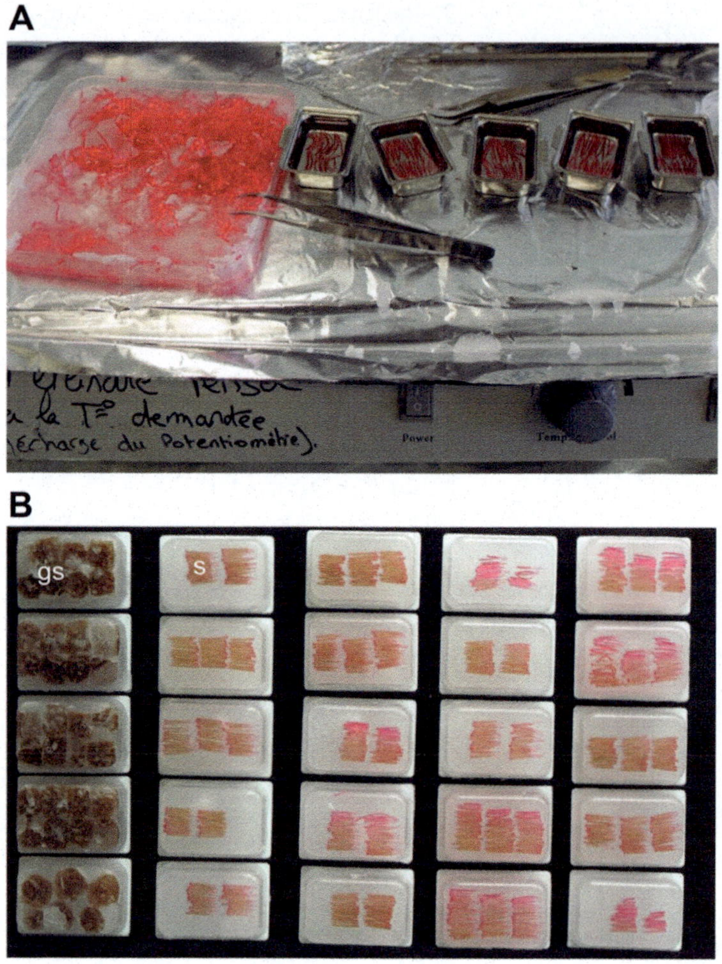

Fig. 3 Tissue microarray paraffin embedding (Subheading 3.3.3). (**a**) Protect the slide warmer (60 °C) with aluminum foil, and transfer the samples in liquid paraffin within square Petri dishes. Compose the tissue microarrays using metallic tweezers and spatula. (**b**) Examples of solidified paraffin-embedded tissue microarrays (stored at 4 °C). *gs* germinating seedlings, *s* developing siliques

position of the samples while the Paraplast starts to solidify (*see* **Note 38**).

6. Transfer the molds at least 1 h to 4 °C for full Paraplast solidification, and complete fixation of plastic ring to the metallic mold. The paraffin tissue microarrays can be stored for years at 4 °C (Fig. 3b).

3.4 Medium-Throughput In Situ Hybridization Protocol

At this stage, all the materials are available to perform an ISH experiment. Starting on Monday (day 1), the results will be available and imaged on Thursday morning (day 4) for 40 slides each corresponding to one probe hybridized with serial sections of tissue arrays. The basis of this protocol comes from [21] with first adaptations by [20, 22, 23] and final simplification using a unique buffer for most of the steps and substituting the tedious charge equilibration step with acetic anhydride/triethanolamine by a more simple carbethoxylation step using DEPC inspired by [10]. It was recently detailed [9].

3.4.1 Microtomy (Few Hours Microtomy on Day 1 and Drying Until Day 2 Morning)

Prepare 40 slides, annotated with a pencil. Dispose them as 2 lines of 20 on a slide warmer set to 30–40 °C. Since each slide will correspond to one probe, the future probe distribution along the 40 slides has to be decided in advance, in order to allow the best morphology comparison for a selection of probes of interest (e.g., antisense and sense probes for the same gene) (*see* **Note 39**).

1. Use a razor blade to trim the edges of the paraffin block with a trapezium shape with a larger base (Fig. 4a) /!\ (*see* **Note 4**).
2. Using a rotary microtome, cut a ribbon of serial sections (10 μm thick) (Fig. 4b) (*see* **Note 40**).
3. Carefully lay down the ribbon of sections on a large Kimwipe (Fig. 4c).

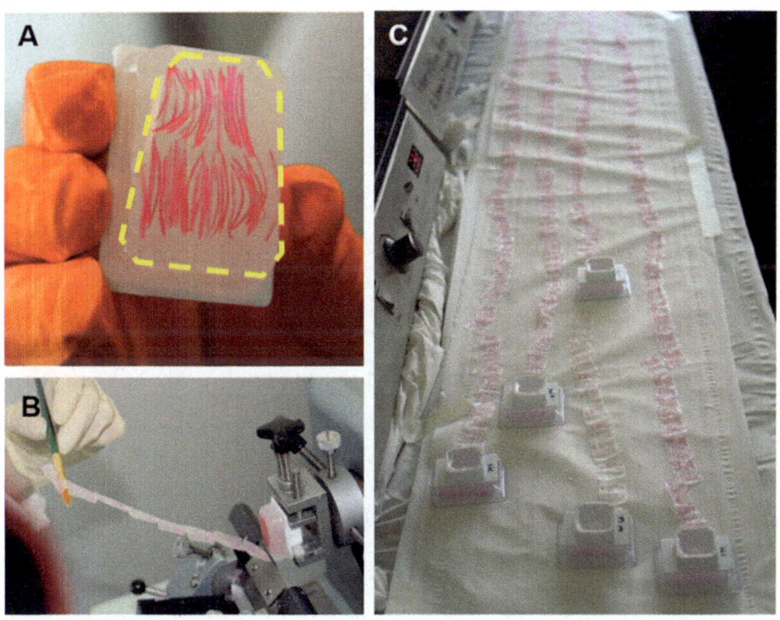

Fig. 4 Microtomy (Subheading 3.4.1). (**a**) A paraffin tissue microarray containing numerous developing siliques was trimmed using a razor blade: edges were removed, forming a "trapezium" with large base (yellow dots). This shape allows obtaining a regular ribbon of serial sections that will be individualized. (**b**) Ribbon of paraffin sections in formation. (**c**) Ribbons of sections on a clean Kimwipe

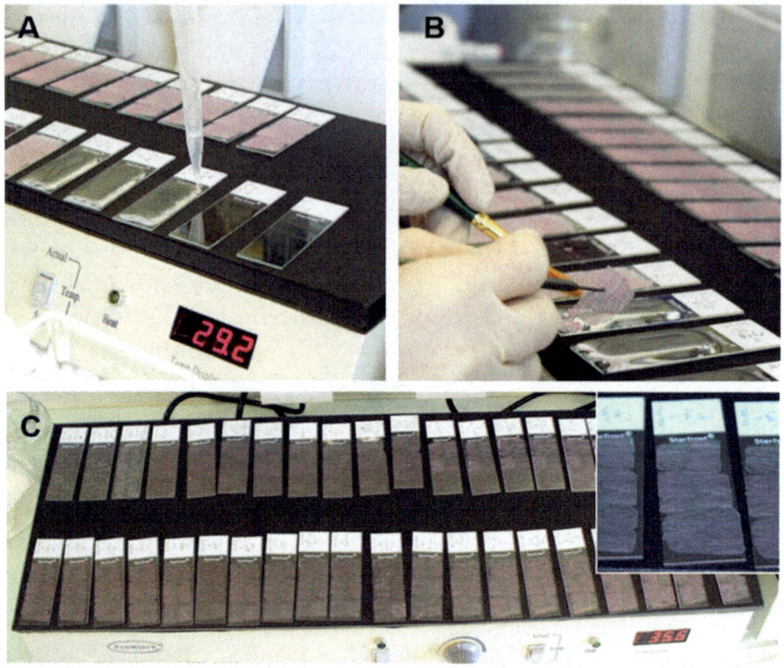

Fig. 5 Serial section positioning on slides (Subheading 3.4.1). (**a**) 3 mL of DEPC-treated water previously degassed. Note the "lotus effect" on silane-coated slides. (**b**) Separated sections are carefully deposited on water with needles and paint brush. (**c**) Excess water was removed, and tissue microarray serial sections are left to dry overnight at 30–40 °C. Inset: zoom on one slide with up to three sections carefully positioned accommodating up to 1000 developing seeds

4. Add 2–3 mL of DEPC-treated water previously degassed for at least 1 h in a vacuum chamber (Fig. 5a) (*see* **Note 41**).
5. Separate sections with needles and paint brush, and carefully lay down individual section on watered slides. Floating sections will spread on the slide. Carefully remove the excess of water with a pipette, and let the slides dry overnight at 30–40 °C on the slide warmer (Fig. 5b, c) (*see* **Note 42**).
6. Dispose the slides in 20-slide racks.

3.4.2 Paraplast Plus Removal and Prehybridization (Day 2 Morning)

Dispose nine glass staining jars under a fume hood for the sequential deparaffinization/rehydration series filled with 200 mL of each solution as follows (*see* **Notes 5** /!\ and **43**).

1. Xylene (two jars, 15 min each).
2. 100% ethanol (two jars, 5 min each).
3. 95%, 70%, and 50% ethanol (three jars 5 min each).
4. DEPC-treated water (two jars, 5 min each).

   During this series prepare two additional glass staining jars filled with 200 mL Tris-EDTA (100 mM Tris–HCl pH 8.0,

50 mM EDTA pH 8.0), and pre-warm at 37 °C in the hybridization oven. At the end of the rehydration series:

5. Equilibrate the first 20-slide rack for 15 min and the second one for 10 min in Tris-EDTA in order to set a 5 min delay between each racks for the following steps.
6. Quickly add and mix concentrated proteinase K to have a 2 μg/mL final concentration in each jar.
7. Digest for 30 min at 37 °C (with the 5 min delay between the two 20-slide racks).

   During this step, prepare 2.6 L of 1× PBS, and organize 14 glass staining jars for the next prehybridization steps to be directly proceeded at the end of the digestion, keeping the 5 min delay between the two racks all over the process. The same jar/solution will be reused for both 20-slide racks for each step equal to or below 5 min. Only six jars will be doubled (one for each 20-slide rack) for the post fixation and the carbethoxylation steps exceeding 5 min.
8. Quickly wash slides in PBS (200 mL, 15 s).
9. Block proteinase K in PBS 2% glycine (200 mL, 2 min).
10. Wash in PBS (200 mL twice, 1 min each).
11. Post-fixate in 10% formalin (3.7% formaldehyde) in PBS (200 mL, 10 min) /!\ (*see* **Note 5**).
12. Wash in PBS (200 mL twice, 5 min each).
13. Carbethoxylation in 0.1% (v/v) DEPC in PBS (200 mL twice, 15 min each) (*see* **Notes 5** /!\ and **44**).
14. Wash in PBS (200 mL twice, 5 min each).
15. Dehydrate (reusing the previous ethanol series kept under the fume hood) as follows: DEPC-treated water (two times), 50% ethanol, 70% ethanol, 95% ethanol, and 100% ethanol 5 min each (**steps 4–2**) (*see* **Note 45**).
16. Air-dry the slides at least 1 h, by storing the 20-slide racks on an RNase-free surface (*see* **Note 46**).

3.4.3 Hybridization

This step requires a few hours to prepare and distribute the mix, followed by an overnight hybridization. Prepare in a 50 mL Falcon tube about 6 mL of "hybridization buffer" by carefully mixing aliquots of the following stock solutions stored at −20 °C (these should be melted on ice a few hours before as for probe stock solutions and should be vortexed and spinned down before use):

1. 1.098 mL DEPC-treated water.
2. 600 μL 10× salts (thoroughly vortex the melted aliquot until no precipitate is visible).
3. 2.4 mL deionized formamide.

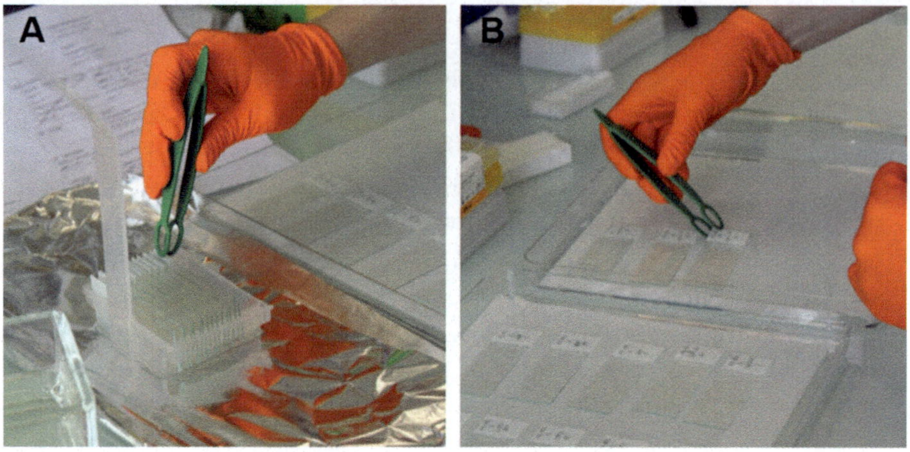

Fig. 6 Slide processing in 20-slide racks and Nunc hybridization boxes (Subheadings 3.4.2–3.4.4). (**a**) Slides transfer from a 20-slide rack to a Nunc box using plastic tweezers. This type of rack fits to 200 mL containing jars for most of the steps. (**b**) Slides are positioned without contact between them on a piece of clean precut Benchkote (soaked with formamide) in Nunc boxes for the riboprobe hybridization and antibody steps

4. 600 μL 10 mg/mL yeast RNA.
5. 1.2 mL 50% dextran sulfate (slowly pipette and distribute since it is highly viscous; it can be heated to 50 °C to reduce the viscosity).
6. 6 μL RNasine (Promega).
7. Mix and distribute 137.76 μL of "hybridization buffer" in a series of 40 microtubes (slowly pipette and distribute since it is relatively viscous).
8. Add 2.24 μL of probes in the dedicated microtubes (one probe per microtube) in order to obtain a concentration of 400 ng/kb/μL hybridization buffer (*see* **Note 47**).
9. Transfer the slides in a moist chamber, containing formamide to ensure good hybridization conditions (Fig. 6). For this, prepare four Nunc square dishes each with a piece of clean precut Benchkote and soaked with a mix of 50% formamide/2× SSC (*see* **Note 48**).
10. Press with the tweezers on the white frame to secure the slide position on the humid Benchkote (*see* **Note 49**).
11. Distribute 135–140 μL of the probe mixtures as a line in the middle of the corresponding slides avoiding air bubbles.
12. Cover each slide with a 25 × 60 mm coverslip. Make sure that the solution mixture does not escape to the white frame before proceeding to the hybridization.
13. Dispose the four covered Nunc square dishes on the top of each other and totally wraps the dishes together with Saran to maintain a saturated atmosphere.

14. Place them overnight (at least 16 h) at 50 °C in the hybridization oven (*see* **Note 50**).

3.4.4 Stringent Washing (Day 3 Morning)

1. Wear clean gloves.
2. Warm three glass staining jars in a 37 °C water bath: one with 200 mL 2× SSC and two with 200 mL TBS-EDTA.
3. Carefully unwrap the pile of covered Nunc square dishes from the hybridization oven.
4. Take the slides individually with tweezers for removal of coverslips, and dip in the 2× SSC bath until the coverslip falls by itself (*see* **Note 51**).
5. Put back the uncovered slides in the 20-slide racks (*see* **Note 52**).
6. Add RNase A (50 μg/mL) to each TBS-EDTA prepared baths, and incubate both racks at the same time, for 30 min at 37 °C.
7. During the RNase A treatment: Prepare three 2 L beakers each containing a magnetic bar for gentle stirring and a glass slide rack (102 × 80 × 24 mm, with 10 slide positions) for raising the slide rack above the stirring magnetic bar that fits within the glass staining jars. Fill them with 1.5 L of 2×, 1×, and 0.1× SSC, respectively (Fig. 7) (*see* **Note 53**).

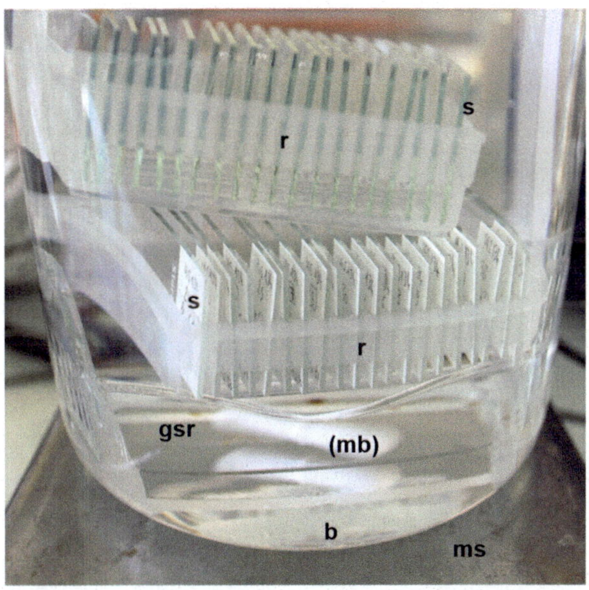

Fig. 7 Stringent washing system (Subheading 3.4.4). Two liter beakers containing each a magnetic bar for gentle stirring and a glass slide rack (102 × 80 × 24 mm, with 10 slide positions) for raising the slide rack above the stirring magnetic bar that fits within the glass slide rack. *b* 2 L beaker, *gsr* glass slide rack, *mb* stirring magnetic bar by transparency, *ms* magnetic stirrer, *r* 20-slide rack, *s* slide

(a) Wash in 2× SSC (1 h, gentle stirring at room temperature).

(b) Wash in 1× SSC (1 h, gentle stirring at room temperature).

(c) Wash in 0.1× SSC (1 h, gentle stirring at 65 °C).

During this series prepare for the Subheading 3.4.5 steps: four additional glass staining jars (two with 200 mL TTBS and two containing 200 mL 2% (w/v) BSA in TTBS). And renew Benchkote (additionally soaked with TTBS) to reuse the same Nunc square dishes as for the hybridization step.

3.4.5 Immunological Detection of Hybridized Probes (Day 3 Afternoon)

At the end of the 65 °C bath, the two slide racks are treated in parallel in two glass staining jars containing the same solutions:

1. Wash in 200 mL TTBS for 10 min.
2. Saturate in 200 mL 2% (w/v) BSA in TTBS for 30 min.
3. As for the hybridization step, take individually the slides from the 20-slide racks with tweezers, and dispose 10 slides per dish. Press with the tweezers on the white frame to secure the slide position on the humid Benchkote, and avoid any contact between them (*see* **Note 54**).
4. Spread on each slide 140 μL anti-digoxigenin-AP Fab fragments diluted 1:200 in 1% (w/v) BSA in TTBS, carefully add a 25 × 60 mm coverslip to each slide, and incubate for 2 h at room temperature.

   During this step, prepare four additional glass staining jars (two with 200 mL TTBS and two with 200 mL TBS-$MgCl_2$).
5. As for the hybridization step, carefully remove the coverslips, using the TTBS-BSA solution (bath from **step 2**). Put back the uncovered slides in the 20-slide racks. Immerse the rack in TTBS-BSA solution every 5–10 slides to avoid section drying.
6. Wash twice 15 min in 200 mL TTBS.
7. Wash twice 10 min in 200 mL TBS-$MgCl_2$.

3.4.6 Color Development of AP Reaction (Overnight Day 4 Morning)

In order to save substrate, we use two rectangular glass staining jars (inside 3″ × 25/16″ × 13/16″).

1. Fill each jar with 80 mL of TBS-$MgCl_2$.
2. Extemporaneously add 250 μL BCIP and 500 μL NBT stock solutions, and mix with a 5 mL pipet.
3. Carefully dispose the 20 slides 2 by 2 and back to back within this 10-slide jars. If needed, add TBS-$MgCl_2$ to fully immerse slides.

4. Cover the jar, wrap in aluminum foil to prevent NBT oxidation by light, and incubate overnight (at least 16 h) at room temperature.
5. Carefully put back slides in the two 20-slide racks, and stop the reaction by washing eight times in 200 mL distilled water, and dry the slides disposed on a slide warmer set to 40–50 °C under a fume hood (*see* **Note 55**).

3.4.7 Slide Permanent Mounting and Scanning (Day 4)

Add three droplets of Eukitt mounting medium along each dry slide (three slides by three slides) /!\ (*see* **Note 5**). Take care to avoid air bubbles, and cover each slide with a 25 × 60 mm coverslip. Let it dry for 1 to 2 h (up to overnight if needed) allowing solvent evaporation (*see* **Note 56**).

The slides are ready to be imaged with classical bright-field microscopy or to be scanned at high throughput using a slide scanner.

3.4.8 Result Analysis and Figure Mounting

Slide analysis is the more time-consuming step and needs organization. For our studies [9], tens of thousands of individual seed sections were viewed. Hybridization signal should appear as a blue to violet staining (Fig. 8). Each hybridization signal needs to be repeatedly observed on various samples on a given tissue microarray. Experimental repeats will ascertain the result. For highly expressed genes, this signal will be strong, and comparing antisense and sense slides will enable localizing the signal. For single-cell expression, or weak expression, antisense and sense slides need to be carefully compared. This comparison work can be performed on bright-field microscope but is more efficient using a slide scanner. We first scan all slides at ×20 focus (we use Nanozoomer HT Hamamatsu, with 200 slides containing racks; ×20 is equivalent of 0.46 μm per pixel on a single z plan). These ×20 scans are used as a first screen to discriminate probes with signals and to localize areas of interest. They are also valued in publication, to prepare figures. Punctually, we also scan at ×40 focus to get better resolution (using several z plans) or for single-cell signals. To improve the analysis throughput, we use the NDP view freeware (Hamamatsu) to load and easily "navigate" the "virtual slides." Despite the file size (×20 scan file, 300–400 Mb), the navigation does not suffer screen lags. We routinely work with two screens, allowing to simultaneously compare 2–8 slides/2–8 different probes on the same exact synchronized zone on serial sections. Image extraction is easy, thanks to selection of area of interest and export at the wished focus (depending of the scanning parameters). We then use Corel Photo-Paint software for figure mounting.

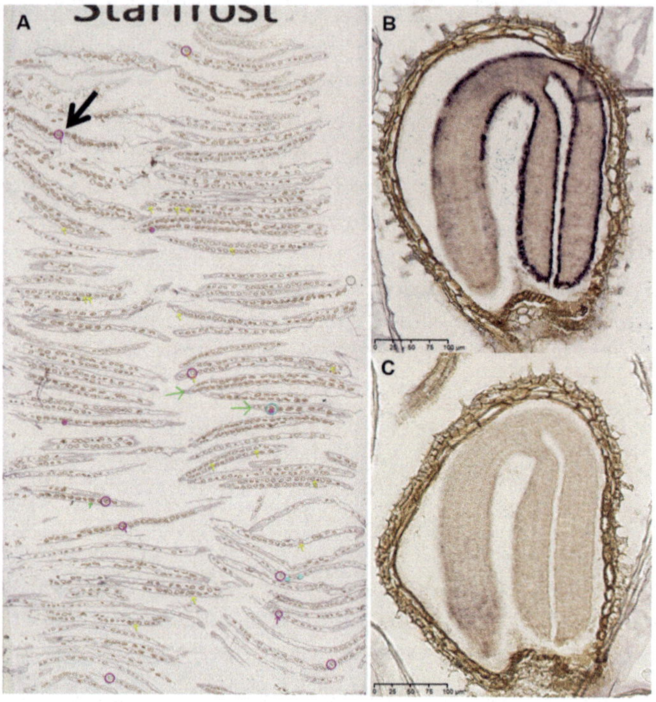

Fig. 8 Example of slides scanned at high resolution (Subheadings 3.4.7 and 3.4.8). (**a**) Example of microscopy slide containing high density of tissues (developing siliques) from several tissue microarray used for ISH experiments. (**b**) Zoom on the embryo marked by the bold black arrow in (**a**) showing hybridization of the antisense probe in the aerial organ epidermis. (**c**) Same exact embryo hybridized with the corresponding sense probe used as a control showing the weak background as well as the natural color of the seed coat, therefore enabling to better see the specific signal in (**b**). All these images were extracted from scans analyzed with NDP viewer (Hamamatsu). The gene analyzed here is LTP2. Complete information is available in [9]

3.5 Integration of ISH and Transcriptomic Results and Concluding Remarks

While ISH results can be valorized solely, we found it interesting to establish ISH-derived expression maps and to establish comparison with published transcriptomic data [9]. Comparison enabled illustrating the complementarity of both datasets. As an example, ISH medium throughput enabled studying dozen of genes on every tissues/cell types at all developmental stages, while transcriptomics provided expression profiles for more than 20,000 genes on 42 selected laser-capture microdissected samples. Interestingly, the sensitivity of ISH overtook those of transcriptomics in some young developing tissues, while transcriptomic sensitivity was better in the more mature tissues. The present chapter does not intend to provide all methods/results of this integrative study, and more details can be found in the original publication [9]. This ISH protocol was also successfully applied on non-model species such as *Madagascar periwinkle* to map the complex multicellular compartmentation of complex anticancer secondary metabolic pathway as recently reviewed [11].

Nowadays, new ISH detection systems such as branched DNA ISH are available [24], but their actual costs prohibit their application to many genes. These newly developed systems promise to be more sensitive, detecting "single transcripts," and to improve signal detection on problematic tissues. Altogether, ISH approaches remain relevant and still have a rosy future to explore cell-specific gene expression.

4 Notes

1. Restriction enzymes producing 3′ protruding ends were avoided to prevent the initiation of transcription of opposite transcription products [18].
2. T3 and SP6 RNA polymerases have a common buffer which is different from T7 RNA polymerase buffer. The Promega RNA polymerase kits also contain 100 mM DTT.
3. The DIG RNA labeling mix contains 10 mM ATP, 10 mM CTP, 10 mM GTP, 6.5 mM UTP, and 3.5 mM DIG, so only one third of UTP is labeled in order to avoid sterical hindrance problem that could occur if successive uridine residues should be incorporated to a given probe.
4. /!\ Safety note: For the microtomy steps, you will have to use razor blades and microtome that could be dangerous. The following sign "/!\" will notify steps requiring to work cautiously.
5. /!\ Safety note: Along this protocol use of solvents, aldehyde-containing solutions, or hazardous solutions requires to carefully work under a fume hood. The following sign "/!\" will notify steps requiring to work cautiously.
6. Slide coating is necessary to have a better adherence of the section to the slide.
7. The same solutions were used for sequential processing of the five racks. Racks were piled in 2 L RNase-free glass beakers, dried overnight at 37 °C, and stored for weeks/months in a clean dedicated place at room temperature.
8. We usually keep them in a separated box from other ISH microtube stocks (especially yeast RNA).
9. The identification of the GSM references can be found on the same webpage to those used to download the "series matrix files."
10. Download the GPL198 [ATH1-121501] Affymetrix Arabidopsis ATH1 Genome Array to be found on the GEO webpage

used to download the "series matrix files." At the bottom of the download page, click on "Download full Table," and save as a .txt file. Open this file with Microsoft Excel and save it as an .xls file. Finally, use the "VLOOKUP" function of Microsoft Excel to establish the correlation.

11. To build a co-expression network: use the Pearson function of Microsoft Excel on the gene of interest, and extend it to all the genes. Sort the genes from the highest Pearson coefficient value (=1) to the lowest (close to −1), and apply a rank to all the genes.
12. We use Z-Competent *E. coli* Transformation Kit (Zymo Research).
13. It is advised to prepare a glycerol stock to be stored at −80 °C using 0.9 mL of this LB culture prior to the miniprep.
14. It is necessary to obtain a high concentration in order to get enough plasmid for the following steps: The two linearization reactions necessitate 3–7 μg of plasmid each, and the two subsequent in vitro transcription reactions use 1 μg of purified linearized plasmid each.
15. Use your favorite sequence analysis software to establish in silico the predicted restriction profile with selected restriction enzymes giving contrasted profiles depending on the orientation. Check which prediction corresponds to your clone by performing in vitro restriction with the selected enzymes. This restriction profile will also enable designing the template for further plasmid linearization necessary for in vitro riboprobe transcription.
16. The plasmid digested at the 5′ end of the insert constitutes the template for in vitro transcription of the antisense probe enabling localizing the mRNA. The same plasmid digested at the 3′ end of the insert constitutes the template for in vitro transcription of the sense probe having the same sequence as the mRNA and routinely used as a negative control.
17. Classical 1% agarose in Tris acetate EDTA 0.5× is suitable. A single band corresponding to the size of the vector + insert should be obtained. We typically keep the digestion at 37 °C during the control gel migration to allow longer incubation in case of incomplete restriction. A few more hours up to overnight incubation are rarely necessary. Incomplete linearization would result in the in vitro transcription of probes using the vector and not only the insert as a template.
18. The samples can be directly processed to the next step or stored at 4 °C for short period or −20 °C for longer period.
19. This step is performed by slightly modifying the protocol for the GeneJET Miniprep Kit (Thermo Fisher Scientific). Directly

start the protocol at the neutralization buffer step, and follow the manufacturer's instructions for the subsequent steps. For the final elution step, add 50 μL of DEPC-treated water, and incubate for 5 min at RT before the final centrifugation.

20. To localize the expression of (m)RNA(s) of interest on tissues, it is necessary to label the riboprobe(s) during in vitro transcription by using uracil residues carrying digoxigenin molecule. The "digoxigenin-tagged" riboprobe will be recognized by anti-digoxigenin antibodies to localize the AS (antisense riboprobe) hybridized to the (m)RNA expressed within tissues. For the synthesis of unlabeled riboprobes used for competitive inhibition, the in vitro transcription step is similar except for the use of uracil residues without digoxigenin. Whenever possible, let the tubes on ice. Before starting, pre-warm a rack for 1.5 mL tubes in a 37° oven.
21. The Vmax = 24 μL for DIG-labeled probes (and Vmax = 18 μL for unlabeled riboprobes) so make sure to avoid working with diluted plasmid solutions.
22. Pay attention: T3 and SP6 RNA polymerases have a common buffer which is different from T7 RNA polymerase buffer.
23. It is important to store these 2 solutions at 4 °C until their use, as cold helps to precipitate and conserve the RNA.
24. First use P1000, and finish with P200 pipettes by pipetting on the opposed side of the pellet. The opalescent pellet is not always clearly visible, but this is not a problem. For drying at 45 °C on a slide warmer, check for the disappearance of the last droplets.
25. Keep 2 μL of this non-hydrolyzed riboprobe (containing the riboprobe + remaining linearized plasmid) before proceeding to the alkaline hydrolysis, as it constitutes a good control to assess transcription and hydrolysis efficiency on 1% agarose gel electrophoresis (see next section).
26. The hydrolysis duration (expressed in min) is determined by the formula set up by [21]: $T\,(\text{min}) = (\text{Li}/\text{Lf})/(\text{k} \times \text{Li} \times \text{Lf})$ with K (constant) = 0.11 kb/min, Li = initial RNA length (kb), and Lf = final RNA length (here 0.4 kb).
27. Pay attention not to lose the pellet that may move.
28. In order to make accurate comparison, we used a homemade lightly colored RNA loading dye so as not to impair the quantification of the probe smear. The commercial RNA ladder can be replaced by a homemade non-labeled riboprobe first quantified with the ladder and later becoming the internal standard for each in vitro transcription.
29. Pay attention to not saturate the band intensity while capturing picture, and exclude the remaining plasmid band from the

selection zone so as not to impair the quantification of the probe smear.

30. This allows to take into consideration the initial probe length (before hydrolysis) and to normalize the probe concentration according to the number of probe molecules and not only to the mass of probes. For example, a 1 kb riboprobe has to be diluted twice as compared to a probe whose initial size was 2 kb. This normalization allows to pipette constant volumes of probes on slides while performing ISH.
31. After this step, the samples can be stored for months at 4 °C in 50% ethanol, without impairing ISH quality.
32. Room temperature has to be higher than 25 °C corresponding to the *tert*-butanol melting point. In winter, work in a 25–28 °C oven.
33. Optionally add 1/40 of 1% (w/v) erythrosine ethanolic solution to transiently stain the depigmented samples. This facilitates positioning of samples in the paraffin. The stain will be eliminated during rehydration of the sections after dewaxing and will not impair the ISH efficiency.
34. Following the first Paraplast change, remove the tube caps to accelerate the evaporation of *tert*-butanol traces. Ideally the 60 °C oven should be placed under a fume hood. If not possible, keep the caps longer, and make sure to discard the solvent Paraplast mixes under a fume hood.
35. Pause point: Samples can be stored for years at 4 °C or directly used for the embedding step.
36. Plastic molds can also be used, but with the hot slide warmer, they lose their flatten base after a while. Metallic molds keep a flat base regardless of the temperature. Spatula and tweezers will help to manipulate the samples in the mold so they need to be kept warm.
37. It is advisable to carefully organize and concentrate the samples in the bloc in order to maximize the sample area on the slides during the ISH. Draw the exact organization of samples present in the mold as a "map" in a dedicated lab book. Just remind that when the paraffin blocs will be unmolded and sectioned, you will get the mirrored image of the organization chosen during the embedding. This is really helpful to choose the blocs for sectioning. In our conditions, 1 bloc can easily comprise 30 siliques, and 1 single tissue array slide can contain up to 90 siliques at different developmental stages (about 1000 developing seeds). The more detailed the drawings are, the easier the final analysis of ISH results will be.

38. We usually use a pack of ice to make this step faster. If samples have a tendency to move while displacing the molds, leave them on the slide warmer and simply shut it down.
39. Use new clean gloves to hold the slides by the side. Do not use a "permanent" ink pen since ink may disappear with solvents such as xylene.
40. The thickness of the section may be adjusted. Thicker sections are sometimes easier to cut and can provide stronger signal but lower resolution. Thinner sections may be useful in providing better morphology and allowing to study more genes (more serial sections) on single small samples, but they should result in weaker signals.
41. Degassing of water is crucial to avoid air bubble production under the tissue during the drying of sections. Alternatively, a RNase-free water bath can be used without degassing. In this case, use a brush to keep the floating sections in the desired orientation on the water bath, and plunge the slide under the section with a 45° angle. The section will spread to the slide while carefully removing the slide.
42. Keep the serial sections in the same orientation and position on each slide using the brush and needles; this is crucial to further permit comparison of serial sections from the same sample on different slides. Depending on the size of the tissue microarrays, we usually position 2–3 sections from 2–3 different tissue arrays on each slide, therefore accommodating a high density of tissues (e.g., up to 1000 developing seeds).
43. The same bath series is used for both 20-slide racks with a 15 min delay. Keep this bath series clean for reuse during the dehydration series at the end of the prehybridization (*see* **step 21**).
44. Note that DEPC will sediment in the jars during the incubation: (1) extemporaneously prepare 800 mL of 0.1% DEPC in PBS in a 1 L Schott bottle, and thoroughly handshake the bottle before each distribution of 200 mL in the first jar. Sequentially repeat the shaking and the distribution in the three other jars just before use. (2) During the 15 min incubation periods of the 20-slide racks in the jars, invert the racks by 180° every 5 min.
45. Do not reuse the first 100% ethanol bath that followed the xylene bath since it is saturated in paraffin.
46. This step could be performed during the preparation of the next steps (Subheading 3.4.3 hybridization) or during lunchtime.
47. Probe stock solutions were set to 25 ng/kb/μL. If not, calculate the volume needed to reach this concentration.

48. Use 12–16 mL/dishes. Wear clean gloves, carefully flatten the fully soaked Benchkote, and eliminate the excess solution before disposing with tweezers ten slides per dish without any contact between them.
49. Think to preserve the racks during the hybridization period by wrapping them in RNAse-free aluminum foil; you will reuse them the next day.
50. Competitive inhibition experiments can be performed similarly except that the hybridization medium contains 400 ng/kb/μL of digoxigenin-labeled antisense probe together with 4000 ng/kb/μL of unlabeled competitive antisense probe (113.1 μL of hybridization buffer mixed with 2.24 μL DIG-labeled probe and 22.4 μL unlabeled probe).
51. Firmly hold the slide using the white frame, being careful to maintain a 90° angle between the tweezers and the slide to dip in the 200 mL bath. Be patient since careless removal of the coverslip would result in section loss.
52. To avoid drying the slides, quickly immerse the rack in the 2× SSC solution every 5–10 slides.
53. Place the 2× SSC and 1× SSC beakers on top of magnetic stirrers and the 0.1× SSC beaker in a pre-warmed 65 °C water bath placed on top of a magnetic stirrer. This homemade system uses this type of classical glass rack only to allow magnetic stirring in its middle while accommodating the two 20-slide racks on its top during subsequent stringent washing steps necessary for hybridization specificity.
54. Alternatively, use humid chambers such as homemade large Petri dishes with tapped Pasteur pipette lines allowing to hold the slides.
55. No need of RNase-free water at this step. Sometimes dark background deposit may appear on the back and sides of the slides. These may be carefully removed with a Kimwipe.
56. Eukitt is a viscous medium, and the coverslip may move. If needed, you can adjust the coverslips, and remove air bubbles by gently pushing them away with two pipette tips before Eukitt dries.

Acknowledgment

Université Paul Sabatier Toulouse 3, Centre National de la Recherche Scientifique (CNRS) and Laboratoire d'Excellence (LABEX) entitled TULIP (ANR -10-LABX-41) granted this work. Nanozoomer virtual microscopy was performed on the Toulouse Réseau Imagerie (TRI, FR3450). We thank Josiane CHOURRE for help in tissue fixation and infiltration.

References

1. Birnbaum K, Shasha DE, Wang JY, Jung JW, Lambert GM, Galbraith DW, Benfey PN (2003) A gene expression map of the Arabidopsis root. Science 302(5652):1956–1960. https://doi.org/10.1126/science.1090022
2. Yadav RK, Girke T, Pasala S, Xie MT, Reddy V (2009) Gene expression map of the Arabidopsis shoot apical meristem stem cell niche. Proc Natl Acad Sci U S A 106(12): 4941–4946. https://doi.org/10.1073/pnas.0900843106
3. Deal RB, Henikoff S (2011) The INTACT method for cell type-specific gene expression and chromatin profiling in *Arabidopsis thaliana*. Nat Protoc 6(1):56–68. https://doi.org/10.1038/nprot.2010.175
4. Belmonte MF, Kirkbride RC, Stone SL, Pelletier JM, Bui AQ, Yeung EC, Hashimoto M, Fei J, Harada M, Munoz MD, Le BH, Drews GN, Brady SM, Goldberg RB, Harada JJ (2013) Comprehensive developmental profiles of gene activity in regions and subregions of the Arabidopsis seed. Proc Natl Acad Sci U S A 110(5):E435–E444. https://doi.org/10.1073/pnas.1222061110
5. Dekkers BJW, Pearce S, van Bolderen-Veldkamp RP, Marshall A, Widera P, Gilbert J, Drost HG, Bassel GW, Muller K, King JR, Wood ATA, Grosse I, Quint M, Krasnogor N, Leubner-Metzger G, Holdsworth MJ, Bentsink L (2013) Transcriptional dynamics of two seed compartments with opposing roles in Arabidopsis seed germination. Plant Physiol 163(1):205–215. https://doi.org/10.1104/pp.113.223511
6. Khan D, Millar JL, Girard IJ, Chan A, Kirkbride RC, Pelletier JM, Kost S, Becker MG, Yeung EC, Stasolla C, Goldberg RB, Harada JJ, Belmonte MF (2015) Transcriptome atlas of the Arabidopsis funiculus - a study of maternal seed subregions. Plant J 82(1):41–53. https://doi.org/10.1111/tpj.12790
7. Le BH, Cheng C, Bui AQ, Wagmaister JA, Henry KF, Pelletier J, Kwong L, Belmonte M, Kirkbride R, Horvath S, Drews GN, Fischer RL, Okamuro JK, Harada JJ, Goldberg RB (2010) Global analysis of gene activity during Arabidopsis seed development and identification of seed-specific transcription factors. Proc Natl Acad Sci U S A 107(18):8063–8070. https://doi.org/10.1073/pnas.1003530107
8. Hejatko J, Blilou I, Brewer PB, Friml J, Scheres B, Benkova E (2006) In situ hybridization technique for mRNA detection in whole mount Arabidopsis samples. Nat Protoc 1 (4):1939–1946. https://doi.org/10.1038/nprot.2006.333
9. Francoz E, Ranocha P, Pernot C, Le Ru A, Pacquit V, Dunand C, Burlat V (2016) Complementarity of medium-throughput in situ RNA hybridization and tissue-specific transcriptomics: case study of Arabidopsis seed development kinetics. Sci Rep 6:24644. https://doi.org/10.1038/srep24644
10. Brewer PB, Heisler MG, Hejatko J, Friml J, Benkova E (2006) In situ hybridization for mRNA detection in Arabidopsis tissue sections. Nat Protoc 1(3):1462–1467. https://doi.org/10.1038/nprot.2006.226
11. Courdavault V, Papon N, Clastre M, Giglioli-Guivarc'h N, St-Pierre B, Burlat V (2014) A look inside an alkaloid multisite plant: the *Catharanthus* logistics. Curr Opin Plant Biol 19:43–50. https://doi.org/10.1016/j.pbi.2014.03.010
12. Geu-Flores F, Sherden NH, Courdavault V, Burlat V, Glenn WS, Wu C, Nims E, Cui YH, O'Connor SE (2012) An alternative route to cyclic terpenes by reductive cyclization in iridoid biosynthesis. Nature 492 (7427):138–142. https://doi.org/10.1038/nature11692
13. Miettinen K, Dong LM, Navrot N, Schneider T, Burlat V, Pollier J, Woittiez L, van der Krol S, Lugan R, Ilc T, Verpoorte R, Oksman-Caldentey KM, Martinoia E, Bouwmeester H, Goossens A, Memelink J, Werck-Reichhart D (2014) The seco-iridoid pathway from *Catharanthus roseus*. Nat Commun 5:3606. https://doi.org/10.1038/ncomms4606
14. Munkert J, Pollier J, Miettinen K, Van Moerkercke A, Payne R, Muller-Uri F, Burlat V, O'Connor SE, Memelink J, Kreis W, Goossens A (2015) Iridoid synthase activity is common among the plant progesterone 5 beta-reductase family. Mol Plant 8(1):136–152. https://doi.org/10.1016/j.molp.2014.11.005
15. Seki M, Carninci P, Nishiyama Y, Hayashizaki Y, Shinozaki K (1998) High-efficiency cloning of Arabidopsis full-length cDNA by biotinylated CAP trapper. Plant J 15(5):707–720. https://doi.org/10.1046/j.1365-313x.1998.00237.x
16. Seki M, Narusaka M, Kamiya A, Ishida J, Satou M, Sakurai T, Nakajima M, Enju A, Akiyama K, Oono Y, Muramatsu M, Hayashizaki Y, Kawai J, Carninci P, Itoh M, Ishii Y, Arakawa T, Shibata K, Shinagawa A, Shinozaki K (2002) Functional annotation of

a full-length Arabidopsis cDNA collection. Science 296(5565):141–145. https://doi.org/10.1126/science.1071006

17. Sambrook J, Fritsh E, Maniatis T (eds) (1989) Molecular cloning. A laboratory manual. Laboratory Press, Cold Spring Harbor
18. Schenborn ET, Mierendorf RC (1985) A novel transcription property of SP6 and T7 RNA-polymerases: dependence on template structure. Nucleic Acids Res 13 (17):6223–6236. https://doi.org/10.1093/nar/13.17.6223
19. Schneider CA, Rasband WS, Eliceiri KW (2012) NIH Image to ImageJ: 25 years of image analysis. Nat Methods 9(7):671–675. https://doi.org/10.1038/nmeth.2089
20. Burlat V, Oudin A, Courtois M, Rideau M, St-Pierre B (2004) Co-expression of three MEP pathway genes and geraniol 10-hydroxylase in internal phloem parenchyma of *Catharanthus roseus* implicates multicellular translocation of intermediates during the biosynthesis of monoterpene indole alkaloids and isoprenoid-derived primary metabolites. Plant J 38(1):131–141. https://doi.org/10.1111/j.1365-313X.2004.02030.x
21. Jackson D (1992) In situ hybridisation in plants. In: Gurr SJ, McPherson MJ, Bowles DJ (eds) Molecular plant pathology: a practical approach. IRL Press at Oxford University, Oxford, pp 163–174
22. Mahroug S, Courdavault V, Thiersault M, St-Pierre B, Burlat V (2006) Epidermis is a pivotal site of at least four secondary metabolic pathways in *Catharanthus roseus* aerial organs. Planta 223(6):1191–1200. https://doi.org/10.1007/s00425-005-0167-y
23. St-Pierre B, Vazquez-Flota FA, De Luca V (1999) Multicellular compartmentation of *Catharanthus roseus* alkaloid biosynthesis predicts intercellular translocation of a pathway intermediate. Plant Cell 11(5):887–900. https://doi.org/10.1105/tpc.11.5.887
24. Baker M (2012) RNA imaging in situ. Nat Methods 9(8):787–790

Chapter 7

Purification and Functional Analysis of Plant Long Noncoding RNAs (lncRNA)

Trung Do, Zhipeng Qu, and Iain Searle

Abstract

More than 70% of eukaryotic genomes are transcribed into RNA transcripts, the majority of these transcripts are noncoding protein, and their biological functions are largely unknown. Over the last decade, the application of high-throughput sequencing technologies has led to the description of almost all cellular coding and noncoding RNA transcripts except perhaps for those transcripts that are lowly abundant or those present only in specific cells that are underrepresented in sampled tissue(s). An often underrepresented class of noncoding are long noncoding RNAs (lncRNAs), and these often play key regulatory functions for many biological processes such as cell identity and cell division. However, the purification and functional characterization in vitro are still a challenge in both animal and plant experimental systems. Here, we describe in detail methodology for purification of specific cell types, bioinformatic annotation of lncRNAs, and investigation of biological function using the reference plant *Arabidopsis thaliana*.

Key words *Arabidopsis thaliana*, Long noncoding RNA, Nuclei purification, RNA-Seq, Functional analysis

1 Introduction

Eukaryotic genomes transcribe genetic information from chromatin to RNA, and subsequently a small portion is translated into proteins. However, the majority of these RNA transcripts are not translated and are described as noncoding RNAs (ncRNAs) [1, 2]. Many ncRNAs are posttranscriptionally processed; examples include introns removed and covalent RNA modifications added or diced into smaller ncRNAs [3–6]. Arbitrarily, ncRNAs have been grouped by size into small sRNAs, those less than 200 nucleotides (nt) and long ncRNAs (lncRNAs) those larger than 200 nt in length [7–9]. To date, lncRNAs have been shown to have diverse functional roles in many fundamental cellular processes, and they also represent important components of ongoing research in many fields [10].

Julia A. Chekanova and Hsiao-Lin V. Wang (eds.), *Plant Long Non-Coding RNAs: Methods and Protocols*, Methods in Molecular Biology, vol. 1933, https://doi.org/10.1007/978-1-4939-9045-0_7,

Often the first step is to identify or sequence the lncRNAs within specific tissue(s) of interest. However, cell-specific expression or low abundance of the lncRNA within a tissue can often make this first step a challenge. In this chapter, we adapt and describe a method called isolation of nuclei tagged in specific cell types (INTACT) to purify a specific cell type [11]. The next step is often application of sequencing technology to sequence the purified RNA and bioinformatic annotation of protein coding and noncoding transcripts. This method has already been applied to a large number of species including *Saccharomyces cerevisiae*, *Schizosaccharomyces pombe*, *Arabidopsis thaliana*, mouse, and human cells [12–21]. We briefly describe these bioinformatic steps within the chapter. Often the next step is functional characterization of novel lncRNAs, and this can involve perturbing transcription by CRISPR-mediated deletion, overexpression of the lncRNA by using a transgene or transcriptional activators, or knockdown using dsRNA or artificial miRNAs [22–25]. In this chapter were detail protocols for the functional characterization of lncRNAs from *Arabidopsis thaliana* siliques (Fig. 1).

2 Materials

Prepare all solutions using RNase-free and DNase-free H_2O and analytical grade reagents. Store and prepare all reagents at room temperature unless indicated otherwise. Prepare and perform RNA extraction, cDNA synthesis, and PCR amplification experiments in an RNase-free area. Follow all state or national safety and waste disposal regulations when performing experiments.

2.1 Plant Growth and Tissue Sampling

1. Transgenic seeds expressing the *E. coli* biotin ligase BirA and tissue-specific nuclear-targeting fusion (NTF) protein in the endosperm.
2. Appropriate plant growth media.
3. Liquid nitrogen (and container).
4. Polystyrene box and/or second liquid nitrogen-proof container.
5. Sharp knife, scalpel, razor blade, tweezers, metal needle/probe, and flame source.
6. Eppendorf tubes (1.5 mL).
7. –80 °C freezer or liquid nitrogen storage container and/or dry ice.
8. RNA*later*™ solution (ThermoFisher Scientific).

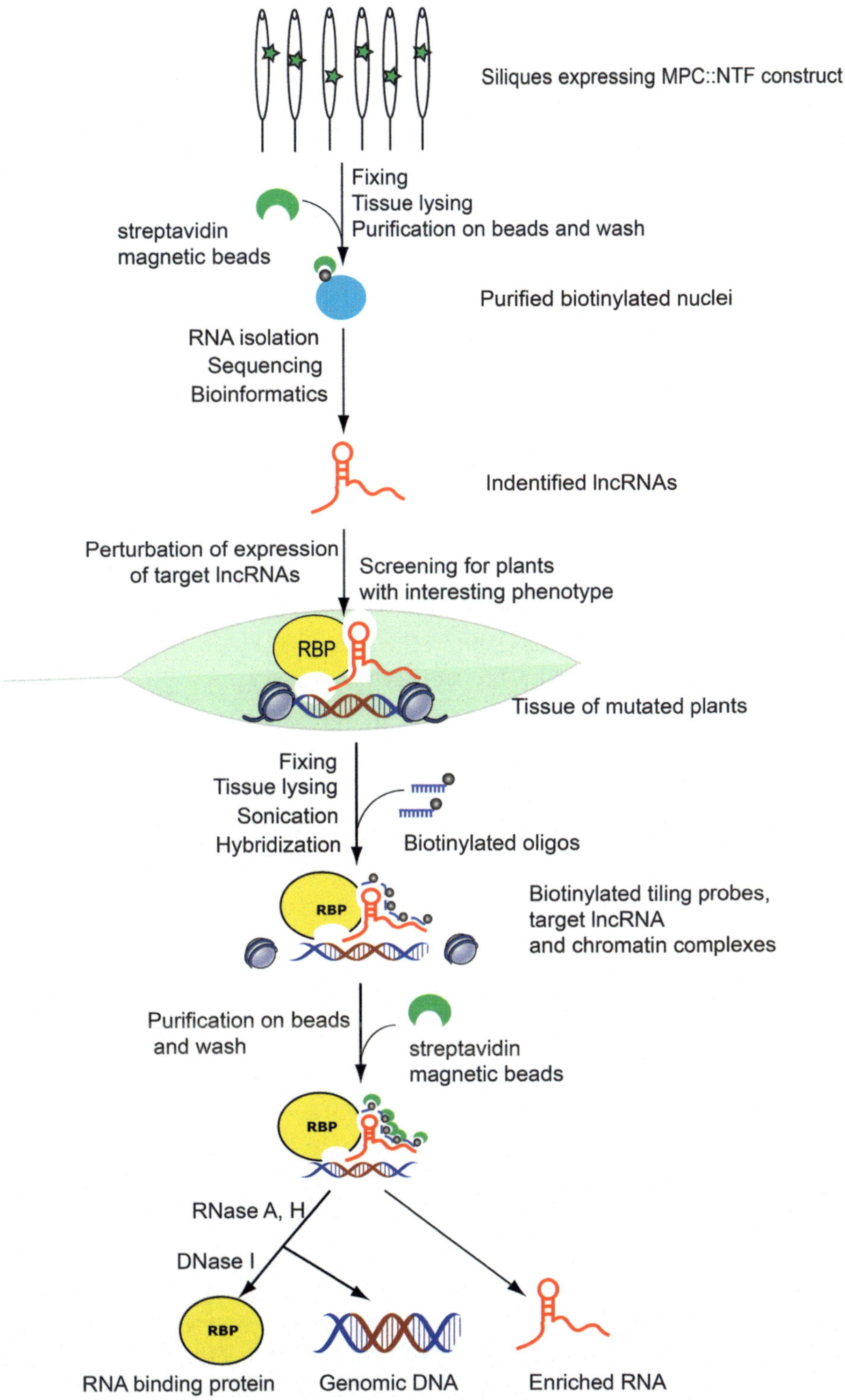

Fig. 1 Overview showing the workflow for this chapter. The siliques of transgenic plants expressing the MPC/NTF construct (green stars) were collected and fixed with the formaldehyde. The fixed samples were then

2.2 Nuclei Purification

2.2.1 Materials and Reagents

1. 37% (w/v) formaldehyde.
2. Glycine.
3. Liquid nitrogen (N_2).
4. MOPS (Merck).
5. NaCl.
6. KCl.
7. EDTA.
8. EGTA.
9. Spermidine.
10. Spermine.
11. cOmplete™, Mini, EDTA-free Protease Inhibitor Cocktail (Merck).
12. Tris–HCl.
13. SDS.
14. 4′,6-Diamidino-2-phenylindole (DAPI).
15. M-280 Streptavidin Dynabeads (ThermoFisher Scientific).
16. Triton X-100.
17. 40-μM cell strainer (BD Falcon).
18. MiniMACS™ separation magnet (Miltenyi Biotec).
19. Refrigerated microcentrifuge (Eppendorf, model 5415R, or equivalent).
20. Refrigerated centrifuge (Sorvall, model RC5C or equivalent).
21. Rotating mixer for 1.5-mL tubes (Labquake; ThermoFisher Scientific, or equivalent).
22. Electronic serological pipetting device (Easypet; Eppendorf, or equivalent).
23. Fluorescence microscope with DAPI and GFP filters.

2.2.2 Nuclei Purification Buffers (NPB)

Spermidine, spermine, and cOmplete™ protease inhibitors are added just before use, and the solution is kept on ice.

1. 20 mM MOPS (pH 7.0).
2. 40 mM NaCl.

Fig. 1 (continued) lysed into purification buffer, and streptavidin magnetic beads were used to enrich the biotinylated nuclei. Next, RNA was extracted from purified nuclei libraries constructed and sequenced using an Illumina NextSeq. The data were analyzed by a bioinformatic pipeline to identify potential long noncoding RNAs (lncRNAs). The target lncRNAs were confirmed and functional analysis undertaken by perturbing their expression using an overexpression or knockdown vector. Mutants with an interesting phenotype were used as material for chromatin isolation by RNA purification (CHIRP) experiments to identify the target genomic DNA and/or target-lncRNA binding proteins

3. 90 mM KCl.
4. 2 mM EDTA.
5. 0.5 mM EGTA.
6. 0.5 mM spermidine.
7. 0.2 mM spermine.
8. 1× cOmplete™ protease inhibitors.

2.2.3 Nuclear Lysis Buffer

This solution should be prepared just before use and kept at RT. Do not store this solution.

1. 50 mM Tris (pH 8.0).
2. 10 mM EDTA.
3. 1% (w/v) SDS.
4. 1× cOmplete™ protease inhibitors.

2.3 RNA Extraction

1. TRIzol™ reagent (ThermoFisher Scientific). Refer to the manufacturer's instructions and guidelines for stability and storage, and handle with eye and glove protection.
2. Chloroform (ThermoFisher Scientific).
3. Isopropyl alcohol.
4. Absolute ethanol.
5. RNase-free water.
6. 0.5% SDS and UV-treated plastic ware, oven-baked sterile glassware.
7. Eppendorf tubes or clean sterile Falcon tubes (conical bottom).
8. Liquid nitrogen, mortar, and pestle.
9. Benchtop centrifuges (refrigerated or access to cold room).

2.4 Generation of RNA-Seq Library

1. Ribo-Zero magnetic kit (Illumina).
2. NEBNext® Ultra™ Directional RNA Library Prep Kit (Illumina).

2.5 Library Sequencing

1. 0.2 M NaOH.
2. 0.2 M Tris–HCl, pH 7.0.
3. NextSeq 500 Kit.
4. Illumina PhiX Control Kit.

2.6 Bioinformatic Analysis

1. FASTQ RNA-Seq files.
2. Pipeline to identify the lncRNAs.

2.7 Functional Analysis of lncRNA

1. Transcripts of target lncRNAs.
2. *Arabidopsis thaliana* seeds.

3. Donor and binary vectors (pENTR/D, pLeela, pJawohl18, pRS300, pGreen II).
4. Appropriate antibiotics.
5. Plant growth facilities.
6. Primers of target lncRNAs (lncRNA_1246_R: 5′-TGACCTG CTGCTCTCATCTCG-3′, lncRNA_1246_F: 5′-GTTGCAC ATCAGGGACATG-3′) and housekeeping genes (Actin1_F: 5′-GTCTCGAGAGATGACTCAGATCATGTTTGAG-3′; Actin1_R: 5′-GGCGCGCCACAATTTCCCGTTCTGCGGT AG-3′).

2.8 Chromatin Isolation by RNA Purification (ChIRP)

2.8.1 Materials and Reagents

1. T_3 seed of target-lncRNA mutants.
2. Agar.
3. Phosphate-buffered saline (PBS).
4. Formaldehyde.
5. 0.125 M glycine.
6. 40 μM strainer (BP Falcon).
7. Vacuum indicator (Sorvall, model RC5C or equivalent).
8. Probes, design an array of probes along the lncRNAs by using Stellaris® Probe Designer version 4.2 at http://www.singlemoleculefish.com.
9. Yeast total RNA (Merck).
10. Bovine serum albumin (BSA).
11. DNase I (ThermoFisher Scientific).
12. RNA extraction kit: RNeasy Mini column extraction (other commercial RNA extraction kits can be used) (Merck).
13. $MgCl_2$.
14. Sucrose.
15. Beta-mercaptoethanol.
16. Dithiothreitol (DTT).
17. Phenylmethylsulfonyl fluoride (PMSF).
18. RNaseOUT™ (ThermoFisher Scientific).
19. $NaHCO_3$.
20. NP-40 (Merck).

2.8.2 Buffers

1. Buffer 1: 0.4 M sucrose, 10 mM Tris–HCl, pH 8.0, 10 mM $MgCl_2$. Before use, add 5 mM beta-mercaptoethanol, 1 mM DTT, 1 mM PMSF, 1 tablet of cOmplete™ protease inhibitor, and 0.1 U/μL RNaseOut™.
2. Buffer 2: 0.25 M sucrose, 10 mM Tris–HCl, pH 8.0, 10 mM $MgCl_2$, 1% Triton X-100. Before use, add 5 mM beta-

mercaptoethanol, 1 mM DTT, 1 mM PMSF, 1 tablet of cOmplete™ protease inhibitor, and 0.1 U/μL RNaseOut™.

3. Buffer 3: 1.7 M sucrose, 10 mM Tris–HCl, pH 8.0, 2 mM $MgCl_2$, 0.15% Triton X-100. Before use, add 5 mM beta-mercaptoethanol, 1 mM DTT, 1 mM PMSF, 1 tablet of cOmplete™ protease inhibitor, and 0.1 U/μL RNaseOut™.
4. Lysis buffer: 50 mM Tris–HCl, pH 8.0, 10 mM EDTA, 1% SDS. Before use, add 5 mM beta-mercaptoethanol, 1 mM DTT, 1 mM PMSF, 1 tablet of cOmplete™ protease inhibitor, and 0.1 U/μL RNaseOut™.
5. Hybridization buffer: 500 mM NaCl, 1% SDS, 100 mM Tris-HCl, pH 7.0, 10 mM EDTA, 15% formamide. Before use, add 5 mM beta-mercaptoethanol, 1 mM DTT, 1 mM PMSF, 1 tablet of cOmplete protease inhibitor, and 0.1 U/μL RNaseOut™.
6. Wash buffer: 2xSSC, 0.5% SDS. Add 1 mM DTT and 1 mM PMSF fresh.
7. RNA elution buffer: Tris–HCl, pH 7.0 and 1% SDS.
8. DNA elution buffer: 50 mM $NaHCO_3$, 1% SDS, 200 mM NaCl.
9. DNase buffer: 100 mM NaCl and 0.1% NP-40.

3 Methods

Carry out all procedures described below at room temperature unless otherwise stated.

3.1 Sampling

Collect plant material (1 day after pollination siliques, DAP) from plant grown in a controlled environment facility, place the siliques in pre-weighed Eppendorf tubes, and immediately immerse in liquid nitrogen. Alternatively, submerge the siliques in RNA*later*™ solution (*see* **Notes 1–5**).

3.2 Purification of Nuclei from Specific Plant Cell Types Using the INTACT Method

1. In a mortar and pestle pre-cooled with liquid N_2, grind 5 g of siliques, and resuspend the tissue powder in 10 mL of cold NPB buffer. Keep on ice for 10 min (*see* **Note 6**).
2. Filter the extract through a 40 μM strainer, and centrifuge down the nuclei at 1200*g* for 15 min at 4 °C.
3. Discard the supernatant and gently resuspend the nuclei in 1 mL of cold NPB, and transfer to a 1.5 mL tube.
4. Wash the appropriate amount (25 μL of beads for each 5 g of root tissue or 10 μL for each 0.5 g of leaf tissue) of Dynabead™ M-280 Streptavidin beads with 1 mL of NPB, and then resuspend the beads with NPB to their original volume. Add the

bead suspension to the nuclei from **step 3**, and rotate at 4 °C for 30 min.

5. Dilute 1 mL of bead-nuclei mixture with 14 mL of four ice-cold NPB containing 0.1% Triton X-100 (NPBt) in a 15 mL tube. Mix gently and place on ice for 30 s. Place the tube in the DynaMag™-5 magnet for 5 min at 4 °C.
6. Carefully remove the supernatant with a serological pipette, and gently resuspend the beads in 14 mL of cold NPBt. Mix gently and place on ice for 30 s. Place the tube in the DynaMag™-5 magnet for 5 min at 4 °C.
7. Repeat **step 5**.
8. Gently remove the supernatant with a serological pipette, and resuspend the beads in 1 mL of cold NPBt. Remove a 25 μL sample for counting the number of captured nuclei on a hemocytometer.
9. Transfer the resuspended beads to a 1.5 mL tube, and capture on a DynaMag™-5 magnet.
10. Remove the supernatant, resuspend the beads in 20 μL of cold NPB, and proceed with downstream processing (RNA isolation or ChIP). Alternatively, nuclei/beads can be stored at −80 °C until further use.
11. To view the purified nuclei under a microscope, add 1 μL of 0.2 μg/μL DAPI to each 25 μL sample (taken at **step 7**), and place on ice for 5 min. Count the number of nuclei using a hemocytometer (*see* **Notes 7–9**) (Fig. 2).

3.3 RNA Extraction, Purification, and DNase Treatment

1. Add 0.5 mL of TRIzol reagent to the beads from Subheading 3.2, and vortex immediately (*see* **Note 10**).
2. Transfer the solution to 1.5 mL tubes.
3. Centrifuge at 12,000 × *g* for 5 min at 2–4 °C.
4. Remove the supernatant to a new 1.5 mL tube.
5. Add 200 μL of chloroform, and shake vigorously by hand for approximately 15 s.
6. Let the tube stand at RT for 3 min.
7. Centrifuge the tube at 12,000 × *g* for 15 min at 4 °C.
8. Carefully transfer the upper aqueous phase to a new 1.5 mL tube (ensure no interface debris are transferred) (*see* **Note 11**).
9. Add 0.5 mL of isopropyl alcohol and mix thoroughly.
10. Let the mix stand at room temperature for 10 min.
11. Centrifuge at 12,000 × *g* for 10 min at 4 °C.
12. Carefully discard the supernatant (tip the tube with the pellet position angled up and away from you, and pipet out the

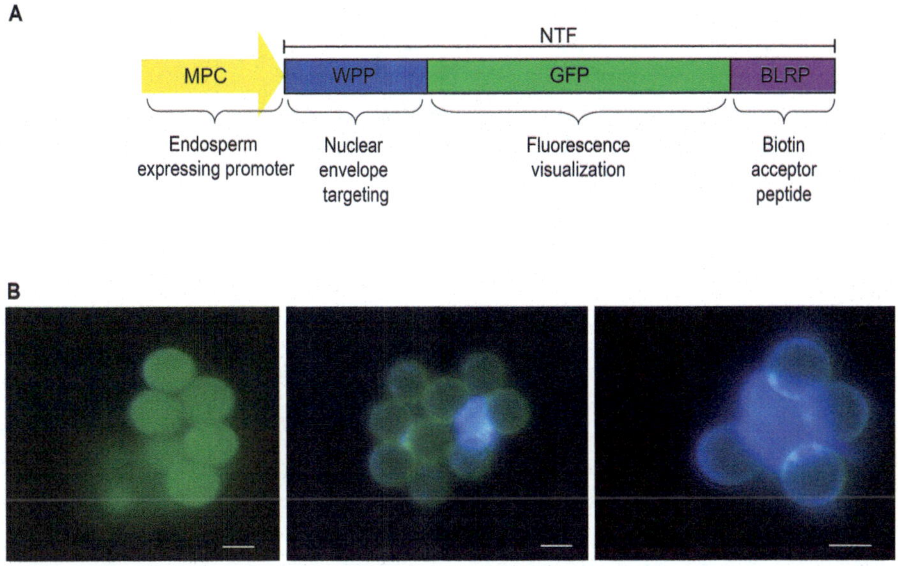

Fig. 2 Purification of tagged nuclei using the isolation of nuclei tagged in specific cell types (INTACT) system [10]. (**a**) Partial cassette of the transgenic vector showing the three-part structure of nuclear-targeting fusion (NTF) protein is shown. The chimeric protein consists of the WPP domain of RanGAP1 for nuclear envelope targeting, the GFP to allow visualization, and the biotin ligase recognition peptide (BLRP), which is biotinylated by BirA. (**b**) Fluorescence microscopy images of NTF-labeled nuclei that have been bound by streptavidin-coated magnetic beads prior to capture on a magnet. Biotinylated nuclei are shown in blue, and GFP fluorescence from the beads themselves is shown in green. Scale bar, ~3 μm in each panel. The right panel is at a higher magnification than the left and center panels

supernatant). The pellet may be slightly glassy and transparent or may not be visible at all.

13. Add 1 mL of 75% ethanol.
14. Vortex briefly and centrifuge at 12,000 × *g* for 5 min at 4 °C.
15. Discard the ethanol and allow the pellet to air-dry for 10 min.
16. Dissolve the pellet in 20 μL of RNase-free water by very gently mixing with a pipette.
17. Check the quantity and purity of the RNA by using a fluorimeter; RNA quality can be assessed by separation of 1 μg on 2.5% agarose gel. Store the RNA at −80 °C until further use.
18. Remove contaminating genomic DNA from the RNA by treating with the Turbo DNA-free kit in a 0.6-mL tube according to the manufacturer's instructions.
19. Remove the treated RNA to a new 0.6-mL tube.

3.4 Generation of RNA-Seq Libraries

1. Remove ribosomal RNA (rRNA) from DNase-treated RNA by treating with the Ribo-Zero™ magnetic kit according to the manufacturer's instructions.

2. Prepare sequencing libraries for the RNA from **step 1** by using the NEBNext® Ultra™ Directional RNA Library Prep Kit for Illumina as per the manufacturer's instructions.
3. Check the prepared libraries for quality and quantity by separation on an Agilent High Sensitivity DNA chip. Store the remaining library at −20 °C until further use.

3.5 NextSeq Sequencing

1. Prepare the sample sheet using the Illumina Experiment Manager by following the manufacturer's protocol (*see* **Note 12**).
2. Dilute the constructed libraries to 4 nM in resuspension buffer (RSB) based on the concentrations determined by the Bioanalyzer. From this point onward, keep the libraries on ice.
3. Dilute the PhiX control library to 4 nM by adding 15 μL of RSB to 10 μL of the 10 nM PhiX control library (*see* **Note 13**).
4. Denature the pooled libraries and PhiX control library separately by adding 5 μL 0.2 M NaOH to 5 μL of the 4 nM libraries (*see* **Note 14**).
5. Vortex thoroughly to mix and incubate at RT for 5 min. Add 5 μL 200 mM Tris–HCl, pH 7.0, and vortex briefly before centrifuge at 280*g* for 1 min.
6. Dilute the denatured pooled libraries and PhiX control library separately to 20 pM by adding 985 μL pre-chilled HT1 to 15 μL denatured libraries.
7. Dilute the 20 pM pooled libraries and PhiX control library separately to 1.8 pM by adding 1183 μL pre-chilled HT1 to 117 μL 20 pM denatured libraries.
8. Combine 13 μL of the 1.8 pM PhiX control library with 1287 μL of the 1.8 pM pooled libraries, and vortex to mix.
9. Load 1300 μL of the final sample into the cartridge. Ensure that any air bubbles are removed by gently tapping the cartridge.
10. Perform the sequencing run according to the manufacturer's protocol.

3.6 Bioinformatic Analysis to Identify lncRNAs

1. Trim adaptors and low-quality sequences from raw reads by using trim_galore with the following parameter:

```
-- stringency 6.
```

2. Align the trimmed reads against the *Arabidopsis thaliana* genome TAIR10 assembly by using TopHat2 with the following parameters:

```
-N 5 -- read-edit- dist 5.
```

3. Merge the aligned reads from all samples by using SAMtools, and assemble the reads into transcripts by using cufflinks by using the parameter:

```
-- library-type fr-firststrand -u.
```

4. Remove transcripts shorter than 200 nt by a custom script.
5. Determine the genomic locations of long transcripts from **step 4** by comparing the genomic coordinates against the reference genes of TAIR10, and annotate the transcripts into either gene, intergenic, intronic, or antisense.
6. Determine the protein-coding potential of the annotated intergenic, intronic, and antisense transcripts by undertaking the following two steps: (1) Sequence similarity search against the SWISS-PROT protein database; and (2) Predict Open Reading Frame(s) (ORF). Transcripts that have no sequence similarity to proteins in SWISS-PROT or no ORFs longer than 20 amino acids are candidate lncRNAs.
7. Save and export the sequences of novel lncRNAs.

3.7 Functional Analysis of lncRNAs

1. Amplify target lncRNAs by using PCR from either genomic DNA or cDNA.
2. To overexpress or knockdown your target lncRNAs by using dsRNA, continue with the following steps (**step 3**). Alternatively, to strand specifically knockdown lncRNA, follow the procedures in **step 4**. Then proceed to **step 5**, transformation using *Agrobacterium tumefaciens.*
3. To overexpress or knockdown your target lncRNAs by using dsRNA, clone the PCR amplicons into pENTR™/D donor vector as per the manufacturer's instructions. Carry out a LR reaction between the recombinant pENTR™/D donor vector and destination binary vector (vector pLEELA for overexpression or pJahwol18 for knockdown) as per the manufacturer's instructions.
4. To strand specifically knockdown a lncRNA, strand specific knockdown of a lncRNA may be desirable, for example, if it is antisense to another transcript, and can be performed by using artificial miRNAs (amiRNAs). Generate, candidate amiRNA sequences that target your lncRNA by using the web app for the automated design of artificial microRNAs, WMD3 (http://wmd3.weigelworld.org/cgi-bin/webapp.cgi). From the list of candidate amiRNA sequences, select an amiRNA sequence, include 5′ and 3′ miRNA319 stem loop and amiRNA* sequences, and order the sequence as a gBLOCK® from IDT. Clone the amiRNA gBLOCK® into vector pRS300 to generate 35S:amiRNA. Digest the vector with *Pvu*II to remove the 35S:

amiRNA cassette, and clone the cassette into binary vector pGreenII.

5. Introduce the recombinant binary vectors into *Agrobacterium tumefaciens* strain AGL 1 by electroporation. Select transformants by using the appropriate antibiotics (rifampicin and vector-conferred antibiotic resistance, e.g., ampicillin) (*see* **Note 15**).
6. Grow 20 plants under long-day conditions for 4 weeks until the inflorescence is approximately 10 cm long. Transform the binary vector into *Arabidopsis thaliana* plants by using the floral dipping method.
7. Grow a 450 mL culture of Agrobacterium containing the binary vector overnight (*see* **Note 16**).
8. Repeatedly dip the flowering plants into the Agrobacterium solution for 60 s. Place the dipped plants horizontally onto a tray such that the Agrobacterium solution does not run down onto the leaves. Cover the tray by using a plastic cover, and place into a growth chamber for 24 h.
9. Remove the plastic cover and place the dipped plants vertically.
10. Collect seeds approximately 3 weeks later.
11. Plant on soil ~2000 dried seeds, and 1 week after germination, select for transformed plants by an aerial application of the herbicide BASTA®. Apply the herbicide to runoff point. Repeat BASTA® application 1 week later (*see* **Note 17**). Identify approximately 20 T_1 plants.
12. From mature T_1 plants, harvest seed; store in a cool, low-humidity place; and dry the seed for 2 weeks.
13. Plant approximately 100 seeds for each of the 20 T_1 lines. One week later, count the number of germinated T_2 seedlings.
14. Spray BASTA® onto the T_2 seedlings, and 5 days later count the number of resistant seedlings. Select T_2 lines that have 3:1 (resistant/sensitive) segregation, and grow 20 plants, and harvest the seed.
15. Perform phenotypic and molecular analysis on T_3 plants after selection with BASTA® (Fig. 3). Homozygous transgenic plants can be identified by progeny screening after BASTA® application.

3.8 Chromatin Isolation by RNA Purification (ChIRP)

1. Sow 1 g of lncRNA mutant and wild-type control seeds on ½ Murashige and Skoog (MS), 1% sucrose agar plates.
2. Seven days after germination, cross-link seedlings with 1% formaldehyde in 1× PBS solution by using vacuum infiltration at 4 °C for 15 min (*see* **Note 18**).

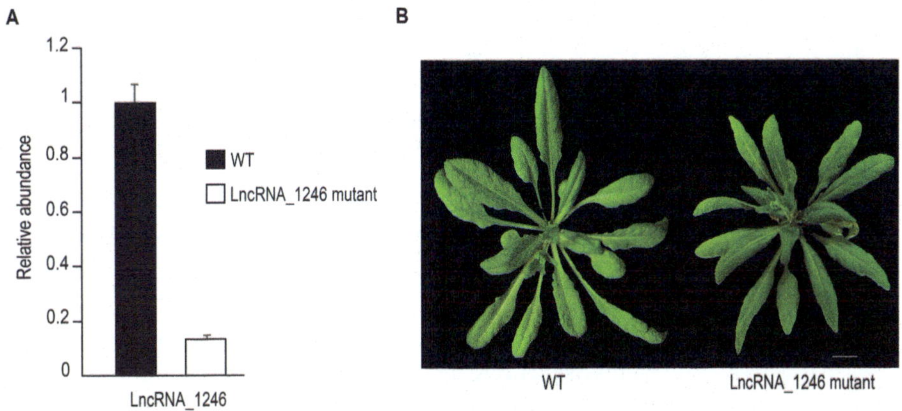

Fig. 3 Mutation of lncRNA_1246 caused smaller rosette leaves than wild type (WT). (**a**) RT_PCR to measure the abundance of lncRNA_1246 in the mutants and WT. Mean ± SD are shown. (**b**) Images of lncRNA_1246 mutant and WT at flowering stage. Scale bar is 1 cm

3. Terminate cross-linking by adding 0.125 M glycine, and vacuum infiltrate for another 5 min.
4. Wash the cross-linked seedlings three times with cold, sterile water.
5. Grind the seedlings to a fine powder in a mortar and pestle that has been pre-chilled with liquid nitrogen. Add 1 mL of ground powder to 10 mL of buffer 1.
6. Keep the samples on ice for 10 min. Filter the samples through two 40 μM strainers, and then centrifuge at 2000 × *g* at 4 °C for 30 min.
7. Remove the supernatant, and resuspend the pellet in 1 mL of buffer 2, centrifuged at 15,000 × *g* at 4 °C for 10 min.
8. Remove the supernatant, and resuspend the pellet in 300 μL of buffer 3, layer on top another 300 μL of fresh buffer 3, centrifuged at 15,000 × *g* at 4 °C for 1 h.
9. Remove the supernatant, and resuspend the pellets in 300 μL of lysis buffer, sonicated (15 s ON/60s OFF) until the DNA is fragmented into 200 ± 500 bp length fragments.
10. Centrifuge at 15,000 × *g* at 4 °C for 15 min. Dilute the chromatin in the supernatant with two volumes of hybridization buffer.
11. Add the Stellaris® designed biotin-labelled lncRNA probes to 150 μL of diluted chromatin to give a final concentration of 100 pmol/μL, and incubate by end-to-end rotation at 4 °C overnight.
12. For each sample, wash 50 μL of streptavidin magnetic C1 beads with six volumes of lysis buffer, and repeat three times. Block the beads by adding 500 ng/μL of yeast total RNA and

1 mg/mL BSA, and incubate for 1 h at RT, and then wash again in lysis buffer before resuspending in the original volume of 50 μL.

13. Mix the samples from **step 11** and the washed beads from **step 12** at 4 °C for 2 h.
14. Wash the captured beads five times with wash buffer. The beads are now ready for different elution protocols depending on the downstream assays.
15. *RNA elution*: Resuspend the beads in 10× the original volume of RNA elution buffer, and boil for 15 min. Purify the RNA by following the RNeasy Mini column procedure according to the manufacturer's protocol. Detect the enriched transcripts by quantitative reverse-transcription PCR (qRT-PCR).
16. DNA elution: Resuspended the beads in 3× the original volume of DNA elution buffer with a cocktail of 100 μg/mL RNase A and 0.1 U/μL RNase H at 37 °C with end-to-end rotation. Reverse-cross-link the chromatin by incubation at 65 °C overnight. Purify the DNA by phenol/chloroform/isoamyl alcohol extraction. Measure the enriched DNA by qPCR or high-throughput sequencing relative to the negative control.
17. DNA elution: Resuspend the beads in 3× original volume of DNase buffer (100 mM NaCl and 0.1% NP-40), and elute the protein with a cocktail of 100 μg/mL RNase A and 0.1 U/μL RNase H and 0.1 U/μL DNase I, and incubate at 37 °C for 30 min.
18. To observe enriched proteins, add 0.2 volume of 5× Laemmli buffer to the sample and negative control, boil for 5 min, and then separate the samples on an acrylamide gel. Silver stain the gel to observe the RNA-binding proteins that are present only in the sample and not in the negative control.

4 Notes

1. If using Eppendorf tubes, prepare the tubes with a small hole in the lid to prevent the tube opening as the tube warms due to residual liquid nitrogen expanding.
2. For storage in RNA*later*™ solution, add approximately 3× the volume of RNA*later*™: 1× tissue.
3. Tissue stored in RNA*later*™ and frozen (−80 °C) was defrosted just enough to remove the tissue from the RNA*later*™ solution prior to extraction. Repeated removal from storage often leads to reduced quality of RNA.

4. It is very important to minimize the time between removal of the tissue from the plant and immersion in liquid nitrogen.
5. The amount of tissue required to achieve acceptable yields of RNA varies according to the material. Tissues with a high-water content require larger amounts of tissue for the same yield of RNA.
6. It is essential to grind the tissue as finely as possible and maintain the samples as cold as possible during grinding to avoid RNA degradation.
7. The actual yield of purified nuclei is generally around 50% of the theoretical yield for the cell types we have examined. Therefore, it is recommended to begin with an amount of tissue that will yield at least double the required number of nuclei. The number of nuclei should yield 100–200 ng of total RNA when purified by using the RNeasy Micro kit.
8. For cross-linked chromatin immunoprecipitation experiments, starting tissue can be treated with formaldehyde, quenched, washed, and used directly in the above protocol without any alterations.
9. The use of this protocol with other types or amounts of tissue may require optimization. The most important parameters seem to be the number of beads used per mass of tissue and the volume of solution when capturing the beads after nuclei binding. For larger scale purifications, the DynaMag-50 magnet can be used to capture beads in volumes up to 40 mL.
10. If the beads were stored at −80 °C, do not let them thaw without being in the presence of extraction buffer. It is important to ensure that the beads do not form a clump, where the outside of the clump is in contact with the buffer, but the inner beads are not.
11. Plant tissues also contain other compounds that interfere with RNA extraction (such as polysaccharides, lipids, proteins). If these compounds are not removed in the first steps (discarded in the aqueous phase), they will remain through the rest of the extraction. Therefore, it is very important not to remove any debris or interface material during the chloroform extraction.
12. The sample sheet is required to insert the sample names and adaptor indices used for each sample. We have selected the "Other" as the category followed by "FASTQ only." This option generates FASTQ files only and also enables the deselection of downstream processing steps like adaptor trimming, allowing trimming and mapping to be performed separately.
13. The prepared PhiX library is added to the pooled amplicon libraries as an internal control for the MiSeq sequencing run.

14. It is best to prepare fresh 0.2 M NaOH for the denaturation of libraries.
15. Other binary vectors may confer different antibiotic resistances.
16. Do not cover the plants for more than 24 h. Excess humidity over a long time can yield low number of transformed plants.
17. Resistant seedlings should be screened for the expression of target lncRNAs by RT-PCR using specific primers. The highest or lowest expression of target lncRNAs for overexpression or knockdown experiments should be used for further analysis, respectively.
18. When the solution starts to boil, stop and slowly release the vacuum. Repeat the vacuum infiltration until the seedlings sink to the bottom after the vacuum is released. Do not exceed 15 min of the total exposure time to the formaldehyde.

Acknowledgment

This work was supported by an Australian Research Council Future Fellowship (FT130100525) awarded to IS and a MOET-VIED Ph. D. scholarship awarded to T.D.

References

1. Chekanova JA, Gregory BD, Reverdatto SV, Chen H, Kumar R, Hooker T, Yazaki J, Li P, Skiba N, Peng Q, Alonso J, Brukhin V, Grossniklaus U, Ecker JR, Belostotsky DA (2007) Genome-wide high-resolution mapping of exosome substrates reveals hidden features in the *Arabidopsis* transcriptome. Cell 131(7):1340–1353
2. Kapranov P, Cheng J, Dike S, Nix DA, Duttagupta R, Willingham AT, Stadler PF, Hertel J, Hackermüller J, Hofacker IL, Bell I, Cheung E, Drenkow J, Dumais E, Patel S, Helt G, Ganesh M, Ghosh S, Piccolboni A, Sementchenko V, Tammana H, Gingeras TR (2007) RNA maps reveal new RNA classes and a possible function for pervasive transcription. Science 316(5830):1484–1488
3. Burgess AL, David R, Searle IR (2015) Conservation of tRNA and rRNA 5-methylcytosine in the kingdom Plantae. BMC Plant Biol 15:199
4. Wang D, Qu Z, Yang L, Zhang Q, Liu ZH, Do T, Adelson DL, Wang ZY, Searle I, Zhu JK (2017) Transposable elements (TEs) contribute to stress-related long intergenic noncoding RNAs in Plants. Plant J 90:133–146
5. Ashby R, Forêt S, Searle I, Maleszka R (2016) MicroRNAs in honey bee caste determination. Sci Rep 6:18794
6. Jin J, Liu J, Wang H, Wong L, Chua N-H (2013) PLncDB: plant long non-coding RNA database. Bioinformatics 29(8):1068–1071
7. David R, Burgess A, Parker B, Li J, Pulsford K, Sibbritt T, Preiss T, Searle I (2017) Transcriptome-wide mapping of RNA 5-methylcytosine in Arabidopsis mRNAs and ncRNAs. Plant Cell 29(3):445–460
8. Wang H, Chung PJ, Liu J, Jang I-C, Kean MJ, Xu J, Chua N-H (2014) Genome-wide identification of long noncoding natural antisense transcripts and their responses to light in *Arabidopsis*. Genome Res 24(3):444–453
9. Zhang Y-C, Liao J-Y, Li Z-Y, Yu Y, Zhang J-P, Li Q-F, Qu L-H, Shu W-S, Chen Y-Q (2014) Genome-wide screening and functional analysis identify a large number of long noncoding RNAs involved in the sexual reproduction of rice. Genome Biol 15(12):512
10. Wang HLV, Chekanova JA (2017) Long noncoding RNAs in plants. In: Rao M (ed) Long non coding RNA biology. Advances in

experimental medicine and biology, vol 1008. Springer, Singapore.

11. Deal RB, Henikoff S (2011) The INTACT method for cell type-specific gene expression and chromatin profiling in *Arabidopsis thaliana*. Nat Protoc 6(1):56–68
12. Nagalakshmi U, Wang Z, Waern K, Shou C, Raha D, Gerstein M (2008) The transcriptional landscape of the yeast genome defined by RNA sequencing. Science 320:1344–1349
13. Wilhelm BT, Marguerat S, Watt S, Schubert F, Wood V, Goodhead I, Penkett CJ, Rogers J, Bähler J (2008) Dynamic repertoire of a eukaryotic transcriptome surveyed at single-nucleotide resolution. Nature 453:1239–1243
14. Mortazavi A, Williams BA, McCue K, Schaeffer L, Wold B (2008) Mapping and quantifying mammalian transcriptomes by RNA-Seq. Nat Methods 5:621–628
15. Lister R, O'Malley RC, Tonti-Filippini J, Gregory BD, Berry CC, Millar AH, Ecker JR (2008) Highly integrated single-base resolution maps of the epigenome in *Arabidopsis*. Cell 133:523–536
16. Cloonan N et al (2008) Stem cell transcriptome profiling via massive-scale mRNA sequencing. Nat Methods 5:613–619
17. Marioni J, Mason C, Mane S, Stephens M, Gilad Y (2008) RNA-seq: an assessment of technical reproducibility and comparison with gene expression arrays. Genome Res 18 (9):1509–1517
18. Morin R, Bainbridge M, Fejes A, Hirst M, Krzywinski M, Pugh T, McDonald H, Varhol R, Jones S, Marra M (2008) Profiling the HeLa S3 transcriptome using randomly primed cDNA and massively parallel short-read sequencing. Biotechniques 45:81–94
19. Djebali S, Davis CA, Merkel A, Dobin A, Lassmann T, Mortazavi A et al (2012) Landscape of transcription in human cells. Nature 489:101–108
20. Liu J, Jung C, Xu J, Wang H, Deng S, Bernad L, Arenas-Huertero C, Chua N-H (2012) Genome-wide analysis uncovers regulation of long intergenic noncoding RNAs in *Arabidopsis*. Plant Cell 24(11):4333–4345
21. Fort A, Hashimoto K, Yamada D, Salimullah M, Keya CA, Saxena A et al (2014) Deep transcriptome profiling of mammalian stem cells supports a regulatory role for retrotransposons in pluripotency maintenance. Nat Genet 46:558–566
22. Guil S, Soler M, Portela A, Carrère J, Fonalleras E, Gómez A et al (2012) Intronic RNAs mediate EZH2 regulation of epigenetic targets. Nat Struct Mol Biol 19:664–670
23. Qiu J, Wang Y, Ding J, Jin H, Yang D, Hua K (2015) The long non-coding RNA HOTAIR promotes the proliferation of serous ovarian cancer cells through the regulation of cell cycle arrest and apoptosis. Exp Cell Res 333 (2):238–248
24. SheikMohamed J, Gaughwin PM, Lim B, Robson P, Lipovich L (2010) Conserved long noncoding RNAs transcriptionally regulated by Oct4 and Nanog modulate pluripotency in mouse embryonic stem cells. RNA 16:324–337
25. Liu SJ, Horlbeck MA, Cho SW, Birk HS, Malatesta M, He D, Attenello FJ, Villalta JE, Cho MY, Chen Y, Mandegar MA, Olvera MP, Gilbert LA, Conklin BR, Chang HY, Weissman JS, Lim DA (2017) CRISPRi-based genome-scale identification of functional long noncoding RNA loci in human cells. Science 355 (6320):eaah7111

Part III

Studying lncRNAs Associated with Response to Abiotic and Biotic Stresses in Plants

Chapter 8

The Involvement of Long Noncoding RNAs in Response to Plant Stress

Akihiro Matsui and Motoaki Seki

Abstract

Plant growth and productivity are greatly impacted by environmental stresses. Therefore, plants have evolved mechanisms which allow them to adapt to abiotic stresses through alterations in gene expression and metabolism. In recent years, studies have investigated the role of long noncoding RNA (lncRNA) in regulating gene expression in plants and characterized their involvement in various biological functions through their regulation of DNA methylation, DNA structural modifications, histone modifications, and RNA–RNA interactions. Genome-wide transcriptome analyses have identified various types of noncoding RNAs (ncRNAs) that respond to abiotic stress. These ncRNAs are in addition to the well-known housekeeping ncRNAs, such as rRNAs, tRNAs, snoRNAs, and snRNAs. In this review, recent research pertaining to the role of lncRNAs in the response of plants to abiotic stress is summarized and discussed.

Key words Long noncoding RNA, Small RNA, RNA metabolism, Epigenetic, Abiotic stress response

1 Introduction

In their natural environment, plants are exposed to a variety of environmental stresses, such as drought, heat, salinity, and low temperature, which are the major limiting factors of plant growth and productivity. In response, plants have evolved several molecular mechanisms that allow them to adapt to and survive abiotic stress [1]. Genome-wide transcriptome analyses have revealed that, in addition to the well-known housekeeping ncRNAs, such as rRNAs, tRNAs, snoRNAs, and snRNAs, various types of other noncoding RNAs (ncRNAs) are also involved in the response of plants to abiotic stress conditions [2–6]. The presence of long noncoding RNA in plants has been known for more than 20 years; however, their function and mechanism of action have not been fully understood [7]. As a result, studies to elucidate the role of lncRNAs in the gene regulation within plants have significantly increased over the past few years and have underscored the

Julia A. Chekanova and Hsiao-Lin V. Wang (eds.), *Plant Long Non-Coding RNAs: Methods and Protocols*, Methods in Molecular Biology, vol. 1933, https://doi.org/10.1007/978-1-4939-9045-0_8,

common mechanisms used to regulate transcription in eukaryotic genomes and highlighted their complexity [8]. Functionally, lncRNAs operate through altering DNA methylation, DNA structural modifications, histone modifications, and RNA-RNA interactions. Even though these regulatory RNAs are grouped together in the category of lncRNAs, they are generated by different generation processes and form mutually related cross-networks of RNA metabolism. For example, some lncRNAs represent primary transcripts of small regulatory RNAs, such as miRNAs and siRNAs [9, 10]. Other lncRNAs are produced through mRNA transcription and mRNA processing. Additionally, miRNA-cleaved mRNA and RNA degradation products can also be processed into functional siRNA [11, 12]. In this review, recent progress in lncRNA research in relation to plant abiotic stress response is summarized and discussed.

2 Plant lncRNA Research

The function of lncRNAs in human beings has been investigated since they were first reported in 1992 [13]. Since then, numerous lncRNAs have been identified in a variety of different organisms. The development of large-scale transcriptome studies, such as tiling arrays and next-generation sequencing (NGS), has also accelerated the identification of intergenic regions and their regulation. More than 35,000 ncRNAs derived from 10,000 independent transcriptional regions have been identified in the mammalian genome [14]. Since that report, genome-wide analyses have now identified more than 50,000 lncRNAs in the human genome [15]. It has been reported that approximately 75% of the human genome, and up to 85% of the yeast genome, is transcribed [16, 17]. In soybean (*Glycine max*), the expression of an lncRNA (GMENOD 40) during nodule development was first reported in 1993 [18]. Additionally, early reports of biologically important plant lncRNA genes, such as MtENOD40 [19], TPS11 [20], and OsPI1 [21], were also previously published. The early discovery of plant lncRNAs was fostered by the full-length cloning of cDNAs and has been subsequently greatly expanded by genome-wide transcriptome analyses such as tiling arrays and RNA-seq utilizing NGS. The first analysis of an *Arabidopsis* tiling array indicated that among 18,093 of 25,540 (70%) annotated genes and 5817 novel transcriptional sequences, approximately 24% were antisense RNA [2]. A similar transcriptome profile has been reported in rice [22]. A subsequent, comprehensive analysis of the *Arabidopsis thaliana* transcriptome identified the presence of 40,000 lncRNAs containing more than 30,000 NATs and more than 6000 lincRNAs [3, 4, 6]. Approximately 6480 lncRNAs were identified in an analysis of 200 *Arabidopsis* transcription data sets containing tissue-specific or stress-inducible expression profiles [4]. Further analysis indicated that 1456 out of 7719 non-annotated transcripts (basically representing lncRNAs) exhibited alterations in their expression in response to abiotic stress [5]. Interestingly, these stress-responsive transcripts included natural, long noncoding antisense transcripts, corresponding to more than 60% of the annotated mRNAs [5, 6]. These results suggest that multiple types of lncRNAs are produced anywhere in the genome. Since that report, lncRNAs have been identified in more than 40 species of plants. Comprehensive information on the numerous plant and/or type-specific lncRNAs is available in several databases, as described in the review by Bhatia et al. [23].

3 Classification of lncRNAs

The term, ncRNA, is a generic term, while other descriptors have been used to define specific types of ncRNA, such as rRNA, tRNA, small nucleolar RNA, and small nuclear RNA. Transcripts without an open reading frame (ORF), or transcripts with an ORF that do not appear to have a function, are generally defined as ncRNA [24]. Genome-wide transcriptome analyses have identified various types of ncRNAs that are generated and transcribed from anywhere in the genome. The molecular function of lncRNAs depends not only on their biogenesis but also on their positional relationship to nearby protein-encoding genes and their association with any interacting proteins. Therefore, the classification of lncRNAs is often overlapping, inconsistent, and problematic. In this review, a commonly used classification scheme based upon size is used to categorize ncRNAs. In this scheme, ncRNAs less than 40 nt in length are classified as small RNAs, including microRNAs (miRNA) and small interference RNAs (siRNA). ncRNAs greater than 200 nt are classified as long noncoding RNA (lncRNA) [25]. Lastly, in *Arabidopsis* and rice, ncRNAs that are 50–300 nt in length are classified as intermediate-sized ncRNAs (im-ncRNAs) [26, 27]. Notably, im-ncRNAs are associated with the 5′UTR and are correlated with a high level of transcripts of their target genes. They are also correlated with the presence of positive histone marks, such as H3K4me3 and H3K9ac, but not with negative marks. Downregulation of several im-ncRNAs induces changes in molecular or developmental programs [26]. A second classification is based on the polyadenylation of the transcript. These novel types of lncRNAs are not covered in the aforementioned classification scheme. Non-polyadenylation is thought to be due to differences in RNA polymerase and differences in the process. Non-polyadenylated lncRNAs are often shorter in length than polyadenylated lncRNAs and have been observed to be more readily induced by several different abiotic stresses [6]. A third classification method is based on the positional relationship of the lncRNA to adjacent protein-coding genes. The various categories in this classification scheme are as follows: (1) long intergenic ncRNA (lincRNA) that is present between two protein-coding genes, (2) intronic ncRNA that is derived from intron substrates, (3) long sense ncRNA that overlaps with exons on the same orientation strand of a protein-coding gene, (4) long antisense ncRNA that overlaps with exons on the opposite strand of a protein-coding gene, and (5) promoter-associated sense RNA and upstream antisense RNA that are located at a position close to the promoter of a protein-coding gene [24]. Recent studies have shown that in many cases the location of an lncRNA is associated to some extent with its molecular function.

4 Transcriptional Forms of lncRNAs

The general expression level of lncRNAs is 60-fold lower than protein-coding RNA [6]. Although it is not clear how all of the different types of lncRNA are transcribed, various mechanisms for producing lncRNA have been reported. One type of lncRNA, mRNA-like ncRNA, which is modified by 5 capping, splicing, and polyadenylation, is produced by RNA polymerase II [14, 28]. Functional lncRNAs, trans-acting siRNA precursors, and miRNA precursors also possess promoter sequences for RNA polymerase II and regulate their expression independently [29]. Many of these potential lncRNAs share structural similarities with known lncRNAs and are distinctly different from protein-coding genes in terms of sequence and potential secondary structure [30]. A subset of SERRATE, CAP-BINDING PROTEIN 20 (CBP 20), and CBP 80 have been identified as regulators of splicing in the biosynthesis of lncRNA, suggesting that RNA processing plays an important role in regulating the expression of mRNA-like ncRNAs [4]. Several important plant lncRNAs are also non-polyadenylated [26, 27, 31]. In this regard, recent studies have identified hundreds of non-polyadenylated lncRNAs in *Arabidopsis* that are induced by specific abiotic stresses [5, 32]. The transcriptional mechanism associated with these lncRNAs remains unclear, but some of them are lncRNAs that are involved in transcriptional gene silencing (TGS). Additionally, several species of a type of lncRNAs are transcribed by the plant-specific RNA polymerases Pol IV and Pol V and lead to the production of TGS. These lncRNAs are double-stranded RNA (dsRNA) produced by RNA-dependent RNA polymerase (RDR) and are subsequently processed into small RNAs. Several transcriptional regions of lncRNAs are also aligned with 24 nt siRNA loci involved in TGS. In *Arabidopsis*, changes in the expression of 22 lncRNAs in response to abiotic stress were identified. Among these, the overexpression of two of these lncRNAs was shown to regulate root growth or leaf morphology in response to salt stress [33].

Another type of ncRNA that is less than 400 nt is transcribed by RNA polymerase III. A genome sequence analysis of *Arabidopsis* predicted the existence of 20 novel, putative ncRNAs [34]. Although their molecular function was unclear, the studies implied that RNA polymerase III-derived lncRNAs have a biological function. A specific Pol III-derived ncRNA, AtR18, responds negatively to hypoxic stress [34]. AtR8 expression characteristically occurs during seed germination and is also induced by salicylic acid (SA) [35]. High concentrations of SA decrease seed germination of wild-type and an AtR8 partial deletion mutant. The decrease in germination, however, was more evident in the AtR8 RNA partial deletion mutant.

5 Molecular Functions of lncRNA in Plants Exposed to Abiotic Stress

As perhaps evident from the previous section, there appears to be insufficient knowledge to accurately classify lncRNAs and understand their molecular function. Developing accurate data on the biological and/or molecular functions of individual lncRNAs will lead to a better understanding of their diversity and a more accurate classification scheme. In this section, recent research on the relationship between lncRNAs and plant abiotic stress response is summarized. Among the numerous reports published on lncRNAs, we will preferentially focus on those in which the molecular mechanism of the lncRNAs is more complete. Although miRNAs and siRNAs are small types of ncRNAs, they are processed from lncRNAs, and a type of lncRNA is closely associated with miRNAs and their function in regulating gene expression.

5.1 microRNA

miRNA precursors are transcribed as lncRNA by RNA polymerase II. In the next process, 21–24 nt mature miRNA is generated through the processing of the hairpin-loop structure of lncRNA which cleaves the target mRNA or inhibits translation [36]. Although the latter process is assumed to occur more rapidly, some miRNA precursors have been identified as or predicted from lncRNA [37]. The biological function of miRNA has been well investigated. Reduced expression of miR398 under condition of low Cu, an element that plays an essential role in photosynthesis and oxidative stress, results in the reduced transcription and translation of superoxide dismutase (SOD) and Cu/Zinc (Zn) superoxide dismutases (CSD1 and CSD2) [38, 39]. The expression of miR400, which is located within the first intron of At1g32583, is reduced by an alternative splicing event in response to heat stress, and this contributes to increased tolerance to heat stress [40]. Importantly, miRNAs also function in the fine-tuning of abiotic stress response by regulating the homeostasis of global miRNA metabolism and phytohormone signaling. The corporation of miR168 into AGO1 complex, which cleaves *AGO1* mRNA, constructs a negative feedback regulatory loop for miRNA activity. Plants overexpressing MIR168a, and its target AGO1 loss-of-function mutants, exhibit ABA hypersensitivity and drought tolerance. In accordance to these observations, the promoter region of *MIR168a* contains four ABREs, which adds further evidence to suggest that its expression is directly induced by ABA [41]. Other examples of miRNA-mediated crosstalk between auxin and ABA in abiotic stress response are evidenced by the regulation of miR160-ARF10 [42] and miR393-TIR1 [43]. Research has also demonstrated that miRNAs function in controlling long-distance transport in response to nutrient stress by limiting the expression of specific targets in specific tissues. The translocation of sulfate in xylem to shoots during sulfate starvation has been shown to be regulated by miR395 through ATP sulfurylases, which catalyze the first step of inorganic sulfate assimilation [44]. The induction of *miR399* expression in response to phosphate starvation regulates Pi homeostasis by controlling the expression of *UBC24* (*PHO2*), which encodes a ubiquitin-conjugating E2 enzyme [45, 46]. Thus, miR399 functions as a positive regulator of Pi uptake and translocation. In addition to miR399 regulating *UBC24* expression, the cleavage activity of miR399 is suppressed by a long, intergenic ncRNA, whose transcription is induced by phosphate starvation 1 (IPS1) [47]. A detailed description of this regulatory process is discussed in a later section of this review.

5.2 Trans-acting siRNA (ta-siRNA)

In plants, trans-acting siRNA (ta-siRNA) can be classified as a specialized form of siRNAs in plants [48, 49]. The production of these siRNAs is triggered by the cleavage of lncRNA by miR173, miR390, and miR828. The miRNA-cleaved 3′ fragments are stabilized by suppressor of gene silencing 3 (SGS3) and synthesized antisense strand RNAs to form dsRNAs with the aid of RDR6 [48]. The dsRNAs are then processed into siRNA by Dicers [48, 49]. Tight feedback regulation between miR390, TAS3 ta-siRNA, and ARF4, which is targeted by TAS3 ta-siRNA, is required for lateral root initiation [50]. Primary and lateral roots both undergo a quiescent phase in response to salt stress, followed by a recovery stage [51, 52]. Structural changes in roots in response to abiotic stress have been reported [53]. Drought and high-salinity stress was reported to repress three groups of ta-siRNAs (TAS1, TAS2, and TAS3) through downregulation in the expression of their precursors [54]. These observations suggest that TAS3-siRNA is also involved in the regulation of the response to abiotic stress. The TAS3-siRNA-mediated auxin signal is required for the proper development of floral architecture in plants subjected to mild drought or salinity stress [54]. In regard to other types of abiotic stress, inhibition of TAS1-siRNA induced the expression of

HTT1 (*HEAT-INDUCED TAS1 TARGET 1*) and *HTT2* expression, which modulates thermotolerance through a HSFA1a-mediated pathway [55]. The downregulation of ta-siRNAs is brought about by the destabilization of SGS3 protein at high temperatures [56]. Based on these results, ta-siRNAs targeting auxin signaling are thought to function as mediators that connect abiotic and developmental signals.

5.3 Pol IV- and Pol V-Derived lncRNAs and Small Interference RNAs (siRNAs)

Plant lncRNAs contribute to epigenetic silencing via RNA-directed DNA methylation (RdDM), which requires the plant-specific RNA polymerases Pol IV and Pol V [57], as well as Pol II-derived RNAs at some loci [58, 59]. DNA methylation functions in maintaining the stability of the genome and in silencing the activity of transposable elements. In a major pathway of RdDM activity, 24 nt siRNAs are generated from Pol IV transcripts through processes mediated by RDR2 and DCL3 (Dicer-like 3) and are then incorporated into Argonaute 4 (AGO4). siRNAs control de novo methylation at CHH sites that are involved in several processes, including transposon gene-mediated regulation of plant development and plant stress response [58, 60]. Pol V transcribes target regions, and the siRNA-AGO4 complex is recruited to the transcript in a sequence-dependent manner, in which the DDR complex includes defective in RNA-directed DNA methylation (DRD1), defective in meristem silencing 3 (DMS3), and RNA-directed DNA methylation 1 (RDM1). Finally, Dnmt3 class de novo methylase domains rearranged methylase 2 (DRM2) is recruited to target regions where it adds a methyl group to cytosines. The lncRNAs produced by Pol IV and Pol V have been difficult to identify, possibly due to their very low abundance or stability. It has been reported that several Pol V-transcribed scaffold lncRNAs are non-polyadenylated and either tri-phosphorylated or capped at their 5′ end [57]. One recent study identified thousands of regions that generate Pol IV/RDR2-dependent transcripts [61]. Most of them are located within intergenic regions, and 65% of them overlap with annotated transposons or repeats, while 9% of them overlap with protein-coding genes. These observations suggest that RdDM regulates gene expression during plant development and in response to abiotic stress through de novo methylation [60]. Several studies have demonstrated epigenetic reprograming in response to abiotic stress [62]. The defective mutant of *NRPD2,* which is the second largest subunit of Pol IV and Pol V, exhibits hypersensitivity to heat stress [63]. Similarly, *rdr2*, *dcl3*, and *ago4*, which represent other types of RdDM mutants, also exhibit mild sensitivity to heat stress. Heat stress increases aberrant read-through transcription which decreases during the recovery process in wild-type plants, but not in *nrpd2* mutants [63]. These data suggest that RdDM supports the transcriptional termination of heat-related genes that are located near transposons in plants exposed to a heat stress and during the recovery stage. Another abiotic stress, low humidity, induces the RdDM-mediated transcriptional suppression of SPCH and FAMA, which positively regulates stomatal development [64]. DNA methylation also alters the expression of stress-responsive genes in several plant species. Oxidative stress in tobacco (*Nicotiana tabacum*) reduces DNA methylation and increases transcription at NtGPDL loci (glycerophosphodiesterase-like protein) [65]. In soybean, salinity stress reduces the level of DNA methylation in portions of the genome encoding salt stress-responsive transcription factors [66].

5.4 Cis-Natural Antisense RNAs (cis-NATs) and the Generation of Small RNAs

Recent analyses have indicated that a large number of sense-antisense NAT pairs are expressed in plant genomes [2, 5, 6]. The expression of sense-antisense NAT pairs in response to abiotic stresses, such as drought, cold, salt, and light, has also been documented. Many light-responsive NATs located between protein-encoding genes and ncRNA exhibit high levels of histone acetylation that is dynamically correlated with changes in *NAT* expression [6]. One hypothesis regarding the origin of NATs is that nat-siRNAs are produced from the RNA interference of sense and antisense transcripts that are independently transcribed from the genome. Nat-siRNAs have been reported to be generated during high-salinity stress from dsRNAs produced from natural cis-antisense gene pairs of *Δ1-pyrroline-5-carboxylate dehydrogenase* (*P5CDH*) and a high-salinity stress-inducible gene of unknown function (*SRO5*) [67]. A study integrating RNA sequencing and smRNA sequencing data reported the accumulation of smRNAs in their overlapping regions, suggesting the occurrence of an RNA interference event [68]. The biological process responsible for generating nat-siRNAs, however, is not fully understood [69, 70].

5.5 Cis-NATs That Enhance Protein Translation

In rice, a long antisense ncRNA of cis-NATs positively assists the translation of mRNA present in the opposite strand [71]. *PHOSPHATE1;2* (*PHO1;2*) encodes a protein in responsible for loading phosphate into xylem conducting elements. A cis-lncNAT (cis--NATPHO1;2) associates with the antisense strand of *PHO1;2.* Although both transcripts have promoters that are expressed in vascular tissue, the *cis-NATPHO1;2* has a promoter whose expression is only induced by phosphate starvation. Knockdown and overexpression of *cis-NATPHO1;2* regulates the protein level of *PHO1;2*, without affecting the expression, or nuclear export of *PHO1;2* mRNA. Interestingly, expression of *cis-NATPHO1;2* is correlated with the transport of sense-antisense pairs to polysomes. Recent studies in mice demonstrated that an antisense ncRNA derived from a repeat sequence, SINEUPs, promotes protein translation by binding the start codon and KOZAK sequence of mRNA present on the opposite strand [72, 73]. These results indicate that initiation of translation is regulated by a type of antisense lncRNA.

5.6 Long Antisense, Noncoding RNA Derived from RNA Degradation

Other functions of long antisense ncRNAs are associated with them being a part of double-strand RNAs that are derived from cleaved mRNA and ncRNA, thus producing siRNAs. The production of a type of antisense RNAs that have complimentary sequences to a portion of mRNAs are produced from degrading mRNA without cap structure and are correlated with RNA quality control and/or RNA decay pathways. Studies of PTGS revealed that certain amounts of aberrant RNAs escape nuclear and cytoplasmic RNA degradation and are transformed into dsRNA by the action of SGS3, SDE5, and RDR6 [74–76]. RDR6-derived dsRNAs are processed into 21 nt and 22 nt primary siRNAs by DCL4 and DCL2. The siRNAs are loaded into AGO1 and serve as a guide in the cleavage of complementary target RNA [77–80]. RNAs with a premature stop codon or an excessive length of 3′ untranslated region are processed by nonsense-mediated decay (NMD) [81]. UPFRAMESHIFT 1 (UPF1), UPF2, and UPF3 are required for the activation of NMD. When RNAs enter into NMD, their 5′ cap structure and/or 3′ poly(A) tails are substantially removed and degraded. DECAPPING1 (DCP1), DCP2, DCP5, VARICOSE (VCS), and DEAD BOX HELICASE HOMOLOG1 (DHH1) form a complex for decapping. A 3′-5′ POLY(A)-SPECIFIC RIBONUCLEASE (PARN) and/or CARBON CATABOLITE REPRESSOR 4 (CCR4) complex shorten the 3′ poly(A) tail [81].

Additionally, miRNA-directed RISC cleavage also produces siRNA, suggesting that the same RNA structure without cap and polyadenylation are target-producing siRNA [11, 12]. These aberrant RNAs are generally degraded from both ends. Nuclear XRN2, nucleoplasmic XRN3, and cytoplasmic XRN4 degrade uncapped RNA [82]. RIBOSOMAL RNA PROCESSING4 (RRP4), RRP40, RRR41, RRR43, RRP42 RRP43, RRP45, RRP46, CENTROMERE ENHANCER OF POSITION EFFECT1 SYNTHETIC LETHAL PROTEIN 3 (CLS4) mRNA, and TRANSPORT REGULATOR 3 (MTR3) comprise a 3′ to 5′ exoribonuclease core complex [83]. In addition, nucleolar MTR4 and nucleoplasmic HEN2, and the SUPERKILLER (SKI) complex in the cytoplasm, work together with the exosome [84–86]. Mutation of UPF1 and UPF3 of the NMD factor, DCP1, DCP2 and VCS of the decapping complex, XRN2, XRN3 and XRN4 of the 5′ to 3′ exoribonuclease, and RRP4, RRP41, RRP6L1, MTR4, HEN2 ad SKI3 of the 3′ to 5′ exoribonuclease complex enhances siRNA-mediated TGS or PTGS [11, 83, 85, 87–90]. These results suggest that when the capacity of RNA degradation is saturated, accumulated aberrant RNA is transferred to the siRNA synthesis pathway. RNA decay mutants, however, only exhibit a small effect on the production of siRNAs [85, 91, 92]. A genome-wide analysis, however, identified a large variety of exosome targets that correspond to housekeeping ncRNAs, cleaved mRNAs, and ncRNAs from centromeric regions, repetitive sequences, and other siRNA-producing regions

[83]. These data suggest that there is a large, unexplained gap in understanding the switch from the degradation of aberrant RNA to siRNAs synthesis.

A genome-wide analysis of plants growing under normal conditions identified a type of long antisense ncRNA in more than 6000 protein-coding loci that were associated with the antisense strand of protein-coding genes and were co-expressed with sense protein-coding RNAs in plants subjected to abiotic stress [5]. Since abiotic stress induces the accumulation of aberrant RNA, mRNA decapping activity increased in plants subjected to abiotic stress, through MPK6- and SnRK2-mediated phosphorylation of DCP1 and VARICOSE [93, 94].

In contrast, XRN activity is reduced by the inhibition of phosphonucleotide (3′-phosphoadenosine 5′-phosphate (PAP)) [95]. These stress-inducible lancRNAs are present in more than 1000 protein-coding genes in *Arabidopsis* and are produced by at least 3 out of 6 *Arabidopsis* RDRs (RDR1/2/6) redundantly to make dsRNA [96]. Long antisense ncRNAs promote mRNA degradation in its opposite strand, corresponding to poly(A) sense RNA degradation without siRNA production. This suggests the existence of a novel RNA decay pathway in which aberrant RNA produces dsRNA by RDR directly. Although the process of making dsRNA and target selectivity is not clear, RNA degradation is strongly involved in the production of long antisense ncRNA in response to abiotic stress.

5.7 Multiple lncRNAs, COLDAIR, COOLAIR, and COLDWRAP Regulate Chromatin Modification at the FLC

It is possible that lncRNA expression could serve as a functional link to the regulation of epigenetic silencing by chromatin. Multiple lncRNAs participate in the gene silencing of FLOWERING LOCUS C (FLC), which suppresses the activation of a set of genes required for floral induction [31, 97]. *COLDAIR* is a 5′ capped, non-polyadenylated RNA that is transcribed by Pol II from the first intron of *FLC* in the same direction [31]. *COLDAIR* has a cold-response promoter, and its expression is induced during the period of vernalization. Accumulation of COLDAIR may recruit polycomb repressive complex 2 (PRC2) by directly interacting with one of its components, CURLY LEAF (CLF), and increasing the level of H3K27me3 at the *FLC* [31], thus maintaining the stable silencing of the locus [98]. These results demonstrate that the interaction of lncRNAs and the chromatin modification complex mediates epigenetic regulation during vernalization.

A recent bioinformatic analysis comparing antisense ncRNA and chromatin status speculated that one type of antisense ncRNA from the genome is suppressed by cytosine methylation and H3K36me [99]. *COOLAIR* is an antisense lncRNA that is derived from and encompasses the whole *FLC*. It also possesses a cold-inducible promoter, and its expression is highly induced by 2 weeks of a cold treatment [97]. The induction of COOLAIR occurs prior to COLDAIR and also before the accumulation of H3K27me3. Thus, it was thought that COOLAIR functions during the early phase of vernalization [97]. Disruption of COOLAIR, however, does not affect vernalization [100]. A recent study suggested that COOLAIR assists the transcriptional repression of FLC independently of PRC2 and H3K27me3 [101]. The loss of COOLAIR disrupts the synchronization between the replacement of H3K36 methylation and the accumulation of H3K27me3 [100]. In addition to the vernalization pathway, COOLAIR also participates in the autonomous pathway to repress FLC. A short splice variant of COOLAIR that is processed by CstF64 and CstF77, along with the spliceosome factor PRP8, affects the recruitment of the histone demethylase FLD (FLOWERING LOCUS D) to FLC. This leads to H3K4me2 demethylation of FLC [102–104]. Although the biological and molecular function of this process is unclear, the R-loop of *COOLAIR* promoter that is stabilized by the *AtDX* suppresses *COOLAIR* transcription, suggesting that another regulatory ncRNA hybridizes to the *FLC* region [105]. Moreover, different types of lncRNA have been isolated from the *FLC* locus. A non-polyadenylated antisense lncRNA, *ASL*, is transcribed from the same *COOLAIR* promoter in an early-flowering *Arabidopsis* ecotype [106]. *COLDWRAP* is a non-polyadenylated lncRNA that associates with the upstream region of *FLC* in the same direction [107]. *COLDWRAP* is induced during vernalization from the chromatin loop formation that forms between the *FLC* promoter and the 3′ end of the first intron together with the polycomb complex and stabilizes FLC repression. Many of these results indicate that several lncRNA mechanisms have been developed to regulate the correct expression of *FLC* by cold stimuli and act as a memory for appropriately adjusting to prevailing environmental conditions.

5.8 ENOD40 and ASCO, mRNA-Like Long Intergenic ncRNAs, Regulate Alternative Splicing Events by Interacting with RNA-Binding Protein

EARLY NODULIN 40 (ENOD40) is involved in the organogenesis of root symbiotic nodules. ENOD40 is highly conserved in legumes and has also been found in other species, such as rice [108]. ENOD40 has sORFs encoding short peptides that can specifically bind to sucrose synthase in vitro. Additionally, ENOD40 has a biological function that requires its structured RNA region. Highly structured *enod40* RNA directly interacts with RNA-binding protein 1 (MtRBP1) that exists in nuclear speckles where splicing machinery is located in plant cells [109]. Interestingly, MtRBP1 is translocated from the nucleus to the cytoplasm specifically in root tissues where *ENOD40* is expressed, suggesting that ENOD40 functions in RNA-binding protein trafficking [109]. A recent report indicated that ENOD40 conserved the translocation activity of AtNSRs, the closest *Arabidopsis* homologs of MtRBP1 [110]. NSRs act as nuclear regulators of alternative splicing (AS). NSRs can interact with an lncRNA designated ALTERNATIVE SPLICING COMPETITOR (ASCO) [110] and a known structured lncRNA npc351 [33]. Expression of *ENOD40/ASCO* RNAs in *Arabidopsis* affects the splicing pattern of target mRNAs. Additionally, the highly structured *ASCO* RNA can also compete in the binding of NSRs to an alternative splicing target in vitro, suggesting that the lncRNA regulation of the splicing system in plant species is conserved.

5.9 HID1 and APOLO, Long Intergenic ncRNAs, Form RNA-DNA Hybrids that Repress TF Activity

lncRNA can regulate the chromatin loop structure formation when there is an interaction between lcRNA and a genomic region. The lncRNA makes scaffolds that recruit chromatin-modifying complexes independently of siRNAs, although how lncRNAs are recognized by protein complexes remains unclear. The intergenic lncRNA, *APOLO*, is transcribed by two RNA polymerases, RNA Pol II and Pol V, from a region about 5 kb upstream of a key regulator of polar auxin transport, PINOID (PID) [111]. *APOLO* was isolated from cDNA subclones as *npc34* [33]. Auxin treatment reduces the H3K27me3 repressive mark, as well as the level of DNA methylation in the portion of the *APOLO-PID* genomic region where the chromatin loop is constructed. The reduction in the number of repressive marks promotes the formation of an open loop structure, allowing *PID* mRNA and *APOLO* lncRNA to be transcribed by Pol II. Subsequently, *APOLO* RNA then gradually recruits LIKE HETEROCHROMATIN PROTEIN 1 (LHP1) to close the loop. The *APOLO* transcripts produced by RNA Pol V trigger RNA-dependent DNA methylation, and PRC2 likely redeposits the repressive marks. Finally, the *APOLO* lncRNA-mediated chromatin loop is reformed, and *PID* expression is downregulated. In another case, *HID1* lncRNA regulates gene expression in trans by being a part of a large nuclear protein-RNA complex, which can bind to the chromatin of its target gene [112]. Although *HID1* is classed as an intermediate ncRNA (236 nt) that is induced by light stress, *HID1* has a potential ORF and two stem-loop structures that are required for photomorphogenesis. These observations suggest that structured ncRNA plays an important role in biological function. HID1 is assembled into a large nuclear protein-RNA complex that becomes associated with the chromatin of the first intron of PHYTOCHROME-INTERACTING FACTOR 3 (PIF3) to repress its transcription. The rice homolog, OsHID1, can rescue the phenotype of the *hid1* mutant in *Arabidopsis*, suggesting that the molecular function of HID1 is conserved across species [112].

5.10 Long Intergenic RNA Mimics miRNA Targets

One type of lincRNA is referred to as an RNA decoy or as an RNA sponge, as they contain a short sequence that shares homology with the miRNA-binding sites of endogenous targets and thus acts as a competitor. A RNA decoy family, induced by phosphate starvation 1 (IPS), AT4.1, and AT4.2, has a highly conserved 22 nt sequence that is not perfectly complementary to miR399 due to a mismatch occurring at the 10–11th position [47, 113]. Although miR399 can bind to an *IPS1* transcript, there is no indication of cleavage via miR399. The miR399 that IPS1 is interfering and competing with is a major regulator of plant phosphate homeostasis and also suppresses *UBC24* expression by cleaving its mRNA [114]. Therefore, induction of IPS1 under phosphate starvation results in the release of *UBC24* mRNA from its miR399 target. Another example, npc536, which is induced by salt and phosphate starvation, has several characteristics, including being located on the antisense strand of a Golgi transport complex-related protein, being an sORF, and also being the complementary sequence of miR319a [33]. Plants overexpressing npc536 exhibit increased root growth under salt stress conditions. Thus, this RNA may fine-tune the stress response in plants through multiple molecular mechanisms. Genome-wide analyses of *Arabidopsis* and rice have identified other conserved lncRNAs that potentially act through target mimics of 20 miRNAs [115]. Interestingly, RNA decoys of miRNAs are well known in animals as competing endogenous RNAs (ceRNAs), suggesting that a common system to prevent miRNA regulation may be present in both plants and animals [116].

5.11 Circular RNA

circRNAs are produced in eukaryotic cells by a noncanonical splicing of pre-mRNA, which are referred to as back splicing [117–119]. The presence of circRNAs in plants was initially confirmed in *Arabidopsis* [120]. A current database contains circRNAs from eight different plant species [121]. Plant circRNAs appear to have shorter complementary sequences in the flanking introns compared to animal circRNAs [122]. Although, the molecular function of circRNAs is largely unknown, there is some evidence that circRNAs play a role in controlling gene expression. A well-known circRNA, ciRS-7, functions as an miRNA sponge in animals [123]. In plants, an exonic circle RNA from the *SEPALLATA 3* (*SEP 3*) transcript binds to the antisense strand of the *SEP 3* locus and promotes exon-skipped alternate splicing of *SEP 3* pre-mRNA [124]. circRNAs are enriched in the GO term “response to osmotic stress” [125] and alter their expression in response to abiotic stress [126, 127].

5.12 Viroids: Subviral Plant-Pathogenic lncRNAs

Viroids are 240–400-nt-long subviral plant-pathogenic lncRNAs composed of a circular, single-stranded molecule. During infection, the viroid is inserted into the host genome and accumulates in a subcellular component through transcription and processing. Subsequently, they then move to neighboring cells and are also transported over a long distance through the vascular tissues [128]. Viroid lncRNA-induced DNA methylation of plant-integrated PSTVd-specific sequences was demonstrated when viroids infected PSTVd-expressing transgenic tobacco plants

[129]. Thus far, the data indicate that viroids can modulate gene expression as miRNA sponges and siRNA precursors.

6 Summary

The biological and molecular function of lncRNAs in plant response to abiotic stress was highlighted in the present review. Studies which demonstrated that lncRNAs act as key regulators of various physiological and metabolic processes were cited and discussed. Indeed, lncRNAs play a role in epigenetic regulation, transcription, processing translational activity, and mRNA stability. Future studies can further clarify the structural and functional characteristics of lncRNA by investigating their biogenesis, processing, accumulation in subcellular components, and their interaction with target proteins. New technological approaches will facilitate these studies. Several studies have indicated that aptamer-tagged RNA sequences may be a useful tool for studying the cellular localization and interaction of lncRNAs. Additionally, it is known that the structure of lncRNA is required for stability and protein binding. Computational analysis of RNA structure has also become an important tool for identifying potential functional parts of lncRNAs. In addition to understanding the molecular function of lncRNAs, there is a critical need to elucidate their biological role as key regulators of gene expression. Sometimes the length and structure of lncRNA and the specific characteristics of different plant species appear to have a relationship, such as COOLAIR and ASL, suggesting that lncRNA may modify the expression of their target expression to facilitate the plant's ability to adapt to a prevailing environmental condition. Therefore, a better understanding of inter-tissue and inter-organ control of lncRNA is needed. The ability to conduct single-cell sequencing and single-molecule sequencing will also provide additional opportunities to increase the number of identified lncRNAs. The accumulation of knowledge from many studies will facilitate a better understanding of the functional significance of new and known lncRNAs.

Acknowledgments

This work was supported by grants from RIKEN, Grants-in-Aid for Scientific Research provided from the Ministry of Education, Culture, Sports, Science, and Technology, (Innovative Areas 18H04791 and 18H04705), the Japan Science and Technology Agency (JST), and the Core Research for Evolutionary Science and Technology (CREST) JPMJCR13B4 to M.S.

References

1. Hirayama T, Shinozaki K (2010) Research on plant abiotic stress responses in the post-genome era: past, present and future. Plant J 61:1041–1052. https://doi.org/10.1111/j.1365-313X.2010.04124.x
2. Yamada K, Lim J, Dale JM et al (2003) Empirical analysis of transcriptional activity in the Arabidopsis genome. Science 31:842–846. https://doi.org/10.1126/science.1088305
3. Jin J, Liu J, Wang H et al (2013) PLncDB: plant long non-coding RNA database. Bioinformatics 29:1068–1071. https://doi.org/10.1093/bioinformatics/btt107
4. Liu J, Jung C, Xu J et al (2012) Genome-wide analysis uncovers regulation of long intergenic noncoding RNAs in Arabidopsis. Plant Cell 24:4333–4345. https://doi.org/10.1105/tpc.112.102855
5. Matsui A, Ishida J, Morosawa T et al (2008) Arabidopsis transcriptome analysis under drought, cold, high-salinity and ABA treatment conditions using a tiling array. Plant Cell Physiol 49:1135–1149. https://doi.org/10.1093/pcp/pcn101
6. Wang H, Chung PJ, Liu IC et al (2014) Genome-wide identification of long noncoding natural antisense transcripts and their responses to light in Arabidopsis. Genome Res 24:444–453
7. Cech TR, Steitz J (2014) The noncoding RNA revolution-trashing old rules to forge new ones. Cell 157:77–94. https://doi.org/10.1016/j.cell.2014.03.008
8. Berretta J, Morillon A (2009) Pervasive transcription constitutes a new level of eukaryotic genome regulation. EMBO Rep 10:973–982. https://doi.org/10.1038/embor.2009.181
9. Chen XM (2009) Small RNAs and their roles in plant development. Annu Rev Cell Dev Biol 25:21–44. https://doi.org/10.1146/annurev.cellbio.042308.113417
10. Lee Y, Kim M, Han J et al (2004) MicroRNA genes are transcribed by RNA polymerase II. EMBO J 23:4051–4060. https://doi.org/10.1038/sj.emboj.7600385
11. Gazzani S, Lawrenson T, Woodward C et al (2004) A link between mRNA turnover and RNA interference in Arabidopsis. Science 306:1046–1048. https://doi.org/10.1126/science.1101092
12. Luo Z, Chen Z (2007) Improperly terminated, unpolyadenylated mRNA of sense transgenes is targeted by RDR6-mediated RNA silencing in Arabidopsis. Plant Cell 19:943–958. https://doi.org/10.1105/tpc.106.045724
13. Lukiw WJ, Handley P, Wong L et al (1992) BC200 RNA in normal human neocortex, non-Alzheimer dementia (NAD), and senile dementia of the Alzheimer type (AD). Neurochem Res 17:591–597
14. Carninci P, Kasukawa T, Katayama S et al (2005) The transcriptional landscape of the mammalian genome. Science 309:1559–1563. https://doi.org/10.1126/science.1112014
15. Iyer MK, Niknafs YS, Malik R et al (2015) The landscape of long noncoding RNAs in the human transcriptome. Nat Genet 47:199–208. https://doi.org/10.1038/ng.3192
16. Djebali S, Davis CA, Merkel A et al (2012) Landscape of transcription in human cells. Nature 489:101–108. https://doi.org/10.1038/nature11233
17. David L, Huber W, Granovskaia M et al (2006) A high-resolution map of transcription in the yeast genome. Proc Natl Acad Sci U S A 103:5320–5325. https://doi.org/10.1073/pnas.0601091103
18. Yang WC, Katinakis P, Hendriks P et al (1993) Characterization of GmENOD40, a gene showing novel patterns of cell-specific expression during soybean nodule development. Plant J 3:573–585
19. Crespi MD, Jurkevitch E, Poiret M et al (1994) enod40, a gene expressed during nodule organogenesis, codes for a non-translatable RNA involved in plant growth. EMBO J 13:5099–5112
20. Liu C, Muchhal US, Raghothama KG (1997) Differential expression of TPS11, a phosphate starvation-induced gene in tomato. Plant Mol Biol 33:867–874
21. Wasaki J, Yonetani R, Shinano T et al (2003) Expression of the OsPI1 gene, cloned from rice roots using cDNA microarray, rapidly responds to phosphorus status. New Phytol 158:239–248. https://doi.org/10.1046/j.1469-8137.2003.00748.x
22. Li L, Wang X, Stolc V et al (2006) Genome-wide transcription analyses in rice using tiling microarrays. Nat Genet 38:124–129. https://doi.org/10.1038/ng1704
23. Bhatia G, Goyal N, Sharma S et al (2017) Present Scenario of Long Non-Coding RNAs in Plants. RNA 3:16. https://doi.org/10.3390/ncrna3020016

24. Ponting CP, Oliver PL, Reik W (2009) Evolution and functions of long noncoding RNAs. Cell 136:629–641. https://doi.org/10.1016/j.cell.2009.02.006
25. Mercer TR, Dinger ME, Mattick JS (2009) Long non-coding RNAs: insights into functions. Nat Rev Genet 10:155–159. https://doi.org/10.1038/nrg2521
26. Wang Y, Wang X, Deng XW et al (2014) Genomic features and regulatory roles of intermediate-sized non-coding RNAs in Arabidopsis. Mol Plant 7:514–527. https://doi.org/10.1093/mp/sst177
27. Liu T-T, Zhu D, Chen W et al (2013) A global identification and analysis of small nucleolar RNAs and possible intermediate-sized non--coding RNAs in Oryza sativa. Mol Plant 6:830–846. https://doi.org/10.1093/mp/sss087
28. Ulitsky I, Bartel DP (2013) lincRNAs, genomics, evolution, and mechanisms. Cell 154:26–46. https://doi.org/10.1016/j.cell.2013.06.020
29. Vannini A, Cramer P (2012) Conservation between the RNA polymerase I, II, and III transcription initiation machineries. Mol Cell 45:439–446. https://doi.org/10.1016/j.molcel.2012.01.023
30. Wen J, Parker BJ, Weiller GF (2007) In Silico identification and characterization of mRNA-like noncoding transcripts in Medicago truncatula. In Silico Biol 7:485–505
31. Heo JB, Sung S (2011) Vernalization-mediated epigenetic silencing by a long intronic noncoding RNA. Science 331:76–79. https://doi.org/10.1126/science.1197349
32. Di C, Yuan J, Wu Y et al (2014) Characterization of stress-responsive lncRNAs in Arabidopsis thaliana by integrating expression, epigenetic and structural features. Plant J 80:848–861. https://doi.org/10.1111/tpj.12679
33. Ben Amor B, Wirth S, Merchan F et al (2009) Novel long non-protein coding RNAs involved in Arabidopsis differentiation and stress responses. Genome Res 19:57–69. https://doi.org/10.1101/gr.080275.108
34. Wu J, Okada T, Fukushima T et al (2012) A novel hypoxic stress-responsive long non-coding RNA transcribed by RNA polymerase III in Arabidopsis. RNA Biol 9:302–313. https://doi.org/10.4161/rna.19101
35. Li D, Huang X, Liu Z et al (2016) Effect of AtR8 lncRNA partial deletion on Arabidopsis seed germination. Mol Soil Biol 7:1–7. https://doi.org/10.5376/msb.2016.07.0007
36. Voinnet O (2009) Origin, biogenesis, and activity of plant microRNAs. Cell 136:669–687. https://doi.org/10.1016/j.cell.2009.01.046
37. Hirsch J, Lefort V, Vankersschaver M et al (2006) Characterization of 43 non-protein-coding mRNA genes in Arabidopsis, including the MIR162a-derived transcripts. Plant Physiol 140:1192–1204. https://doi.org/10.1104/pp.105.073817
38. Sunkar R, Kapoor A, Zhu JK (2006) Posttranscriptional induction of two Cu/Zn superoxide dismutase genes in Arabidopsis is mediated by downregulation of miR398 and important for oxidative stress tolerance. Plant Cell 18:2051–2065. https://doi.org/10.1105/tpc.106.041673
39. Yamasaki H, Abdel-Ghany SE, Cohu CM et al (2007) Regulation of copper homeostasis by micro-RNA in Arabidopsis. J Biol Chem 282:16369–16378. https://doi.org/10.1074/jbc.M700138200
40. Yan K, Liu P, Wu CA et al (2012) Stress-induced alternative splicing provides a mechanism for the regulation of microRNA processing in Arabidopsis thaliana. Mol Cell 48:521–531. https://doi.org/10.1016/j.molcel.2012.08.032
41. Li W, Cui X, Meng Z et al (2012) Transcriptional regulation of Arabidopsis MIR168a and ARGONAUTE1 homeostasis in abscisic acid and abiotic stress responses. Plant Physiol 158:1279–1292. https://doi.org/10.1104/pp.111.188789
42. Liu PP, Montgomery TA, Fahlgren N et al (2007) Repression of AUXIN RESPONSE FACTOR10 by microRNA160 is critical for seed germination and post-germination stages. Plant J 52:133–146. https://doi.org/10.1111/j.1365-313X.2007.03218.x
43. Blomster T, Salojärvi J, Sipari N et al (2011) Apoplastic reactive oxygen species transiently decrease auxin signaling and cause stress-induced morphogenic response in Arabidopsis. Plant Physiol 157:1866–1883. https://doi.org/10.1104/pp.111.181883
44. Kawashima CG, Matthewman CA, Huang S et al (2011) Interplay of SLIM1 and miR395 in the regulation of sulfate assimilation in Arabidopsis. Plant J 66:863–876. https://doi.org/10.1111/j.1365-313X.2011.04547.x
45. Aung K, Lin SI, Wu CC et al (2006) pho2, a phosphate overaccumulator, is caused by a nonsense mutation in a microRNA399 target

gene. Plant Physiol 141:1000–1011. https://doi.org/10.1104/pp.106.078063

46. Fuji H, Chiou TJ, Lin SI et al (2005) A miRNA involved in phosphate-starvation response in Arabidopsis. Curr Biol 15:2038–2043. https://doi.org/10.1016/j.cub.2005.10.016
47. Franco-Zorrilla JM, Valli A, Todesco M et al (2007) Target mimicry provides a new mechanism for regulation of microRNA activity. Nat Genet 39:1033–1037. https://doi.org/10.1038/ng2079
48. Allen E, Xie Z, Gustafson AM et al (2005) MicroRNA-directed phasing during trans-acting siRNA biogenesis in plants. Cell 121:207–221. https://doi.org/10.1016/j.cell.2005.04.004
49. Ronemus M, Vaughn MW, Martienssen RA (2006) MicroRNA-targeted and small interfering RNA-mediated mRNA degradation is regulated by argonaute, dicer, and RNA-dependent RNA polymerase in Arabidopsis. Plant Cell 18:1559–1574. https://doi.org/10.1105/tpc.106.042127
50. Marin E, Jouannet V, Herz A et al (2010) miR390, Arabidopsis TAS3 tasiRNAs, and their AUXIN RESPONSE FACTOR targets define an autoregulatory network quantitatively regulating lateral root growth. Plant Cell 22:1104–1117. https://doi.org/10.1105/tpc.109.072553
51. Duan L, Dietrich D, Ng CH et al (2013) Endodermal ABA signaling promotes lateral root quiescence during salt stress in Arabidopsis seedlings. Plant Cell 25:324–341. https://doi.org/10.1105/tpc.112.107227
52. Geng Y, Wu R, Wee CW et al (2013) A spatio-temporal understanding of growth regulation during the salt stress response in Arabidopsis. Plant Cell 25:2132–2154. https://doi.org/10.1105/tpc.113.112896
53. Matsui A, Mizunashi K, Tanaka M et al (2014) tasiRNA-ARF pathway moderates floral architecture in Arabidopsis plants subjected to drought stress. Biomed Res Int 2014:303451. https://doi.org/10.1155/2014/303451
54. Bartels D, Sunkar R (2005) Drought and salt tolerance in plants. Crit Rev Plant Sci 24:23–58. https://doi.org/10.1080/07352680590910410
55. Li S, Liu J, Liu Z et al (2014) Heat-induced TAS1 TARGET1 mediates thermotolerance via heat stress transcription factor A1a-directed pathways in Arabidopsis. Plant Cell 26:1764–1780. https://doi.org/10.1105/tpc.114.124883
56. Zhong SH, Liu JZ, Jin H et al (2013) Warm temperatures induce transgenerational epigenetic release of RNA silencing by inhibiting siRNA biogenesis in Arabidopsis. Proc Natl Acad Sci U S A 110:9171–9176. https://doi.org/10.1073/pnas.1219655110
57. Wierzbicki AT, Haag JR, Pikaard CS (2008) Noncoding transcription by RNA polymerase Pol IVb/Pol V mediates transcriptional silencing of overlapping and adjacent genes. Cell 135:635–648. https://doi.org/10.1016/j.cell.2008.09.035
58. Matzke MA, Mosher RA (2014) RNA-directed DNA methylation: an epigenetic pathway of increasing complexity. Nat Rev Genet 15:394–408. https://doi.org/10.1038/nrg3683
59. Zheng B, Wang Z, Li S et al (2009) Intergenic transcription by RNA Polymerase II coordinates Pol IV and Pol V in siRNA-directed transcriptional gene silencing in Arabidopsis. Genes Dev 23:2850–2860. https://doi.org/10.1101/gad.1868009
60. Matzke M, Kanno T, Daxinger L et al (2009) RNA-mediated chromatin-based silencing in plants. Curr Opin Cell Biol 21:367–376. https://doi.org/10.1016/j.ceb.2009.01.025
61. Li S, Vandivier LE, Tu B et al (2015) Detection of Pol IV/RDR2-dependent transcripts at the genomic scale in Arabidopsis reveals features and regulation of siRNA biogenesis. Genome Res 25:235–245. https://doi.org/10.1101/gr.182238.114
62. Chinnusamy V, Zhu JK (2009) Epigenetic regulation of stress response in plant. Curr Opin Plant Biol 12:133–139. https://doi.org/10.1016/j.pbi.2008.12.006
63. Popova OV, Dinh HQ, Aufsatz W et al (2013) The RdDM pathway is required for basal heat tolerance in Arabidopsis. Mol Plant 6:396–410. https://doi.org/10.1093/mp/sst023
64. Tricker PJ, Gibbings JG, Rodríguez-López CM et al (2012) Low relative humidity triggers RNA-directed de novo DNA methylation and suppression of genes controlling stomatal development. J Exp Bot 63:3799–3814. https://doi.org/10.1093/jxb/ers076
65. Choi CS, Sano H (2007) Abiotic-stress induces demethylation and transcriptional activation of a gene encoding a glycerophosphodiesterase-like protein in tobacco plants. Mol Genet Genomics 277:589–600. https://doi.org/10.1007/s00438-007-0209-1

66. Song Y, Ji D, Li S et al (2012) The dynamic changes of DNA methylation and histone modifications of salt responsive transcription factor genes in soybean. PLoS One 7:e41274. https://doi.org/10.1371/journal.pone.0041274
67. Borsani O, Zhu J, Verslues PE et al (2005) Endogenous siRNAs derived from a pair of natural cis-antisense transcripts regulate salt tolerance in Arabidopsis. Cell 123:1279–1291. https://doi.org/10.1016/j.cell.2005.11.035
68. Zhang X, Lii Y, Wu Z et al (2013) Mechanisms of small RNA generation from cis-NATs in response to environmental and developmental cues. Mol Plant 6:704–715. https://doi.org/10.1093/mp/sst051
69. Henz SR, Cumbie JS, Kasschau KD et al (2007) Distinct expression patterns of natural antisense transcripts in Arabidopsis. Plant Physiol 144:1247–1255. https://doi.org/10.1104/pp.107.100396
70. Zhan S, Lukens L (2013) Protein-coding cis--natural antisense transcripts have high and broad expression in Arabidopsis. thaliana Plant Physiol 161:2171–2180. https://doi.org/10.1104/pp.112.212100
71. Jabnoune M, Secco D, Lecampion C et al (2013) A rice cis-natural antisense RNA acts as a translational enhancer for its cognate mRNA and contributes to phosphate homeostasis and plant fitness. Plant Cell 25:4166–4182. https://doi.org/10.1105/tpc.113.116251
72. Carrieri C, Cimatti L Biagioli M et al (2012) Long non-coding antisense RNA controls Uchl1 translation through an embedded SINEB2 repeat. Nature 491:454–457. https://doi.org/10.1038/nature11508
73. Takahashi H, Kozhuharova A, Sharma H et al (2018) Identification of functional features of synthetic SINEUPs, antisense lncRNAs that specifically enhance protein translation. PLoS One 13:e0183229. https://doi.org/10.1371/journal.pone.0183229
74. Jauvion V, Elmayan T, Vaucheret H (2010) The conserved RNA trafficking proteins HPR1 and TEX1 are involved in the production of endogenous and exogenous small interfering RNA in Arabidopsis. Plant Cell 22:2697–2709. https://doi.org/10.1105/tpc.110.076638
75. Mourrain P, Beclin C, Elmayan T et al (2000) Arabidopsis SGS2 and SGS3 genes are required for posttranscriptional gene silencing and natural virus resistance. Cell 101:533–542. https://doi.org/10.1016/S0092-8674(00)80863-6
76. Parent JS, Jauvion V, Bouché N et al (2015) Post-transcriptional gene silencing triggered by sense transgenes involves uncapped antisense RNA and differs from silencing intentionally triggered by antisense transgenes. Nucleic Acids Res 43:8464–8475. https://doi.org/10.1093/nar/gkv753
77. Baumberger N, Baulcombe DC (2005) Arabidopsis ARGONAUTE1 is an RNA Slicer that selectively recruits microRNAs and short interfering RNAs. Proc Natl Acad Sci U S A 102:11928–11933. https://doi.org/10.1073/pnas.0505461102
78. Dalmay T, Horsefield R, Braunstein TH et al (2001) SDE3 encodes an RNA helicase required for post-transcriptional gene silencing in Arabidopsis. EMBO J 20:2069–2078. https://doi.org/10.1093/emboj/20.8.2069
79. Garcia D, Garcia S, Pontier D et al (2012) Ago hook and RNA helicase motifs underpin dual roles for SDE3 in antiviral defense and silencing of nonconserved intergenic regions. Mol Cell 48:109–120. https://doi.org/10.1016/j.molcel.2012.07.028
80. Parent JS, Bouteiller N, Elmayan T et al (2015) Respective contributions of Arabidopsis DCL2 and DCL4 to RNA silencing. Plant J 81:223–232. https://doi.org/10.1111/tpj.12720
81. Belostotsky DA, Sieburth LE (2009) Kill the messenger: mRNA decay and plant development. Curr Opin Plant Biol 12:96–102. https://doi.org/10.1016/j.pbi.2008.09.003
82. Kastenmayer JP, Green PJ (2000) Novel features of the XRN-family in Arabidopsis: evidence that AtXRN4, one of several orthologs of nuclear Xrn2p/Rat1p, functions in the cytoplasm. Proc Natl Acad Sci U S A 97:13985–13990. https://doi.org/10.1073/pnas.97.25.13985
83. Chekanova JA, Gregory BD, Reverdatto SV et al (2007) Genome-wide high-resolution mapping of exosome substrates reveals hidden features in the Arabidopsis transcriptome. Cell 131:1340–1353. https://doi.org/10.1016/j.cell.2007.10.056
84. Lange H, Zuber H, Sement FM et al (2014) The RNA helicases AtMTR4 and HEN2 target specific subsets of nuclear transcripts for degradation by the nuclear exosome in

Arabidopsis thaliana. PLoS Genet 10: e1004564. https://doi.org/10.1371/journal.pgen.1004564

85. Yu A, Saudemont B, Bouteiller N et al (2015) Second-site mutagenesis of a hypomorphic argonaute1 allele identifies SUPERKILLER3 as an endogenous suppressor of transgene posttranscriptional gene silencing. Plant Physiol 169:1266–1274. https://doi.org/10.1104/pp.15.00585
86. Zhang X, Zhu Y, Liu X et al (2015) Plant biology. Suppression of endogenous gene silencing by bidirectional cytoplasmic RNA decay in Arabidopsis. Science 348:120–123. https://doi.org/10.1126/science.aaa2618
87. Gy I, Gasciolli V, Lauressergues D et al (2007) Arabidopsis FIERY1, XRN2, and XRN3 are endogenous RNA silencing suppressors. Plant Cell 19:3451–3461. https://doi.org/10.1105/tpc.107.055319
88. Thran M, Link K, Sonnewald U (2012) The Arabidopsis DCP2 gene is required for proper mRNA turnover and prevents transgene silencing in Arabidopsis. Plant J 72:368–377. https://doi.org/10.1111/j.1365-313X.2012.05066.x
89. Moreno AB, Martinez de Alba AE, Bardou F et al (2013) Cytoplasmic and nuclear quality control and turnover of single-stranded RNA modulate post-transcriptional gene silencing in plants. Nucleic Acids Res 41:4699–4708. https://doi.org/10.1093/nar/gkt152
90. Hematy K, Bellec Y, Podicheti R et al (2016) The Zinc-finger protein SOP1 is required for a subset of the nuclear exosome functions in Arabidopsis. PLoS Genet 12:e1005817. https://doi.org/10.1371/journal.pgen.1005817
91. Martínez de Alba AE, Moreno AB et al (2015) In plants, decapping prevents RDR6-dependent production of small interfering RNAs from endogenous mRNAs. Nucleic Acids Res 43:2902–2913. https://doi.org/10.1093/nar/gkv119
92. Branscheid A, Marchais A, Schott G et al (2015) SKI2 mediates degradation of RISC 5′-cleavage fragments and prevents secondary siRNA production from miRNA targets in Arabidopsis. Nucleic Acids Res 43:10975–10988. https://doi.org/10.1093/nar/gkv1014
93. Xu J, Chua NH (2012) Dehydration stress activates Arabidopsis MPK6 to signal DCP1 phosphorylation. EMBO J 31:1975–1984. https://doi.org/10.1038/emboj.2012.56
94. Soma F, Mogami J, Yoshida T et al (2017) ABA-unresponsive SnRK2 protein kinases regulate mRNA decay under osmotic stress in plants. Nat Plants 3:16204. https://doi.org/10.1038/nplants.2016.204
95. Estavillo GM, Crisp PA, Pornsiriwong W et al (2011) Evidence for a SAL1-PAP chloroplast retrograde pathway that functions in drought and high light signaling in Arabidopsis. Plant Cell 23:3992–4012. https://doi.org/10.1105/tpc.111.091033
96. Matsui A, Iida K, Tanaka M et al (2017) Novel stress-inducible antisense RNAs of protein-coding loci are synthesized by RNA-Dependent RNA Polymerase. Plant Physiol 175:457–472. https://doi.org/10.1104/pp.17.00787
97. Liu F, Marquardt S, Lister C et al (2010) Targeted 3′ processing of antisense transcripts triggers Arabidopsis FLC chromatin silencing. Science 327:94–97. https://doi.org/10.1126/science.1180278
98. Buzas DM, Robertson M, Finnegan EJ et al (2011) Transcription-dependence of histone H3 lysine 27 trimethylation at the Arabidopsis polycomb target gene FLC. Plant J 65:872–881. https://doi.org/10.1111/j.1365-313X.2010.04471.x
99. Luo C, Sidote D, Zhang Y et al (2013) Integrative analysis of chromatin states in Arabidopsis identified potential regulatory mechanisms for natural antisense transcript production. Plant J 73:77–90. https://doi.org/10.1111/tpj.12017
100. Helliwell CA, Robertson M, Finnegan EJ et al (2011) Vernalization-repression of Arabidopsis FLC requires promoter sequences but not antisense transcripts. PLoS One 6:e21513. https://doi.org/10.1371/journal.pone.0021513
101. Csorba T, Questa JI, Sun Q et al (2014) Antisense COOLAIR mediates the coordinated switching of chromatin states at FLC during vernalization. Proc Natl Acad Sci U S A 111:16160–16165. https://doi.org/10.1073/pnas.1419030111
102. Hornyik C, Terzi LC, Simpson GG (2010) The spen family protein FPA controls alternative cleavage and polyadenylation of RNA. Dev Cell 18:203–213. https://doi.org/10.1016/j.devcel.2009.12.009

103. Marquardt S, Raitskin O, Wu Z et al (2014) Functional consequences of splicing of the antisense transcript COOLAIR on FLC transcription. Mol Cell 54:156–165. https://doi.org/10.1016/j.molcel.2014.03.026
104. Liu F, Quesada V, Crevillén P et al (2007) The Arabidopsis RNA-binding protein FCA requires a lysinespecific demethylase 1 homolog to downregulate FLC. Mol Cell 28:398–407. https://doi.org/10.1016/j.molcel.2007.10.018
105. Sun Q, Csorba T, Skourti-Stathaki K et al (2011) R-loop stabilization represses antisense transcription at the Arabidopsis FLC locus. Science 340:619–621. https://doi.org/10.1126/science.1234848
106. Shin JH, Chekanova JA (2014) Arabidopsis RRP6L1 and RRP6L2 function in FLOWERING LOCUS C silencing via regulation of antisense RNA synthesis. PLoS Genet 10: e1004612. https://doi.org/10.1371/journal.pgen.1004612
107. Kim DH, Sung S (2017) Vernalization-triggered intragenic chromatin loop formation by long noncoding RNAs. Dev Cell 40:302–312.e4. https://doi.org/10.1016/j.devcel.2016.12.021
108. Gultyaev AP, Roussis A (2007) Identification of conserved secondary structures and expansion segments in enod40 RNAs reveals new enod40 homologues in plants. Nucleic Acids Res 35:3144–3152. https://doi.org/10.1093/nar/gkm173
109. Campalans A, Kondorosi A, Crespi M (2004) Enod40, a short open reading frame—containing mRNA, induces cytoplasmic localization of a nuclear RNA binding protein in Medicago truncatula. Plant Cell 16:1047–1059. https://doi.org/10.1105/tpc.019406
110. Bardou F, Ariel F, Simpson CG et al (2014) Long noncoding RNA modulates alternative splicing regulators in Arabidopsis. Dev Cell 30:166–176. https://doi.org/10.1016/j.devcel.2014.06.017
111. Ariel F, Jegu T, Latrasse D et al (2014) Noncoding transcription by alternative RNA polymerases dynamically regulates an auxin-driven chromatin loop. Mol Cell 55:383–396. https://doi.org/10.1016/j.molcel.2014.06.011
112. Wang Y, Fan X, Lin F et al (2014) Arabidopsis noncoding RNA mediates control of photomorphogenesis by red light. Proc Natl Acad Sci U S A 111:10359–10364. https://doi.org/10.1073/pnas.1409457111
113. Shin H, Shin HS, Chen R et al (2006) Loss of At4 function impacts phosphate distribution between the roots and the shoots during phosphate starvation. Plant J 45:712–726. https://doi.org/10.1111/j.1365-313X.2005.02629.x
114. Lin SI, Chiang SF, Lin WY et al (2008) Regulatory network of microRNA399 and PHO2 by systemic signaling. Plant Physiol 147:732–746. https://doi.org/10.1104/pp.108.116269
115. Wu HJ, Wang ZM, Wang M et al (2013) Wide-spread long non-coding RNAs (lncRNAs) as endogenous target mimics (eTMs) for microRNAs in plants. Plant Physiol 161:1875–1884. https://doi.org/10.1104/pp.113.215962
116. Kartha RV, Subramanian S (2014) Competing endogenous RNAs (ceRNAs): new entrants to the intricacies of gene regulation. Front Genet 5:8. https://doi.org/10.3389/fgene.2014.00008
117. Zhang Y, Zhang XO, Chen T et al (2013) Circular intronic long noncoding RNAs. Mol Cell 51:792–806. https://doi.org/10.1016/j.molcel.2013.08.017
118. Starke S, Jost I, Rossbach O et al (2015) Exon circularization requires canonical splice signals. Cell Rep 10:103–111. https://doi.org/10.1016/j.celrep.2014.12.002
119. Chen LL (2016) The biogenesis and emerging roles of circular RNAs. Nat Rev Mol Cell Biol 17:205–211. https://doi.org/10.1038/nrm.2015.32
120. Wang PL, Bao Y, Yee MC et al (2014) Circular RNA is expressed across the eukaryotic tree of life. PLoS One 9:e90859. https://doi.org/10.1371/journal.pone.0090859
121. Chu Q, Zhang X, Zhu X et al (2017) PlantcircBase: a database for plant circular RNAs. Mol Plant 10:1126–1128. https://doi.org/10.1016/j.molp.2017.03.003
122. Ye CY, Chen L, Liu C et al (2015) Widespread noncoding circular RNAs in plants. New Phytol 208:88–95. https://doi.org/10.1111/nph.13585
123. Hansen TB, Jensen TI, Clausen BH et al (2013) Natural RNA circles function as efficient microRNA sponges. Nature 495:384–388. https://doi.org/10.1038/nature11993
124. Conn VM, Hugouvieux V, Nayak A et al (2017) A circRNA from SEPALLATA3 regulates splicing of its cognate mRNA through R-loop formation. Nat Plants 3:17053. https://doi.org/10.1038/nplants.2017.53

125. Lu T, Cui L, Zhou Y et al (2015) Transcriptome-wide investigation of circular RNAs in rice. RNA 21:2076–2087. https://doi.org/10.1261/rna.052282.115
126. Pan T, Sun X, Liu Y et al (2018) Heat stress alters genome-wide profiles of circular RNAs in Arabidopsis. Plant Mol Biol 96:217–229. https://doi.org/10.1007/s11103-017-0684-7
127. Zuo J, Wang Q, Zhu B et al (2016) Deciphering the roles of circRNAs on chilling injury in tomato. Biochem Biophys Res Commun 479:132–138. https://doi.org/10.1016/j.bbrc.2016.07.032
128. Ding B (2009) The biology of viroid–host interactions. Annu Rev Phytopathol 47:105–131
129. Wassenegger M, Heimes S, Riedel L et al (1994) RNA-directed de novo methylation of genomic sequences in plants. Cell 76:567–576. https://doi.org/10.1016/0092-8674(94)90119-8

Chapter 9

Subcellular Localization and Functions of Plant lncRNAs in Drought and Salt Stress Tolerance

Tao Qin and Liming Xiong

Abstract

Plant lncRNAs are expected to play important roles in various plant processes including response to abiotic stress, although the functions of most plant lncRNAs are still unknown. To investigate these potential functions, integrated approaches that employ genetic, transgenic, and molecular biology methods are required. Here, we describe general methods to study the function of lncRNAs in plant response to drought and salt stresses. First, the expression pattern and subcellular localization of lncRNA are analyzed by GUS staining and RNA fluorescence in situ hybridization (FISH). Then the responses of lncRNA mutant and overexpressing transgenic plants to drought and salt stress treatments are characterized, and their sensitivities to ABA are also assayed. To understand the molecular mechanism of lncRNAs' function in stress response, transcriptome sequencing (RNA-seq) and real-time quantitative PCR are performed to analyze altered expression of stress-related genes. Finally, proline content and ROS content are measured to reveal the accumulation of osmolytes and second messengers in these plants in response to drought and salt stress treatments.

Key words Plant lncRNAs, Drought stress, Salt stress, GUS staining, FISH, Transcriptome sequencing, Stress response genes, Proline content, ROS content

1 Introduction

Long noncoding RNAs (lncRNAs) are classes of transcripts longer than 200 nt without or with little protein-coding potential. Increasing evidence suggests that plant lncRNAs are important regulators of plant development and response to the environment. For instance, lncRNAs are shown to regulate plant vegetative growth [1, 2], reproduction [3–5], and symbiosis between bacteria or fungi and leguminous plants [6]. Plant lncRNAs have also been suggested to play roles in response to abiotic stress [7–10], including drought and salt stress [11–15]. Since lncRNAs do not encode proteins, revealing the potential functions of lncRNAs in plant stress response may lead to innovative methods to genetically engineer plant stress tolerance, which previously was mainly restricted

Julia A. Chekanova and Hsiao-Lin V. Wang (eds.), *Plant Long Non-Coding RNAs: Methods and Protocols*, Methods in Molecular Biology, vol. 1933, https://doi.org/10.1007/978-1-4939-9045-0_9,

to use protein-coding genes. Here, we present strategies to study the functions of plant lncRNAs in response to drought or salt stress.

The function of lncRNAs is closely related to their subcellular localization. Nucleus-localized lncRNAs can affect chromatin topology and regulate the transcription of their neighboring genes [16, 17]; they may also interact with splicing factors to regulate alternative splicing of target RNA [1]. Nuclear lncRNAs may regulate DNA methylation [18] and the degradation of their complementary mRNA through directly binding [9, 19]. On the other hand, cytoplasm-localized lncRNAs may function as target mimicry of miRNAs and thus allow normal translation of these miRNA-targeted mRNAs [7, 20]; cytoplasm lncRNA can also promote the translation of overlapping mRNAs by recruiting them to polysomes [8]. Determining the subcellular localization of lncRNAs is thus important to clarify their functions. Therefore, for a newly identified lncRNA, we will firstly use the lncRNA promoter-driving GUS reporter gene to determine where the lncRNA is expressed and then use RNA fluorescence in situ hybridization (FISH) to investigate the subcellular localization of the lncRNA transcript.

To determine the functions of the lncRNA, knockout mutants and/or transgenic plants overexpressing the lncRNA should be generated; and these plants will then be tested for their drought and salt stress tolerance. Since ABA plays critical roles in plant response to drought and salt stress [21, 22], the response of these plants to ABA will also be analyzed. Transcriptome sequencing (RNA-seq) and real-time quantitative PCR are performed with these plants to elucidate the molecular mechanism of the lncRNA in affecting plant response to drought and salt stresses. Since proline is an important osmolyte with cellular protection functions in plants [23, 24], and reactive oxygen species (ROS) are critical second messengers in plant response to drought and salt stress [25], proline and ROS contents in these plants under stress treatments are measured to reveal how the lncRNA may exert its role in stress responses.

2 Materials

Unless otherwise stated, all solutions are prepared with ultrapure water (prepared with Direct-Q® Water Purification System from deionized water to attain a resistivity of 18.2 MΩ cm at 25 °C) and analytical grade reagents.

2.1 GUS Staining

1. Plant material: *Arabidopsis thaliana* Columbia-0 seeds.
2. GUS staining buffer (30 ml): 100 mM sodium phosphate solutions (pH 7.0), 1 mM EDTA (pH 8.0), 0.5 mM $K_3[Fe(CN)_6]$ (protect from light), 0.5 mM $K_4[Fe(CN)_6]$ (protect

Table 1
GUS staining buffer (30 mL)

Concentration of the stock reagents	Reagent volume
0.2 M sodium phosphate solutions (pH 7.0)	15 mL
0.5 M EDTA (pH 8.0)	60 μL
20 mM $K_3[Fe(CN)_6]$ (protect from light)	750 μL
20 mM $K_4[Fe(CN)_6]$ (protect from light)	750 μL
20 mM X-gluc	1.5 mL
Triton X-100	300 μL
Water	to 30 mL

from light), 1 mM X-gluc, 1% Triton X-100. Refer to Table 1 for the complete buffer recipe (*see* **Note 1**).

3. Destaining buffer: 70% ethanol and 30% acetic acid.

2.2 RNA In Situ Hybridization

1. Plant material: *Arabidopsis thaliana* Columbia-0 seeds (*see* **Note 2**).
2. Fixation buffer: 120 mM NaCl, 7 mM Na_2HPO_4, 3 mM NaH_2PO_4, 2.7 mM KCl, 80 mM EGTA (pH 8.0), 0.1% Tween 20, 5% paraformaldehyde, 10% DMSO. Refer to Table 2 for the complete buffer recipe (*see* **Note 3**).
3. 2 × SSC, 0.1% SDS buffer (100 ml): 0.3 M NaCl, 30 mM sodium citrate, 0.1% SDS. Refer to Table 3 for the complete buffer recipe (*see* **Note 4**).
4. 0.2 × SSC, 0.1% SDS (100 mL): 0.03 M NaCl, 3 mM sodium citrate, 0.1% SDS. Refer to Table 4 for the complete buffer recipe (*see* **Note 4**).
5. PerfectHyb™ Plus Hybridization Buffer (Sigma-Aldrich; H-7033).
6. ULYSiS Nucleic Acid Labeling Kit (Molecular Probes company; U21650) (*see* **Note 5**).
7. BioRad® Micro Bio-Spin® P-30 (Bio Rad; 7326250).
8. Zeiss LSM 710 confocal microscope.

2.3 Drought Tolerance Assay

Plant materials: wild-type (WT), lncRNA mutant, and overexpressing seedlings grown in soil

2.4 Salt Tolerance Assay

1. Plant materials: WT, lncRNA mutant, and overexpressing seeds; WT, mutant, and overexpressing seedlings in soil.
2. ½ MS medium plates containing 0, 100, 150, or 200 mM NaCl. MS (Murashige and Skoog) salt (1/2 strength), 3%

Table 2
Fixation buffer (30 mL)

Reagents for the buffer and their concentrations	Concentration of the stock reagents	Reagent volume
120 mM NaCl	3 M	1.20 mL
7 mM Na_2HPO_4	200 mM	1.05 mL
3 mM NaH_2PO_4	200 mM	0.45 mL
2.7 mM KCl	1 M	81 μL
80 mM EGTA (pH 8.0)	500 mM	4.8 mL
0.1% Tween 20	n/a	30 μL
5% paraformaldehyde	n/a	1.5 g
10% DMSO	n/a	3 mL
Water	n/a	to 30 mL

Table 3
2 × SSC, 0.1% SDS buffer (100 mL)

Reagents for the buffer and their final concentrations	Concentration of the stock reagents	Reagent volume
0.3 M NaCl	3 M	10 mL
30 mM sodium citrate	200 mM	15 mL
0.1% SDS	10%	1 mL
Water		74 mL

Table 4
0.2 × SSC, 0.1% SDS (100 mL)

Reagents for the buffer and their final concentrations	Concentration of the stock reagents	Reagent volume
0.03 M NaCl	3 M	1 mL
3 mM sodium citrate	200 mM	1.5 mL
0.1% SDS	10%	1 mL
Water		96.5 mL

(w/v) sucrose, and NaCl are dissolved in water to the indicated final concentrations. Either 0.6% (w/v) (for general growth assay where roots can penetrate the medium) or 1.2% (w/v) (for root growth assay where seedlings are on the agar surface with the plate placed vertically and roots will not penetrate the medium) agar is added into the medium before autoclaving.

2.5 ABA Treatment

1. Plant materials: WT, lncRNA mutant, and overexpressing seedlings.
2. ½ MS plates containing 0, 2, or 5 μM ABA.
3. Stomata open solution (5 mM KCl, 50 μM $CaCl_2$, and 10 mM MES-Tris, pH 5.6).
4. 20 mM ABA (dissolved in absolute ethanol).
5. Microscope.

2.6 Transcriptome Sequencing

1. Plant materials: WT, lncRNA mutant, and overexpressing seedlings.
2. RNeasy Mini Kit (Invitrogen).
3. DNase I (New England Biolabs).
4. Illumina Whole Transcriptome Analysis Kit (Illumina, HiSeq system).
5. HiSeq 2000 platform.

2.7 Real-Time Quantitative PCR

1. Plant materials: WT, lncRNA mutant, and overexpressing seedlings.
2. Plant RNeasy Kit (Qiagen).
3. DNase I (New England Biolabs).
4. Superscript III Reverse Transcriptase Kit (Invitrogen).
5. Fast SYBR Green Master Mix (Thermo Fisher Scientific).
6. ABI 7900HT Fast Real-Time PCR System (Thermo Fisher Scientific).

2.8 Proline Content Assay

1. Plant materials: WT, lncRNA mutant, and overexpressing seedlings.
2. L-Proline: 0.2, 0.5, 1.0, 2.0, and 5.0 mM proline (for making standard curve).
3. Sulfosalicylic acid (3%): dissolve 3.0 g 5-sulfosalicylic acid (2-hydroxy-5-sulfobenzoic acid) in distilled water, and make up to 100 mL.
4. Acidic ninhydrin: dissolve 1.25 g ninhydrin (1,2,3-indantrione monohydrate) in 30 mL glacial acetic acid and 20 mL of 6 M orthophosphoric acid by vortexing and gentle warming.
5. Glacial acetic acid.
6. Toluene.
7. Cuvette spectrophotometer.

2.9 ROS Content Assay

1. Plant material: WT, lncRNA mutant, and overexpressing seedlings.
2. H2DCF-DA (2′,7′-dichlorofluorescin diacetate, Sigma).
3. Zeiss LSM 710 confocal microscope.

3 Methods

3.1 Analysis of Promoter: GUS Activity

1. PCR-amplify the promoter fragment of the lncRNA, and insert into the pMDC162 vector using the GATEWAY cloning technology.
2. After sequence confirmation, the construct was transformed into *Arabidopsis* using *Agrobacterium tumefaciens* (strain GV3101)-mediated flower-dipping method.
3. Screen for the transgenic seedlings.
4. Seedling samples are stained in GUS staining buffer at 37 °C overnight followed by decoloring with destaining buffer (*see* **Note 6**).
5. Examine and photograph with a microscope.

3.2 RNA In Situ Hybridization

1. Prepare the Alexa Fluor® 488 labeled probes. Find the specific nucleotide sequence (for labeling) of the lncRNA by using Basic Local Alignment Search Tool: Standard Nucleotide BLAST (https://blast.ncbi.nlm.nih.gov/Blast.cgi?PROGRAM=blastn&PAGE_TYPE=BlastSearch&LINK_LOC=blasthome). Copy and paste the lncRNA transcript sequence to "Enter Query Sequence." In "Choose Search Set," set "database" as "Reference RNA sequences (refseq_rna)" and "Organism" as the target plant species. Set "Optimize for" as "Somewhat similar sequences (blastn)." Then run the search. The BLAST search result would look like Fig. 1. Find out the most specific region of the gene, and design the probe as the reverse complement of the sequence of this specific region (*see* **Note 7**).
2. Design a scramble probe sequence. Design a scramble probe with the same sequence length as the lncRNA probe. Then BLAST search the scramble probe sequence as described above. If the blast result would show "No significant similarity found. For reasons why, click here," the scramble probe can be used (*see* **Note 8**).
3. Synthesize the specific probe and scramble probe by a DNA/RNA synthesis company.
4. Before labeling probes with Alexa Fluor 488, resuspend 1 μg probe in 20 μL of the labeling buffer (Component C of ULYSiS Nucleic Acid Labeling Kit).

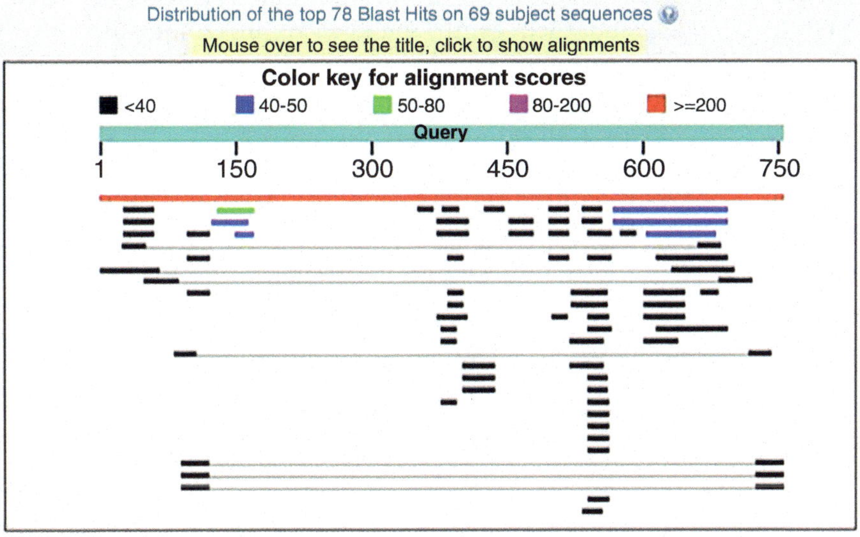

Fig. 1 BLAST search result of the lncRNA *DRIR* transcript sequence as an example. The sequence alignment shown in red is the RNA sequence of *DRIR*. The most specific sequence of *DRIR* is between 171 and 351 since there is no other sequence significantly matching this region

5. Label the probes as per ULYSiS Nucleic Acid Labeling Kit manual (*see* **Note 9**).
6. Roots (or other tissues with higher expression of the lncRNA) from 7-day-old seedlings are fixed in a glass vial by adding 10 mL of fixation buffer. The samples are gently shaken for 2 h at room temperature. Discard the fixation buffer carefully (*see* **Note 10**).
7. Add 10 mL absolute methanol. Gently shake for 5 min at room temperature. Discard absolute methanol carefully.
8. Repeat **step** 7 one more time.
9. Add 10 mL absolute ethanol. Gently shake for 5 min at room temperature. Discard ethanol carefully.
10. Repeat **step 9** two more times.
11. Add 10 mL 1:1 ethanol/xylene. Gently shake at room temperature for 30 min.
12. Discard ethanol/xylene buffer, and add 10 mL absolute ethanol to wash the samples for two times, each time for 5 min at room temperature. Discard ethanol.
13. Add 10 mL absolute methanol to wash the samples for two times, each for 5 min at room temperature. Discard methanol.
14. Add 10 mL 1:1 methanol/fixation buffer without 5% paraformaldehyde. Gently shake at room temperature for 5 min.

15. Discard methanol/fixation buffer, add 10 mL fixation buffer, and then gently shake at room temperature for 30 min to postfix the samples.
16. Rinse with fixation buffer without 5% paraformaldehyde to wash the samples for two times.
17. Transfer samples from glass vials to Eppendorf tubes. Wash the samples with 1 mL of PerfectHyb™ Plus Hybridization Buffer for three times, each for 1 min (*see* **Note 11**).
18. Add 1 mL hybridization buffer to each tube, and keep the tubes in an incubator for 1 h at 50 °C for pre-hybridization.
19. Add 5 pmol specific probes or scramble probes into the tubes, and hybridize at 50 °C in darkness for more than 8 h.
20. After hybridization, wash the samples with 2 × SSC, 0.1% SDS buffer at 50 °C in darkness for 60 min.
21. Wash the samples with 0.2 × SSC, 0.1% SDS at 50 °C in darkness for 20 min.
22. Observe the samples immediately using a Zeiss LSM 710 confocal microscope.

3.3 Drought Tolerance Assay

1. Measure water loss rate of detached leaves. Detach leaves from 4-week-old plants, place immediately in weighing boats at room temperature, and weigh at 0, 20, 40, 60, 80, 100, and 120 min, respectively (*see* **Note 12**). Calculate the weight loss rate over time.
2. Assess drought tolerance in soil. Three-week-old well-watered WT, lncRNA mutant, and overexpressing seedlings grown in soil are treated with drought stress by stopping watering for 20 days. Upon the end of the treatment, WT seedlings clearly wither. Then rewater the seedlings and allow recovery for 2 days (*see* **Note 13**). Assess the damage or survival rate, and document the phenotypes by taking photographs of the plants.

3.4 Salt Tolerance Assay

1. NaCl treatment of plants on agar plates. Seeds of WT, lncRNA mutant, and overexpressing plants are planted onto ½ MS medium plates (with 0.6% agar) containing 0, 100, 150, or 200 mM NaCl and stratified at 4 °C for 3 days. The plates are then incubated in a growth chamber for germination and growth for 8 days. The survival rate or damage is then scored (*see* **Note 14**). This measures both germination and growth sensitivity to NaCl. To exclude the effect of salt stress on seed germination, seedlings that are germinated and grown on ½ MS plates (1.2% agar) without NaCl are transferred individually onto ½ MS plates (1.2% agar) supplemented with 0, 100, 150, or 200 mM NaCl. The position of root tips of the seedlings is marked on the back of the plates. The plates with

seedlings are incubated in a growth chamber for 4 more days, and new root growth (length) after the seedling transfer and the survival rate are measured (*see* **Note 15**).

2. NaCl treatment for soil-grown seedlings. Irrigate 4-week-old WT, lncRNA mutant, and overexpressing seedlings with 200 mM NaCl solution. Take photographs 2 weeks later (*see* **Note 16**). For more comprehensive examination of salt tolerance, seedlings can be treated with multiple concentrations of NaCl, e.g., 50, 100, 150, and 200 mM. Seedlings can be irrigated with NaCl solution with stepwise increase from 50 mM every 4 days to reach (and then maintain) at the intended concentration with a total treatment time of 16 days. Plant damage can be assessed at the end of the treatments.

3.5 ABA Treatment

1. Seedling ABA sensitivity. Transfer 4-day-old WT, mutant, and overexpressing seedlings that are germinated and grown on ½ MS agar plates to ½ MS agar plates containing 0, 2, or 5 μM ABA. After an 8-day incubation in a growth chamber, seedlings are photographed, and plants with green leaves are scored. Root length can also be documented (*see* **Note 17**).
2. ABA-induced stomata closure assays. The fifth and sixth rosette leaves of 4-week-old WT, mutant, and overexpressing plants are detached and incubated in a stomata open solution under cool-white light for 2 h. To induce stomata closure, the leaves are transferred to the stomata open solution supplemented with 20 μM ABA. After incubation under cool-white light for another 2 h, leaf epidermis is carefully peeled and observed under a microscope, and images are captured. Stomata apertures are measured using the ImageJ software (*see* **Note 18**). If desired, more comprehensive investigation of stomata response can be similarly performed with different concentrations of ABA (e.g., 5, 10, 15, and 20 μM) and at different time points after ABA treatment.

3.6 Transcriptome Sequencing Analysis

1. Dehydration treatment of seedlings. Ten-day-old WT, mutant, and overexpressing seedlings grown on ½ MS agar plates are harvested and placed on dry filter paper in Petri dishes to lose 40% fresh weight and then incubated for 2 h in sealed plastic bags to prevent further water loss.
2. Extract total RNA using the RNeasy Mini Kit (Invitrogen) as per the manual provided by the manufacturer.
3. Clean DNA using DNase I (New England Biolabs).
4. Construct RNA-seq libraries with 2–4 μg of cleaned total RNA by using an Illumina Whole Transcriptome Analysis Kit following the standard protocol (Illumina, HiSeq System).

5. Sequence on the HiSeq 2000 or another platform.
6. Determine the gene expression levels (FPKM).

3.7 Real-Time Quantitative PCR

1. For dehydration treatment, 10-day-old WT, mutant, and overexpressing seedlings grown on ½ MS agar plates are harvested and placed on dry filter paper in Petri dishes to lose 40% fresh weight and then incubated for 2 h in sealed plastic bags to prevent further water loss. For salt stress treatment, 10-day-old WT, mutant, overexpressing seedlings grown on ½ MS plates are transferred onto filter paper saturated with ½ MS solution supplemented with 200 mM NaCl and incubated under light for 3 h before harvesting by immediately freezing in liquid nitrogen.
2. Extract total RNA by using the Plant RNeasy Kit with DNase I treatment (Qiagen).
3. Synthesize cDNAs from total RNA by using Superscript III Reverse Transcriptase Kit (Invitrogen) as per the manufacturer's manual.
4. Perform real-time quantitative PCR on the ABI 7900HT Fast Real-Time PCR System using 18S rRNA as a control.

3.8 Proline Content Assay

1. For dehydration treatment, approximately 0.5 g of 10-day-old WT, mutant, and overexpressing seedlings grown on ½ MS agar plates are transferred onto dry filter paper in Petri dishes to lose 40% fresh weight and then incubated for 2 h in sealed plastic bags to prevent further water loss. For salt stress treatment, approximately 0.5 g of 10-day-old WT, mutant, overexpressing seedlings grown on ½ MS plates are transferred onto filter paper saturated with ½ MS solution supplemented with 200 mM NaCl and incubated under light for 3 h.
2. Approximately 0.5 g of plant materials are homogenized in 5 mL of 3% aqueous sulfosalicylic acid and the homogenate filtered through filter paper (*see* **Note 19**).
3. Prepare the reaction mixture by mixing 1 mL of 3% sulfosalicylic acid, 2 mL glacial acetic acid, and 2 mL acidic ninhydrin.
4. Add 5 mL reaction mixture to 1 mL of filtrate in a test tube. Incubate the tubes at 96 °C for 60 min.
5. Terminate the reaction on ice.
6. Add 1 mL toluene to the reaction mixture, vortex the samples, and leave on the bench for 5 min to allow the separation of the organic and water phases.
7. Transfer the chromophore-containing toluene into another tube.
8. Measure the absorbance at 520 nm and determine the proline concentration.

3.9 ROS Content Assay

1. Immerse detached leaves from 4-week-old WT, mutant, and overexpressing seedlings in H_2O with or without 20 μM ABA under cool-white light for 1 h.
2. Peel epidermises and stain with 2.5 μM H2DCF-DA (Sigma) for 30 min.
3. Wash three times with H_2O, and examine the fluorescent signal of guard cells using Zeiss LSM 710 confocal microscope with excitation at 488 nm and emission at 525 nm.

4 Notes

1. X-gluc (5-bromo-4-chloro-3-indolyl-beta-D-glucuronic acid) should be dissolved in dimethylformamide and protected from light.
2. Since most plant lncRNAs are expressed at low levels, the fluorescent signal usually is very weak. Using transgenic plants, overexpressing the lncRNA is another option to obtain stronger fluorescent signal in in situ hybridization assays.
3. Prepare the fixation buffer before use. Paraformaldehyde is difficult to dissolve. When preparing the buffer, heat the buffer in 60 °C water bath to dissolve all the paraformaldehyde.
4. Prepare before use.
5. Upon receipt, store the kit at ≤–20 °C in a desiccator and protect from light. Before using, add 5 μL Component B to the vial containing the Alexa Fluor® 488 ULS® reagent (Component A). Vortex until all the ULS® labeling reagent has dissolved and no particulate material remains. *Do not heat* the solution. After use, ULS® reagent stock solutions should be stored at 4 °C.
6. While overnight staining is generally recommended, the suitable staining time varies dependent on the promoters. For strong promoters, the staining time should be reduced accordingly to allow differences among different tissues to be clearly observed.
7. While designing the probe, several factors should be considered regarding the length of the probe. First is the length of the unique sequence of the target gene that may restrict the overall length of the probe. In addition, short probe is easier to get into cells and nuclei and could thus hybrid with more RNA molecules during fluorescent in situ hybridization. Further, when labeling probes by using the ULYSiS Nucleic Acid Labeling Kit, longer probes would have more fluorescent molecules at the same labeling ratio and result in stronger fluorescent signal after in situ hybridization.

8. Scramble probe is used as a negative control. Another negative control probe usually used is the sense sequence of the lncRNA.
9. The labeled DNA must be purified to remove the remaining ULS® labeling reagents. A gel filtration-based spin column, such as BioRad® Micro Bio-Spin® P-30, is recommended to purify the labeled DNA.
10. Since the expression of plant lncRNAs may be confined to specific cell types, it is recommended to learn the expression pattern of lncRNAs in different tissues using the promoter-GUS reporter before doing fluorescent in situ hybridization, so that a tissue optimal for fluorescent in situ hybridization could be identified.
11. HybPlus hybridization buffer usually produces a solid precipitate when stored at room temperatures. The buffer should be heated in a water bath to dissolve the precipitate before use.
12. Detach and combine five to six pieces of 5th and 6th rosette leaves from different plants to weigh to avoid plant difference. Three replicates for each genotype.
13. Monitor plant conditions daily during drought treatment. If leaves of any genotype show withered or death phenotype, then stop drought treatment. The treatment time may be longer or shorter than 20 days depending on pot size and plant conditions.
14. Monitor seedling phenotypes daily during the treatment. If seedlings of any genotype show great number of death, then take photographs. The treatment time may be longer or shorter than 8 days.
15. To facilitate the measure of root elongation rate, the plates are placed vertically so that seedlings are grown on the surface of the agar. If seedlings of any genotype show great number of death, then take photographs.
16. Monitor seedling conditions closely after treatment with 200 mM NaCl solution. If seedlings of any genotype show damage phenotype, then take photographs. The treatment time may be longer or shorter than 2 weeks.
17. To facilitate the measure of root elongation rate, the plates are placed vertically so that seedlings are grown on the surface of the agar. If seedlings of any genotype show great number of yellow leaves, then take photographs. Root elongation can also be measured. The treatment duration may be longer or shorter than 8 days.
18. Avoid dark conditions, as darkness would induce stomata closure.

19. The sample weight (0.5 g) is the fresh weight of plant materials before dehydration or salt treatment.

Acknowledgment

This work was supported by Texas A&M AgriLife Research Center at Dallas and by King Abdullah University of Science and Technology to L.X.

References

1. Bardou F, Ariel F, Simpson CG, Romero-Barrios N, Laporte P, Balzergue S, Brown JW, Crespi M (2014) Long noncoding RNA modulates alternative splicing regulators in Arabidopsis. Dev Cell **30**(2):166–176
2. Wang Y, Fan X, Lin F, He G, Terzaghi W, Zhu D, Deng XW (2014) Arabidopsis noncoding RNA mediates control of photomorphogenesis by red light. Proc Natl Acad Sci U S A 111(28):10359–10364
3. Csorba T, Questa JI, Sun Q, Dean C (2014) Antisense *COOLAIR* mediates the coordinated switching of chromatin states at FLC during vernalization. Proc Natl Acad Sci U S A 111(45):16160–16165
4. Ding J, Lu Q, Ouyang Y, Mao H, Zhang P, Yao J, Xu C, Li X, Xiao J, Zhang Q (2012) A long noncoding RNA regulates photoperiod-sensitive male sterility, an essential component of hybrid rice. Proc Natl Acad Sci U S A 109 (7):2654–2659
5. Ietswaart R, Wu Z, Dean C (2012) Flowering time control: another window to the connection between antisense RNA and chromatin. Trends Genet 28(9):445–453
6. Röhrig H, Schmidt J, Miklashevichs E, Schell J, John M (2002) Soybean *ENOD40* encodes two peptides that bind to sucrose synthase. Proc Natl Acad Sci U S A 99(4):1915–1920
7. Franco-Zorrilla JM, Valli A, Todesco M, Mateos I, Puga MI, Rubio-Somoza I, Leyva A, Weigel D, García JA, Paz-Ares J (2007) Target mimicry provides a new mechanism for regulation of microRNA activity. Nat Genet 39(8):1033–1037
8. Jabnoune M, Secco D, Lecampion C, Robaglia C, Shu Q, Poirier Y (2013) A rice cis-natural antisense RNA acts as a translational enhancer for its cognate mRNA and contributes to phosphate homeostasis and plant fitness. Plant Cell 25(10):4166–4182
9. Wunderlich M, Groß-Hardt R, Schöffl F (2014) Heat shock factor HSFB2a involved in gametophyte development of *Arabidopsis thaliana* and its expression is controlled by a heat-inducible long non-coding antisense RNA. Plant Mol Biol 85(6):541–550
10. Xin M, Wang Y, Yao Y, Song N, Hu Z, Qin D, Xie C, Peng H, Ni Z, Sun Q (2011) Identification and characterization of wheat long non-protein coding RNAs responsive to powdery mildew infection and heat stress by using microarray analysis and SBS sequencing. BMC Plant Biol 11(1):61
11. Amor BB, Wirth S, Merchan F, Laporte P, d'Aubenton-Carafa Y, Hirsch J, Maizel A, Mallory A, Lucas A, Deragon JM (2009) Novel long non-protein coding RNAs involved in Arabidopsis differentiation and stress responses. Genome Res 19(1):57–69
12. Chung PJ, Jung H, Jeong D-H, Ha S-H, Do Choi Y, Kim J-K (2016) Transcriptome profiling of drought responsive noncoding RNAs and their target genes in rice. BMC Genomics 17(1):563
13. Qi X, Xie S, Liu Y, Yi F, Yu J (2013) Genome-wide annotation of genes and noncoding RNAs of foxtail millet in response to simulated drought stress by deep sequencing. Plant Mol Biol 83(4–5):459–473
14. Qin T, Zhao H, Cui P, Albesher N, Xiong L (2017) A nucleus-localized long non-coding rna enhances drought and salt stress tolerance. Plant Physiol 175(3):1321–1336
15. Zhang W, Han Z, Guo Q, Liu Y, Zheng Y, Wu F, Jin W (2014) Identification of maize long non-coding RNAs responsive to drought stress. PLoS One 9(6):e98958
16. Heo JB, Sung S (2011) Vernalization-mediated epigenetic silencing by a long intronic noncoding RNA. Science 331 (6013):76–79
17. Zhu Y, Rowley MJ, Böhmdorfer G, Wierzbicki AT (2013) A SWI/SNF chromatin-remodeling complex acts in noncoding

RNA-mediated transcriptional silencing. Mol Cell 49(2):298–309

18. Wierzbicki AT (2012) The role of long non-coding RNA in transcriptional gene silencing. Curr Opin Plant Biol 15(5):517–522
19. Zubko E, Meyer P (2007) A natural antisense transcript of the *Petunia hybrida Sho* gene suggests a role for an antisense mechanism in cytokinin regulation. Plant J 52(6):1131–1139
20. Wu H-J, Wang Z-M, Wang M, Wang X-J (2013) Widespread long noncoding RNAs as endogenous target mimics for microRNAs in plants. Plant Physiol 161(4):1875–1884
21. Munemasa S, Hauser F, Park J, Waadt R, Brandt B, Schroeder JI (2015) Mechanisms of abscisic acid-mediated control of stomatal aperture. Curr Opin Plant Biol 28(6):154–162
22. Yamaguchi-Shinozaki K, Shinozaki K (2006) Transcriptional regulatory networks in cellular responses and tolerance to dehydration and cold stresses. Annu Rev Plant Biol 57:781–803
23. Ábrahám E, Rigó G, Székely G, Nagy R, Koncz C, Szabados L (2003) Light-dependent induction of proline biosynthesis by abscisic acid and salt stress is inhibited by brassinosteroid in Arabidopsis. Plant Mol Biol 51(3):363–372
24. Sharma S, Villamor JG, Verslues PE (2011) Essential role of tissue-specific proline synthesis and catabolism in growth and redox balance at low water potential. Plant Physiol 157(1):292–304
25. Kwak JM, Mori IC, Pei ZM, Leonhardt N, Torres MA, Dangl JL, Bloom RE, Bodde S, Jones JD, Schroeder JI (2003) NADPH oxidase *AtrbohD* and *AtrbohF* genes function in ROS-dependent ABA signaling in Arabidopsis. EMBO J 22(11):2623–2633

Chapter 10

Discovery, Identification, and Functional Characterization of Plant Long Intergenic Noncoding RNAs After Virus Infection

Ruimin Gao, Peng Liu, Nadia Irwanto, De Rong Loh, and Sek-Man Wong

Abstract

Long intergenic noncoding RNAs (lincRNAs), which possess diverse features such as remodeling chromatin and genome architecture, RNA stabilization, and genome architecture, are important regulatory factors in plant genomes. They serve to fine-tune the expression of neighboring genes. Here, we describe a procedure of discovery, identification, and functional characterization of plant lincRNAs after virus infection. From high-throughput RNA-Seq transcriptome analysis, the noncoding RNA transcripts with significant fold changes (upregulation or downregulation) will be discovered and identified. The lincRNA of interest will be further confirmed and validated using rapid amplification of cDNA ends (RACE). In addition, functional characterization of the lincRNA will be followed up through overexpression and knockdown strategies.

Key words Long noncoding RNAs, Long intergenic noncoding RNAs (lincRNAs), RNA-Seq, Transcriptome analysis, Rapid amplification of cDNA ends (RACE)

1 Introduction

Long noncoding RNAs (lncRNAs) have emerged as important regulatory factors in mammalian and plant genomics [1]. In many diverse and substantial biological processes, lncRNAs are involved in remodeling chromatin and genome architecture; RNA processing machinery; small RNA pathways, as well as transcriptional and posttranscriptional regulation of gene expression; genomic imprinting; modulation of protein activity; and cellular structural maintenance [2–4]. In order to simplify analyses by avoiding complications arising from overlapping with other types of genes, and to identify lncRNAs and their corresponding genes clearly, long "intergenic" noncoding RNAs (lincRNAs) have been focused recently. Some lincRNAs derived from genes, and it does not overlap exons of either protein coding or other non-lincRNA

Julia A. Chekanova and Hsiao-Lin V. Wang (eds.), *Plant Long Non-Coding RNAs: Methods and Protocols*,
Methods in Molecular Biology, vol. 1933, https://doi.org/10.1007/978-1-4939-9045-0_10,

types of genes [5]. LincRNAs also contain promoter- or enhancer-associated RNAs that are gene proximal with either in the sense or antisense orientation to its associated protein-coding gene [6]. A comprehensive characterization of lincRNAs function study reveals that long-range chromatin interactions play crucial roles in transcriptional regulation of lincRNAs [7].

Genome-wide analysis has becoming a promising tool to discover global gene transcript change profiles in pathogenesis. For example, RNA-Seq was used to find the correlation of aberrant expression of lincRNAs with cancer among different breast cancer tissue and respective adjacent normal tissue [8]. An RNA-sequencing-based transcriptome analysis of compatible interaction of tomato (*Solanum lycopersicum*) and *Verticillium dahliae* facilitates researchers to identify key susceptible genes and understand the mechanism of tomato susceptible response to *V. dahliae* [9]. Analysis of the tomato leaf transcriptome during successive hemibiotrophic stages of a compatible interaction with the oomycete pathogen *Phytophthora infestans* will provide a broad view of the hemibiotrophic infection process from the plant perspective [10].

In this study, NGS-based genome-wide transcriptome analysis is used to identify all the RNA transcripts (including the noncoding RNA transcripts) with significant fold changes after virus infection. It mainly focuses on the discovery, identification, and functional characterization of noncoding RNA transcripts in plants after virus infection. For selected noncoding RNA transcripts, RACE PCR are used to further validate its existence. Finally, to functionally characterize this noncoding RNA transcript, overexpressing and knocking down of this lncRNA transcripts are performed. Successfully characterizing lincRNA provides us more knowledge in understanding the noncoding RNA function in plant development.

2 Materials

2.1 Plant Materials and Reagents

1. *Arabidopsis* ecotypes Col-0 plants.
2. *Agrobacterium* strain(s): GV3101, EHA105, LBA 4404, or others.
3. 0.05% agarose (wt/vol). Autoclaved and stored at room temperature.
4. Autoclave MS medium: 4.3 g Murashige and Skoog salts, 10 g sucrose, 0.5 g MES, 8 g agar per liter; pH 5.7, cool to approximately 50 °C before pouring into Petri dishes (MS solid medium).
5. Liquid LB medium: 10 g tryptone, 5 g yeast extract, 10 g NaCl per liter.
6. 5% (wt/vol) sucrose, prepared fresh.

7. Silwet L-77.
8. 70% (vol/vol) ethanol (with DEPC water prepared for RNA extraction).
9. 50% Clorox bleach/50% sterile double-distilled water/0.05% Tween, prepared fresh.
10. Restriction enzyme.
11. T4 ligase.

2.2 Equipment and Used Software Programs

1. Growth chambers or light room or greenhouse, adjustable to long-day condition of 16 h light/8 h dark, 23 °C, humidity 70%.
2. Eppendorf or other microcentrifuge capable of centrifugation at 16,000 × *g*, at 4 °C.
3. Tissue homogenizer or pestle and mortar.
4. Laminar flow hood.
5. NanoDrop 2000C (Thermo Scientific).
6. 2100 Bioanalyzer (Agilent).
7. Linux-based bioinformatics analysis server; TopHat, Cufflinks-based NGS transcriptome analysis.

2.3 Commercial Kit, Solutions, and Chemicals

1. TRIzol reagents for RNA extraction (Invitrogen).
2. SMARTer® RACE 5′/3′ Kit (Clontech).

3 Methods

3.1 Plant Materials and Generation of Transgenic Arabidopsis

1. Geminate and grow *Arabidopsis* ecotypes Col-0 plants under long-day conditions (16 h light/8 h dark cycles) at 23 °C and humidity 70%. Keep the plants growing for 3–4 weeks until they begin to bolt and produce floral inflorescences.
2. Prepare the *Agrobacterium* strain that harbors the gene of interest in a binary vector for floral-dipping transformation.
3. Collect *Agrobacterium* cells by centrifugation at 4000 × *g* for 10 min at room temperature, and gently resuspend cells in 1 volume of freshly made 5% (wt/vol) sucrose solution with a stirring bar.
4. Add Silwet L-77 to a concentration of 0.02% (vol/vol) (100 μl per 500 mL of solution), and mix well immediately before dipping.
5. Invert plants and dip aerial parts of plants in the *Agrobacterium* cell suspension for 10 s with gentle agitation.

6. Cover dipped plants with a plastic cover or wrap them with plastic film. Lay down the treated plants on their sides for 16–24 h to maintain high humidity.
7. Remove the cover next day, and send the treated plants back to the greenhouse or the growth chamber for growing normally for 1 month. Then collect dry seeds.

3.2 Total RNA Extraction Using TRIzol Reagents (See Note 1)

1. Homogenize tissue samples using pestle and mortar or power homogenizer. Use 1 mL of TRIzol® reagent for 50–100 mg tissue (*see* **Note 2**).
2. Incubate the homogenized samples for 5 min at 15–30 °C to permit the complete dissociation of nucleoprotein complexes. Add 0.2 mL of chloroform per 1 mL of TRIzol® reagent. Mix completely and incubate them at 15–30 °C for 2–3 min (*see* **Note 3**). Centrifuge the samples at 12,000 × *g* for 15 min at 4 °C.
3. Transfer 0.4 mL aqueous phase to a fresh tube and mix 0.5 mL of isopropyl alcohol. Incubate samples at 15–30 °C for 10 min, and centrifuge at 12,000 × *g* for 10 min at 4 °C.
4. Wash the RNA pellet once with 75% ethanol.
5. Briefly dry the RNA pellet (air-dry or vacuum-dry for 5–10 min) (*see* **Note 4**), and dissolve RNA in RNase-free water.

3.3 NGS-Based Transcriptome Analysis for Discovering Noncoding RNA

3.3.1 Library Construction and Sequencing

1. Use total RNA extracted from desired plant tissues using TRIzol reagents.
2. Before library construction, measure the total RNA amount using NanoDrop 2000C (Thermo Scientific), and assess the quality of RNA samples by agarose gel electrophoresis. Then assess the integrity of RNA samples using Agilent Bioanalyzer 2100 RNA Nano chip.
3. Construct an RNA library using the Illumina TruSeq mRNA library preparation protocol.
4. The poly(A) RNA from 500 ng (or optimized amount) of total RNA was enriched with oligo dT beads and then fragmented to convert to double-stranded cDNA.
5. One microliter of each of the final RNA libraries was loaded on a Bioanalyzer 2100 DNA High Sensitivity chip (Agilent Technologies) to check for size.
6. The RNA libraries were quantified by qRT-PCR using the Kapa Library Quantification Illumina/ABI Prism Kit protocol (KAPA Biosystems).
7. Pool a few libraries in one lane with equimolar quantities and sequenced on an Illumina HiSeq sequencer.

3.3.2 Analysis of the RNA-Seq Dataset

1. For obtained raw reads, carry out QC analyses for all the raw reads using FastQC program.
2. Raw sequence reads are 5′ trimmed on quality score ($Q > 30$), adapter sequences removed, and short/unpaired reads dropped.
3. Map the filtered and properly paired reads for each sample to plant genome using TopHat, the used command is:

```
tophat -p 8 -g all.gff3 -r 60 -o 1th R1_001.fastq R2_001.
fastq
```

4. Use fragment alignments generated by TopHat as input files to analyze the differentially expressed RNA transcripts using Cufflink packages. Assemble transcripts for each sample with reference gene structure:

```
cufflinks -g ../all.gff3 -p 8 -o mock_cl mock_tophat/
accepted_hits.bam
```

5. Run Cuffcompare on all assemblies to create a single merged transcriptome annotation:

```
cuffcompare -o TCV -s At.fa -r ../all.gff3 -R -C mock_cl/
transcripts.gtf TCV_cl/transcripts.gtf
```

6. Run Cuffdiff by using the merged transcriptome assembly along with the BAM from TopHat for each replicate:

```
cuffdiff -o diff1 -b At.fa -p 8 -L mock,TCV -u 1_cuffcom-
pare/\1.combined.gtf \./mock_tophat/accepted_hits.bam \./
TCV_tophat/accepted_hits.bam
```

7. All the significant differentially expressed (both upregulation and downregulation) noncoding RNA transcripts are retrieved from the above mentioned Cuffdiff output, where the transcript ID is labeled with XLOC_xxxxxx. The interested noncoding gene transcript with specific ID can be further identified (its location and neighboring genes in the genome) in the Arabidopsis long noncoding RNA database (http://chualab.rockefeller.edu/gbrowse2/homepage.html), which is a good resource for further illustrating this long noncoding RNA [11].

Based on the predicted sequences of interest LINC RNA, SMARTer® RACE 5′/3′ Kit (Clontech) or any other available RACE kits are used to amplify the full length of LINC RNA transcript with its respective 5′- and 3′-sequences.

3.4 RACE PCR to Amplify the Complete Sequences of LINC RNA

1. To design gene-specific primers (GSPs) (*see* **Note 5**), the length should be 23–28 nt to ensure specific annealing, be 50–70% GC, and have a Tm ≥65 °C; best results are obtained if Tm >70 °C, which enables the use of touchdown PCR, not be complementary to the 3′-end of the Universal Primer Mix (long primer = 5′–TAATACGACTCACTATAGGGCAAGCAGTGG-TATCAACGCAGAGT–3′, short primer = 5′–CTAATACGAC-TCACTATAGGGC–3′) and be specific to the gene of interest; both have 15 bp overlaps with the vector at their 5′ ends.
2. Total RNA or poly A^+ RNA is isolated from the established protocol following the manufacture's protocols (i.e., TRIzol®) (*see* **Note 6**). The quality of the RNA template can be assessed by Agilent 2100 Bioanalyzer (RNA integrity number should be 7 or higher), or you can visualize the RNA on a denaturing formaldehyde agarose gel under UV light. The theoretical 28S:18S ratio for eukaryotic RNA is approximately 2:1. If the 28S:18S ratio of the RNA is less than 1, the RNA template is not suitable for SMARTer RACE.
3. The two 20 μL reactions described in the protocol convert 10 ng to 1 μg of total or poly A+ RNA into RACE-Ready first-strand cDNA (*see* **Note 7**).
4. Carry out 5′-RACE and 3′-RACE PCR reactions which generate the 5′ and 3′ transcript fragments.
5. Based on the DNA fragments obtained from RACE PCR reactions. Primers of interest which can amplify the full-length transcript are designed (*see* **Note 8**). Finally the PCR products obtained were cloned and sequenced for sequence verification.

3.5 Generation of LINC-AP2 OE and KO Transgenic Plants

For functional characterization of LincRNA, similar to coding RNA, overexpression and knockout strategies are two crucial steps to investigate them.

1. Similar to protein-coding gene sequences, the functional lncRNA transcript sequences are amplified.
2. Based on the selected specific overexpression vector, restriction enzyme digestion and T4 ligation steps are set up for overexpression construct generation.
3. Depending on the overexpression vector of interest, gateway cloning strategy can be used to generate overexpression constructs.
4. For knocking down/knocking out the specific lincRNA of interest, the CRISPR/Cas9-mediated genome editing method can be used to precisely edit plant genome at specific loci.
5. The commercial or customized CRISPR/Cas9 vector with its unique promoters is used for construct generation.

6. The sgRNAs are designed using the web-based tool CRISPR-P (http://cbi.hzau.edu.cn/cgi-bin/CRISPR) [12] or any other available resources for sgRNAs design (*see* **Note 9**).
7. The sgRNAs with the highest score are selected and cloned into all-in-one CRISPR plasmid.
8. Finally, the lincRNA of interest obtained for overexpression or CRISPR/Cas9 constructs was transformed into *Arabidopsis* plants for transgenic plant generation.

4 Notes

1. Have a separate bench and/or pipette set dedicated to RNA work, free of RNase contamination. Wear gloves throughout to protect RNA samples from nucleases; use DEPC-treated water or freshly deionized (e.g., Milli-Q) H_2O.
2. To obtain good quality of RNA, the sample volume should not exceed 10% of the volume of used TRIzol®. Too much tissue used will cause incomplete dissociation of nucleoprotein complexes and result in low quality of RNA.
3. During phase separation step for RNA extraction, shake the tubes vigorously to make sure complete mixing.
4. It is crucial not to let the RNA pellet dry completely as this will greatly decrease its solubility. Partially dissolved RNA samples have an O.D. A260/280 ratio < 1.6.
5. The requirement for SMARTer RACE cDNA amplification is to know at least 23–28 nucleotides (nt) of sequence information in order to design gene-specific primers (GSPs) for the 5′- and 3′-RACE reactions.
6. For carrying out RACE PCR to amplify the full-length transcript of lncRNA, the integrity and purity of total RNA starting material are the important elements to obtain high-quality DNA.
7. Poly A+ RNA is recommended for cDNA synthesis. However, if less than 50 μg of total RNA is available, do not bother to purify poly A+ RNA because the final yield will be too small to analyze the RNA quantity and quality.
8. To characterize RACE products, first verify the amplified product of interest because multiple transcriptional start sites create a number of different transcripts.
9. In order to increase genome editing efficiency, the selection criteria for the interested sgRNAs such as the secondary structure for the folded RNA and GC content need to be evaluated. It is also crucial to validate the quality of the selected sgRNAs

using a range of calculation algorithms that are used in different programs, instead of relying on only one.

Acknowledgments

This work was supported by the Ministry of Education Tier 1 research grant R-154-000-A34-114 through the National University of Singapore (NUS) and NUS High School of Mathematics and Science.

References

1. Wang KC, Chang HY (2011) Molecular mechanisms of long noncoding RNAs. Mol Cell 43(6):904–914. https://doi.org/10.1016/j.molcel.2011.08.018
2. Au PC, Zhu QH, Dennis ES, Wang MB (2011) Long non-coding RNA-mediated mechanisms independent of the RNAi pathway in animals and plants. RNA Biol 8(3):404–414
3. Liu J, Wang H, Chua NH (2015) Long noncoding RNA transcriptome of plants. Plant Biotechnol J 13(3):319–328. https://doi.org/10.1111/pbi.12336
4. Ransohoff JD, Wei Y, Khavari PA (2018) The functions and unique features of long intergenic non-coding RNA. Nat Rev Mol Cell Biol 19:143–157. https://doi.org/10.1038/nrm.2017.104
5. Ulitsky I, Bartel DP (2013) lincRNAs: genomics, evolution, and mechanisms. Cell 154 (1):26–46. https://doi.org/10.1016/j.cell.2013.06.020
6. Deniz E, Erman B (2017) Long noncoding RNA (lincRNA), a new paradigm in gene expression control. Funct Integr Genomics 17 (2–3):135–143. https://doi.org/10.1007/s10142-016-0524-x
7. Cai L, Chang H, Fang Y, Li G (2016) A comprehensive characterization of the function of LincRNAs in transcriptional regulation through long-range chromatin interactions. Sci Rep 6:36572. https://doi.org/10.1038/srep36572
8. Ding X, Zhu L, Ji T, Zhang X, Wang F, Gan S, Zhao M, Yang H (2014) Long intergenic non-coding RNAs (LincRNAs) identified by RNA-seq in breast cancer. PLoS One 9(8): e103270. https://doi.org/10.1371/journal.pone.0103270
9. Tan G, Liu K, Kang J, Xu K, Zhang Y, Hu L, Zhang J, Li C (2015) Transcriptome analysis of the compatible interaction of tomato with Verticillium dahliae using RNA-sequencing. Front Plant Sci 6:428. https://doi.org/10.3389/fpls.2015.00428
10. Zuluaga AP, Vega-Arreguin JC, Fei Z, Matas AJ, Patev S, Fry WE, Rose JK (2016) Analysis of the tomato leaf transcriptome during successive hemibiotrophic stages of a compatible interaction with the oomycete pathogen Phytophthora infestans. Mol Plant Pathol 17 (1):42–54. https://doi.org/10.1111/mpp.12260
11. Jin J, Liu J, Wang H, Wong L, Chua NH (2013) PLncDB: plant long non-coding RNA database. Bioinformatics 29(8):1068–1071. https://doi.org/10.1093/bioinformatics/btt107
12. Lei Y, Lu L, Liu HY, Li S, Xing F, Chen LL (2014) CRISPR-P: a web tool for synthetic single-guide RNA design of CRISPR-system in plants. Mol Plant 7(9):1494–1496. https://doi.org/10.1093/mp/ssu044

Part IV

Identification and Functional Analysis of lncRNAs

Chapter 11

Bioinformatics Approaches to Studying Plant Long Noncoding RNAs (lncRNAs): Identification and Functional Interpretation of lncRNAs from RNA-Seq Data Sets

Hai-Xi Sun and Nam-Hai Chua

Abstract

Long noncoding RNAs (lncRNAs) play important roles in regulating various biological processes including growth and stress responses in plants. RNA-seq data sets provide a good resource to exploring the noncoding transcriptome and studying their comprehensive interactions with the coding transcriptome. Here, we describe computational procedures for studying plant lncRNAs including long intergenic noncoding RNAs (lincRNAs) and long noncoding natural antisense transcripts (lncNATs). Bioinformatics tools for transcriptome assembly, lncRNA identification, and functional interpretations are included. Finally, we also introduce PLncDB, a user-friendly database that provides comprehensive information of plant lncRNAs for researchers to compare their own data sets to those in public database.

Key words lincRNAs, lncNATs, RNA-seq, *cis*-regulation, lncRNA-miRNA interaction, RNA-DNA triplex

1 Introduction

With the advance of omics technology, more and more layers of the "dark matter" within the genomic DNA sequence are being explored. One such layer is the presence of long noncoding RNAs (lncRNAs) that are transcribed from the intergenic region (lincRNAs) or from the opposite DNA strand encoding sense transcripts (lncNATs) [1, 2]. In plants, lncRNAs have been shown to be involved in the regulation of diverse biological processes including vegetative growth [3, 4], flowering time [5–7], reproductive process [8, 9], biotic stress responses [10, 11], and abiotic stress responses [12, 13]. In addition, like coding genes, lncRNA genes are also associated with epigenetic marks indicating that their expression is precisely regulated [2, 11]. As transcriptional regulators, lncRNAs can regulate their target genes by altering their histone modification status and chromatin structures [5–7,

Julia A. Chekanova and Hsiao-Lin V. Wang (eds.), *Plant Long Non-Coding RNAs: Methods and Protocols*, Methods in Molecular Biology, vol. 1933, https://doi.org/10.1007/978-1-4939-9045-0_11,

14]. Furthermore, some lncRNAs can also interact with microRNAs (miRNAs) through imperfect pairing at the 5′ end 9th to 11th positions of miRNAs, a negative regulatory mechanism of miRNA functions termed target mimicry which was first described in *Arabidopsis* under phosphate starvation [12, 15].

In recent years following the accumulated discoveries of plant lncRNAs, several databases have been constructed to store and share information relating to plant lncRNAs [16–22]. Among them, PLncDB [18] provides not only lncRNA sequences and annotations but also a genome browser view for researchers to upload, visualize, and compare their own data to public data sets including lncRNAs expressed in different organs [1]; lncRNAs responsive to ABA, cold, drought, and salinity treatments [23]; lncRNAs associated with different histone modifications [24–26]; and lncRNAs embedded in conserved noncoding sequences of crucifer species [27]. All the above data sets can be downloaded for free from PLncDB.

The fast development and extensive application of high-throughput sequencing technology facilitate the exploration of the lncRNA world. Strand-specific RNA-sequencing provides not only the genomic coordinates of all expressed transcripts but also their directional information of the originating strand [28]. This information together with their transcript abundance allows us to systematically identify lncRNAs, determine their expression changes in different conditions, and study their interactions with other coding/noncoding transcripts. In this chapter we describe a step-by-step guide to studying lncRNAs including mapping and transcriptome assembly of RNA-seq data using HISAT2 and StringTie [29–31], identification of lncRNAs from assembled transcripts using getorf and CPC [32, 33], co-expression analysis of lncRNAs and their putative targets using R, investigating their possible interactions with miRNAs using psRobot [34] and RNA-DNA triplex prediction using Triplexator [35], and, finally, comparison to the public lncRNA data sets in PLncDB [18].

2 Materials

1. HISAT2: HISAT2 [29] is a fast and sensitive alignment program for mapping next-generation sequencing reads (both DNA and RNA) to a reference genome. HISAT2 is available for download from https://ccb.jhu.edu/software/hisat2/index.shtml.
2. StringTie: StringTie [30] is a fast and highly efficient assembler of RNA-seq alignments into potential transcripts. StringTie is available for download from https://ccb.jhu.edu/software/stringtie/.

3. Getorf: This program finds and outputs the sequences of open reading frames (ORFs) in one or more nucleotide sequences [32]. Getorf is attached from the EMBOSS package which is available for download from http://emboss.sourceforge.net/download/.
4. R: R is a free software environment for statistical computing and graphics. R is available for download from https://www.r-project.org/.
5. psRobot: psRobot [34] is a plant small RNA analysis software package. The module psRobot_tar is designed to find potential small RNA targets on a large scale. psRobot is available for download from http://omicslab.genetics.ac.cn/psRobot/index.php.
6. psMimic: psMimic is developed to predict plant miRNA endogenous target mimics in a genome-wide scale. psMimic is available for download from http://omicslab.genetics.ac.cn/psMimic/.
7. PLncDB: Plant long noncoding RNA database (PLncDB) attempts to provide comprehensive functions related to plant long noncoding RNAs [18]. PLncDB is available for access at http://chualab.rockefeller.edu/gbrowse2/homepage.html.
8. SAMtools: SAMtools [36] is a suite of programs for interacting with high-throughput sequencing data. SAMtools is available for download from http://www.htslib.org/.
9. Gffread: Gffread is used to generate a FASTA file with the DNA sequences for all transcripts in a GFF file. gffread is available for download from https://github.com/gpertea/gffread.
10. Triplexator: Triplexator [35] is a computational framework for RNA-DNA triplex prediction. Triplexator is available for download from http://bioinformatics.org.au/tools/triplexator/.
11. CPC: Coding potential calculator (CPC) is a support vector machine-based classifier to assess the protein-coding potential of a transcript based on six biologically meaningful sequence features [33]. CPC is available for download from http://cpc.cbi.pku.edu.cn/.
12. miRBase: The miRBase database [37] is a searchable database of published miRNA sequences and annotation. miRBase is available for access at http://www.mirbase.org/.

3 Methods

3.1 Transcriptome Assembly

3.1.1 Building Index of Reference Genome (hisat2-Build)

```
hisat2_extract_splice_sites.py annotation.gtf > ss.exon.txt
hisat2_extract_exons.py annotation.gtf > exon.exon.txt
hisat2-build --ss ss.exon.txt --exon exon.exon.txt -f genome.
fa index
```

3.1.2 Mapping Reads to Reference Genome (hisat2, See Note 1)

Single-end reads with strand-specific library:

```
hisat2 -p 12 -x index -S accepted_hits.NC.sam -q --no-unal --
dta --rna-strandness R -U SE_NC.fastq.gz
hisat2 -p 12 -x index -S accepted_hits.ABA.sam -q --no-unal --
dta --rna-strandness R -U SE_ABA.fastq.gz
```

Paired-end reads with strand-specific library:

```
hisat2 -p 12 -x index -S accepted_hits.NC.sam -q --no-unal --
dta --rna-strandness RF -1 PE_NC_1.fq.gz -2 PE_NC_2.fq.gz
      hisat2 -p 12 -x index -S accepted_hits.ABA.sam -q --no-
unal --dta --rna-strandness RF -1 PE_ABA_1.fq.gz -2 PE_ABA_2.
fq.gz
```

Single-end reads with nondirectional library:

```
hisat2 -p 12 -x index -S accepted_hits.NC.sam -q --no-unal --
dta -U SE_NC.fastq.gz
      hisat2 -p 12 -x index -S accepted_hits.ABA.sam -q --no-
unal --dta -U SE_ABA.fastq.gz
```

Paired-end reads with nondirectional library:

```
hisat2 -p 12 -x index -S accepted_hits.NC.sam -q --no-unal --
dta -1 PE_NC_1.fq.gz -2 PE_NC_2.fq.gz
      hisat2 -p 12 -x index -S accepted_hits.ABA.sam -q --no-
unal --dta -1 PE_ABA_1.fq.gz -2 PE_ABA_2.fq.gz
```

3.1.3 Sort the SAM File (SAMtools)

```
samtools view -Su accepted_hits.NC.sam | samtools sort -o
accepted_hits.sorted.NC.sam -O SAM
samtools view -Su accepted_hits.ABA.sam | samtools sort -o
accepted_hits.sorted.ABA.sam -O SAM
```

3.1.4 Assembling Transcripts of Each Sample (StringTie, See Note 2)

1. Strand-specific library:

```
stringtie -p 12 --rf -o transcripts.NC.gtf accepted_hits.sorted.NC.sam
stringtie -p 12 --rf -o transcripts.ABA.gtf accepted_hits.sorted.ABA.sam
```

2. Nondirectional Library:

```
stringtie -p 12 -o transcripts.NC.gtf accepted_hits.sorted.NC.sam
stringtie -p 12 -o transcripts.ABA.gtf accepted_hits.sorted.ABA.sam
```

3.1.5 Merging Transcripts from all Samples (Gffcompare, See Note 3)

```
gffcompare -r annotation.gtf -R -s genome.fa -i assembly_GTF_-list.merged.txt
```

3.1.6 Extracting Sequences of all Assembled Transcripts (Gffread)

```
gffread -w transcript.fa -g genome.fa gffcmp.combined.gtf
```

3.2 Identification of lncRNAs from Assembled Transcripts

3.2.1 Calculating Expression Levels of Assembled Transcripts (StringTie)

1. Strand-specific library:

```
stringtie --rf -e -B -A gene_abund.NC.tab -p 12 -G gffcmp.combined.gtf -o transcripts.NC.gtf accepted_hits.sorted.NC.sam
stringtie --rf -e -B -A gene_abund.ABA.tab -p 12 -G gffcmp.combined.gtf -o transcripts.ABA.gtf accepted_hits.sorted.ABA.sam
```

2. Nondirectional Library:

```
stringtie -e -B -A gene_abund.NC.tab -p 12 -G gffcmp.combined.gtf -o transcripts.NC.gtf accepted_hits.sorted.NC.sam
stringtie -e -B -A gene_abund.ABA.tab -p 12 -G gffcmp.combined.gtf -o transcripts.ABA.gtf accepted_hits.sorted.ABA.sam
```

3.2.2 Identifying lincRNAs and lncNATs (Getorf and CPC, See Note 4)

All assembled intergenic transcription units are collected as lincRNA candidates. For strand-specific RNA-seq, lncNAT candidates are defined as transcription units transcribing from the opposite strand of an mRNA coding gene and having at least 50-nt-long overlapping region with the sense transcript. Candidates with length $\geq$ 200 nt, predicted ORF $\leq$100 amino acids and without coding potential (CPC score $<$ 0) are defined as lincRNAs and lncNATs. ORFs are predicted using getorf (getorf -sequence lncRNA.fa -outseq ORF.fa). Coding potential is predicted using CPC (run_predict.sh lncRNA.fa cpc.txt).

3.3 Functional Interpretation of lncRNAs

3.3.1 Identifying Putative Cis-Regulatory lncRNAs (R)

To identify putative *cis*-regulatory lincRNAs, the Pearson correlation coefficients (PCCs) are calculated using FPKMs of the lincRNAs vs. mRNAs of their neighboring upstream and downstream genes. Only neighboring genes with $r^2 \geq 0.6$ are considered as putative targets of the *cis*-regulatory lincRNAs. Similarly, putative *cis*-regulatory lncNATs are identified using the PCCs of these lncNATs vs. the corresponding sense mRNAs, and only mRNAs with $r^2 \geq 0.6$ are considered as putative targets of the *cis*-regulatory lncNATs.

3.3.2 Identifying lncRNA's Function as Putative Endogenous Target Mimics (psMimic, See Note 5)

```
psRobot_mim -s miRNA.fa -t lncRNA.fa -p 12 -o mimic.pTM
```

3.3.3 Identifying Putative lncRNAs Targeted by miRNAs (psRobot, See Note 5)

```
psRobot_tar -s miRNA.fa -t lncRNA.fa -o target.gTP -ts 2.5 -fp
2 -tp 17 -gl 17 -p 12 -gn 1
```

3.3.4 Identifying Putative Triplex-Forming lncRNAs (Triplexator)

```
triplexator -e 10 -l 13 -g 50 -ss lncRNA.fa -ds double_-
strand_DNA.fa -o triplex.txt -of 1
```

3.4 Comparison to Public Data Sets in PLncDB (See Note 6)

1. Uploading custom tracks: Upload your own data to PLncDB at http://chualab.rockefeller.edu/cgi-bin/gb2/gbrowse/arabidopsis/.
2. Comparing to public data sets: By clicking the "Select Tracks" button, you can compare your lncRNAs to (a) known lncRNA data sets including RepTAS lncRNAs, EST lncRNAs, Okamoto2010 lncRNAs, RNA-seq lncRNAs, Matsui2008 lncRNAs, and NATs [1]; (b) public abiotic stress data sets including ABA, cold, drought, and salinity treatments [23]; (c) public histone modification data sets in order to identify lncRNAs associated with histone modifications [24–26]; and (d) conserved noncoding sequences of crucifer species [27].

4 Notes

1. When mapping RNA-seq reads to reference genome, it is better to refine the maximum intron length (default, 500,000 bp) based on the annotated transcripts to avoid false positives. For

example, in *Arabidopsis* because 99.9% of TAIR10 annotated intron sequences are 40–5000 bp in length, we therefore specify the intron size to 40–5000 bp in order to improve the mapping accuracy of splice junctions (--min-intronlen 40 --max-intronlen 5000).

2. When assembling transcripts using StringTie, you can add "-t" to disable trimming at the ends of the assembled transcripts in order to get the full length of the transcripts especially lncRNAs.
3. After merging transcripts from all samples using gffcompare, it is better to remove the transcript whose class code is "s" because this code means an intron of the transcript overlaps a reference intron on the opposite strand, which is likely due to incorrect direction of the reads.
4. We do not think nondirectional RNA-seq data are useful for investigations on lncNATs because these data do not allow us to distinguish between the Watson and the Crick strand-originated reads.
5. The two input files (lncRNA.fa and miRNA.fa) of psRobot_mim and psRobot_tar are sequences of lncRNAs and mature miRNAs. The mature miRNA sequences are available for download from miRBase.
6. Integrative Genomics Viewer (IGV, http://software.broadinstitute.org/software/igv/) is a high-performance visualization tool for interactive exploration of large, integrated genomic data sets [38]. Using IGV you can visualize your RNA-seq data in your local computer. To do this, you need to convert your sam file to bam file and then build bam index using SAMtools (samtools view -bS accepted_hits.sorted.sam > accepted_hits.sorted.bam && samtools index accepted_hits.sorted.bam). You can also see your assembled transcripts by loading the GTF file (gffcmp.combined.gtf).

Acknowledgment

We thank Jun Liu and Huan Wang for developing the above bioinformatics methods for lincRNA and lncNAT analyses and Jingjing Jin for constructing PLncDB. This work was funded in part by Singapore NRF RSSS Grant NRF-RSSS-002.

References

1. Liu J, Jung C, Xu J, Wang H, Deng S, Bernad L, Arenas-Huertero C, Chua NH (2012) Genome-wide analysis uncovers regulation of long intergenic noncoding RNAs in Arabidopsis. Plant Cell 24(11):4333–4345. https://doi.org/10.1105/tpc.112.102855
2. Wang H, Chung PJ, Liu J, Jang IC, Kean MJ, Xu J, Chua NH (2014) Genome-wide

identification of long noncoding natural antisense transcripts and their responses to light in Arabidopsis. Genome Res 24(3):444–453. https://doi.org/10.1101/gr.165555.113

3. Ariel F, Jegu T, Latrasse D, Romero-Barrios N, Christ A, Benhamed M, Crespi M (2014) Noncoding transcription by alternative RNA polymerases dynamically regulates an auxin-driven chromatin loop. Mol Cell 55(3):383–396. https://doi.org/10.1016/j.molcel.2014.06.011
4. Bardou F, Ariel F, Simpson CG, Romero-Barrios N, Laporte P, Balzergue S, Brown JW, Crespi M (2014) Long noncoding RNA modulates alternative splicing regulators in Arabidopsis. Dev Cell 30(2):166–176. https://doi.org/10.1016/j.devcel.2014.06.017
5. Swiezewski S, Liu F, Magusin A, Dean C (2009) Cold-induced silencing by long antisense transcripts of an Arabidopsis Polycomb target. Nature 462(7274):799–802. https://doi.org/10.1038/nature08618
6. Heo JB, Sung S (2011) Vernalization-mediated epigenetic silencing by a long intronic noncoding RNA. Science 331 (6013):76–79. https://doi.org/10.1126/science.1197349
7. Kim DH, Sung S (2017) Vernalization-triggered intragenic chromatin loop formation by long noncoding RNAs. Dev Cell 40 (3):302–312 e304. https://doi.org/10.1016/j.devcel.2016.12.021
8. Zhang YC, Liao JY, Li ZY, Yu Y, Zhang JP, Li QF, Qu LH, Shu WS, Chen YQ (2014) Genome-wide screening and functional analysis identify a large number of long noncoding RNAs involved in the sexual reproduction of rice. Genome Biol 15(12):512. https://doi.org/10.1186/s13059-014-0512-1
9. Ding J, Lu Q, Ouyang Y, Mao H, Zhang P, Yao J, Xu C, Li X, Xiao J, Zhang Q (2012) A long noncoding RNA regulates photoperiod-sensitive male sterility, an essential component of hybrid rice. Proc Natl Acad Sci U S A 109 (7):2654–2659. https://doi.org/10.1073/pnas.1121374109
10. Seo JS, Sun HX, Park BS, Huang CH, Yeh SD, Jung C, Chua NH (2017) ELF18-INDUCED LONG-NONCODING RNA associates with mediator to enhance expression of innate immune response genes in Arabidopsis. Plant Cell 29(5):1024–1038. https://doi.org/10.1105/tpc.16.00886
11. Zhu QH, Stephen S, Taylor J, Helliwell CA, Wang MB (2014) Long noncoding RNAs responsive to Fusarium oxysporum infection in Arabidopsis thaliana. New Phytol 201 (2):574–584. https://doi.org/10.1111/nph.12537
12. Franco-Zorrilla JM, Valli A, Todesco M, Mateos I, Puga MI, Rubio-Somoza I, Leyva A, Weigel D, Garcia JA, Paz-Ares J (2007) Target mimicry provides a new mechanism for regulation of microRNA activity. Nat Genet 39(8):1033–1037. https://doi.org/10.1038/ng2079
13. Di C, Yuan J, Wu Y, Li J, Lin H, Hu L, Zhang T, Qi Y, Gerstein MB, Guo Y, Lu ZJ (2014) Characterization of stress-responsive lncRNAs in Arabidopsis thaliana by integrating expression, epigenetic and structural features. Plant J 80(5):848–861. https://doi.org/10.1111/tpj.12679
14. Liu F, Marquardt S, Lister C, Swiezewski S, Dean C (2010) Targeted 3′ processing of antisense transcripts triggers Arabidopsis FLC chromatin silencing. Science 327 (5961):94–97. https://doi.org/10.1126/science.1180278
15. Wu HJ, Wang ZM, Wang M, Wang XJ (2013) Widespread long noncoding RNAs as endogenous target mimics for microRNAs in plants. Plant Physiol 161(4):1875–1884. https://doi.org/10.1104/pp.113.215962
16. Berardini TZ, Reiser L, Li D, Mezheritsky Y, Muller R, Strait E, Huala E (2015) The Arabidopsis information resource: making and mining the "gold standard" annotated reference plant genome. Genesis 53(8):474–485. https://doi.org/10.1002/dvg.22877
17. Cheng CY, Krishnakumar V, Chan AP, Thibaud-Nissen F, Schobel S, Town CD (2017) Araport11: a complete reannotation of the Arabidopsis thaliana reference genome. Plant J 89(4):789–804. https://doi.org/10.1111/tpj.13415
18. Jin J, Liu J, Wang H, Wong L, Chua NH (2013) PLncDB: plant long non-coding RNA database. Bioinformatics 29(8):1068–1071. https://doi.org/10.1093/bioinformatics/btt107
19. Zhao Y, Li H, Fang S, Kang Y, Wu W, Hao Y, Li Z, Bu D, Sun N, Zhang MQ, Chen R (2016) NONCODE 2016: an informative and valuable data source of long non-coding RNAs. Nucleic Acids Res 44(D1):D203–D208. https://doi.org/10.1093/nar/gkv1252
20. Szczesniak MW, Rosikiewicz W, Makalowska I (2016) CANTATAdb: a collection of plant long non-coding RNAs. Plant Cell Physiol 57 (1):e8. https://doi.org/10.1093/pcp/pcv201

21. Yi X, Zhang Z, Ling Y, Xu W, Su Z (2015) PNRD: a plant non-coding RNA database. Nucleic Acids Res 43(Database issue): D982–D989. https://doi.org/10.1093/nar/gku1162
22. The RC (2017) RNAcentral: a comprehensive database of non-coding RNA sequences. Nucleic Acids Res 45(D1):D128–D134. https://doi.org/10.1093/nar/gkw1008
23. Matsui A, Ishida J, Morosawa T, Mochizuki Y, Kaminuma E, Endo TA, Okamoto M, Nambara E, Nakajima M, Kawashima M, Satou M, Kim JM, Kobayashi N, Toyoda T, Shinozaki K, Seki M (2008) Arabidopsis transcriptome analysis under drought, cold, high-salinity and ABA treatment conditions using a tiling array. Plant Cell Physiol 49(8):1135–1149. https://doi.org/10.1093/pcp/pcn101
24. Oh S, Park S, van Nocker S (2008) Genic and global functions for Paf1C in chromatin modification and gene expression in Arabidopsis. PLoS Genet 4(8):e1000077. https://doi.org/10.1371/journal.pgen.1000077
25. Zhang X, Bernatavichute YV, Cokus S, Pellegrini M, Jacobsen SE (2009) Genome-wide analysis of mono-, di- and trimethylation of histone H3 lysine 4 in Arabidopsis thaliana. Genome Biol 10(6):R62. https://doi.org/10.1186/gb-2009-10-6-r62
26. Charron JB, He H, Elling AA, Deng XW (2009) Dynamic landscapes of four histone modifications during deetiolation in Arabidopsis. Plant Cell 21(12):3732–3748. https://doi.org/10.1105/tpc.109.066845
27. Haudry A, Platts AE, Vello E, Hoen DR, Leclercq M, Williamson RJ, Forczek E, Joly-Lopez Z, Steffen JG, Hazzouri KM, Dewar K, Stinchcombe JR, Schoen DJ, Wang X, Schmutz J, Town CD, Edger PP, Pires JC, Schumaker KS, Jarvis DE, Mandakova T, Lysak MA, van den Bergh E, Schranz ME, Harrison PM, Moses AM, Bureau TE, Wright SI, Blanchette M (2013) An atlas of over 90,000 conserved noncoding sequences provides insight into crucifer regulatory regions. Nat Genet 45(8):891–898. https://doi.org/10.1038/ng.2684
28. Levin JZ, Yassour M, Adiconis X, Nusbaum C, Thompson DA, Friedman N, Gnirke A, Regev A (2010) Comprehensive comparative analysis of strand-specific RNA sequencing methods. Nat Methods 7(9):709–715. https://doi.org/10.1038/nmeth.1491
29. Kim D, Langmead B, Salzberg SL (2015) HISAT: a fast spliced aligner with low memory requirements. Nat Methods 12(4):357–360. https://doi.org/10.1038/nmeth.3317
30. Pertea M, Pertea GM, Antonescu CM, Chang TC, Mendell JT, Salzberg SL (2015) StringTie enables improved reconstruction of a transcriptome from RNA-seq reads. Nat Biotechnol 33 (3):290–295. https://doi.org/10.1038/nbt.3122
31. Pertea M, Kim D, Pertea GM, Leek JT, Salzberg SL (2016) Transcript-level expression analysis of RNA-seq experiments with HISAT, StringTie and Ballgown. Nat Protoc 11 (9):1650–1667. https://doi.org/10.1038/nprot.2016.095
32. Rice P, Longden I, Bleasby A (2000) EMBOSS: the European molecular biology open software suite. Trends Genet 16 (6):276–277
33. Kong L, Zhang Y, Ye ZQ, Liu XQ, Zhao SQ, Wei L, Gao G (2007) CPC: assess the protein-coding potential of transcripts using sequence features and support vector machine. Nucleic Acids Res 35(Web Server issue): W345–W349. https://doi.org/10.1093/nar/gkm391
34. Wu HJ, Ma YK, Chen T, Wang M, Wang XJ (2012) PsRobot: a web-based plant small RNA meta-analysis toolbox. Nucleic Acids Res 40 (Web Server issue):W22–W28. https://doi.org/10.1093/nar/gks554
35. Buske FA, Bauer DC, Mattick JS, Bailey TL (2012) Triplexator: detecting nucleic acid triple helices in genomic and transcriptomic data. Genome Res 22(7):1372–1381. https://doi.org/10.1101/gr.130237.111
36. Li H, Handsaker B, Wysoker A, Fennell T, Ruan J, Homer N, Marth G, Abecasis G, Durbin R, Genome Project Data Processing Subgroup (2009) The sequence alignment/map format and SAMtools. Bioinformatics 25 (16):2078–2079. https://doi.org/10.1093/bioinformatics/btp352
37. Kozomara A, Griffiths-Jones S (2014) miRBase: annotating high confidence microRNAs using deep sequencing data. Nucleic Acids Res 42(Database issue):D68–D73. https://doi.org/10.1093/nar/gkt1181
38. Robinson JT, Thorvaldsdottir H, Winckler W, Guttman M, Lander ES, Getz G, Mesirov JP (2011) Integrative genomics viewer. Nat Biotechnol 29(1):24–26. https://doi.org/10.1038/nbt.1754

Chapter 12

Identification of Novel lincRNA and Co-Expression Network Analysis Using RNA-Sequencing Data in Plants

Song Qi, Shamima Akter, and Song Li

Abstract

Long intergenic noncoding RNA (lincRNA) plays important biological functions in plants. Identification and annotation of lincRNA in plants largely rely on RNA sequencing followed by computational analysis. In this protocol, we describe a multistep computational pipeline for lincRNA identification using RNA-sequencing data. This pipeline can also construct co-expression network that is made of both lincRNA and mRNA genes. The co-expression network generated by this pipeline can be used to provide putative annotation of lincRNAs that have no known biological functions.

Key words Co-expression network, lincRNA, Noncoding RNA, Plant

1 Introduction

Long intergenic noncoding RNAs (lincRNAs) are commonly found in multicellular organisms including both plants and animals [1–4]. In plants, except for a few cases, most lincRNAs have no known biological function. Therefore, it is important to systematically predict the function of all known lincRNAs in plants to guide experimental characterization. In recent years, both short- and long-read sequencing technologies have been widely applied in identification of lincRNAs in many plant species. Bioinformatic algorithms and computational tools play indispensable roles in the analysis of high-throughput sequencing (HTS) data and whole-genome tilling arrays to predict lincRNAs [1, 4]. Because HTS can be applied in species with and without reference genomes, different computational strategies have to be applied to identify lincRNAs in these situations. Short-read sequencing typically generates reads (50–150 base pairs) that are much shorter than full-length transcripts; therefore, multiple reads must be used to reconstruct a full-length lincRNA. Long-read sequencing can produce reads that are much longer than a typical transcript; however,

Julia A. Chekanova and Hsiao-Lin V. Wang (eds.), *Plant Long Non-Coding RNAs: Methods and Protocols*,
Methods in Molecular Biology, vol. 1933, https://doi.org/10.1007/978-1-4939-9045-0_12,

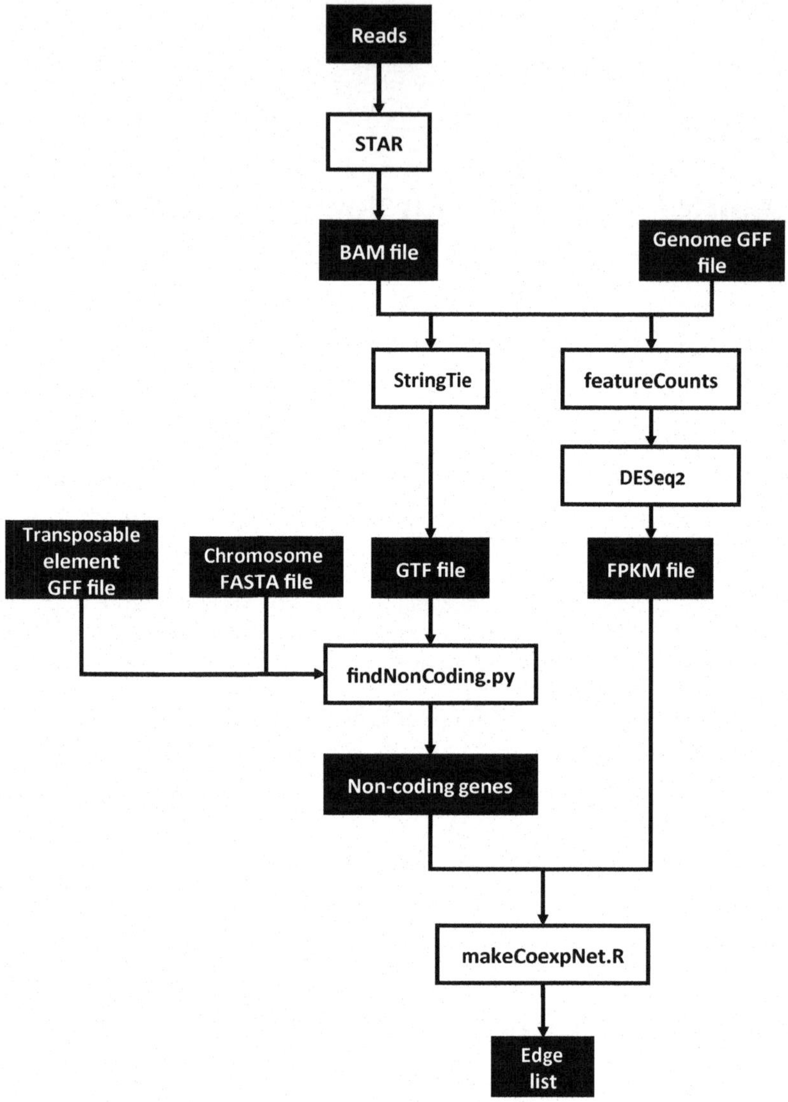

Fig. 1 The analysis pipeline for identification of lincRNA from RNA-seq data

currently, the sequencing depth of long-read technologies is still lower as compared to short reads. This could reduce the sensitivity of detecting lowly expressed lincRNAs using long-read technologies. In this book chapter, we focus on the computational pipeline to identify new lincRNAs using RNA-seq data and to generate co-expression networks of lincRNAs and mRNAs (*see* Fig. 1).

This pipeline for lincRNA identification and co-expression analysis includes three major steps. The first step is mapping of sequencing reads to the reference genome and to reconstruct transcripts using mapped reads. The second step is to identify novel lincRNA from these reconstructed transcripts, following published method [1]. Finally, the expression matrix of lincRNA and protein-coding genes will be compared using co-expression analysis. A co-expression network will be generated for visualization. This pipeline was developed in the model plant *Arabidopsis*, using root cell-type-specific gene expression data. The same pipeline can be used to identify lincRNAs in other plant species such as soybean and maize, where RNA-seq has been generated [5–7]. Multiple genomic features can be used to further refine the prediction of lincRNAs [8], but this is not the focus of this chapter. This pipeline also relies on an annotated reference genome. For species without reference genome, the Python script provided in this protocol can be easily modified to identify long noncoding transcripts.

2 Materials

All scripts used in this analysis can be obtained from GitHub using the following command (*see* **Note 1**). The "$" means the command is executed under a Linux terminal (*see* **Note 2**).

```
$ git clone https://github.com/LiLabAtVT/LincRNA_MIMB.gi-
tLincRNA_ATH
```

The user can replace "LincRNA_ATH" with another folder name that better represents your project. All scripts in this project are tested under the project folder created by the "git clone" command (default LincRNA_ATH).

This protocol was tested under CentOS 7 (Linux), which supports command-line interface that is commonly found in UNIX-type operating system. The steps described in this protocol can be used in most Linux operating systems and MacOSX. For Windows users, we suggest the user to install lightweight virtual machine such as Docker (https://www.docker.com). A Linux system can be installed in the virtual machine and perform the analysis under the virtual environment. In this protocol, we will install multiple software packages for the analysis, including STAR [9] for read mapping, featureCounts [10] for counting reads, and StringTie [11] for transcript reconstruction. We will require R (*see* **Note 3**) and Python (*see* **Note 4**) programming languages to be installed for running the scripts for lincRNA identification and co-expression network analysis.

2.1 Setup Folder Structure

There are three steps for this analysis. The first step is processing of RNA-seq data, which includes mapping reads, reconstructing isoforms, and generating a consensus gene transfer format (GTF) file for all transcripts identified. The second step is lincRNA identification. We have implemented the rules for identification of lincRNAs using a Python script. The third step is to perform co-expression analysis using expression data from both protein-coding genes and lincRNAs. We suggest the user to set up three folders. All data and scripts for performing the analysis in each step will be included in one folder. Using the "git clone" command will automatically create the three folders for the user.

The folder structure can also be easily created using simple "mkdir" command. We will create three folders for the three steps of the analysis and one folder to install software for this analysis.

```
$ cd LincRNA_ATH
$ mkdir step01 step02 step03
$ mkdir software
```

2.2 Download and Install Software for Analysis

2.2.1 Download and Install STAR, featureCounts, StringTie, and SRA Toolkit

In this analysis, we will use STAR for read mapping, featureCounts for counting number of reads mapped to each gene, StringTie for transcript reconstruction, and DESeq2 [12] and edgeR [13] for calculation of gene expression levels (*see* **Note 5**). To obtain the raw data to repeat the analysis in this chapter, the user also needs to download the SRA Toolkit which provides fast download of sequencing data from NCBI short-read archive (SRA) database [14].

1. To download and install STAR, run the following scripts (*see* **Note 6**):

```
$ cd LincRNA_ATH/software
$ wget https://github.com/alexdobin/STAR/archive/2.5.2b.
tar.gz
$ tar -xzf 2.5.2b.tar.gz
$ STAR-2.5.2b/bin/Linux_x86_64_static/STAR --version
```

2. To download and install featureCounts, run the following scripts (*see* **Note 7**):

```
$ wget https://sourceforge.net/projects/subread/files/sub-
read-1.5.1/subread-1.5.1-Linux-x86_64.tar.gz/download
$ tar -zxvf download
$ subread-1.5.1-Linux-x86_64/bin/featureCounts -v
```

3. To download and install StringTie, run the following scripts:

```
$ wget http://ccb.jhu.edu/software/stringtie/dl/stringtie-
1.3.3b.Linux_x86_64.tar.gz
$ tar -zxvf stringtie-1.3.3b.Linux_x86_64.tar.gz
$ stringtie-1.3.3b.Linux_x86_64/stringtie --version
```

4. To download and install SRA Toolkit, run the following scripts:

```
$ wget http://ftp-trace.ncbi.nlm.nih.gov/sra/sdk/current/
sratoolkit.current-centos_linux64.tar.gz
$ tar -xzf sratoolkit.current-centos_linux64.tar.gz
$ ./sratoolkit.2.8.2-1-centos_linux64/bin/fastq-dump --
version
```

2.2.2 Installation of Required R Packages

Unlike the last four software tools, DESeq2 and edgeR are R packages and can be installed in the R interactive programming environment. To install DESeq2 and edgeR, use the following command under R environment. You can start R environment in Linux command line by typing "r" followed by hitting the "enter" key. DESeq2 and edgeR are part of the bioconductor system [15] and should be installed using the installer provided by bioconductor.

```
> source("https://bioconductor.org/biocLite.R")
> biocLite("DESeq2")
> biocLite("edgeR")
```

2.2.3 Installation of Required Python Packages

Before running the script, two Python packages (*see* **Note 8**), Pandas and Biopython, need to be installed. The two packages can be conveniently installed using pip or conda. Please refer to https://pip.pypa.io/en/stable/installing/ for installing or upgrading pip and https://conda.io/docs/user-guide/install/index.html for installing conda.

To install Pandas and Biopython through pip, use the two commands below:

```
$ pip install pandas.
$ pip install biopython.
```

Alternatively, if you prefer using conda to install and manage the packages, use the two commands below:

```
$ conda install pandas.
$ conda install -c anaconda biopython.
```

All package dependencies will be automatically handled by pip or conda.

2.3 Download Genome Annotation and TE Annotation

We will use the genomic sequence of *Arabidopsis thaliana* for read mapping (for other species, *see* **Note 9**). Genomic sequence for *Arabidopsis* can be downloaded from the Araport website (www.araport.org). Araport is a data portal for *Arabidopsis* genomic research that hosts the latest genomic sequences and genome annotations for this model organism [16]. The latest version of the genome sequence of *Arabidopsis* is "TAIR10_Chr.all.fasta.gz." This file is unlikely to change because the genome assembly of

Arabidopsis is likely to remain the same in the future. The latest version of the gene annotation file is "Araport11_GFF3_genes_-transposons.201606.gtf.gz."

2.4 Download Raw Data for Analysis

RNA-seq reads for root cell type-specific gene expression (*see* **Note 10**) can be downloaded from NCBI, SRA database, using this link: https://www.ncbi.nlm.nih.gov/bioproject/323955. A list of sample IDs can be downloaded into a text file. To download one file at a time, one can use the following command. Simple bash script with a for loop can be used to download a list of files. Using ascp (http://downloads.asperasoft.com) can significantly speed up the downloading process.

```
$ fastq-dump -I -split-files SRR3664408
```

3 Methods

3.1 Read Mapping and Isoform Reconstruction

There are many established protocols and pipelines for read mapping and isoform reconstruction. Our current pipeline uses STAR for read mapping and StringTie for isoform reconstruction. Other pipelines may produce different results. Comparing and merging results from multiple pipelines may provide a better representation of the actual underlying transcriptome [17].

The command for mapping reads with STAR and reconstructing transcripts using StringTie is provided in the GitHub repository. In brief, mapping reads with STAR require two steps. First, a genome index has to be constructed using STAR. Second, STAR will map the reads to the genome index and generate read mapping file in the bam format. Each sample will generate one single bam file. We sequenced 15 root cells with three biological replicates for each cell type, which will produce 45 bam files. For transcript reconstruction, there are also two steps. First, StringTie will be performed on each sample from the mapped reads. Each sample (bam file) will generate one GTF file that represents the transcripts detected in the sample. Second, StringTie merge command will be used to merge all reconstructed transcript models into a single output file, which is also a GTF file. This GTF file will be used as input file for the analysis in the next step.

1. Command for genome index creation using STAR (*see* **Note 11**):

```
$ STAR --runThreadN 16 \
 --runMode genomeGenerate \
 --genomeDir Gnmdir \
 --genomeFastaFiles TAIR10_Chr.all.fasta \
 --sjdbGTFfile Araport11_GFF3_genes_transposons.201606.gtf
```

In this command, "\" is a special Linux operator, which indicates all the white space after this symbol should be ignored. This means the command is not finished in the current line and will continue in the next line. The user should replace "Gnmdir" with a target directory name for the genomic index file. The user should also make sure the "TAIR10_Chr.all.fasta" and "Araport11_GFF3_genes_transposons.201606.gtf" files are in the current working directory.

2. Command for read mapping using STAR:

```
$ STAR --runThreadN 16 \
 --genomeDir Gnmdir \
 --readFilesIn SRR3664408.1.fastq SRR3664408.2.fastq \
 --outSAMstrandField intronMotif \
 --outFileNamePrefix output/bam/SRR3664408 \
 --outSAMtype BAM SortedByCoordinate;
```

In this command, "Gnmdir" should be the directory where the previous command has used to store genomic index. This command is using SRR3664408 as an example. We used paired-end sequencing, such that there will be two fastq files associated with this SRR ID.

3. Command for transcript reconstruction (GTF file creation):

```
$ stringtie SRR3664408.bam \
 -o SRR3664408.gtf \
 -p 4 \
 -G Araport11_GFF3_genes_transposons.201606.gtf
```

4. Command for merging GTF files:

```
$ stringtie --merge \
 -G Araport11_GFF3_genes_transposons.201606.gtf \
 -o merged.Arath.gtf ../gtf_list.txt
```

5. In the gtf_list.txt file, all files assembled using the previous StringTie command should be included. This file can be easily generated using Linux command:

```
$ ls SRR*.gtf > gtf_list.txt
```

3.2 Identification of lincRNA

To identify lincRNA from reconstructed transcript annotation GTF files, the user needs to download the two Python scripts (myGTF.py and findNonCoding.py) from the GitHub repository. The myGTF.py defines functions for parsing the GTF file; the findNonCoding.py uses these functions to parse input file and find noncoding genes.

1. The findNonCoding.py requires four input files (some of these files have been downloaded in Subheading 2.3):

   A GTF file containing annotations for genes of interest

   BED format file that stores annotations for protein-coding genes

   BED format file that stores annotations for transposable elements

   FASTA format file for assembled chromosome sequences (*see* **Note 12**)

2. The findNonCoding.py script will ignore any transcript ID with the format of "ATXGXXXXX" and take other transcripts to check if they are the noncoding transcripts (*see* **Note 13**). This is because the purpose of this script is to identify new lincRNAs. In the downloaded annotation file, all known lincRNA and protein-coding genes will have an "ATXGXXXXX" ID. Other transcripts that are identified by StringTie but do not overlap with the existing annotated genes will not have this type of ID.

3. The output file of findNonCoding.py is a list of identified noncoding gene IDs. Gene is considered as a noncoding gene when all isoforms of that gene are noncoding transcripts. The noncoding transcripts are selected from the input GTF file using the four criteria as described in our previous publication [1]:

   Mature transcript length > 200 bps (including UTRs and exons, excluding introns).

   Longest open reading frame does not encode more than 100 amino acids.

   Transcribed regions do not overlap with any transposable elements.

   Transcribed regions are 500 bp away from any protein-coding gene.

4. Before running the script, make sure myGTF.py and findNonCoding.py are put into the same directory; otherwise findNonCoding.py will fail to load the required functions. To use the script, the user can use -h / --help option that can be used to display usage and help information for findNonCoding.py:

```
$ python findNonCoding.py -help
usage: python findNonCoding.py [-h] -o OUTFILE GTF_file
Pro_BED Transpos_BED FASTA
```

Identify noncoding genes
Positional arguments:
GTF_file GTF file for transcripts of interest
Pro_BED BED file for protein-coding genes
Transpos_BED BED file for transposable elements
FASTA FASTA file for chromosome sequences
Optional arguments:
-h, --help show this help message and exit
-o OUTFILE, --out OUTFILE Output file name

5. An example of running the script also provided here:

```
$ python findNonCoding.py -o out \
 merged.Arath.gtf Araport11_protein_coding.201606.bed Ara-
port11_transposable_element_gene.201606.bed TAIR10_Chr.all.fasta
```

In this command, -o\--out specifies the output file name. In this example, the output gene list was saved into a file named "out."

3.3 Read Counting and FPKM Calculation

There are two steps in generating gene expression levels from RNA-seq data. The first step is to count how many reads mapped to each genomic feature, which, in this analysis, includes both transcribed regions for both protein-coding genes (mRNA) and lincRNA genes. The second step is to calculate gene expression level in the format of FPKM.

3.3.1 Read Counting

The command to count reads using featureCounts is the following (*see* **Note 14**):

```
$ featureCounts -T 8 \
 -t exon \
 -g gene_id \
 -p \
 -a $GTF \
 -o out.readcount.txt \
 $mappingdir/$file 1
```

The command to calculate FPKM expression is included in the edgeR and DESeq2 package. This chapter was developed when FPKM function is not included in the DESeq2 package; therefore, the example code is provided using an edgeR function. However, it is straightforward to implement an R script to perform FPKM calculation using DESeq2.

3.3.2 FPKM Calculation

The full script to generate FPKM from read count files is included in the GitHub repository. Here we only show the key commands (*see* **Note 15**):

```
dds = DESeqDataSetFromMatrix(countData = InputDF2,
colData = sampleInfo,
design = ~Treatment)
dds<-DESeq(dds)
dds.EstSizFac <- estimateSizeFactors(dds)
norm.count=counts(dds.EstSizFac, normalized=TRUE)
AnnoData <- read.table(GeneidLengthFile,
sep='\t',
as.is=T,
header=T)
norm.count.DGEList <- DGEList(counts=norm.count,
genes = AnnoData[,c("Geneid","Length")])
norm.rpkm <- rpkm(norm.count.DGEList,
norm.count.DGEList$genes$Length)
```

These commands are executed in the R interactive programming environment. "dds" is an R object which contains DESeq2 normalized read counts. "AnnoData" is a table which contains the information of gene length. "norm.count.DEGList" is an R object which contains read count data as well as gene length information. This function "DEGlist" is from edgeR package, which is to generate the DEGlist object (norm.count.DEGList). "rpkm" function is then used to calculate normalized FPKM values from gene length and read count table.

3.4 Co-expression Analysis

Co-expression network can be generated using an R script provided in the GitHub repository. The script, makeCoexpNet.R, needs two input files: [1] a FPKM file containing the expression level of lncRNAs and [2] a FPKM file containing the expression level of protein-coding genes. Additionally, one parameter k, which specifies the number of desired clusters, and one argument giving the output file name should be provided. The script, makeCoexpNet.R, will process the input files and construct co-expression network by the following four steps (*see* **Note 16**):

1. Remove those having expression <0.05 across all columns (conditions) in lncRNA FPKM file.
2. For protein-coding gene FPKM file, calculate variance of each gene, and keep the top 5000 genes with greatest variance.

3. Perform K-means clustering using filtered data set of lncRNAs and protein-coding genes; the number of cluster is specified by command line argument.
4. Find one cluster with at least one lncRNA and one protein-coding gene with a total number of nodes <100. Calculate Pearson correlation coefficient (PCC) for each gene pair and p values for the PCC using Student's t-distribution. Select gene pairs with p values <0.05. An edge list of these gene pairs will then be generated and saved into the output file.

   Below is an example of running the Rscript:

```
$ Rscript makeCoexpNet.R AllLincAveFPKM.v4.1.csv \
 aveFPKM.pro.v4.csv \
 60 \
 edgeList.csv
```

This will generate the co-expression network and save the edge list into edgeList.csv. This edgeList.csv file can be used to visualize the co-expression network using software such as Cytoscape [18].

4 Notes

1. Software installation: The git software is installed in most Linux systems by default. If git is not installed in your system, please refer to https://git-scm.com for installation instructions.
2. Code blocks: All code blocks started with "$" are command line scripts that should be executed under a Linux terminal. All code blocks started with ">" are command line scripts that should be executed under an interactive R programming language console.
3. For Linux and MacOSX users, R can be installed by downloading the installation package from the following website (https://cran.r-project.org/mirrors.html). The reader can select a mirror site that is closest to the reader's location and follow the download link to download the installation package for the corresponding platform. For Windows users, if the reader prefers to perform analysis in Windows, the R program language provides native installer and interface for Windows. The reader can also perform the analysis within the Docker virtual machine, which requires R installation following a standard Linux installation procedure.
4. For Linux and MacOSX users, Python can be installed by downloading the installation package from the following website (https://www.python.org/downloads/). For Windows users, similar to R installation, if the reader prefers to perform

analysis in Windows, the Python program language provides native installer and interface for Windows. The reader can also perform the analysis within the Docker virtual machine, which requires Python installation following a standard Linux installation procedure.

5. In this protocol, STAR, featureCounts, StringTie, and SRA Toolkit are installed under Linux command line. Installing this package under MacOSX requires almost identical command with minor modifications. For example, MacOSX does not include "wget" by default. To download a file from command line, the reader can use "curl -O" followed by the URL of the target package. The reader can also type the link provided in this protocol in a web browser, and the download should start automatically. After manual download, the reader should move the downloaded package to the folder that has been created for data analysis. DESeq2 and edgeR are R packages, which can be installed under R interactive session.

6. In this section, after installing each software, a testing command is provided. The testing command usually is to display the version information for the package that was installed. If the program is installed properly, the testing command should not invoke any error message. Some command line options are preceded by single dash (-), and others are preceded by double dash (--). To read the "help" page for each program, the reader can usually use "--help" or "-help," following the program name.

7. In the following sections, a single-line command usually starts with "$." If a line does not start with "$," that means this line is a continuation of the previous line. For example, in the command "$ wget https://sourceforeg.net/projects/subread ...... Linux-x86_64. . ." appears in two rows. However, in the actual command line, this should be in the same row. This line appears in two rows because of the page width limit.

8. In this protocol, we use Python 2.7.x. However, the code should be functional under Python 3.x.x with minimum modifications.

9. To apply this protocol in other plant species, the user needs to download the corresponding genomic sequence and gene annotation files. There are many excellent resource websites available, which include Ensembl Plant (https://plants.ensembl.org/index.html) and Gramene (http://www.gramene.org/). Many species-specific websites are also available. In this protocol, we used GTF file for genome annotation. Ensembl Plant and Gramene usually provide both GTF and GFF annotation for any plant genome that are hosted in these websites. If a GTF annotation is not available, a GFF file can be

converted to GTF file. There are many free scripts online that can convert GFF to GTF file.

10. In this protocol, we use our published root cell expression data for co-expression analysis. However, this protocol can be used for any number of expression data from any plant species where sufficient data are available. For co-expression analysis, we recommend that the reader use at least ten different data points under different conditions. Each data point can be either a single replicate or the average of multiple replicates (recommended).

11. In these commands, we use "$ STAR --runThreadN 16." The reader should check how many cores are available. For a typical laptop, there are either two or four cores. For a typical computing server, there could be 12–48 or even more cores. In Linux, the command to check the number of cores available is "lscpu." Using more cores usually can speed up the analysis process; however, the reader is also advised to not use more cores than the actual available number of cores.

12. Annotation BED files can be either downloaded from website or converted from other formats such as GTF or GFF files. Unlike *Arabidopsis*, many newly sequenced plant genomes usually include contigs that are not included in the assembled chromosome pseudo-molecules. Our suggestions would be to only include the assembled chromosomes in this analysis to reduce the complexity of the analysis. The reader can also map the reads to all chromosomes including unassembled contigs and determine how many reads can map to contigs and how many reads can map to chromosomes. The reader can decide whether or not reads that map outside the assembled chromosomes are providing important biological information.

13. The gene ID "ATXGXXXXX" with X as numerical numbers is specific for *Arabidopsis* genome annotation. When using annotation GTF as input for StringTie to assemble new transcripts, annotated transcripts will inherit ATXGXXXXX ID from the annotation file. Newly assembled transcripts that are different from the existing annotation will be labeled by StringTie with different IDs. For different species, there is typically a different convention for name genes. For example, soybean genes are usually named as "GLYMAXXGXXXXXX." The script can be simply modified to recognize these IDs that are species-specific.

14. Explanation of the parameters. "-T 8" means use eight threads (cores). "-t exon" means only count reads mapped to exons. "-g gene_id" means summarize the number of reads by gene_id. This is important because there are multi-exonic genes and lincRNAs. This option allows all reads mapped to all exons

of a gene to be added together. "-p" means this dataset is generated using paired-end reads. "-a $GTF" provides annotation to the program. $GTF should be the location and file name of the GTF annotation file. "-o out.readcount.txt" is the output file name. "$mappingdir/$file 1" provides the directory where the mapped reads are located and the file name for the mapping file.

15. For R program language, a multi-line command can be used if the first line of the command is not finished but ends with a ",." For example, in this code block, the command "dds=DESeq......InputDF2" is not complete. This command is followed by "colData=sampleInfo," which is also not finished. The next line is "design =~Treatment)," which finished this command. These three lines are considered as a single command by R.

16. The filtering and clustering methods are flexible and should be determined by the reader when analyzing other datasets. The first step filtering of expression <0.05 removes lowly expressed genes. The second step keeps only highly variable genes for clustering. The actual number should be determined by the reader. However, we do not recommend to use all genes for expression clustering. The third step we used is K-means clustering. However, any clustering methods can be used in this step to determine co-expressed genes. The fourth step is to produce an edge list for visualization. This step can be skipped if the reader prefers other methods for visualization. For example, igraph is an R package which can provide network visualization within R environment.

Acknowledgment

This work is supported by the Virginia Agricultural Experiment Station (Blacksburg) and the National Institute of Food and Agriculture, US Department of Agriculture (Washington, DC).

References

1. Li S, Yamada M, Han X et al (2016) High resolution expression map of the Arabidopsis root reveals alternative splicing and lincRNA regulation. Dev Cell 39(4):508–522
2. Ulitsky I, Bartel DP (2013) lincRNAs: genomics, evolution, and mechanisms. Cell 154:26–46
3. Shuai P, Liang D, Tang S et al (2014) Genome-wide identification and functional prediction of novel and drought-responsive lincRNAs in Populus trichocarpa. J Exp Bot 65 (17):4975–4983
4. Liu J, Jung C, Xu J et al (2012) Genome-wide analysis uncovers regulation of long intergenic noncoding RNAs in Arabidopsis. Plant Cell 24:4333–4345
5. Shen Y, Zhou Z, Wang Z et al (2014) Global dissection of alternative splicing in paleopolyploid soybean. Plant Cell 26:996–1008

6. Thatcher SR, Danilevskaya ON, Meng X et al (2016) Genome-wide analysis of alternative splicing during development and drought stress in maize. Plant Physiol 170:586–599
7. Zhang W, Han Z, Guo Q et al (2014) Identification of maize long non-coding RNAs responsive to drought stress. PLoS One 9: e98958
8. Aghamirzaie D, Batra D, Heath LS et al (2015) Transcriptome-wide functional characterization reveals novel relationships among differentially expressed transcripts in developing soybean embryos. BMC Genomics 16:928
9. Dobin A, Davis CA, Schlesinger F et al (2013) STAR: ultrafast universal RNA-seq aligner. Bioinformatics 29:15–21
10. Liao Y, Smyth GK, Shi W (2014) FeatureCounts: an efficient general purpose program for assigning sequence reads to genomic features. Bioinformatics 30:923–930
11. Pertea M, Pertea GM, Antonescu CM et al (2015) StringTie enables improved reconstruction of a transcriptome from RNA-seq reads. Nat Biotechnol 33:290–295
12. Love MI, Huber W, Anders S (2014) Moderated estimation of fold change and dispersion for RNA-seq data with DESeq2. Genome Biol 15:1–34
13. Robinson MD, McCarthy DJ, Smyth GK (2010) edgeR: a Bioconductor package for differential expression analysis of digital gene expression data. Bioinformatics 26:139–140
14. Leinonen R, Sugawara H, Shumway M (2011) International nucleotide sequence database collaboration. The sequence read archive. Nucleic Acids Res 39:D19–D21
15. Gentleman RC, Carey VJ, Bates DM et al (2004) Bioconductor: open software development for computational biology and bioinformatics. Genome Biol 5:R80
16. Krishnakumar V, Hanlon MR, Contrino S et al (2015) Araport: the Arabidopsis information portal. Nucleic Acids Res 43:D1003–D1009
17. Cheng CY, Krishnakumar V, Chan AP et al (2017) Araport11: a complete reannotation of the Arabidopsis thaliana reference genome. Plant J 89:789–804
18. Smoot ME, Ono K, Ruscheinski J et al (2011) Cytoscape 2.8: new features for data integration and network visualization. Bioinformatics 27:431–432

Chapter 13

An Easy-to-Follow Pipeline for Long Noncoding RNA Identification: A Case Study in Diploid Strawberry *Fragaria vesca*

Chunying Kang and Zhongchi Liu

Abstract

Long noncoding RNAs (lncRNAs), defined as transcripts longer than 200 nucleotides without coding potential, are a new class of regulatory molecules with roles in diverse biological processes. New lncRNAs can readily be identified by mining RNA-seq data from a wide range of plant species. However, challenges remain as to how one can distinguish functional lncRNAs from mRNAs coding for small peptides or products of pseudogenes without any function. In this chapter, stepwise instruction is provided using RNA-seq datasets of developing wild strawberry fruit to illustrate each step. The workflow can be divided into three parts. Part I concerns standard RNA-seq data processing and analysis; part II describes lncRNA identification; part III describes several approaches aimed at shedding lights on lncRNA function. The description is intended for beginners with easy-to-follow steps. Text boxes provide codes and explanations. While it is relatively easy to identify lncRNAs, it is difficult to infer their function in the absence of coding information. Multiple RNA-seq libraries across tissues and stages are useful resources for deducing possible function of lncRNAs based on their expression and co-regulation.

Key words lncRNA, RNA-seq, Strawberry, Identification, Correlation analysis

1 Introduction

lncRNAs represent a new class of regulatory molecules, whose function and importance are just beginning to be unraveled in a wide range of organisms including both plants and animals. The increasing realization and identification of lncRNAs owe much to the next-generation sequencing (NGS) technology that has led to the rise of RNA-seq data output and the massive datasets in the public databases. NGS is particularly powerful for studying economically important non-model organisms as it doesn't rely on availability of genome sequence or gene annotations, unlike the older technologies such as microarray. Further, even for model

Julia A. Chekanova and Hsiao-Lin V. Wang (eds.), *Plant Long Non-Coding RNAs: Methods and Protocols*,
Methods in Molecular Biology, vol. 1933, https://doi.org/10.1007/978-1-4939-9045-0_13,

organisms, novel transcripts could be discovered by analyzing RNA-seq data obtained from new tissue types, new treatments, or higher depth of sequencing. Hence, lncRNAs can be readily identified not only in model organisms such as *Arabidopsis* and rice but also in non-model species such as strawberry [1–4].

As the name implies, noncoding RNAs are transcripts that do not appear to code for polypeptides as they harbor either none or only very short open reading frames. The noncoding transcripts can be further grouped into short and long ones. Some short noncoding RNAs perform housekeeping functions, such as tRNAs, rRNAs, and small nuclear RNAs, while others such as microRNAs and small interfering RNAs (siRNA) play critical regulatory roles during development or stress responses. In contrast, noncoding RNAs with a length greater than 200 bp, named as long noncoding RNA (lncRNA), played diverse and varied roles ranging from structural and regulatory to unknown function [5]. This chapter is aimed at identifying and characterizing the lncRNAs.

According to where a lncRNA resides in the genome, it can be classified as intergenic lncRNAs (residing in the intergenic region of annotated genes), intronic lncRNAs (from intronic region of annotated genes), and long noncoding natural antisense transcripts (lnc-NATs) which overlap with annotated genes but are transcribed in the antisense orientation. However, the definition of lncRNAs based on length is somewhat arbitrary. While miRNA precursors and transcripts from tranposons are usually ruled out, transcripts that code for peptides fewer than 100 residues are generally included as lncRNAs even that some short peptides may potentially possess function [4, 5]. In animals, a majority of lncRNAs are transcribed by RNA polymerase II with a 5′ cap and a 3′ poly-A tail, just like protein-coding genes. By contrast, lncRNAs in plants can also be transcribed by PolIV and V and therefore may lack poly-A tails [6, 7]; most of these plant lncRNAs participate in siRNA-mediated gene silencing at the transcriptional level. As the majority of RNA-seq data were generated from libraries involving poly-A selection, this chapter is focused on the identification pipeline of lncRNAs that have poly-A tails, whose roles are more diverse than the transcripts lacking poly-A tails.

A major challenge in the study of lncRNA function is that lncRNAs do not produce protein products, leaving no clue as to what the transcripts may do in the absence of conserved protein domains or protein localization signals. Further, in the absence of coding constraints, lncRNAs evolve fast and are often not conserved, making it hard to deduce function based on homologs in other species. Finally, lncRNA could play roles in multiple ways [8]; some may rely on secondary structural features, while others may rely on homology-based features. Therefore, lncRNAs are more difficult to study than protein-coding genes. Here, we show that one could gain functional insights by co-expression analysis. If an lncRNA shows a high degree of positively correlated co-expression with a protein-coding gene across diverse experimental conditions, it may suggest that the two genes have some functional connections, such as involving in the same biological processes or promoting each other's expression. If an lncRNA shows a high degree of negative correlation in expression with a protein-coding gene across diverse experimental conditions, they might negatively regulate each other's expression or function. For example, the lncRNA could potentially antagonize or silence the protein-coding gene's expression [6]. Considering that lncRNAs could function *in cis*, an analysis scheme is introduced to identify correlation in the expression between lncRNAs and their adjacent genes.

This chapter used the datasets derived from the wild strawberry *Fragaria vesca* as an example to illustrate our analysis pipeline. *F. vesca* is a new model for studying fruit development as its edible flesh, the receptacle, is enlarged from the stem tip rather than from the ovary [9]. A comprehensive RNA-seq dataset has been previously generated in *F. vesca* including 37 distinct *F. vesca* floral and fruit tissues at different developmental stages [10, 11]. For simplicity, only a time series of embryo samples were used to illustrate the pipeline in this chapter. The entire pipeline could be divided into three parts: Part I is the general RNA-seq processing and analysis, including read quality control, alignment, assembly, and obtaining gene expression data. Part II is the lncRNA identification step, including a series of filtering steps. Part III is the co-expression analysis and transcript visualization, which helps to identify potential function of lncRNAs (summarized in Fig. 1).

2 Materials

2.1 RNA-Seq Datasets

The RNA-seq reads from six embryo samples will be taken from a larger dataset [10]. The six embryo samples are a developmental time series of embryos at fruit developmental stages 3, 4, and 5, respectively. Two biological replicates are there for each stage, and the samples are named Embryo 3-1, 3-2, 4-1, 4-2, 5-1, and 5-2. Each RNA-seq library was made from poly-A-selected RNA and sequenced using Illumina HiSeq2000 (*see* **Note 1**). About 20–40 million single-end, 51 bp reads were obtained per sample (*see* **Notes 2** and **3**). It is available at Sequence Read Archive (SRA) at NCBI (http://www.ncbi.nlm.nih.gov/sra) with the accession number SRP035308 [10].

The diploid strawberry genome sequence file "Fvesca_226.fa" and genome annotation file "Fvesca_226_gene.gff3" were downloaded from Phytozome (http://www.phytozome.net), which is the initial version of *F. vesca* genome ($2n = 2\times = 14$; version 1.1)

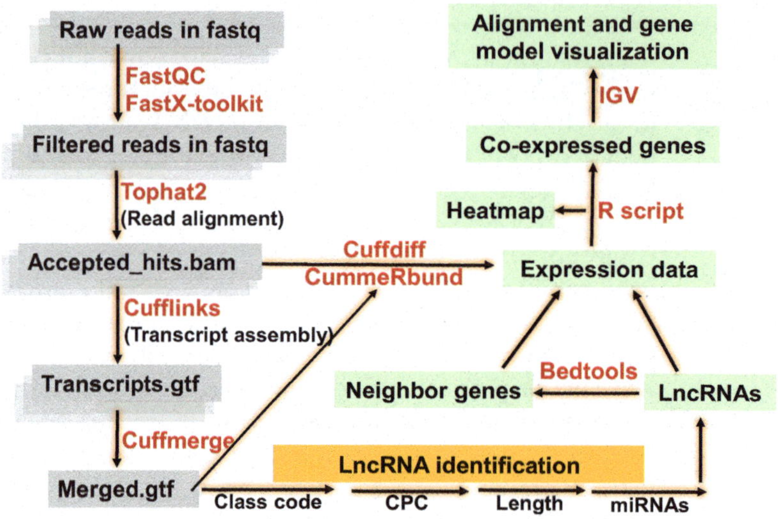

Fig. 1 The workflow of lncRNA analysis presented in this chapter. The workflow could be separated into three parts. Part I (gray box) is the general RNA-seq data processing and analysis, including read filtering, mapping, and assembly. Part II (yellow box) is the lncRNA identification with a series of filters. Part III (green box) is the investigation of lncRNA expression and function by co-expression analysis

obtained from the accession "Hawaii 4" but curated by this website [12]. The strawberry RNA-seq datasets were derived from the *F. vesca* accession "Yellow Wonder 5AF7"; its genome is highly similar to that of "Hawaii 4," albeit with some accession-specific SNPs and InDels [13].

2.2 Tools and Websites

As TopHat and Cufflinks are both operated through the UNIX shell, the following tools would be run on a Mac computer, which is much easier for a beginner to use the command lines. Most tools will be run in the Terminal; the commands on expression analysis will be run in R environment (prefixed with ">"). The required tools and websites are listed as follows (*see* **Note 4**):

1. FastQC (http://www.bioinformatics.babraham.ac.uk/projects/download.html#fastqc) for examining the overall read quality.
2. FastX-Toolkit (hannonlab.cshl.edu/fastx_toolkit/) for read trimming.
3. Bowtie2 (http://sourceforge.net/projects/bowtie-bio/files/bowtie2/2.2.6/), the command "bowtie2-build" was used for indexing the genome sequence.

4. Tophat2 (https://ccb.jhu.edu/software/tophat/index.shtml, "Mac OS X x86_64 binary") for aligning RNA-seq reads to the genome and identifying splice junctions between exons.
5. Cufflinks (http://cole-trapnell-lab.github.io/cufflinks/install/) for transcript assembly, combination, and comparison, which includes a few function modules, such as "cufflinks," "cuffmerge," and "cuffdiff."
6. Coding potential calculator (CPC) server (http://cpc.cbi.pku.edu.cn/) for evaluating the coding potential [14].
7. Stand-alone blast executables (ftp://ftp.ncbi.nlm.nih.gov/blast/executables/blast+/) for rapidly blasting the mature miRNA sequences against the lncRNA candidates.
8. Bedtools (https://github.com/arq5x/bedtools2/releases) for looking for neighboring genes [15].
9. RStudio (https://www.rstudio.com/products/rstudio/download2/), a friendly interface for using the R language.
10. R package CummeRbund, a toolkit for extracting the gene expression data, differentially expressed genes, gene annotations, etc. from the Cuffdiff output.
11. Samtools (https://sourceforge.net/projects/samtools/files/samtools/) for indexing the genome sequence file and the .bam file.
12. IGV (http://software.broadinstitute.org/software/igv/download) for RNA-seq read alignment and gene model visualization [16].

3 Methods

3.1 General RNA-Seq Processing and Analysis

3.1.1 Initial Quality Control and Trimming of the Raw RNA-Seq Data

The first thing to do with the raw RNA-seq data in the .fastq format is to evaluate the data quality. FastQC is an easy-to-use tool for a quick evaluation of the raw read data. To start the interactive graphical application of FastQC, just double-click the .dmg file (called "fastqc_v0.11.5.dmg" in the current version), and then double-click the QC icon; a window will be shown (Fig. 2). To evaluate one sequence file, simply choose "File > Open" to select any file you want to analyze. It will take a few minutes to finish the evaluation depending on the size of the file. In the result window, each item will be shown on the left with graphic symbols and clickables for details (Fig. 2).

A proportion of the reads may be of low quality. Filtering and trimming of the reads before mapping will relieve the computing load and yield better results. The FASTX-Toolkit, a collection of command line tools, is suitable for such read preprocessing. To use

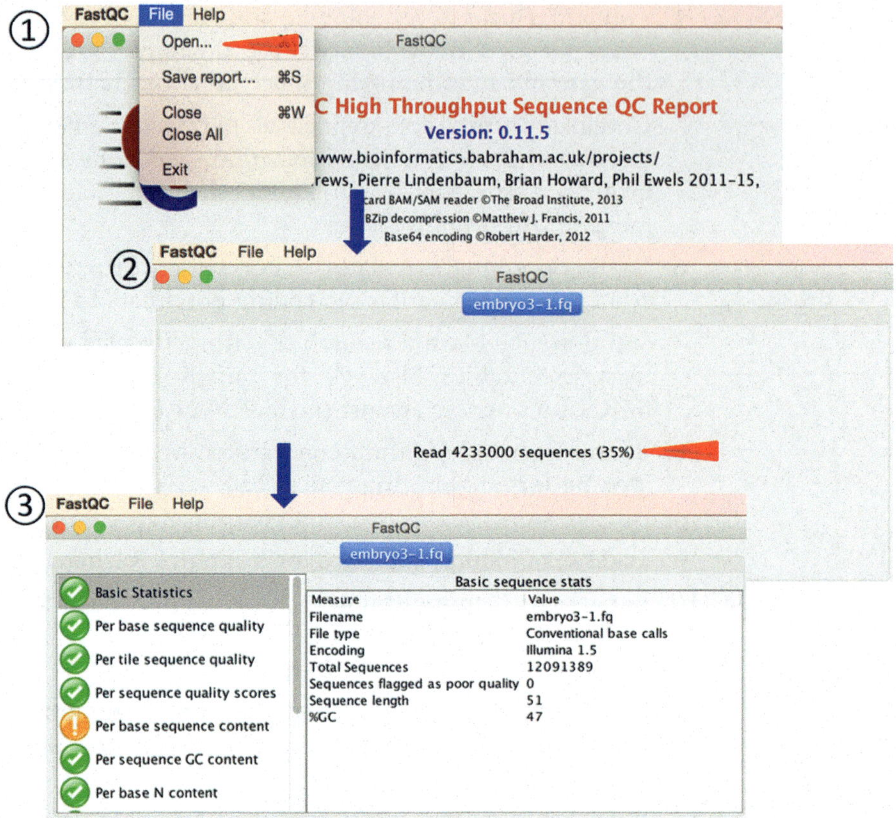

Fig. 2 The steps of using FastQC to evaluate raw RNA-seq data quality. Step 1: open the file to be analyzed. Step 2: wait for the job to be done. Step 3: check the data quality summary in each item one by one

it, first download the binary file, and then use the command line in the Terminal to move all the binaries in the FASTX-Toolkit folder to one directory in the PATH, usually "/usr/local/bin/," which will enable calling the binaries when the working directory is not in the PATH (*see* **Note 5**). The command "fastx_quality_stats" will do a similar job as FastQC (Text Box 1). The command "fastq_quality_filter" will filter out the low-quality reads as required by -q (minimum quality score to keep, 20 herein) and -p (minimum percent of bases that must have [-q] quality, 90 herein). The command "fastx_trimmer" will trim off the low-quality nucleotides in each read based on the FastQC evaluation results; for example, "-f 7" here means that the first 7 nucleotides will be trimmed off. The processed .fastq file, called "embryo3-1-processed.fq" in the example, will be used for read mapping in the next step.

Text Box 1

```
cd /Path/to/the bin folder in the FASTX-Toolkit file
sudo cp * /usr/local/bin
#Move all the binaries into PATH to call them in any working directory
cd /usr/local/bin
ls
#To confirm that all the binaries exist in the folder

cd /Path/to/the FastQ files
fastx_quality_stats -i embryo3-1.fq -o embryo3-1.txt
#Quality evaluation
fastq_quality_filter -q 20 -p 90 -i embryo3-1.fq -o embryo3-1-filtered.fq

#Filter out the low quality read
fastx_trimmer -f 7 -i embryo3-1-filtered.fq > embryo3-1-processed.fq
#Trim off the first 7 nucleotides in each read
fastx_trimmer -h
#To show the help page of the command with descriptions about the parameters and the
results
```

Note (applicable to all textboxes in this article): the file names in bold could be changed in other analyses. The text after "#" explains the purpose of the script.

3.1.2 Transcript Assembly by the Tuxedo Suite

A number of bioinformatic tools have been developed for the purpose of RNA-seq read alignment and transcript assembly [17]. The Tuxedo suite (TopHat, Cufflinks, and CummeRbund) is among the most popular ones for species with sequenced genome [18]. First, TopHat aligns reads to the genome and discovers transcript splice sites. Next, Cufflinks uses the alignment to assemble the reads into transcripts. Then, Cuffdiff (a function module from Cufflinks) takes the aligned reads from two or more conditions to identify genes and transcripts that are differentially expressed using a rigorous statistical analysis. Lastly, the R package "CummeRbund" renders Cuffdiff output into publication-ready figures and plots. The Tuxedo suite provides a complete RNA-seq analysis workflow, which is a great option for beginners to learn the RNA-seq analysis because of the straightforward installation, detailed tutorials, and broad user base from which one can easily find help.

In this step, three tools (Bowtie, TopHat, and Cufflinks) should be installed. The binaries will be placed into the PATH for easy calling in any working directory. To go through the entire pipeline smoothly, the genome sequence file (Fvesca_226.fa), the genome annotation file (Fvesca_226_gene.gff3), and each raw RNA-seq read file (embryo3-1-processed.fq, etc.) should all be placed into one folder. All the commands will be run under this folder.

To map the RNA-seq reads by Tophat2, "bowtie2-build" should be used first to index the genome (Text Box 2), and six files with the same prefix and different suffixes (.1.bt2, .2.bt2, .3.

bt2, .4.bt2, .rev.1.bt2, and .rev.2.bt2.) will be generated. The key is to use the same names for the genome file and indexed files. Second, each .fastq file is individually aligned to the genome by "Tophat2" that preserves junction reads. The default settings were used except that the thread number (-p) was designated 4 to accelerate the processing, and the maximum intron size was set at 5000 bp (-I) (*see* **Note 6**). The result files will be saved in the folder "embryo3-1_thout" (-o). Third, the file "accepted_hits.bam" in the folder "embryo3-1_thout" (the "TopHat2" output) is fed to "cufflinks" for both known and novel transcript assembly, and the output files will be saved in the folder "embryo3-1_clout." Only the file "transcripts.gtf" in the folder "embryo3-1_clout" is needed for the next step.

Text Box 2

```
#Download the Bowtie2 binary file
cd /Path/to/Bowite2
sudo cp bowtie2* /usr/local/bin
#To access the binaries in any working directory

#Index the genome sequence file using bowtie2
bowtie2-build Fvesca_226.fa Fvesca_226
#The two names in bold should be exactly the same
bowtie2-inspect -s Fvesca_226

#Download the tophat2 binary and unpack it (double-click the file)
export PATH="/Path/to/tophat-2.1.0.OSX_x86_64:$PATH"
#Add the tophat2 directory into the PATH to access its binaries
echo $PATH
#To show the current PATH

#Align the reads using tophat2
cd /Path/to/FastQ folder/
tophat2 -p 4 -I 5000 -G Fvesca_226_gene.gff3 -o embryo3-1_thout Fvesca_226 embryo3-1-
processed.fq

#Download the cufflinks binary and unpack it (double-click the file)
export PATH="/Path/to/cufflinks-2.2.1.OSX_x86_64:$PATH"
#Add the cufflinks directory into the PATH to access its binaries
echo $PATH #To show the current PATH

#Assemble the transcripts using cufflinks
cufflinks -p 4 -g Fvesca_226_gene.gff3 -I 5000 -o embryo3-1_clout embryo3-1_thout/
accepted_hits.bam

#Merge the transcripts assembled from individual libraries into one master file
find . -name transcripts.gtf > assemblies.txt
```

(continued)

```
#Put the .gtf file directories together
cuffmerge -p 4 -g Fvesca_226_gene.gff3 -s Fvesca_226.fa assemblies.txt

#Run cuffdiff to get the expression level of each locus and transcript
cuffdiff -o diff_out -b Fvesca_226.fa -p 8 -T -L embryo3,embryo4 -u merged_asm/merged.
gtf embryo3-1_thout/accepted_hits.bam,embryo3-2_thout/accepted_hits.bam embryo4-
1_thout/accepted_hits.bam,embryo4-2_thout/accepted_hits.bam
```

Obviously, there will be as many "transcripts.gtf" files as the raw .fastq files. However, the transcripts in each individual .gtf file were randomly named by numbers according to the order in the genome. To compare the expression levels of those transcripts among samples, each of them should have a consensus name. To do this, "cuffmerge" could be utilized to make a merged annotation file (merged.gtf) including all the assembled transcripts. Finally, "cuffdiff" calculated the expression level of each transcript present in "merged.gtf" using the file "accepted_hits.bam" in the "TopHat2" output. In the "cuffdiff" command, to be noted, the label for each sample is defined by "-L"; the .bam files for replicates should be separated by a comma following the same order as the labels; the option "-T" allows making comparisons between successive samples rather than between all pairs of samples to relieve the computing load. The output files will be saved in the folder "diff_out."

3.2 lncRNA Identification

In "merged.gtf" obtained by the codes in Text Box 2, each transcript was assigned 1 of the 12 class codes according to their locations relative to the annotation [18], such as "j," "=," "u," "o," "x," "i," etc. To identify lncRNAs, one can select and keep the transcripts with class codes "u," "o," "i," and "x," which are most likely to be noncoding. Next, coding potential calculator (CPC) will be used to identify those transcripts that have low coding potential scores. In the meantime, transcripts shorter than 200 bp will be filtered out. At last, miRNAs will be blasted against candidate lncRNA transcripts; miRNA precursors will be identified and removed. These filtering steps will lead to the final list of lncRNAs.

3.2.1 Select u, o, i, and x Classes of Transcripts Using Class Code

The annotations in the "merged.gtf" consist of all the assembled transcripts from each sample, irrespective of coding or noncoding and known or unknown. The main task here is to distinguish long noncoding transcripts from others. Noticeably, 12 class_codes were designated to each transcript describing the transcript feature when the transcript was compared with the annotations in "merged.gtf." For instance, class code "u" means unknown/intergenic transcript, whereas class code "i" means a transcript residing entirely within an intron. A majority of transcripts are either partially (class_code "j") or completely (class_code "=") matched with the annotation. As the strawberry version 1.1 annotation includes only protein-coding genes, these two categories (j and =) likely code for genes and thus are excluded from further analysis. The transcripts with class_code "u" (unknown or intergenic regions), "o" (generic exonic overlap with a reference transcript), "x" (natural antisense transcript, NAT) (*see* **Note 6**), and "i" (located entirely within the intronic regions) may include noncoding transcripts and will be subjected to further filtering (Text Box 3).

Text Box 3

```
#Select the transcripts with the class_codes "u," "i," "o," and "x" from "merged.gtf"
cat merged.gtf | grep'class_code "[ uiox] "'> selected.gtf

#Obtain the selected transcript sequences in the fasta format
gffread -w selected.fa -g Fvesca_226.fa selected.gtf

#Count the total number of selected transcripts (see Note 7)
cat selected.fa | grep ">" | wc -l
```

3.2.2 Use CPC to Calculate Coding Potential of Putative lncRNAs

Long noncoding RNAs are defined as transcripts without coding potential and longer than 200 bp in length. The *coding potential calculator* (CPC) is a good tool for assessing the coding capacities with a high accuracy, as it takes into account six sequence features obtained from the open reading frame quality assessment and the BLASTX search results [18]. There is no sequence length limit in this server (http://cpc.cbi.pku.edu.cn), but it is better to split a large size file into multiple small size files if the large file includes too many transcripts. To start a job, just submit the transcript sequences (the .fasta file), and wait to download the results, which usually takes hours depending on the work load. In this chapter, a transcript was classified as noncoding when it has a coding potential score (CPC score) lower than "−1," which is a relatively stringent standard (*see* **Note 8**). A one-line command is sufficient to filter the CPC score as well as the transcript length (>200 nt) listed in "cpc.txt," one of the tables in the CPC output (Text Box 4).

Text Box 4

```
#Find the noncoding transcripts longer than 200 nt from the CPC output
cat cpc.txt | awk '$4 < -1 && $2 > 200{ print $0} ' | cut -f1 > non-coding-transcript.txt
```

(continued)

```
#Get the noncoding transcript sequences
sed 's/^/transcript_id "&/g' non-coding-transcript.txt > non-coding-transcript-new.txt
cat merged.gtf | fgrep -f non-coding-transcript-new.txt > non-coding-transcript.gtf
gffread -w non-coding-transcript.fa -g Fvesca_226.fa non-coding-transcript.gtf
```

3.2.3 Removing Transcripts Coding for miRNAs

So far, the two primary filters, coding potential and transcript length, have been applied for lncRNA identification. However, some of the lncRNA candidates may encode small peptides, be transcribed from transposons, or generate small RNAs. At the present, it is difficult to determine which small peptides may actually been made and have functions. However, transposon transcripts could be determined by checking if their genomic loci overlap with the transposon loci. Small RNA-sequencing data are needed for detecting and then removing small RNA-producing transcripts. Since the version1.1 strawberry genome annotation doesn't include miRNAs, some of the putative lncRNAs could be miRNA precursors. To remove such miRNA precursors, a blast search can be easily carried out in a stand-alone manner. First, download and move the stand-alone BLAST binary to the PATH, then an *F. vesca* lncRNA (FvlncRNA) database was established, and finally the mature miRNA sequences in "FvmiRNA.fasta" (a file with the names and sequences of miRNAs in the fasta format) were used to blast against the FvlncRNA database in local computer (Text Box 5). The lncRNAs in "miRNA-lncRNA.txt" are potential miRNA precursors and will be excluded from further analysis.

Text Box 5

```
#Download the zipped file "ncbi-blast-2.2.28+-universal-macosx.tar.gz"
cd /Path/to/bin
sudo cp * /usr/local/bin

#Make the lncRNA database
makeblastdb -in non-coding-transcript.fa -input_type fasta -parse_seqids -dbtype
nucl -out FvlncRNA

#Blast the mature miRNA sequences against the FvlncRNA database
blastn -db FvlncRNA -query FvmiRNA.fasta -out miRNA-lncRNA.txt -evalue 0.001 -outfmt
6
-word_size 7
```

The lncRNAs do not code for any protein products. Hence lncRNA sequences are not well conserved due to a lack of constraint by

3.3 Investigate lncRNA Expression and Potential Function

coding potential. As a result, the function of lncRNA is difficult to infer based on sequence homology between genes in the same or different species (*see* **Note 9**). Therefore, several analysis schemes are proposed here aiming at a better understanding of lncRNAs using existing resources such as RNA-seq data. The co-expression analysis between lncRNAs and neighbor protein-coding genes is shown below. The co-expression analysis between lncRNAs and all protein-coding genes is more straightforward and could be similarly determined (codes not shown).

3.3.1 Find the Neighboring Protein-Coding Genes of lncRNAs

One option to make a functional connection is to identify co-expression pairs between lncRNAs and protein-coding genes. Such co-expression study is only possible when RNA-seq data from multiple tissues, developmental stages, or stress conditions are available. Since some lncRNAs may function *in cis* to regulate the transcription of neighboring protein-coding genes [6], the correlation analysis was carried out between lncRNAs and their neighboring protein-coding genes. Figure 3 shows an example of such negatively correlated gene pairs [1]. Herein, one should first find lncRNAs and their neighboring genes. The "windowBed" function from Bedtools [15], a software suite for comparing genomic features, could be used to identify both upstream and downstream neighboring genes of all lncRNAs at a distant of 10Kb (-w) and will save the output in "genepair.gtf" with all the columns from the two input .gtf files (Text Box 6). In the input .gtf files generated by the Tuxedo suite, the names of gene loci were prefixed by "XLOC" and could be parsed using the arguments -f (columns) and -b (bytes) in the command line (Text Box 6). As the expression level of each locus is more reliable, the correlation coefficients between lncRNAs and neighboring protein-coding genes will be calculated only at the locus level. The locus names of each gene pair were saved in "genepair.txt" with the lncRNA locus name in the first column and the neighboring protein-coding gene name in the second column.

Text Box 6

```
#Obtain the expressed encoding genes
cat merged.gtf | grep 'class_code "="' > coding.gtf

#Download the bedtools and put the executables into the PATH (/usr/local/bin)

#Obtain the neighbor gene pairs between lncRNAs and encoding genes
windowBed -a non-coding-transcript.gtf -b coding.gtf -w 10000 > genepair.gtf
cat genepair.gtf | cut -f9 | cut -b 10-20 > lncRNA.txt
cat genepair.gtf | cut -f18 | cut -b 10-20 > coding.txt
paste lncRNA.txt coding.txt | sort | uniq | awk '$1 != $2{print $0}' > genepair.txt
```

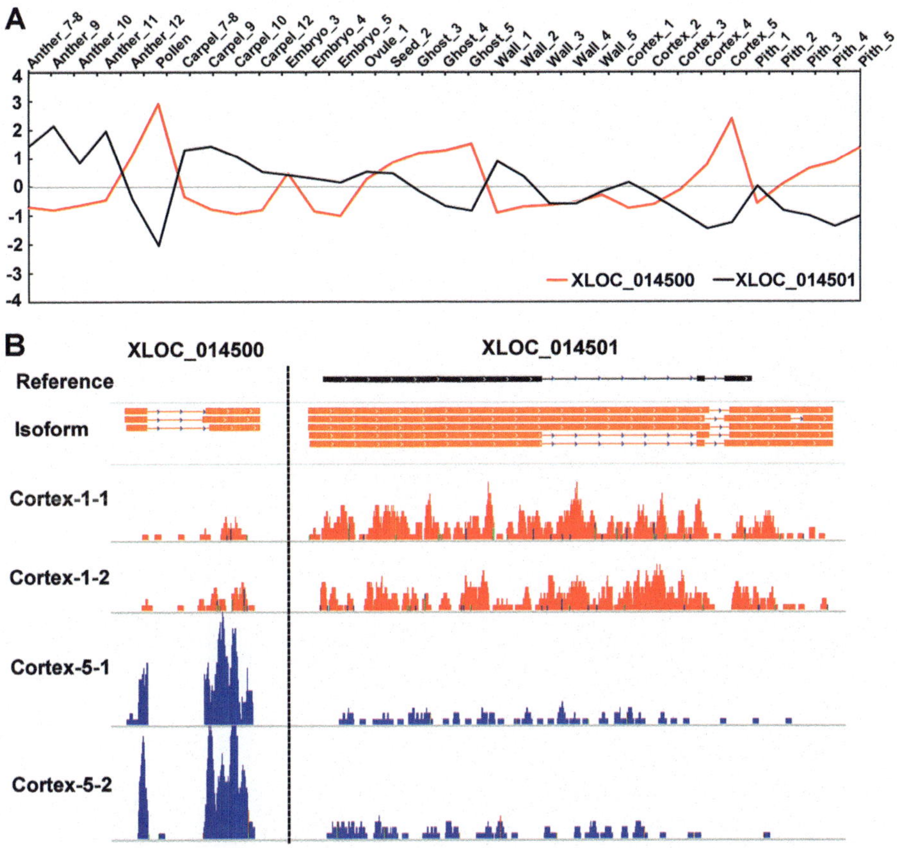

Fig. 3 An example of negative correlation in expression between an lncRNA and its neighboring protein-coding gene (Reproduced from Kang and Liu (2015) [1]) (**a**) The expression of lncRNA XLOC_014500 (red line) and neighboring coding gene XLOC_014501 (black line, gene22438) is negatively correlated. Y-axis shows the expression level by Z-score obtained from averaged FPKM of two replicates (**b**) IGV view of aligned RNA-seq read counts for XLOC_014500 and XLOC_014501 based on the fruit cortex tissue at two stages: Cortex-1 (pre-fertilization) and Cortex-5 (postfertilization). The "Reference" row shows the gene structure based on *F. vesca* genome annotation version1.1. Thin line indicates intron and thick line denotes exon. The "Isoform" row shows transcript variants predicted by Cufflinks. The bottom four rows show the RNA-seq read counts in respective tissues. The two replicates are shown with identical color

3.3.2 Retrieve the Transcript Level of lncRNAs and the Neighboring Protein-Coding Genes

The outputs from the "cuffdiff" run at Subheading 3.1.2 were saved in the "diff_out" folder, which preserved the data on gene expression, gene annotation, and differentially expressed genes in separate files. To parse those data, the R package "cummeRbund" could be run in RStudio, a powerful and productive user interface for R. As shown in Text Box 7, import the gene names in lncRNA.txt and coding.txt, respectively, into RStudio by "read.table()," find the genes in the cuffdiff data by "getGenes()," and extract their expression data by "fpkmMatrix()" which will only provide the average expression level in FPKM of each gene in the two biological replicates. Finally, the R function "write.table()" was used to save the expression data into a .csv file.

Text Box 7

```
> library(cummeRbund)
> setwd("/Path/to/diff_out")
#Set up the working directory

#Read the data
> cuff <- readCufflinks()
> cuff

#Get the expression data for the lncRNAs
> lncRNA <- read.table("lncRNA.txt", header=FALSE)
> lncRNA <- lncRNA[ ,1]
> lncRNA1 <- getGenes(cuff, lncRNA)
> lncRNA2 <- fpkmMatrix(lncRNA1)
> write.table(lncRNA2, "Expression of noncoding genes.csv", sep="\t", quote=FALSE)

#Get the expression data for the neighbor coding genes
> coding <- read.table("coding.txt", header=FALSE)
> coding <- coding[ ,1]
> coding1 <- getGenes(cuff, coding)
> coding2 <- fpkmMatrix(coding1)
> write.table(coding2, "Expression of coding genes.csv", sep="\t", quote=FALSE)
```

3.3.3 Expression Correlation Between lncRNAs and the Neighboring Protein-Coding Genes

The correlation coefficients (Pearson's r) between protein-coding genes and lncRNAs can be easily computed by R language. Pearson's r ranges from -1 to 1. Values closer to 0 indicate no association; values closer to 1 indicate positive co-expression; values closer to -1 indicate negative co-expression. The negative co-expression is of most interest as it may indicate possible repression of the neighboring protein-coding gene by the lncRNA or vice versa. The gene expression data in "lncRNA2" and "coding2" obtained from Text Box 7 is a data frame with the libraries/tissues in columns while genes in rows. To calculate the correlation coefficients, the columns and rows should be transposed first by "t()" (Text Box 8); then, cor() in R will generate a data matrix of correlation coefficients with the genes from the first file in rows and the genes from the second file in columns (Text Box 8). Say there are 10 lncRNA genes and 100 coding genes, the output will be a 10×100 data frame with lncRNAs in rows and coding genes in columns. The function "melt()" in R package "reshape2" was used to transform the data into a matrix with three columns, and the last column is the correlation coefficient r. The top rows of the matrix were shown in Text Box 8 under "> cor.m = melt(cor)." Next, the adjacent gene pairs with $r > 0.7$ were selected for further analysis by the function "subset()." The opposite expression trend between a lncRNA and its protein-coding neighbor could be found by filtering the r with a number close to −1, such as $r < -0.7$. In the case of strawberry lncRNAs [1], there are 409 correlated gene pairs when $r < -0.8$, while there are 14,953 gene pairs when $r < -0.7$. In our opinion, the complexity of samples will largely affect the correlation coefficient. A greater number of gene pairs would be identified in fewer samples than more samples. Therefore, the coefficient cutoffs are flexible and should be determined according to the specific context of study.

Text Box 8

```
> lncRNA2trans <- t(lncRNA2)
> coding2trans <- t(coding2)
> cor = cor(lncRNA2trans, coding2trans)
> library(reshape2)
> cor.m = melt(cor)
       Var1        Var2          value
1 XLOC_000002 XLOC_000026 -0.96899322
2 XLOC_000004 XLOC_000026 -0.80749945
3 XLOC_000012 XLOC_000026 0.99611275
4 XLOC_000015 XLOC_000026 -0.95241608

> genepair = read.table ("lncRNAs and the neighbor encoding genes.txt", header=FALSE)
> cor.positive = subset(cor.m, cor.m$Var1 %in% genepair[ ,1] & cor.m$Var2 %in% gene-
pair[ ,2] & cor.m$value > 0.7)
```

3.3.4 Visualize lncRNA Expression Pattern with Heatmap to Gain Functional Insights

It is reported that a greater portion of lncRNAs are expressed in a tissue-specific manner than protein-coding genes [1]. Heatmap with hierarchical clustering is frequently used to show the expression patterns of a large number of genes among a series of tissues. To draw the heatmap, the matrix (lncRNA2) of the lncRNA expression data in FPKM obtained from Text Box 7 was used. Some lncRNAs are highly expressed, whereas others are lowly expressed. To show their expression patterns in the same heatmap, Z-score normalization was carried out first by the function "znorm" in the R package "survJamda" (Text Box 9). Then, the color scheme of the heatmap is defined by the function "colorpanel" in the R package "gplots." At last, the function "heatmap.2" was used to draw the heatmap, which would look like the image shown in Fig. 4 (borrowed from the reference paper [1]) and can be exported as a PDF file or a JPG file. From the heatmap, we can see that the expression of some lncRNAs is tissue-specific, and thus they may play roles in the development of the tissue.

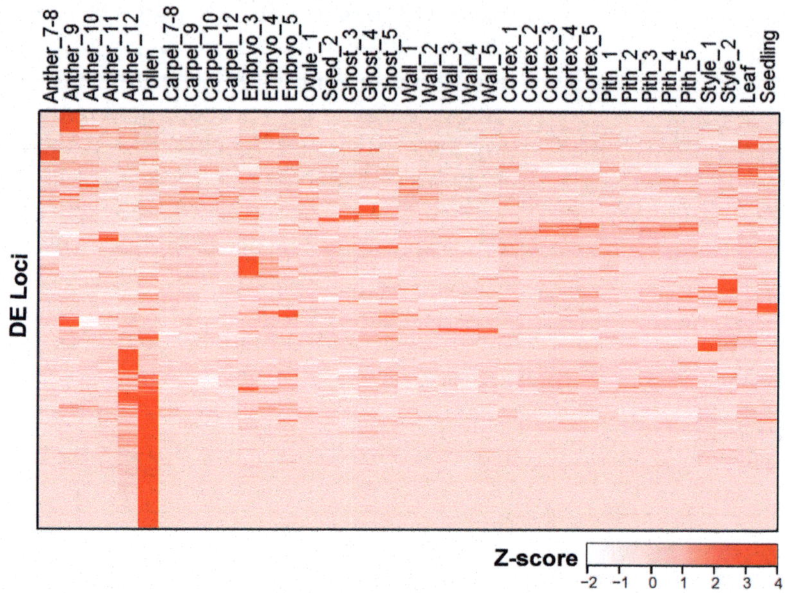

Fig. 4 Heatmap showing tissue-specific expression of lncRNAs across 37 different tissues and stages of wild strawberry (Reproduced from Kang and Liu (2015) [1]). Z-scores were obtained from averaged FPKM of two replicates and used to make the heatmap. A large number of lncRNA genes are shown to be specifically expressed in pollen.

Text Box 9

```
> library(survJamda)
> lncRNA3 <- znorm(lncRNA2)
#Z-score normalization of the original expression data of lncRNAs
> library(gplots)
> mycol <- colorpanel(n=99,low="white",mid="pink",high="red")
> heatmap.2(as.matrix(lncRNA3), dendrogram = "row", Rowv=TRUE, Colv=FALSE, scale="none", col=mycol, margins=c(5, 1), density.info="none",cexCol=1, trace="none", labRow=FALSE)
```

3.3.5 Read Alignment and lncRNA Gene Model Visualization in IGV

A large gene annotation file and gene expression table will be obtained from the above analysis. Yet they are by no means fully capturing the information in the RNA-seq data. This is because genes are often incorrectly annotated, and novel gene models are often inaccurate. Hence, it is extremely helpful to see the RNA-seq reads aligned against the gene annotations. Further, if specific lncRNAs are of particular interests, their read alignment and gene model should be checked at their genomic loci. IGV is an excellent visualization tool for interactive exploration of large, integrated genomic datasets [16, 19]. To use IGV, one should first download and unzip the Mac App archive and then double-click the IGV application to launch it. The genome sequence and aligned .bam files should be indexed by Samtools (Text Box 10) before opening them in IGV. Follow the description in Fig. 5 to open the files in IGV. Some great features are available in IGV, such as search by chromosome coordinates or gene IDs, zoom in/out, change colors and fonts, and fetch the sequence in the region of interest (http://software.broadinstitute.org/software/igv/).

Text Box 10

```
#Index the genome file by samtools
samtools faidx Fvesca_226.fa

#Index the .bam file to be displayed in IGV
cd /path/to/embryo3-1_thout/
samtools index accepted_hits.bam
```

3.4 Conclusions

To conclude, we have provided detailed stepwise instructions on lncRNA identification and subsequent characterization based on gene expression information across a large number of tissues and stages. However, true functional insights will depend on mutational analysis such as CRISPR/CAS9-mediated deletion or over expression via transgenic approaches.

4 Notes

1. The lncRNAs transcribed by PolII have poly-A tails, whereas those transcribed by PolIV or V don't have poly-A tails in plants. The RNA-seq libraries used in this chapter were made from poly-A-selected RNA; therefore nonpolyadenylated lncRNAs were missed from the results. The other strategy is to use rRNA depletion protocol to construct RNA-seq libraries, which will preserve non-poly-A RNA transcripts. However, rRNA depletion is usually incomplete and therefore requires a greater sequencing depth.
2. It is known that a lot of lncRNAs are expressed at a low level. The sequencing depth is an important factor to be considered, which will largely affect the power of identifying novel lncRNAs.
3. Given the low abundance and the noncoding nature of lncRNAs, it is hard to predict their full-length transcript sequences by the short-read sequencing technology. Recently, the single-molecule, real-time (SMRT) sequencing technology,

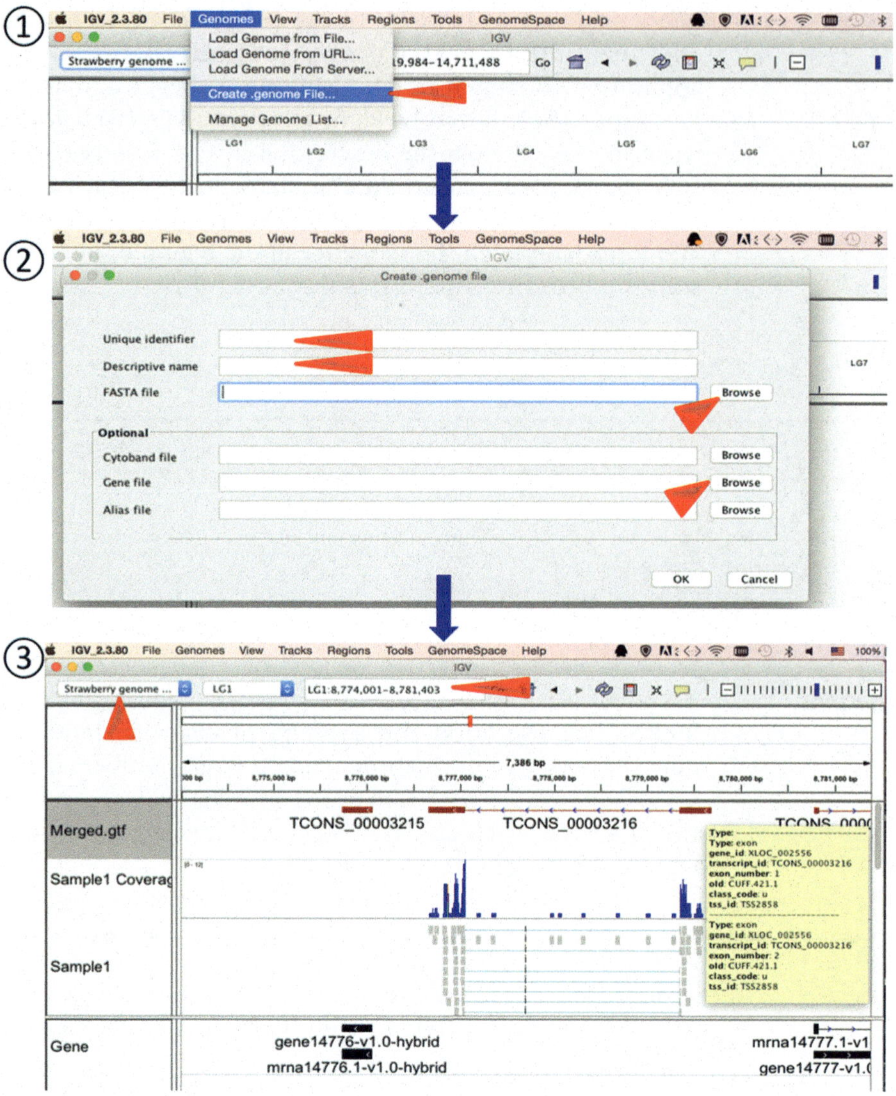

Fig. 5 Using IGV to visualize RNA-seq reads aligned against the genome annotation. Step 1: initiate IGV and click "creat .genome file." Step 2: in the popped window, fill in the "Unique identifier" and "Descriptive name" with whatever you like, and choose the genome file (FASTA) and gene file (.gtf). Step 3: open the created genome ("Descriptive name" is shown on the top left, and gene annotation is shown on the bottom (black bars indicate exons)), and drag the new annotation file (merged.gtf, in red) and indexed .bam file (read coverage in blue, short read in gray) into the middle part. When the mouse points to a transcript, some details will be displayed in a yellow-shaded box

such as the PacBio system, yields kilobase-sized sequence reads usually representing full-length mRNA molecules [20, 21]. It will complement the Illumina strategy.

4. The best practice to install the software tools is to download the precompiled binaries suitable for your computer system. It is usually tedious to compile the tools from the source code.

5. When running the commands, always make sure that the binaries are in the PATH and the working directory is correct.
6. lnc-NATs (long noncoding natural antisense transcripts) usually overlap with the protein-coding genes at their genomic loci. If a library is not made strand-specific, it will be difficult to determine which strand a specific transcript is derived from. As a result, a majority of lnc-NATs will be missed or mis-annotated. To identify lnc-NATs, strand-specific RNA-seq library is recommended. The entire analysis pipeline should be the same as above, except that TopHat2 should be set to treat each read as strand-specific (add the library type tag "fr-firststrand").
7. Always use "less" or "wc -l" to make sure the output files are right in the Terminal.
8. There is no clear rule on what is the proper CPC score cutoff for the determination of lncRNAs. During the filtering of coding potential, the CPC score < -1 was used. However, it is possible that some of the weak coding transcripts ($0 > \text{CPC} > -1$) are bona fide lncRNAs. On the other hand, some lncRNAs defined by the pipeline might encode functional short peptides. Short peptides derived from either pri-miRNAs or 5′ UTR regions of protein-coding genes have been demonstrated to play important regulatory roles [22]. In particular, 5′ UTR-derived short peptides tend not to use ATG as the START codon [23], which really complicates the analyses.
9. Since lncRNAs are much less conserved than the protein-coding genes, one may only find homologous genes between closely related species [1].

Acknowledgments

This work was supported by the National Natural Science Foundation of China (31572098 and 31772274) to C.K., US National Science Foundation Grant (IOS1444987) to Z.L., and the Scientific and Technological Self-innovation Foundation of Huazhong Agricultural University (2014RC005 to Z.L. and 2014RC017 to C.K.).

References

1. Kang C, Liu Z (2015) Global identification and analysis of long non-coding RNAs in diploid strawberry Fragaria vesca during flower and fruit development. BMC Genomics 16 (1):1–15. https://doi.org/10.1186/s12864-015-2014-2
2. Liu J, Jung C, Xu J, Wang H, Deng S, Bernad L (2012) Genome-wide analysis uncovers regulation of long intergenic noncoding RNAs in Arabidopsis. Plant Cell 24:4333–4345. https://doi.org/10.1105/tpc.112.102855

3. Zhang YC, Liao JY, Li ZY, Yu Y, Zhang JP, Li QF (2014) Genome-wide screening and functional analysis identify a large number of long noncoding RNAs involved in the sexual reproduction of rice. Genome Biol 15:512. https://doi.org/10.1186/s13059-014-0512-1
4. Li L, Eichten SR, Shimizu R, Petsch K, Yeh CT, Wu W, Chettoor AM, Givan SA, Cole RA, Fowler JE, Evans MM, Scanlon MJ, Yu J, Schnable PS, Timmermans MC, Springer NM, Muehlbauer GJ (2014) Genome-wide discovery and characterization of maize long non-coding RNAs. Genome Biol 15(2):R40. https://doi.org/10.1186/gb-2014-15-2-r40
5. Chekanova JA (2015) Long non-coding RNAs and their functions in plants. Curr Opin Plant Biol 27:207–216. https://doi.org/10.1016/j.pbi.2015.08.003
6. Ariel F, Jegu T, Latrasse D, Romero-Barrios N, Christ A, Benhamed M (2014) Noncoding transcription by alternative RNA polymerases dynamically regulates an auxin-driven chromatin loop. Mol Cell 55:383–396. https://doi.org/10.1016/j.molcel.2014.06.011
7. Wierzbicki AT, Haag JR, Pikaard CS (2008) Noncoding transcription by RNA polymerase pol IVb/pol V mediates transcriptional silencing of overlapping and adjacent genes. Cell 135(4):635–648. https://doi.org/10.1016/j.cell.2008.09.035
8. Sana J, Faltejskova P, Svoboda M, Slaby O (2012) Novel classes of non-coding RNAs and cancer. J Transl Med 10:103. https://doi.org/10.1186/1479-5876-10-103
9. Hollender CA, Geretz AC, Slovin JP, Liu Z (2012) Flower and early fruit development in a diploid strawberry, Fragaria vesca. Planta 235:1123–1139. https://doi.org/10.1007/s00425-011-1562-1
10. Kang C, Darwish O, Geretz A, Shahan R, Alkharouf N, Liu Z (2013) Genome-scale transcriptomic insights into early-stage fruit development in woodland strawberry Fragaria vesca. Plant Cell 25(6):1960–1978. https://doi.org/10.1105/tpc.113.111732
11. Hollender CA, Kang C, Darwish O, Geretz A, Matthews BF, Slovin J (2014) Floral transcriptomes in woodland strawberry uncover developing receptacle and anther gene networks. Plant Physiol 165. https://doi.org/10.1104/pp.114.237529
12. Shulaev V, Sargent DJ, Crowhurst RN, Mockler TC, Folkerts O, Delcher AL (2011) The genome of woodland strawberry (Fragaria vesca). Nat Genet 43:109–116. https://doi.org/10.1038/ng.740
13. Hawkins C, Caruana J, Schiksnis E, Liu Z (2016) Genome-scale DNA variant analysis and functional validation of a SNP underlying yellow fruit color in wild strawberry. Sci Rep 6:29017. https://doi.org/10.1038/srep29017
14. Kong L, Zhang Y, Ye ZQ, Liu XQ, Zhao SQ, Wei L (2007) CPC: assess the protein-coding potential of transcripts using sequence features and support vector machine. Nucleic Acids Res 35:W345–W349. https://doi.org/10.1093/nar/gkm391
15. Quinlan AR, Hall IM (2010) BEDTools: a flexible suite of utilities for comparing genomic features. Bioinformatics 26(6):841–842. https://doi.org/10.1093/bioinformatics/btq033
16. Thorvaldsdottir H, Robinson JT, Mesirov JP (2013) Integrative genomics viewer (IGV): high-performance genomics data visualization and exploration. Brief Bioinform 14(2):178–192. https://doi.org/10.1093/bib/bbs017
17. Conesa A, Madrigal P, Tarazona S, Gomez-Cabrero D, Cervera A, McPherson A, Szczesniak MW, Gaffney DJ, Elo LL, Zhang X, Mortazavi A (2016) A survey of best practices for RNA-seq data analysis. Genome Biol 17:13. https://doi.org/10.1186/s13059-016-0881-8
18. Trapnell C, Roberts A, Goff L, Pertea G, Kim D, Kelley DR, Pimentel H, Salzberg SL, Rinn JL, Pachter L (2012) Differential gene and transcript expression analysis of RNA-seq experiments with TopHat and Cufflinks. Nat Protoc 7(3):562–578. https://doi.org/10.1038/nprot.2012.016
19. Robinson JT, Thorvaldsdottir H, Winckler W, Guttman M, Lander ES, Getz G, Mesirov JP (2011) Integrative genomics viewer. Nat Biotechnol 29(1):24–26. https://doi.org/10.1038/nbt.1754
20. Eid J, Fehr A, Gray J, Luong K, Lyle J, Otto G, Peluso P, Rank D, Baybayan P, Bettman B, Bibillo A, Bjornson K, Chaudhuri B, Christians F, Cicero R, Clark S, Dalal R, Dewinter A, Dixon J, Foquet M, Gaertner A, Hardenbol P, Heiner C, Hester K, Holden D, Kearns G, Kong X, Kuse R, Lacroix Y, Lin S, Lundquist P, Ma C, Marks P, Maxham M, Murphy D, Park I, Pham T, Phillips M, Roy J, Sebra R, Shen G, Sorenson J, Tomaney A, Travers K, Trulson M, Vieceli J, Wegener J, Wu D, Yang A, Zaccarin D, Zhao P, Zhong F, Korlach J, Turner S (2009) Real-time DNA sequencing from single polymerase molecules. Science 323(5910):133–138. https://doi.org/10.1126/science.1162986

21. Sharon D, Tilgner H, Grubert F, Snyder M (2013) A single-molecule long-read survey of the human transcriptome. Nat Biotechnol 31 (11):1009–1014. https://doi.org/10.1038/nbt.2705
22. Waterhouse PM, Hellens RP (2015) Plant biology: coding in non-coding RNAs. Nature 520 (7545):41–42. https://doi.org/10.1038/nature14378
23. Laing WA, Martinez-Sanchez M, Wright MA, Bulley SM, Brewster D, Dare AP, Rassam M, Wang D, Storey R, Macknight RC, Hellens RP (2015) An upstream open reading frame is essential for feedback regulation of ascorbate biosynthesis in Arabidopsis. Plant Cell 27 (3):772–786. https://doi.org/10.1105/tpc.114.133777

Chapter 14

Reference-Based Identification of Long Noncoding RNAs in Plants with Strand-Specific RNA-Sequencing Data

Xiao Lin, Meng Ni, Zhixia Xiao, Ting-Fung Chan, and Hon-Ming Lam

Abstract

Long noncoding RNAs (lncRNAs) have been shown to play important roles in various organisms, including plant species. Several tools and pipelines have emerged for lncRNA identification, including reference-based transcriptome assembly pipelines and various coding potential calculating tools. In this protocol, we have integrated some of the most updated computational tools and described the procedures step-by-step for identifying lncRNAs from plant strand-specific RNA-sequencing datasets. We will start from clean RNA-sequencing reads, followed by reference-based transcriptome assembly, filtering of known genes, and lncRNA prediction. At the end point, users will obtain a set of predicted lncRNAs for downstream use.

Key words Plant long noncoding RNA, Computational identification, Software pipeline, Strand-specific RNA-sequencing, Reference-based transcriptome assembly

1 Introduction

Protein-coding genes have always been the research focuses in molecular biology. In recent years, noncoding RNAs have been identified in human, animal, and plant based on expressed sequence tag (EST), microarrays, and RNA-seq data. Long noncoding RNAs (lncRNAs) are conventionally defined as RNAs longer than 200 nucleotides and not being translated into protein [1]. In plants, the emerging roles of lncRNAs have been revealed by recent studies [2–4]. More than 6000 lncRNAs were identified from 200 *Arabidopsis thaliana* transcriptome datasets, with either organ-specific or stress-induced expression profile [5]. The same group systematically identified long noncoding natural antisense transcripts (lncNATs) responding to light in *Arabidopsis* [6]. More than 20,000 salt-responsive lncRNAs were identified from root and leaf samples of *Medicago truncatula* [7]. Over the past decade, various strategies have been developed to predict lncRNAs,

Julia A. Chekanova and Hsiao-Lin V. Wang (eds.), *Plant Long Non-Coding RNAs: Methods and Protocols*, Methods in Molecular Biology, vol. 1933, https://doi.org/10.1007/978-1-4939-9045-0_14,

including homology-based, features-based, and transcriptome-based ones [8]. Due to the lack of sequence conservation of lncRNAs among species, the sensitivity of sequence homology-based methods are hampered, while structure homology-based methods can complement other methods [9]. Features-based methods, which mainly predict lncRNAs based on known features of protein-coding and lncRNA transcripts using machine learning, have been actively developed by the community [10]. Among the three strategies, the transcriptome-based approach is of higher sensitivity and lower false-positive rate in detecting the de facto transcribed portion of lncRNAs [8]. These lncRNA prediction methods are usually integrated into a computational pipeline, though they can be used independently [8].

In this book chapter, we will go through a protocol starting from trimmed or cleaned RNA-sequencing reads. A reference-based transcriptome assembly pipeline [11] will be used for identifying a set of unannotated transcripts. Afterward, FEELnc [12] and blastx will be used to identify the potential set of lncRNAs. At the end of this protocol, the sequences of a set of predicted lncRNAs will be obtained. The schematic of this protocol is described in Fig. 1.

2 Materials

2.1 Linux Computational Environment

Linux is a family of free and open-source operating system, where all software mentioned in this protocol can be run in its terminal environment. At least 8 GB RAM is suggested according to some of the software's minimum system requirements. The number of CPUs and storage space depend on the users' needs and data size, respectively.

2.2 RNA-Sequencing Datasets and Genome Annotations

1. Clean RNA-sequencing data: RNA-sequencing data are usually in FASTQ format and compressed by gzip (http://www.gzip.org/). Raw RNA-sequencing data should be first cleaned by removing sequencing adapters and low-quality bases (*see* **Note 1**). Strand-specific RNA-sequencing data [13] is preferred for lncRNA identification, and it will be assumed that the data is strand-specific throughout this protocol.
2. Reference genome, gene annotation, and transcript sequences of the target organism: The reference genome is usually in FASTA format. It can be either in chromosome or contig levels. The gene annotation is usually in general feature format (GFF) or general transfer format (GTF) (*see* **Note 2**). It mainly contains protein coding mRNA, rRNA, and tRNA and other non-coding RNA genes of the organism (*see* **Note 3**). The file of

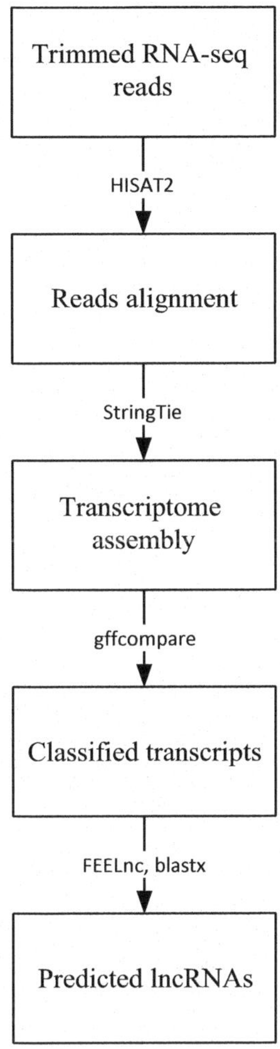

Fig. 1 Schematic of the protocol for computational prediction of lncRNAs from RNA-seq data

transcript sequences in FASTA format is also available, together with the same source of gene annotation files.

2.3 Software Requirements

1. HISAT2 [14]: HISAT2 is an efficient aligner, capable for mapping RNA-sequencing reads to a reference genome. It is available at https://ccb.jhu.edu/software/hisat2/.
2. SAMtools [15]: SAMtools is a tool package for easy manipulation of sequence alignment files. It is available at http://samtools.sourceforge.net/.
3. StringTie [16]: StringTie is a transcriptome assembler based on the mapping results of sequencing reads to a reference genome. It is available at https://ccb.jhu.edu/software/stringtie/.
4. Gffcompare [11]: Gffcompare is a program for comparing different sets of genomic features in GTF or GFF format. It is

available at http://ccb.jhu.edu/software/stringtie/gffcompare.shtml.

5. FEELnc [12]: FEELnc is a flexible tool for lncRNA prediction. It can work with either transcript sequences or gene annotation. It also works when the training set of lncRNA is absent for the organism. It is available at https://github.com/tderrien/FEELnc.
6. Perl: Perl is a scripting language, which is easy to use. It is available at https://www.perl.org/. Some of the codes are written in Perl for customized manipulation of intermediate files.
7. BioPerl: BioPerl is a collection of Perl modules for bioinformatics applications. It is available at http://bioperl.org/.

   Bio::Seq: Bio::Seq is a Perl module in BioPerl for manipulation of biological sequences in different formats.

   Bio::SeqIO: Bio::SeqIO is a Perl module in BioPerl for input and output of biological sequences in different formats.

2.4 Perl Scripts

These scripts are written in Perl for customized manipulation of intermediate files. Users can have their own choice of programming languages, in order to achieve the same purposes.

1. extract_transcript_gtf.pl: This script is to extract entries in a GTF file based on the provided list of transcript IDs.

```
# extract_transcript_gtf.pl
#!/bin/perl
use strict;
my $in = shift @ARGV;
my $out = shift @ARGV;
my $id = shift @ARGV;
open(IN, $in) or die "Cannot open $in!\n";
open(OUT, ">$out") or die "Cannot create $out!\n";
open(ID, $id) or die "Cannot open $id!\n";
my %id_list = ();
while(<ID>){
        chomp;
        my $temp_id = $_;
        $id_list{$temp_id} = 1;
}
close ID;
my $new_line_flag = 0;
while(<IN>){
        chomp;
        my $line = $_;
        $line =~ /transcript_id \"([^\"]+)\"/;
        my $temp_id = $1;
```

```
        if(exists $id_list{$temp_id}){
                if($new_line_flag == 0){
                      $new_line_flag = 1;
                }
                else{
                      print OUT "\n";
                }
                print OUT "$line";
        }
}
close IN;
close OUT;
```

2. extract_transcript_fasta.pl: This script is to extract transcript sequences from a reference genome, both in FASTA format, based on an annotation in GTF format.

```
# extract_transcript_fasta.pl
#!/bin/perl
use strict;
use Bio::Seq;
use Bio::SeqIO;
my $genome = shift @ARGV;
my $gtf = sfhit @ARGV;
my $out = shift @ARGV;
open(GTF, $gtf) or die "Cannot open $gtf!\n";
my $seq_in = Bio::SeqIO->new(-file => $genome, -format => 'fasta',);
my $seq_out = Bio::SeqIO->new(-file => ">$out", -format => 'fasta',);
my %genome = ();
while(my $seq_obj = $seq_in->next_seq){
      my $id = $seq_obj->primary_id;
      $genome{$id} = Bio::Seq->new();
      $genome{$id} = $seq_obj;
}
my %pos = ();
my %strands = ();
my %chrs = ();
my %gid = ();
my %exon_count = ();
while(<GTF>){
       chomp;
       my $line = $_;
       my @line = split('\t', $line);
       my $chr = $line[0];
       my $type = $line[2];
       if($type ne "exon"){
             next;
   }
```

```
        my $start = $line[3];
        my $end = $line[4];
        my $strand = $line[6];
        my $attri = $line[8];
        my $gene_id = "";
        my $transcript_id = "";
       $attri =~ /gene_id\"([^\;]+)\"\; transcript_id\"([^\;]+)\"/;
        $gene_id = $1;
        $transcript_id = $2;
        unless(exists $exon_count{$transcript_id}){
               $exon_count{$transcript_id} = -1;
        }
        $exon_count{$transcript_id}++;
        unless(exists $pos{$transcript_id}){
               @{$pos{$transcript_id}} = ();
        }
        @{$pos{$transcript_id}[$exon_count{$transcript_id}]} =
        ($start, $end);
        unless(exists $chrs{$transcript_id}){
               $chrs{$transcript_id} = $chr;
        }
        unless(exists $strands{$transcript_id}){
               $strands{$transcript_id} = $strand;
        }
        unless(exists $gid{$transcript_id}){
               $gid{$transcript_id} = $gene_id;
        }
}
close GTF;
foreach my $transcript_id (keys %pos){
        my @sorted_pos = sort {$a->[0] <=> $b->[0]} @{$pos
        {$transcript_id}};
        my $num = @sorted_pos;
        my $desc =
"GeneID=$gid{$transcript_id};Strand=$strands{$transcrip-
t_id};Pos=$chrs{$transcript_id}:";
        my $seq = "";
        for(my $i = 0; $i < $num; $i++){
               $seq .= $genome{$chrs{$transcript_id}}->subseq
               ($sorted_pos[$i][0], $sorted_pos[$i][1]);
                $desc .= "$sorted_pos[$i][0]-$sorted_pos[$i]
               [1],";
           }
           if($strands{$transcript_id} eq "-"){
                  $seq = reverse_complement($seq);
           }
           my $seq_obj = Bio::Seq->new();
           $seq_obj->id($transcript_id);
```

```
        $seq_obj->seq($seq);
        $seq_obj->desc($desc);
        $seq_out->write_seq($seq_obj);
}
        sub reverse_complement{
        my $dna = shift;
        my $revcomp = reverse($dna);
        $revcomp =~ tr/ACGTacgt/TGCAtgca/;
        return $revcomp;
}
```

3 Methods

3.1 Read Mapping to Reference Genome

HISAT2 is used to map RNA-sequencing data to the reference genome. An indexed reference genome is needed prior to mapping RNA-sequencing data to the reference genome. The following code assumes paired-end sequencing data are used. Single-end sequencing data can also be handled. The output of HISAT2 should be piped to SAMtools for generating an alignment file in binary alignment map (BAM) format, which will be convenient for downstream process. Users may specify options for HISAT2 according to the manual of HISAT2. If there are multiple sets of RNA-sequencing data for different samples, reads alignment with HISAT2 should be performed separately with each sample, which will generate alignment file <alignment1.bam>, <alignment2.bam> ... <alignmentN.bam>.

```
#Generate the reference genome index.
hisat2-build <reference_genome.fasta> <index>
#Mapping RNA-seq data to the reference genome.
hisat2 [options] <index> -1 <read1.fq.gz> -2 <read2.fq.gz>
| samtools view -Sb - > <alignment.bam>
```

3.2 Transcriptome Assembly

After getting the alignment file in BAM format, it will be used as the input file for transcriptome assembly with StringTie. If there are multiple alignment files, <alignment1.bam>, <alignment2.bam> ... <alignmentN.bam>, from different samples, transcriptome assembly should be performed with each alignment file separately, which will generate transcript assemblies <output1.gtf>, <output2.gtf> ... <outputN.gtf>.

```
#Assemble transcriptome on one sample.
stringtie alignment.bam -rf -o <output.gtf>
```

3.3 Transcript Classification

In this step, assembled transcripts for all samples will be merged and compared with gene annotation of the reference genome using gffcompare, in order to classify transcripts in different classes. The reference gene annotation should include protein-coding genes. If possible, rRNA, tRNA, and other noncoding RNA genes (*see* **Note 3**) can also be included, in order to exclude known noncoding transcripts in the unannotated set. The transcriptome assemblies of different samples, <output1.gtf>, <output2.gtf> ... <outputN.gtf>, should be listed in a file, e.g., <output_gtf_list.txt>, with one file name at a row. The following code is to merge assembled transcripts of all samples and compare them with the reference genome annotation.

```
#Merge assembled transcripts for all samples and compare them
with reference gene annotation.
gffcomapare -r <reference_gene_annotation.gtf> -i
<output_gtf_list.txt> -o <merged>
```

The output transcriptome assembly will be named as <merged.combined.gtf>, with class code for each transcript specified. The details of class codes can be found at http://ccb.jhu.edu/software/stringtie/gffcompare.shtml. A set of unannotated transcripts can be selected based on the class code. Potential class codes indicating unannotated transcripts include “x,” “i,” “y,” “p,” and “u” (*see* **Note 4**). The set of unannotated transcripts can be extracted using the following codes.

```
#Extract transcript IDs of the set of unannotated transcripts.
grep 'class_code \"x\"\|class_code \"i\"\|class_code \"y\"\|
class_code \"p\"\|class_code \"u\"' <merged.combined.gtf> |
cut -d'"' -f2 > unannotated_transcript_id.txt
#Extract the set of unannotated transcripts with the Perl
script in Subheading 2.4, step 1.
perl extract_transcript_gtf.pl <merged.combined.gtf>
<unannotated_transcript.gtf>
<unannotated_transcript_id.txt>
#Extract the sequences of unannotated transcripts with the
Perl script in Subheading 2.4, step 2.
perl extract_transcript_fasta.pl <reference_genome.fa>
<unannotated_transcript.gtf> <unannotated_transcript.fa>
```

3.4 lncRNA Identification

In the previous step, all assembled transcripts have been classified as different classes, and a set of unannotated transcript sequences have been selected based on the class code. FEELnc will be used to further infer potential lncRNAs from these unannotated transcripts. In this protocol, it is assumed that the training set of lncRNAs is absent, where the shuffle mode (-m “shuffle”) of FEELnc can be used to generate a set of noncoding nucleotides by randomly shuffling mRNA sequences as a training set, as suggested by the authors of FEELnc.

```
#Use FEELnc_codpot.pl in the FEELnc software package to
identify lncRNAs.
FEELnc_codpot.pl -i <unannotated_transcript.fa> -a
<mRNA_sequence.fa> -g <reference_genome.fa> -m 'shuffle' -o
<unannotated_transcript>
```

The predicted lncRNA transcripts will be in the files <unannotated_transcript.lncRNA.fa> and <unannotated_transcript.noORF.fa>. The sequences from the two files are then merged to generate a collection of lncRNA transcript sequences, which can be searched against the NCBI nonredundant (NR) protein database by blastx to further remove transcripts with significant homology to known proteins (e.g., e-value <1e-10, coverage of target >80%, and identity >90%) (*see* **Note 5**).

```
#Merge files <unannotated_transcript.lncRNA.fa> and < unan-
notated_transcript.noORF.fa>.
cat <unannotated_transcript.lncRNA.fa> < unannotated_tran-
script.noORF.fa> > <predicted_lncRNA_transcript.fa>
#Search the predicted lncRNA transcript sequences against the
NR protein database for further filtering.
blastx -db </path/to/nr> -query
<predicted_lncRNA_transcript.fa> -strand plus -out <
predicted_lncRNA_transcript.lnRNA_blastx.txt> -evalue 1E-10
-outfmt 6
```

4 Notes

1. Raw sequencing data can be cleaned by various tools, such as Trimmomatic [17] and Trim Galore! (https://www.bioinformatics.babraham.ac.uk/projects/trim_galore/). Optionally, quality controls tools, such as FastQC (https://www.bioinformatics.babraham.ac.uk/projects/fastqc/), can be used to check the quality of the sequencing data.
2. GFF and GTF are interconvertible tab-delineated file formats describing sequence features. The first eight columns are identical for both formats, describing (1) the name of the sequence containing the feature, (2) the source of the feature being generated, (3) the feature type, (4) the start position of the feature in the sequence, (5) the end position of the feature in the sequence, (6) the score of the feature as a floating point value, (7) at which strand the feature locates (+ or −), and (8) the frame information for protein-coding sequence feature. The last column describes additional information of the feature and differs for both formats. Users may refer to https://

ensembl.org/info/website/upload/gff.html for detailed description of the two formats.

3. The categories of genes in the gene annotation may vary from source to source. For example, gene annotations from Ensembl Plants [18] include protein-coding and noncoding genes, while those from Phytozome [18] include only protein-coding genes. It is recommended that both protein-coding and non-coding, such as rRNA and tRNA, genes are included in the gene annotation. If the gene annotation only contain protein-coding genes, RNAmmer [19], tRNAscan-SE [20], and Infernal [21, 22] should be first used to generate annotations for rRNA, tRNA, and other noncoding RNA genes, respectively. The gene annotation of these noncoding RNAs should be merged to the protein-coding gene annotation.
4. The mentioned class codes represent the following:

   "x": the transcript overlapping the exon(s) of an annotated gene at the opposite strand

   "i": the transcript fully contained in an intron of an annotated gene

   "y": the transcript containing an annotated gene within its intron(s)

   "p": the transcript adjacent to the 5′ end of an annotated gene at the same strand

   "u": the transcript locating in an intergenic region.

   Users may specify a different set of class codes according to the types of transcripts they would like to obtain. The code for extracting transcript IDs should be modified accordingly.
5. In this protocol, the result from blastx search is in tab-delineated table format. Users can filter the result by coverage and identify, using table editing software, such as Microsoft Excel. Users can also adjust the output format and contents according to the manual of blastx according to their own needs. Please notice that it is difficult to find a cutoff value of e-value, coverage of target, or identity that perfectly separates all protein-coding and noncoding sequences based on the blastx result. The choice of different cutoff values will imply different false-positive rates.

Acknowledgments

This work was supported by grants from the Hong Kong Research Grants Council Area of Excellence Scheme (AoE/M-403/16); CUHK VC Discretionary Fund (VCF2014004); National Key Research and Development Program–Key Innovative and

Collaborative Science and Technology Scheme for Hong Kong, Macau, and Taiwan (2017YFE0191100); CUHK Direct Grant (3132782); and the Lo Kwee-Seong Biomedical Research Fund to H.-M.L.

References

1. Marchese FP, Raimondi I, Huarte M (2017) The multidimensional mechanisms of long noncoding RNA function. Genome Biol 18 (1):206
2. Bazin J, Bailey-Serres J (2015) Emerging roles of long non-coding RNA in root developmental plasticity and regulation of phosphate homeostasis. Front Plant Sci 6:400
3. Chekanova JA (2015) Long non-coding RNAs and their functions in plants. Curr Opin Plant Biol 27:207–216
4. Zhao J, He Q, Chen G et al (2016) Regulation of non-coding RNAs in heat stress responses of plants. Front Plant Sci 7:1213
5. Liu J, Jung C, Xu J et al (2012) Genome-wide analysis uncovers regulation of long intergenic noncoding RNAs in Arabidopsis. Plant Cell 24 (11):4333–4345
6. Wang H, Chung PJ, Liu J et al (2014) Genome-wide identification of long noncoding natural antisense transcripts and their responses to light in Arabidopsis. Genome Res 24(3):444–453
7. Wang T-Z, Liu M, Zhao M-G et al (2015) Identification and characterization of long non-coding RNAs involved in osmotic and salt stress in Medicago truncatula using genome-wide high-throughput sequencing. BMC Plant Biol 15(1):131
8. Zhang Y, Huang H, Zhang D et al (2017) A review on recent computational methods for predicting noncoding RNAs. Biomed Res Int 2017:1–14
9. Johnsson P, Lipovich L, Grandér D et al (2014) Evolutionary conservation of long non-coding RNAs; sequence, structure, function. Biochim Biophys Acta 1840 (3):1063–1071
10. Han S, Liang Y, Li Y et al (2016) Long non-coding RNA identification: comparing machine learning based tools for long noncoding transcripts discrimination. Biomed Res Int 2016:1–14
11. Pertea M, Kim D, Pertea GM et al (2016) Transcript-level expression analysis of RNA-seq experiments with HISAT, StringTie and Ballgown. Nat Protoc 11(9):1650–1667
12. Wucher V, Legeai F, Hédan B et al (2017) FEELnc: a tool for long non-coding RNA annotation and its application to the dog transcriptome. Nucleic Acids Res 45(8):gkw1306
13. Parkhomchuk D, Borodina T, Amstislavskiy V et al (2009) Transcriptome analysis by strand-specific sequencing of complementary DNA. Nucleic Acids Res 37(18):e123
14. Kim D, Langmead B, Salzberg SL (2015) HISAT: a fast spliced aligner with low memory requirements. Nat Methods 12(4):357–360
15. Li H, Handsaker B, Wysoker A et al (2009) The sequence alignment/map format and SAMtools. Bioinformatics 25(16):2078–2079
16. Pertea M, Pertea GM, Antonescu CM et al (2015) StringTie enables improved reconstruction of a transcriptome from RNA-seq reads. Nat Biotechnol 33(3):290–295
17. Bolger AM, Lohse M, Usadel B (2014) Trimmomatic: a flexible trimmer for Illumina sequence data. Bioinformatics 30 (15):2114–2120
18. Bolser DM, Staines DM, Perry E et al (2017) Ensembl plants: integrating tools for visualizing, mining, and analyzing plant genomic data. Methods Mol Biol 1374:115–140
19. Lagesen K, Hallin P, Rødland EA et al (2007) RNAmmer: consistent and rapid annotation of ribosomal RNA genes. Nucleic Acids Res 35 (9):3100–3108
20. Schattner P, Brooks AN, Lowe TM (2005) The tRNAscan-SE, snoscan and snoGPS web servers for the detection of tRNAs and snoRNAs. Nucleic Acids Res 33(Web Server issue): W686–W689
21. Nawrocki EP, Eddy SR (2013) Infernal 1.1: 100-fold faster RNA homology searches. Bioinformatics 29(22):2933–2935
22. Kalvari I, Argasinska J, Quinones-Olvera N et al (2018) Rfam 13.0: shifting to a genome-centric resource for non-coding RNA families. Nucleic Acids Res 46(D1):D335–D342

Chapter 15

NAMS: Noncoding Assessment of long RNAs in Magnoliophyta Species

Gaurav Sablok, Kun Sun, and Hao Sun

Abstract

Regulation of plant transcriptional machinery has been recently demonstrated to be widely regulated by a class of long noncoding RNAs (lncRNAs) with size larger than 200 nt. The lncRNAs have been demonstrated to play key roles in abiotic stress. Taking into account the rapid pace in the development of the sequencing technologies, accelerated identification of lncRNAs with potential involvement in regulating the gene expression has been witnessed. Although progress has been made to identify the lncRNAs, however, accurate classification of the lncRNAs particularly in the case of plants is still challenging. In this protocol chapter, we present *NAMS*, which provides large-scale noncoding assessment of the lncRNAs specifically designed for Magnoliophyta species. We describe the approach and the usage of *NAMS* with potential applications for the lncRNA discovery.

Key words Noncoding assessment, lncRNAs, Magnoliophyta

1 Introduction

Regulation of gene expression has played a major role in understanding the genotypic changes induced either as a consequence of environmental stress or due to the altered physiological state. Altered gene expression is under the control of several factors among which noncoding RNAs often described as regulatory RNAs play a major role. Although the majority of the regulatory RNAs till now have been classified to the short class of RNAs such as microRNAs, isomiRs, and recently circular RNAs [1, 2], increasing evidences can be seen, which reflect the role of long noncoding RNAs (lncRNAs, defined as RNAs larger than 200 nt and lack coding potential) as potential new regulators. In plants, transcription of these lncRNAs takes place by RNA polymerase II–V and is involved in regulating the splicing of the coding mRNAs

Gaurav Sablok and Kun Sun contributed equally to this chapter.

Julia A. Chekanova and Hsiao-Lin V. Wang (eds.), *Plant Long Non-Coding RNAs: Methods and Protocols*, Methods in Molecular Biology, vol. 1933, https://doi.org/10.1007/978-1-4939-9045-0_15,

[3, 4]. These classes of lncRNAs have been shown to be widely distributed with as much as 6840 lncRNAs in *Arabidopsis thaliana* and have been shown to be involved in biotic and abiotic stress responses [5]. Realizing the potential implications of lncRNAs in plants and their role as key regulators in regulating the expression profiles in stress either as *cis-* or *trans-* regulators, efforts have been leveraged computationally to identify and profile lncRNAs to estimate the distinct role of lncRNAs [6, 7]. In the present protocol, we present *NAMS* (Noncoding Assessment in Magnoliophyta Species). The compiled package uses a decision classifier approach coupled with homology detection to identify and classify the lncRNAs. We benchmarked the *NAMS* algorithm against the previously developed classification approach, which uses SVM to classify lncRNAs [8] and demonstrate high accuracy of the NAMS over the previously developed approaches.

2 Materials

The source package and testing datasets for *NAMS* are freely available from http://sunlab.cpy.cuhk.edu.hk/NAMS/. *NAMS* is implemented in Perl (http://www.perl.org/) and runs on most GNU/Linux machines. To test the efficiency, prediction, and accuracy measures such as sensitivity and specificity of NAMS, *NAMS* has been tested on various Magnoliophyta gene annotation datasets available from previous studies. The source and prediction results of these datasets are listed in Table 1. To illustrate the usage of *NAMS* coupled with transcriptome assembly for lncRNA profiling, we further provided a complete workflow starting from the raw RNA-seq data. Multiple bioinformatics tools were used in the workflow: StringTie can be obtained from https://ccb.jhu.edu/software/stringtie/ [9, 10]; Trimmomatic was obtained from http://www.usadellab.org/cms/?page=trimmomatic [11]; TopHat can be downloaded from http://ccb.jhu.edu/software/tophat/ [12]; cuffcompare and gffread programs were obtained from the Cufflinks suite at http://cole-trapnell-lab.github.io/cufflinks/ [13]. In addition, the reference *Arabidopsis thaliana* genome and gene annotation were both obtained from TAIR: https://www.arabidopsis.org/ [14].

3 Methods

3.1 The Approaches Employed in NAMS

NAMS employs a decision-tree approach to assess the coding potential of the transcripts, which contains multiple steps as illustrated in Fig. 1.

Table 1
Source and prediction results of the Magnoliophyta annotation datasets from major plant databases

						Prediction			Performance	
Data source	Reference	Species	Category	Total items	Length ≥200	Coding	Noncoding	Uncertain	Acc%[a]	Sen%[b]
lncRNAdb	[22]	Magnoliophyta	Noncoding	6	6	0	6	0	100.0	100.0
Phytozome	[23]	*Brachypodium distachyon*	Coding	31,029	30,908	29,027	1398	483	95.4	93.9
		Oryza sativa	Coding	49,061	48,538	41,734	5257	1547	88.8	86.0
		Zea mays	Coding	63,540	63,251	52,348	7611	3292	87.3	82.8
PlantGDB	[24]	*Brachypodium distachyon*	Coding	30,991	27,601	12,222	10,290	5089	54.3	44.3
		Oryza sativa	Coding	44,644	42,204	9529	29,120	3555	24.7	22.6
PNRD	[25]	*Arabidopsis thaliana*	Noncoding	2577	2543	594	1392	557	70.1	54.7
		Oryza sativa	Noncoding	752	752	101	461	190	82.0	61.3
		Populus trichocarpa	Noncoding	538	538	1	536	1	99.8	99.6
		Zea mays	Noncoding	1704	1686	5	1636	45	99.7	97.0
TAIR	[14]	*Arabidopsis thaliana*	Coding	35,386	34,714	31,750	2180	784	93.6	91.5
		Arabidopsis thaliana	Noncoding	480	477	23	294	120	92.7	61.6

[a]Accuracy is defined as the correct predictions versus the number of solid predictions (i.e., coding plus noncoding)
[b]Sensitivity is defined as the correct predictions versus all the transcripts with a length ≥200 bp

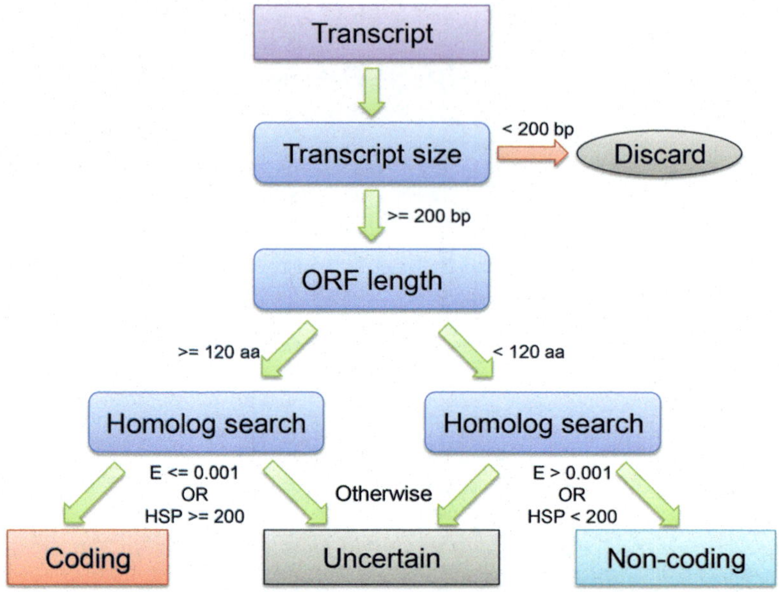

Fig. 1 Schematic overview of *NAMS*. To assess the coding potential of the transcripts, *NAMS* uses a decision-tree method that combines the ORF (open reading frame) size and homolog search result

1. Transcript size filter: Since the primary goal of *NAMS* is to differentiate the long noncoding RNAs from the protein-coding genes, therefore, we only focus on the transcripts that are longer than 200 bp (base pair) (*see* **Note 1**). Any transcripts that do not fulfill this criterion will be ignored for prediction and will be filtered off.
2. ORF size filter: Open reading frame (ORF) size is a widely used parameter to assess the coding potential of the transcript. For any transcript to have a coding potential, it must have accurately marked start codons (ATG) and termination codon (UAG, UGA, and UGG). These codons correspond to the standard genetic code = 1. Additionally, any transcript to be classified as protein coding should have an amino acid stretch more than 120 to be classified as a functional peptide. The rational is that for a protein-coding transcript, it must contain an ORF which is long enough to be translated into a functional peptide. In this task, the *txCdsPredict* program (*see* **Note 2**), which is a utility from the UCSC genome Browser [15], is employed. This program searches the whole transcript and reports the size of the longest ORF in the transcript which takes into account all the possible frameshifts. In addition, *NAMS* also provides inbuilt support for another program named GeneMarkS-T [16], which is an accurate ORF predictor which has been proved in the plant RNAs and especially suits for the reference-guided assembled transcripts [16]. The users

have the option to specify *txCdsPredict* or *GeneMarkS-T* for this task.

3. Homolog filter: Homolog information by *blastx* analysis is another widely used parameter for accessing the coding potential and has been widely implemented in previously developed algorithm such as CPC (coding potential calculator) [17]. Protein-coding genes are much more conserved than the noncoding ones during evolution selection. For a given transcript, the *blastx* program [18] is employed to search the homolog genes against the known protein-coding genes in the *UniProtKB* (Universal Protein Resource Knowledgebase) database [19] (*see* **Note 3**). For each transcript, the *blastx* looks for protein sequences with high similarities in the database; and for each match, *blastx* reports a high-scoring segment pairs (HSP) score as well as an E-value assessing the significance of the HSP score. The higher the HSP score means the higher the similarity of the queried transcript to the matched protein sequence. In our implementation, only the best match is considered: the HSP score and E-value for this match are extracted and used in the following procedures.
4. Noncoding assessment: *NAMS* then combine the ORF size, HSP score, and *E*-value to do the noncoding assessment. As shown in Fig. 1, a decision-tree approach was employed: If the ORF size is longer than or equal to 120 aa (amino acid) and then if either the HSP score is larger than (or equal to) 200 or E-value is smaller than (or equal to) 0.001 is satisfied, the transcript will be classified as "coding"; otherwise it will be classified as "uncertain" (*see* **Note 4**). If the ORF size is shorter than 120 aa and then if the HSP score is smaller than 200 or the E-value is larger than 0.001, the transcript will be classified as "noncoding"; otherwise it will be classified as "uncertain."
5. How to run *NAMS*: There are three parameters that *NAMS* requires: the first is the input RNA sequence in FASTA format; the second is the output directory; and the third is the prefix for the output files. Suppose that the user has an assembled transcript sets in FASTA format named "assembled_v1.fa" and wants to write the prediction results in a directory named "NAMS_pred" with a prefix of "assembled_v1," then the user could use the following command:

```
/path/to/NAMS assembled_v1.fa NAMS_pred assembled_v1
```

The result will be recorded in the file "NAMS_pred/assembled_v1.classification." For more information, the user could refer to README file in the source package.

3.2 Transcriptome Data Analysis Workflow

Here we provided a complete workflow of transcriptome data analysis starting from the raw RNA-seq data.

1. Define your filename for the transcriptome assembly. For example: Forward reads as RNAseq_R1.fastq and reverse reads as RNAseq_R2.fastq:

```
$FILE 1=RNAseq_R1.fastq
$FILE 2=RNAseq_R2.fastq
```

2. Clean the raw reads using the Trimmomatic:

```
$ java -jar trimmomatic-0.36.jar PE -phred33 -threads
8 $FILE 1 $FILE 2 \
$FILE 1.P.fastq $FILE 1.U.fastq $FILE 2.P.fastq $FILE 2.U.
fastq \
ILLUMINACLIP:TruSeq3-PE-2.fa:2:30:10 LEADING:3 TRAILING:3 \
SLIDINGWINDOW:4:15 MINLEN:50
```

3. Make bowtie indices:

```
$ bowtie2-build TAIR10_chr_all.fas TAIR10
```

4. Map the cleaned reads to bowtie indices:

```
$ tophat2 --library-type fr-firststrand -p 8 -G
TAIR10_GFF3_genes.gff \
TAIR10 RNAseq_R1.P.fastq RNAseq_R2.P.fastq
```

5. Use stringtie to assemble the transcriptome reads:

```
$ stringtie -p 8 -G TAIR10_GFF3_genes.gff -o RNAseq.gtf
accepted_hits.bam
```

6. Compare the assembled transcripts with the reference transcriptome:

```
$ cuffcompare -r Athaliana_167_TAIR10.gene.gtf -o RNAseq.
vs.TAIR \
RNAseq.gtf
```

7. Pick up the transcripts with class code "x" (cis antisense), "i" (intronic), "u" (intergenic) or "j" (alternatively spliced):

```
cat RNAseq.vs.TAIR.combined.gtf | perl -ne "print if /
class_code '[xij]';/" \
> RNAseq.vs.TAIR.unannotated.gtf
```

8. Get the sequences of the unannotated transcripts:

```
gffread RNAseq.vs.TAIR.unannotated.gtf -r TAIR10_chr_all.fas \
-w RNAseq.vs.TAIR.unannotated.fa
```

9. Run *NAMS* on the unannotated transcripts to get lncRNAs

```
$ NAMS RNAseq.vs.TAIR.unannotated.fa RNAseq.NAMS \
RNAseq.vs.TAIR.unannotated
```

The noncoding assessment result was then written into the output file "SRA.NAMS/SRA.vs.TAIR.unannotated.classification," and those with statuses of "noncoding" were putative lncRNAs for further downstream analyses.

4 Notes

1. The definition of lncRNAs requires the ncRNA to be longer than 200 bp. Short noncoding RNAs, for instance, the microRNAs and transfer RNAs, are different from lncRNAs in many aspects, including structure and conservation [20]. Since *NAMS* is designed for differentiating lncRNAs from protein-coding genes, therefore, the size filter is essential.
2. For the ORF prediction, there are many other programs available such as ORF Finder (Open Reading Frame Finder) from NCBI (National Center for Biotechnology Information) [21]. The performances are quite similar between these tools since the principle is very straightforward and the implementation is not complex.
3. For homolog search, the *blastx* program supports multi-thread mode to increase the speed which could significantly shorten the running time. Meanwhile, the UniProtKB was used as the protein database based on its comprehensiveness and high quality. Another choice is the protein database constructed by NCBI (the nr database) [21] which shows a similar performance.
4. The "uncertain" category means that the coding status of the transcript cannot be classified because these transcripts show both coding and noncoding characteristics. To avoid the high false prediction rate of these transcripts, *NAMS* leaves them as unpredictable. In this manner, the confidence of the valid predictions (either "coding" or "noncoding") could be largely guaranteed, while as a compromise, some transcripts (around 8% based on the data in Table 1) cannot be predicted.

References

1. Sablok G, Srivastva AK, Suprasanna P, Baev V, Ralph PJ (2015) isomiRs: increasing evidences of isomiRs complexity in plant stress functional biology. Front Plant Sci 6:949
2. Sablok G, Zhao H, Sun X (2016) Plant circular RNAs (circRNAs): transcriptional regulation beyond miRNAs in plants. Mol Plant 9 (2):192–194
3. Dinger ME, Pang KC, Mercer TR, Mattick JS (2008) Differentiating protein-coding and noncoding RNA: challenges and ambiguities. PLoS Comput Biol 4(11):e1000176
4. Wierzbicki AT, Haag JR, Pikaard CS (2008) Noncoding transcription by RNA polymerase Pol IVb/Pol V mediates transcriptional silencing of overlapping and adjacent genes. Cell 135 (4):635–648
5. Di C, Yuan J, Wu Y, Li J, Lin H, Hu L, Zhang T, Qi Y, Gerstein MB, Guo Y, Lu ZJ (2014) Characterization of stress-responsive lncRNAs in Arabidopsis thaliana by integrating expression, epigenetic and structural features. Plant J 80(5):848–861
6. Sun K, Chen X, Jiang P, Song X, Wang H, Sun H (2013) iSeeRNA: identification of long intergenic non-coding RNA transcripts from transcriptome sequencing data. BMC Genomics 14(Suppl 2):S7
7. Sun K, Zhao Y, Wang H, Sun H (2014) Sebnif: an integrated bioinformatics pipeline for the identification of novel large intergenic noncoding RNAs (lincRNAs)—application in human skeletal muscle cells. PLoS One 9(1):e84500
8. Boerner S, McGinnis KM (2012) Computational identification and functional predictions of long noncoding RNA in Zea mays. PLoS One 7(8):e43047
9. Pertea M, Pertea GM, Antonescu CM, Chang TC, Mendell JT, Salzberg SL (2015) StringTie enables improved reconstruction of a transcriptome from RNA-seq reads. Nat Biotechnol 33 (3):290–295
10. Pertea M, Kim D, Pertea GM, Leek JT, Salzberg SL (2016) Transcript-level expression analysis of RNA-seq experiments with HISAT, StringTie and Ballgown. Nat Protoc 11 (9):1650–1667
11. Bolger AM, Lohse M, Usadel B (2014) Trimmomatic: a flexible trimmer for Illumina sequence data. Bioinformatics 30 (15):2114–2120
12. Trapnell C, Pachter L, Salzberg SL (2009) TopHat: discovering splice junctions with RNA-Seq. Bioinformatics 25(9):1105–1111
13. Trapnell C, Williams BA, Pertea G, Mortazavi A, Kwan G, van Baren MJ, Salzberg SL, Wold BJ, Pachter L (2010) Transcript assembly and quantification by RNA-Seq reveals unannotated transcripts and isoform switching during cell differentiation. Nat Biotechnol 28(5):511–515
14. Berardini TZ, Reiser L, Li D, Mezheritsky Y, Muller R, Strait E, Huala E (2015) The Arabidopsis information resource: making and mining the "gold standard" annotated reference plant genome. Genesis 53(8):474–485
15. Kent WJ, Sugnet CW, Furey TS, Roskin KM, Pringle TH, Zahler AM, Haussler D (2002) The human genome browser at UCSC. Genome Res 12(6):996–1006
16. Tang S, Lomsadze A, Borodovsky M (2015) Identification of protein coding regions in RNA transcripts. Nucleic Acids Res 43(12):e78
17. Kong L, Zhang Y, Ye ZQ, Liu XQ, Zhao SQ, Wei L, Gao G (2007) CPC: assess the protein-coding potential of transcripts using sequence features and support vector machine. Nucleic Acids Res 35(Web Server issue):W345–W349
18. Altschul SF, Gish W, Miller W, Myers EW, Lipman DJ (1990) Basic local alignment search tool. J Mol Biol 215(3):403–410
19. Consortium U (2015) UniProt: a hub for protein information. Nucleic Acids Res 43(Database issue):D204–D212
20. Liu J, Wang H, Chua NH (2015) Long noncoding RNA transcriptome of plants. Plant Biotechnol J 13(3):319–328
21. Wheeler DL, Church DM, Federhen S, Lash AE, Madden TL, Pontius JU, Schuler GD, Schriml LM, Sequeira E, Tatusova TA, Wagner L (2003) Database resources of the National Center for Biotechnology. Nucleic Acids Res 31(1):28–33
22. Amaral PP, Clark MB, Gascoigne DK, Dinger ME, Mattick JS (2011) lncRNAdb: a reference database for long noncoding RNAs. Nucleic Acids Res 39(Database issue):D146–D151
23. Goodstein DM, Shu S, Howson R, Neupane R, Hayes RD, Fazo J, Mitros T, Dirks W, Hellsten U, Putnam N, Rokhsar DS (2012) Phytozome: a comparative platform for green plant genomics. Nucleic Acids Res 40(Database issue):D1178–D1186
24. Dong Q, Schlueter SD, Brendel V (2004) PlantGDB, plant genome database and analysis tools. Nucleic Acids Res 32(Database issue): D354–D359
25. Yi X, Zhang Z, Ling Y, Xu W, Su Z (2015) PNRD: a plant non-coding RNA database. Nucleic Acids Res 43(Database issue): D982–D989

Chapter 16

De Novo Plant Transcriptome Assembly and Annotation Using Illumina RNA-Seq Reads

Stephanie C. Kerr, Federico Gaiti, and Milos Tanurdzic

Abstract

The ability to identify and quantify transcribed sequences from a multitude of organisms using high-throughput RNA sequencing has revolutionized our understanding of genetics and plant biology. However, a number of computational tools used in these analyses still require a reference genome sequence, something that is seldom available for non-model organisms. Computational tools employing de Bruijn graphs to reconstruct full-length transcripts from short sequence reads allow for de novo transcriptome assembly. Here we provide detailed methods for generating and annotating de novo transcriptome assembly from plant RNA-seq data.

Key words RNA-seq, Long noncoding RNA, Trinity, De novo transcriptome assembly

1 Introduction

Plant functional genomics has entered an era in which contemporary uses of next-generation sequencing (NGS) have been fully adopted by the research community. However, the required availability of assembled and annotated reference genomes for functional studies, such as gene expression profiling, limits access to these experimental approaches for a large number of plant species, which are often plagued by large genomes with high percentages of repetitive DNA. Another issue hampering quantitative and qualitative interrogations of transcriptomic changes arises from the fact that the vast majority of current transcriptomic analyses relies on generating large numbers of short sequence reads from fragmented RNA transcripts using Illumina high-throughput RNA-seq methods. This has created a need for computational tools to align the RNA reads to a reference and tools used to re-create the RNA

Electronic supplementary material: The online version of this chapter (https://doi.org/10.1007/978-1-4939-9045-0_16) contains supplementary material, which is available to authorized users.

Julia A. Chekanova and Hsiao-Lin V. Wang (eds.), *Plant Long Non-Coding RNAs: Methods and Protocols*, Methods in Molecular Biology, vol. 1933, https://doi.org/10.1007/978-1-4939-9045-0_16,

transcripts where those reads came from, or in other words, to assemble the transcripts into a transcriptome collection.

Availability of a reference genome sequence enables sophisticated and precise quantification of gene expression through the use of bioinformatics tools like TopHat2 [1] and RNA-STAR [2] that rely on reference-based transcriptome assembly and transcript quantification (including the ability to identify differences in gene expression between different transcript isoforms) using tools like DESeq2 [3], edgeR [4], or Cufflinks [5]. Standard computational analysis pipelines incorporating read mapping and transcript quantification are well represented in the literature and are easily streamlined for inexperienced users through the use of graphical user interface access and preset workflow procedures, for example, as Galaxy bioinformatics server instances [6].

With the development of de novo transcriptome assembly algorithms implemented in several different software packages, researchers are now able to perform most types of transcriptome analyses even in the absence of a reference genome for the species they work on [7–10]. The vast majority of de novo assemblers operate on a similar conceptual basis: individual sequence reads (e.g., RNA-seq) are overlapped until a contig of short sequence reads assembles (i.e., reconstructs) a full-length transcript where those reads originally came from. This computationally intensive process most commonly uses a strategy based on de Bruijn graphs created from concatamerization of sequence reads sharing common k-mers (sequence strings shorter than the sequence reads) [8, 11, 12]. Being graph-based, this approach, in principle, allows for identification of multiple transcript isoforms arising from the same locus (e.g., alternatively spliced transcripts). By the same token, however, it also suffers from mis-assembly errors, caused by shared k-mers due to close sequence similarity of gene homologues (as is often the case with gene family members or in individuals with high levels of heterozygosity). A trial-and-error approach and stringent quality control are needed for each de novo assembled transcriptome. The completeness of any transcriptome, i.e., the number of genes whose transcripts can be detected in the transcriptome assembly, is directly proportionate to the RNA sample complexity; the higher the number of genes expressed in the sample, the higher is the likelihood those transcripts will be represented in the assembly. Normalization protocols are available [13], although normalization steps tend to produce transcriptome assemblies with higher complexities but fewer true full-length transcript contigs (T. Chabikwa, F. Barbier, and M. Tanurdzic, pers. obs.).

In this protocol, we provide a detailed description of a computational pipeline that can be used to generate a de novo transcriptome assembly from any single- or paired-end Illumina-generated RNA-seq dataset, using Trinity [11]. This de novo assembler has garnered wide adoption in de novo transcriptome assembly but is far from being the only de novo assembler available [7]. As with any bioinformatics protocol, new and updated versions of the software packages used here may be available after the publication. More appropriate computational tools may be substituted within the pipeline to better suit the research project.

2 Materials

2.1 Hardware

A computer with Internet connection.

A multi-CPU cluster may be necessary for some memory-intensive data analysis steps (e.g., assembly and annotation).

2.2 Software

A program capable of running Unix command line, such as Cygwin for Windows (https://www.cygwin.com/) or Terminal preinstalled on Mac.

FASTX-Toolkit (http://hannonlab.cshl.edu/fastx_toolkit/index.html).

DeconSeq [14] (http://deconseq.sourceforge.net).

Trinity [11, 15] (https://github.com/trinityrnaseq/trinityrnaseq/wiki).

BLAST® Command Line Application (https://www.ncbi.nlm.nih.gov/books/NBK279690/).

HMMER [16] (http://hmmer.org).

SignalP [17] (http://www.cbs.dtu.dk/services/SignalP/).

Getorf tool [18] (http://emboss.sourceforge.net/apps/cvs/emboss/apps/getorf.html).

Coding Potential Calculator (CPC) [19] (http://cpc.cbi.pku.edu.cn/).

Infernal [20] (http://eddylab.org/infernal/).

Python (https://www.python.org/about/).

2.3 RNA-Seq Data

Fastq files obtained from Illumina sequencing platform.

2.4 Databases

UniProtKB (https://www.uniprot.org/uniprot/).

Swiss-Prot (https://www.uniprot.org/uniprot/).

UCOs [21] (http://cgpdb.ucdavis.edu/cgpdb2/).

3 Methods

3.1 Read Quality Control

1. Download and install FASTQC from Babraham Bioinformatics (http://www.bioinformatics.babraham.ac.uk/projects/fastqc/).
2. Select "File -> Open" to open each raw read fastq file individually (read files do not need to be unzipped). Click on each tab

on the left-hand side to examine the FASTQC output. Select "File -> Save report" to save the report with an appropriate name (*see* **Note 1**).

3. Download and install the FASTX-Toolkit from the Hannon Lab (http://hannonlab.cshl.edu/fastx_toolkit/download.html).
4. Remove the appropriate adaptor sequences from the reads:

   fastx_clipper –a *adaptorsequence* –v –Q33 –n –i *input.fastq* –o *output.fastq* (*see* **Notes 2–4**).
5. Next remove the first 10–15 bps from the 5′end of the read (*see* **Note 5**):

   fastx_trimmer –v –Q33 –f10 –i *input.fastq* –o *output.fastq* (*see* **Note 6**).
6. Remove any bases that have a read quality less than Q30 (*see* **Note 7**):

   fastq_quality_trimmer –v –Q33 –t30 –l15 –i *input.fastq* –o *output.fastq* (*see* **Note 8**)
7. Download and install DeconSeq (https://sourceforge.net/projects/deconseq/files/).
8. Download contaminant databases:

   wget ftp://edwards.sdsu.edu:7009/deconseq/db/ (*see* **Note 9**).
9. Finally, use DeconSeq to remove common contaminants such as bacterial and viral, that may be present in your sample (*see* **Note 10**):

   deconseq.pl -i 95 -c 90 -f *input.fastq* -id *output.fq* -dbs bact (*see* **Note 11**).

3.2 De Novo Transcriptome Assembly

1. Download and install python (https://www.python.org/downloads/).
2. Download the python script "fastqCombinePairedEnd.py" (https://github.com/enormandeau/Scripts/blob/master/fastqCombinePairedEnd.py).
3. Remove any unpaired reads from the forward and reverse read files for each paired-end library (*see* **Note 12**):

   python fastqcombinepairedend.py "@M01" "/" *R1.fastq R2.fastq* (*see* **Note 13**).
4. Download and install Trinity (https://github.com/trinityrnaseq/trinityrnaseq/releases).
5. Use Trinity with the default settings to assemble the reads into a de novo transcriptome assembly:

   Trinity.pl --seqType fa --SS_lib_type RF --max_memory 50G --left *R1_paired.fa* --right *R2_paired.fa* --CPU 8 (*see* **Note 14**).

3.3 Assessment of Assembly (See Note 15)

1. Download and install BLAST® Command Line Application (https://www.ncbi.nlm.nih.gov/books/NBK279671/).
2. Create a BLAST database of your transcriptome:

   makeblastdb -in *transcriptome.fa* -input_type fasta -dbtype nucl -out *transcriptome_blastdb*
3. Run BLASTN to determine the redundancy of the transcriptome (*see* **Note 16**):

   blastn -db *transcriptome_blastdb* -query *transcriptome.fa* -evalue 1e-10 -outfmt 6 -out *output*
4. Extract any protein sequences that have already been sequenced from your species of interest from the UniProtKB and Swiss-Prot databases. Go to https://www.uniprot.org/uniprot/, and type your species name in the search box, and press enter. To ensure that the list only contains protein sequences from your species, type your species name into the search box at the left-hand side under "Popular organisms -> Other organisms," and select the appropriate species from the dropdown menu, and press go. Tick the box in the first row first column of the output table to select all of the protein sequences. Click Download, and change the Format to "FASTA (canonical & isoform)," and then press go. A new web page will open with the fasta output, select all, and copy and paste into a simple text program such as TextEdit (Mac) or Notepad (Windows). Save the file using an appropriate name. Repeat the above steps for Swiss-Prot, but instead filter the results by selecting "Reviewed Swiss-Prot" under "Filter by" on the left-hand side of the results table.
5. Create BLAST databases of the UniProtKB and Swiss-Prot protein sequences:

   makeblastdb -in *UniProtKB.fa* -input_type fasta -dbtype prot -out *UniProtKB_blastdb*

   makeblastdb -in *UniProtKB.fa* -input_type fasta -dbtype prot -out *SwissProt_blastdb*
6. Use BLASTX to see how many of these protein sequences from your species have been assembled in your transcriptome:

   blastx -db *UniProtKB_blastdb* -query *transcriptome.fa* -evalue 1e-10 -out *output* -outfmt 6

   blastx -db *Swissprot_blastdb* -query *transcriptome.fa* -evalue 1e-10 -out *output* -outfmt 6
7. Download the protein sequences of the 357 ultra-conserved orthologs (UCOs) A_thaliana_ATGC_2006_08_24.protein.COS_ULTRA.fasta from http://cgpdb.ucdavis.edu/cgpdb2/ (*see* **Note 17**).
8. Create a BLAST database of the UCO protein sequences:

makeblastdb -in A_thaliana_ATGC_2006_08_24.protein.COS_ULTRA.fasta -input_type fasta -dbtype prot -out *UCO_blastdb*

9. Use BLASTX to identify how well your transcriptome has assembled these UCOs:

   blastx -db *UCO_blastdb* -query *transcriptome.fa* -evalue 1e-10 -out *output* -outfmt 6

10. Research the literature to identify if any single copy genes have been sequenced in your species. Obtain the mRNA or CDS sequence for each single copy gene from NCBI (https://www.ncbi.nlm.nih.gov). Assess whether your transcriptome contains these single copy genes and how well the genes have been assembled using BLASTN:
    blastn -db *transcriptome_blastdb* -query *single_copy_gene.fa* -evalue 1e-100 -out *output* -outfmt 5 -max_target_seqs 20″

3.4 Gene Annotation and Ontology

1. Download the nr protein database and unzip:

   wget ftp://ftp.ncbi.nlm.nih.gov/blast/db/FASTA/nr.gz
   gunzip nr.gz

2. Create a BLAST database of the nr protein database for searching:

   makeblastdb -in nr -input_type fasta -dbtype prot -out *nr_blastdb*

3. Annotate the sequences assembled in the transcriptome using the nr protein database:

   blastx -db *nr_blastdb* -query *transcriptome.fa* -evalue 1e-10 -out *output* -outfmt 5 -max_target_seqs 20 (*see* **Note 18**).

4. Download and install BLAST2GO from https://www.blast2go.com/blast2go-pro/download-b2g. Register for free BLAST2GO at https://www.blast2go.com/b2g-register-basic.

5. Load the transcriptome into BLAST2GO by selecting "File -> Load -> Load Sequences -> Load Fasta File (.fasta)."

6. Upload BLAST annotation output files from **step 3** by selecting "File -> Load -> Load BLAST Results -> XML Files."

7. Map the BLAST Hits for each sequence with the Gene Ontology (GO) terms by selecting "Mapping" and then "R" (*see* **Note 19**).

8. Once the mapping is completed, annotate the sequences with GO and KEGG terms by selecting "annot" with the default options. Change the "Filter GO by Taxonomy" option to an appropriate taxon for your species. Select "Next", "Next", then "Run".

9. At the same time, run InterProScan to annotate the sequences with GO terms by selecting "interpro" with the default options, and then select "Next", then "Run."
10. Once InterProScan and annotation is completed, select the dropdown arrow next to "interpro," and then select "Merge InterProScan GOs to Annotation" to merge InterProScan annotation with GO term annotations from the BLAST results.

3.5 Long Noncoding RNA Prediction

1. Download and install HMMER (http://hmmer.org/download.html).
2. Download PfamA database and unzip:

   wget ftp://ftp.ebi.ac.uk/pub/databases/Pfam/current_release/Pfam-A.hmm.gz
   gunzip Pfam-A.hmm.gz
3. Download and install SignalP (http://www.cbs.dtu.dk/cgi-bin/nph-sw_request?signalp).
4. Download and install EMBOSS containing getorf software (http://emboss.sourceforge.net/download/).
5. Translate the isoforms and keep only the longest unique ORF for each isoform:

   getorf -sequence *transcriptome.fa* -find 0 *ORFs.pep*
   get_longest_ORF_per_transcript.pl *ORFs.pep* > *longest_ORF_per_transcript.pep*
   The Perl script (get_longest_ORF_per_transcript.pl) is provided as Supplementary File 1.
6. Run BLASTP query to identify protein coding sequences:

   blastp -query *longest_ORF_per_transcript.pep* -evalue 0.0001 -max_target_seqs 100 -db nr -num_threads 8 -outfmt "7 qseqid qlen sseqid pident length mismatch gapopen qstart qend sstart send ppos evalue bitscore score" -out *BlastP.csv.*
7. Run BLASTX query to identify protein coding sequences:

   blastx -query *transcriptome.fa* -evalue 0.0001 -max_target_seqs 100 -db nr -num_threads 8 -outfmt "7 qseqid qlen sseqid pident length mismatch gapopen qstart qend sstart send ppos evalue bitscore score" -out *BlastX.csv.*
8. Run HMMER query to identify protein coding sequence:

   hmmscan -o PfamA --tblout *PfamA_table* --incE 0.0001 --incdomE 0.0001 --cpu 8 PfamA.hmm *longest_ORF_per_transcript.pep*
9. Use output files from **steps 5–8**, and run bash script "lncRNA_pipeline.sh" to remove any sequences deemed as protein coding. The bash script is provided as Supplementary File 2.

10. Go to http://cpc.cbi.pku.edu.cn/programs/run_cpc.jsp. Copy and paste the sequences from lncRNAs_StartStop50.fa into the "fasta sequence data input" box, and press run. Remove the sequences defined as "coding" to produce the final list of putative lncRNAs.

3.6 MicroRNA Precursor Annotation

1. Download and install Infernal (http://eddylab.org/infernal/).
2. Download RFAM.cm database and unzip the file:

   wget ftp://ftp.ebi.ac.uk/pub/databases/Rfam/CURRENT/Rfam.cm.gz
   gunzip Rfam.cm.gz.

3. Run Infernal to annotate sequences with Rfam:
   cmsearch -A hits --tblout *output* -E 0.01 Rfam.cm *transcriptome.fa*.

4 Notes

1. Ideally, this step should be repeated after each read quality control step to assess the read quality. However, at a minimum, this step should be performed before any read quality control steps have been performed and then after the final read quality control step (e.g., after **step 9** here).
2. -Q33 is needed to specify that the quality score is Sanger (offset 33) and not the old Illumina (ASCII offset 64). This specification will need to be applied for all newer Illumina reads and can also be used for other sequencing technology reads such as Ion Torrent and PacBio.
3. For all command lines, if an object is italicized, then you are required to input your own data or file name. For example, in **step 4**, *adaptor sequence* will need to be replaced with the nucleotide sequence of the appropriate adaptor used for that sample, and also the *input* and *output* file names will also need to be altered to match the file names you have used. If you are using paired-end reads, then both the forward and reverse read files will need to be processed independently for each step.
4. We recommend carefully considering the choice of output file name for each step to ensure that the contents of the file are clear for future reference. For the remainder of the methods, unless the output file was used as an input file in a subsequent step, we have named all output files with the generic term "output."
5. The first 10–15 base pairs of Illumina sequencing are generally of lower quality, but the exact number will vary depending on the sequence error rate of the technology used. Refer to the FASTQC output to assess how many bps at the beginning of

each read are ~<Q30. Specify the number of base pairs to remove using the –f parameter, e.g., to remove 10 bps specify –f10.

6. For all command lines, if an object is underlined, then this indicates that you may wish to adjust this parameter to suit your project, e.g., –f10 will result in 10 bps being removed from the beginning of the read, but if you wish to remove, for example, 20 bps, then you would use –f20.
7. The read quality may also decline toward the end of the read, especially for longer reads. It is therefore recommended to trim the end of each read once the base pairs fail to reach a minimum quality score. To specify the minimum quality score below which the remaining base pairs of a read will be removed use the -t parameter, e.g., to set a minimum quality score of 30, specify -t30.
8. You can also specify to remove any reads less than a certain length after the read quality control step using the –l parameter, e.g., –l15 specifies that any reads less than 15 bps long after the trimming function has been performed will be removed.
9. The database link provided was inaccessible at the time of publishing. If this is still the case, then the databases can be temporarily downloaded from https://osf.io/egn6m/.
10. Specify what database to search using the –dbs parameter. Depending on your RNA source, you may want to remove different contaminating sequences, e.g., bacterial or viral sequences. You can also use the –dbs-retain function to retain certain sequences that may be present in both your sample and the contaminating database, e.g., plant-specific sequences.
11. Specify the percentage identity and percentage coverage using the –i and –c parameters, e.g., –i 95 will specify that the sequences have at least 95% identity matches and –c 90 will specify that sequences have at least 90% coverage.
12. Trinity can only assemble either paired-end or single-end reads. When assembling paired-end reads, Trinity requires each read in the forward read file to have a matching pair in the reverse read file; however, the read quality control steps may result in the loss of some paired-end reads. You can use the fastqCombinePairedEnd.py script to remove any reads that are unpaired after the quality control steps have been performed.
13. "@M01" refers to the beginning characters of each read name and will be different depending on how the Illumina (or other) technology labelled the reads. "/" refers to the delimiter before the numbering that defines R1 and R2, e.g., "/" specifies that the read name ended in either /1 or /2 referring to R1 or R2. Three output files are produced: R1_paired.fastq,

R2_paired.fastq, and Singles.fastq. Use the R1_paired.fastq and the R2_paired.fastq files for the Trinity assembly.

14. --seqType refers to the format of the input files, e.g., fa refers to fasta format and fq refers to fastq format. If you are using strand-specific read libraries, then you need to specify which of the paired-end reads is forward and which is reverse using --SS_lib_type; RF specifies that R1 is reverse and R2 is forward, while FR specifies that R1 is forward and R2 is reverse. If you are not using strand-specific read libraries, then remove this parameter altogether. You may also wish to specify the maximum amount of memory and CPUs for Trinity to use; depending on the computing system being used, these can be specified with --max_memory and --CPU.
15. There are a number of different methods that can be used to assess the quality of the assembled transcriptome in terms of both how comprehensiveness and representative it is. We recommend as many of the methods suggested below to increase the confidence in your assembly.
16. The number of matches between different numbered comps should be low. If there is a high amount of redundancy in your transcriptome, then you may need to use a tool to remove redundant sequences, e.g., CAP3 (web version, http://biosrv.cab.unina.it/webcap3/, or command line version, http://seq.cs.iastate.edu/cap3.html).
17. The ultra-conserved orthologs (UCOs) are a list of 357 single-copy genes that are shared between *Arabidopsis thaliana*, humans, mice, yeast, fruit flies, and *Caenorhabditis elegans* [21]. Determining how many UCOs are present in your transcriptome can give an indication of how comprehensive the assembly is.
18. Only use –outfmt 5 if you will be loading the BLAST results into a program that requires that output such as BLAST2GO in **step 5**. Otherwise use –outfmt 6 which will run faster and produce the results in a tabular format.
19. Running GO and KEGG annotations can take a long time, so only do it if you require these annotations for your transcriptome. You might also want to use an online server such as GOanna (http://agbase.arizona.edu/cgi-bin/tools/GOanna.cgi) for GO term annotation or KAAS (http://www.genome.jp/kaas-bin/kaas_main) for KEGG annotations. However, these servers usually restrict how many sequences can be uploaded at once, so you may need to split your transcriptome into smaller files.

References

1. Kim D, Pertea G, Trapnell C et al (2013) TopHat2: accurate alignment of transcriptomes in the presence of insertions, deletions and gene fusions. Genome Biol 14:R36
2. Dobin A, Davis CA, Schlesinger F et al (2013) STAR: ultrafast universal RNA-seq aligner. Bioinformatics 29:15–21
3. Love MI, Huber W, Anders S (2014) Moderated estimation of fold change and dispersion for RNA-seq data with DESeq2. Genome Biol 15:550
4. Robinson MD, McCarthy DJ, Smyth GK (2010) edgeR: a bioconductor package for differential expression analysis of digital gene expression data. Bioinformatics 26:139–140
5. Trapnell C, Williams BA, Pertea G et al (2010) Transcript assembly and abundance estimation from RNA-Seq reveals thousands of new transcripts and switching among isoforms. Nat Biotechnol 28:511–515
6. Afgan E, Sloggett C, Goonasekera N et al (2015) Genomics virtual laboratory: a practical bioinformatics workbench for the cloud. PLoS One 10:e0140829
7. Ungaro A, Pech N, Martin J-F et al (2017) Challenges and advances for transcriptome assembly in non-model species. PLoS One 12: e0185020
8. Moreton J, Izquierdo A, Emes RD (2016) Assembly, assessment, and availability of de novo generated eukaryotic transcriptomes. Front Genet 6:361
9. Atallah NM, Vitek O, Gaiti F et al (2018) Sex determination in Ceratopteris richardii is accompanied by transcriptome changes that drive epigenetic reprogramming of the young gametophyte. G3 Genes Genomes Genet 8:2205–2214
10. Kerr SC, Gaiti F, Beveridge CA et al (2017) De novo transcriptome assembly reveals high transcriptional complexity in Pisum sativum axillary buds and shows rapid changes in expression of diurnally regulated genes. BMC Genomics 18:221
11. Grabherr MG, Haas BJ, Yassour M et al (2011) Full-length transcriptome assembly from RNA-Seq data without a reference genome. Nat Biotechnol 29:644–652
12. Schulz MH, Zerbino DR, Vingron M et al (2012) Oases: robust de novo RNA-seq assembly across the dynamic range of expression levels. Bioinformatics 28:1086–1092
13. Bogdanov EA, Shagina I, Barsova EV, et al (2010) Normalizing cDNA libraries. Curr Protoc Mol Biol Chapter 5:Unit 5.12.1-27
14. Schmieder R, Edwards R (2011) Fast identification and removal of sequence contamination from genomic and metagenomic datasets. PLoS One 6:e17288
15. Haas BJ, Papanicolaou A, Yassour M et al (2013) *De novo* transcript sequence reconstruction from RNA-seq using the Trinity platform for reference generation and analysis. Nat Protoc 8:1494–1512
16. Finn RD, Clements J, Eddy SR (2011) HMMER web server: interactive sequence similarity searching. Nucleic Acids Res 39: W29–W37
17. Petersen TN, Brunak S, von Heijne G et al (2011) SignalP 4.0: discriminating signal peptides from transmembrane regions. Nat Methods 8:785–786
18. Rice P, Longden I, Bleasby A (2000) EMBOSS: the European molecular biology open software suite. Trends Genet 16: 276–277
19. Kong L, Zhang Y, Ye Z-Q et al (2007) CPC: assess the protein-coding potential of transcripts using sequence features and support vector machine. Nucleic Acids Res 35: W345–W349
20. Nawrocki EP, Burge SW, Bateman A et al (2015) Rfam 12.0: updates to the RNA families database. Nucleic Acids Res 43:D130–D137
21. Lai Z, Kane NC, Kozik A et al (2012) Genomics of Compositae weeds: EST libraries, microarrays, and evidence of introgression. Am J Bot 99:209–218

Part V

Purification of lncRNAs and Ribonucleoprotein Complexes, RNA Secondary Structure, and RNA Modifications

Chapter 17

Identification of Long Noncoding RNA-Protein Interactions Through In Vitro RNA Pull-Down Assay with Plant Nuclear Extracts

Jun Sung Seo and Nam-Hai Chua

Abstract

Recent advances in next-generation sequencing have revealed that majority of the plant genome is transcribed into long noncoding RNA (lncRNA). Many lncRNAs function by interacting with proteins and forming regulatory complexes. RNA-protein interactions are vital in controlling core cellular processes like transcription and translation. Therefore, identifying proteins that interact with lncRNAs is the first step to deciphering lncRNA functions. Here, we describe an RNA-protein pull-down assay, which enables the identification of proteins that interact with an RNA under study. As an example, we describe pull-down of proteins interacting with lncRNA ELENA1, which promotes the enrichment of MED19a on *PR1* promoter to activate *PR1* expression.

Key words Noncoding RNA, RNA-protein interactions, Biotin, In vitro transcription, ELENA1, Pull-down, MED19a

1 Introduction

Majority of the plant genome is transcribed into long noncoding RNA (lncRNA), but only a few lncRNAs have been functionally characterized until now [1, 2]. These few examples support the notion that many lncRNAs function through their interactions with protein partners [2, 3]. Therefore, identification of novel RNA-protein interactions could be highly important in determining unknown functions of lncRNAs. Most of molecular biology tools are protein-centric because of our lack of understanding about functional roles of many ncRNAs. Fewer RNA-centric methods are available for biologists to identify the whole repertoire of RNA-interacting proteins. One of such methods is RNA-protein pull-down assay where one or multiple proteins binding to RNA can be identified. This method (Fig. 1) utilizes in vitro transcription to produce RNA or a part of the RNA of interest which is then

Julia A. Chekanova and Hsiao-Lin V. Wang (eds.), *Plant Long Non-Coding RNAs: Methods and Protocols*, Methods in Molecular Biology, vol. 1933, https://doi.org/10.1007/978-1-4939-9045-0_17,

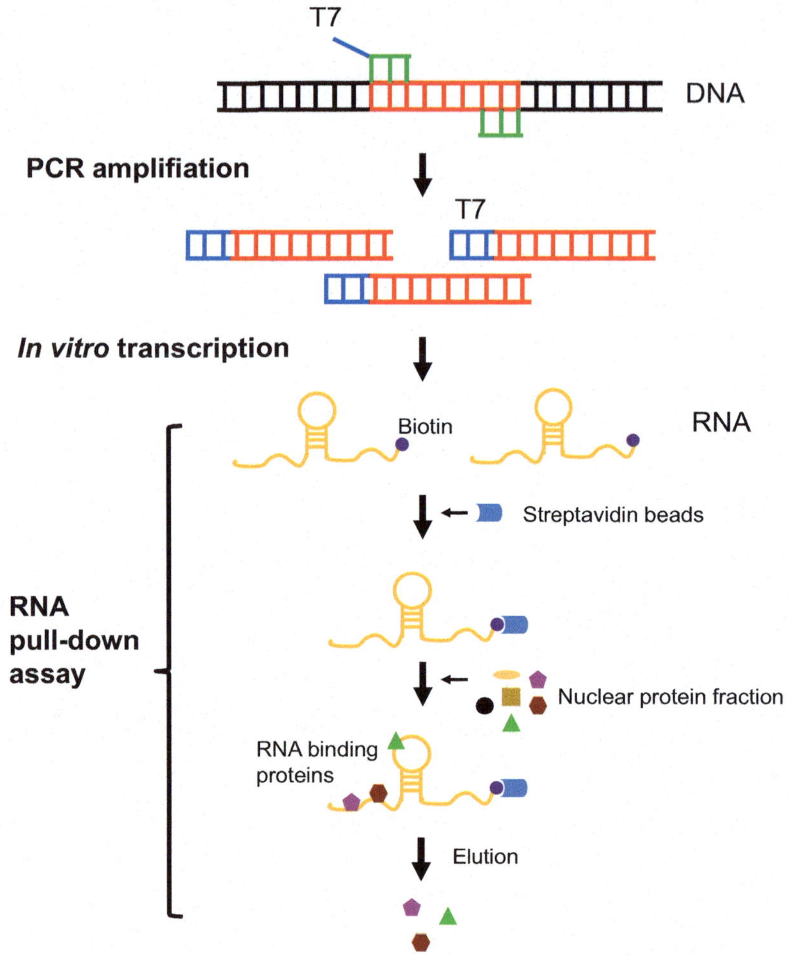

Fig. 1 Work flow of in vitro transcription and RNA-protein pull-down method. The DNA region of interest (shown in red) was amplified and used for in vitro transcription to produce RNA for biotin labeling. Biotin-labeled RNA was captured by streptavidin beads, incubated with nuclear fraction extract to allow for RNA-protein interactions, and the RNA-bound protein was eluted for detection by Western blotting

labeled using biotin. Proteins interacting with the biotin-labeled RNA are retrieved from whole-cell or protein extracts using streptavidin beads. These proteins can be then identified using mass spectrometry or Western blot. In the first step, a DNA template encoding the RNA sequence is produced with an added 5′ T7 RNA polymerase promoter sequence (5′-TAATACGACTCACTATA-3′) [4]. The T7 RNA polymerase promoter sequence is introduced into the DNA template using a polymerase chain reaction (PCR) step and specially designed primers. The T7 promoter is used to transcribe and produce copies of RNA in vitro. The biggest advantage of using in vitro transcription is that it allows synthesis of different lengths of RNAs with a specified sequence. We used this pull-down assay to identify the interaction between the GFP-tagged target protein, Mediator subunit 19a, and in vitro transcribed ELENA1 lncRNA [5].

2 Materials

Maintain RNase-free environment, and prepare all solutions using ultrapure water, RNase-free disposal materials, and analytical grade reagents.

2.1 DNA Template Preparation Components

1. Genomic or cDNA (100 ng/μL) from *Arabidopsis*.
2. 5 μM stock of primers for PCR: We ordered primers for the amplification of *ELENA1* template (forward, 5′-GAAAT TAATACGACTCACTATAGGG ACAAGTTGACATTTCGAGAG- 3′; reverse, 5′-GCAGTATAAGCCTGTGCTGACTTG-3′) from IDT. But other oligonucleotide-synthesizing companies can also be used (*see* **Note 1**).
3. KOD Hot Start (Novagen) or a suitable alternative DNA polymerase.
4. SYBR Safe DNA gel stain (Invitrogen) or a suitable alternative.
5. TBE buffer: For 1 L of a 10× stock, mix 108 g Tris base, 55 g boric acid, 80 mL 0.5 M EDTA, and 700 mL of dH_2O. Adjust pH to 8.0 and bring to 1 L volume with dH_2O.
6. Agarose gel: For 100 mL 1% gel, 1 g agarose in 100 mL 1× TBE buffer.
7. DNA ladders (1 kb and 100 bp).
8. QIAquick gel extraction kit (Qiagen) or a suitable alternative.
9. Thermocycler.
10. Horizontal gel electrophoresis equipment.
11. UV transilluminator.

2.2 T7 In Vitro Transcription Components

1. T7 RNA polymerase (Roche).
2. 10× Transcription buffer supplied with T7 RNA polymerase (Roche).
3. dNTP mix (dATP, dUTP, dGTP, and dCTP; each 10 mM solution) for unlabeled RNA transcription.
4. RNase inhibitor (RiboLock RNase Inhibitor from Thermo Fisher Scientific).
5. TURBO DNase and 10× TURBO DNase reaction buffer (Ambion).

6. Biotin RNA Labeling Mix, 10× conc. (Roche) for labeled RNA transcription.
7. 5× RNA structure buffer (RSB): 100 mM Tris–HCl (pH 7.0), 500 mM KCl, and 50 mM $MgCl_2$.
8. RNeasy Plant Mini Kit (Qiagen).
9. NanoDrop spectrometer.

2.3 Nuclear Protein Extraction Components

1. Homogenization buffer (HB): 20 mM Tris–HCl (pH 7.4), 25% glycerol, 20 mM KCl, 2 mM EDTA, 2.5 mM $MgCl_2$, 250 mM sucrose, 1% plant protease inhibitors (Sigma), and add H_2O to 100 mL.
2. Nuclei resuspension buffer (NRB): 20 mM Tris–HCl (pH 7.4), 25% glycerol, 2.5 mM $MgCl_2$, 1% plant protease inhibitors (Sigma), and add H_2O to 100 mL.
3. Nuclei resuspension buffer with 0.2% Triton X-100 (NRBT): 20 mM Tris–HCl (pH 7.4), 25% glycerol, 2.5 mM MgCl2, 0.2% Triton X-100, and add H_2O to 100 mL.
4. Nuclear storage buffer (NSB): 20 mM Tris–HCl (pH 7.4), 25% glycerol, 2.5 mM $MgCl_2$, sucrose 15.1 g, 1% plant protease inhibitors (Sigma), and add H_2O to 100 mL.
5. Quick start Bradford protein assay (bio-rad).
6. Bovine serum albumin (BSA, Sigma-Aldrich).
7. Mortar, pestle, and liquid nitrogen for grinding of plant materials.
8. Bioruptor® Plus sonication device (Diagenode).
9. Benzonase® Nuclease (Sigma).
10. Refrigerated centrifuge (Eppendorf).

2.4 RNA-Protein Pull-Down Components

1. RNA pull-down buffer (RPDB): 50 mM Tris (pH 7.5), 150 mM NaCl, 2 mM DTT, 0.2% Nonidet P-40, 8 U/mL Ribolock RNase Inhibitor (Thermo Fisher), and 1% plant protease inhibitors (Sigma) at the day of experiment.
2. Streptavidin Mutein Matrix (Roche).
3. 4× NuPAGE™ LDS Sample Buffer (Thermo Fisher).

2.5 Polyacrylamide Gel Electrophoresis and Western Blot Components

1. Precision Plus Protein™ Dual Color Standards Ladder (Bio-Rad).
2. PVDF membrane (0.45 μm) (Millipore).
3. Filter paper.
4. 10× Tris/glycine/SDS buffer; running buffer (Bio-Rad).
5. 10 × Tris/glycine buffer; transfer buffer (Bio-Rad).
6. TBS: 15 mL 5 M NaCl and 10 mL 1 M Tris (pH 7.5), bring to 500 mL with distilled H_2O.

7. TBST solution: 40 mL 1 M Tris, 60 mL 5 M NaCl, and 1 mL Tween-20, bring to 2 L with distilled H_2O.
8. 7.5% (w/v) blocking solution: Dissolve 7.5 g blotting-grade blocker (Bio-Rad) in 100 mL TBST, and adjust pH 7.6 with NaOH.
9. 5% (w/v) blocking solution: Dissolve 5 g blotting-grade blocker (Bio-Rad) in 100 mL TBST, and adjust pH 7.6 with NaOH.
10. 4% stacking gel: 425 μL 30% polyacrylamide, 625 μL 0.5 M Tris (pH 6.8), 25 μL 10% (w/v) SDS, 1.34 mL H_2O, 25 μL 10% (w/v) APS, and 5 μL TEMED.
11. 10% polyacrylamide gel (1.5 mm) with 4% stacking gel: 3.33 mL 30% polyacrylamide, 3 mL 1.5 M Tris (pH 8.8), 125 μL 10% (w/v) SDS, 4.96 mL H_2O, 125 μL 10% (w/v) APS, and 10 μL TEMED.
12. Primary and secondary antibodies: For example, GFP polyclonal rabbit antibody (Santa Cruz) for primary and ECL Rabbit IgG, HRP-linked whole antibody (Amersham) for secondary antibody.
13. Vertical gel electrophoresis equipment (Bio-Rad).
14. Wet/tank blotting systems (Bio-Rad).
15. Clarity™ ECL Western Blotting Substrate (Bio-Rad) or a suitable alternative.
16. ChemiDoc™ Imaging System (Bio-Rad).

3 Methods

3.1 DNA Template Preparation

1. Order the forward and reverse deoxyribo-oligonucleotide primers for PCR from companies such as IDT. The primer with the same sequence as RNA strand should include T7 RNA polymerase promoter sequence (5′-TAATACGACTCAC TATAGGG- 3′) at its 5′ end enabling its incorporation at the 5′ of the RNA sequence of interest (*see* **Note 1**).
2. Prepare 2× 50 μL PCR reactions for each primer pair as follows: 4 μL each of forward and reverse primers (5 μM stock), 5 μL 10× KOD Hot Start reaction buffer, 1 μL KOD Taq, 100 ng of cDNA or genomic DNA, and RNase-free H_2O (to make a final volume of 50 μL) (*see* **Note 2**).
3. Amplify the DNA through PCR as follows (or in case of using alternative DNA polymerase, according to the vendor's instructions): (a) denaturation, 95 °C, 2 min, and (b) amplification (40 cycles), 95 °C, 20 s; annealing temperature, 20 s; 70 °C, 15 s. (c) Final extension and storage: 70 °C, 5 min; 4 °C ∞.

4. Check the size and purity of the PCR product on a 1% agarose gel containing 0.3 μg/mL SYBR safe DNA gel stain (load 5–10 μL of PCR product). Use a standard UV illuminator to view the gel. If only one band of the desired size is observed, then proceed to **step 6**.
5. If multiple products are observed in **step 4**, you will require purification from a gel (*see* **Note 3**); run the remaining PCR product on an agarose gel, and cut the desired band with a clean scalpel (use suitable UV protection).
6. PCR product cleanup using a Qiagen QIAquick Gel Extraction Kit (*see* **Note 4**).

3.2 T7 RNA Polymerase In Vitro Transcription

1. Set up a 20 μL T7 polymerase reaction by incubating 2 μL of T7 RNA polymerase, 2 μL of 10× reaction buffer, 2 μL of 10× biotin RNA labeling mix, 100–200 ng of purified DNA (PCR product containing T7 promoter sequence), and RNase-free H_2O up to 20 μL (*see* **Note 5**).
2. Incubate the reaction for up to 3 h at 37 °C (*see* **Note 6**).
3. Add 1 μL TURBO DNase to each tube and incubate for 15 min at 37 °C.
4. Purify RNA with RNeasy Plant Mini Kit (Qiagen) (*see* **Note 7**), and elute with 40 μL RNase-free water.
5. Add 10 μL 5× RNA structure buffer to purified RNA.
6. Total 50 μL labeled RNA was heated to 90 °C for 2 min and quenched in the ice for 2 min. Then keep the treated RNA on the bench for 20–30 min to allow proper secondary structure formation.
7. Measure RNA quantity by using NanoDrop spectrometer.

3.3 Nuclear Protein Extraction

1. Grow *Arabidopsis* seedlings (*35S:GFP-MED19a*) on MS media for 2 weeks. Collect approximately 1 g of *Arabidopsis* seedlings, and freeze in liquid nitrogen.
2. Grind the tissue to a fine powder in liquid nitrogen using a cold mortar and pestle. Then collect the powder into a 50 mL conical tube.
3. Add 5 mL cold HB buffer into the powder, and homogenize the mixture by gentle shaking or pipetting (*see* **Note 8**).
4. Filter the homogenate through double layer of Miracloth (*see* **Note 9**).
5. Centrifuge the filtered homogenate at 1500 × *g* at 4 °C for 10 min to pellet the nuclei.
6. Discard the supernatant and add 3 mL NRBT to the pellet. Resuspend the nuclei by pipetting.
7. Centrifuge the sample at 1500 × *g* at 4 °C for 10 min. Repeat **steps 5** and **6** twice more.

8. Discard the supernatant and add 3 mL NRB to the pellet. Resuspend the nuclei by pipetting.
9. Centrifuge at 1500 × *g* at 4 °C for 10 min to pellet the nuclei, and discard the supernatant (*see* **Note 10**).
10. Thaw the pellet on ice. Add 300 μL RPDB to the pellet. Resuspend first with an inoculation loop, and then gently pipet the mix up and down. Avoid foaming at all costs.
11. Sonicate the sample five times, 10 s each, with a sonicator set at low intensity at 4 °C by using Bioruptor® Plus sonication device (*see* **Note 11**).
12. (Optional) Add 2 μL of Benzonase to each sample. Incubate on ice for 1 h (*see* **Note 12**).
13. Centrifuge with 12,000 × *g* for 30 min at 4 °C. Save the supernatant.

3.4 RNA-Protein Pull-Down Assay

1. Mix the streptavidin-agarose beads well before aliquoting 50 μL of beads per reaction.
2. Remove supernatant and wash beads two times with 500 μL H_2O.
3. Remove supernatant, and wash beads three times with 500 μL RNA pull-down buffer.
4. Add nuclear proteins (Subheading 3.3) and RNA pull-down buffer (total reaction volume 200–500 μL).
5. Add 500 ng of labeled RNA and mix gently by pipetting.
6. Incubate at 25 °C for 30 min with rotation or at 4 °C for 2 h (*see* **Note 13**).
7. Add the prepared streptavidin-agarose beads and incubate at 4 °C for 2 h.
8. Remove supernatant and wash twice with 50 μL of 20 mM Tris (pH 7.5).
9. Wash five times with 500 μL of RNA pull-down buffer (dilute 10× in RNase-free H_2O).
10. Remove supernatant completely, and add 25 μL of RNA pull-down buffer, and vortex well (*see* **Note 14**).
11. Add 6.25 μL of 4× NuPAGE™ LDS Sample Buffer, mix well, and boil for 10 min (*see* **Note 15**).
12. Keep cold on ice if proceeding to the next step immediately. For later use, store at −20 °C.

3.5 Western Blot

1. Prepare a 1.5 mm thick 10% polyacrylamide gel with a 4% stacking gel (*see* **Note 16**).
2. Load protein ladder and 30 μL of each protein sample from RNA pull-down assay (positive and negative controls) onto 10% gel.

3. Run the gel at 100 V for approximately 1 h or until the blue front reaches the end of the gel. Remove the gel from the plates and put in a clean plastic container.
4. Soak gel in 1× transfer buffer for 10 min before performing the wet transfer with a PVDF membrane (*see* **Note 17**) by running at 100 V for 1 h (*see* **Note 18**).
5. Block the membrane at room temperature for 1 h with 7.5% blocking solution on an orbital shaker (*see* **Note 19**).
6. Remove 7.5% (w/v) blocking solution, and incubate with the primary antibody, polyclonal anti-GFP, diluted (1:1000–1:5000) in 5% (w/v) blocking solution for 1 h.
7. Wash the membrane three times (each for 10 min) with TBST.
8. Incubate for 1 h at room temperature with secondary antibody diluted (1:5000–1:10,000) in 5% (w/v) blocking solution.
9. Wash the membrane three times (each for 10 min) with TBST.
10. Wash the membrane once with PBS for 5 min.
11. Incubate the membrane with Clarity™ ECL Western Blotting Substrate for 2 min.
12. Scan chemiluminescence of the blot by using ChemiDoc™ Imaging System.

4 Notes

1. When designing primers for the PCR, first design the oligonucleotide sequences (excluding the T7 promoter sequence) to have similar melting temperatures before adding T7 promoter sequence at one end. T7 RNA polymerase has increased efficiency when the T7 promoter sequence is preceded by several nucleotides and immediately followed by GG.
2. cDNA should be used if primers are designed across different exons; however, if the primers are designed to amplify a region of RNA which is not spliced, we have found that genomic DNA works sufficiently well. 2× PCR reactions containing 100 ng genomic DNA should be sufficient to produce 1 μg template DNA.
3. In some cases it might not be possible to produce a single, discrete band after PCR. In such a scenario, the correct size band must be extracted from the gel. For those that do produce a unique band, direct purification of PCR product can be used.
4. Follow the manual provided by Qiagen.

5. To prepare unlabeled lncRNA for competition assay, we used dNTP mix for in vitro transcription of unlabeled lncRNA instead of biotin-labeling kit.
6. For RNA with a strong secondary structure, lower the reaction temperature to 16 °C or 4 °C, and incubate for a longer period (24–72 h) to improve the reaction yield.
7. Follow the manual provided by Qiagen.
8. If the sample is frozen from excess liquid nitrogen, wait until it is thawed enough that it can be homogenized.
9. Once all the homogenate has been filtered, squeeze out the remaining homogenate into the membranes. Always keep the sample on ice.
10. The purpose of this step is to remove the Triton X-100. The nuclei can now be used for any purpose. For example, they can be used to detect nuclear protein using a Western blot. If the nuclei cannot be used immediately, they should be resuspended in 400 μL NSB buffer, quickly frozen in liquid nitrogen, and stored at −80 °C. The nuclear preparation can last for at least half a year.
11. Sonication disrupts the nuclei. The intensity and duration of sonication need to be optimized.
12. Benzonase removes DNA and RNA, thereby releasing DNA-binding proteins. One can infer the relative distribution of proteins between chromatin-bound and chromatin-unbound states based on the Benzonase treatment. Also, the resulting nuclear extracts will not be complicated by the presence of genomic DNA and mRNA. Normally skip this step.
13. To maximize the amount of RNA that binds to the beads, we recommend the use of rotation rather than agitation. And use freshly transcribed RNA for the best result.
14. We found it more efficient to elute in a smaller volume to concentrate the protein samples.
15. To maximize the elution, it is preferable to heat the streptavidin beads with loading buffer.
16. To increase the resolution of the Western band, adjust the polyacrylamide concentration depending on the protein size.
17. Place membranes in methanol for 20 s, and wash with water for 2 min, and then put it in 1× transfer buffer before transferring.
18. We have optimized transfer for the Bio-Rad wet transfer system.
19. Otherwise, you can block the membrane with 5% (w/v) blocking solution overnight.

Acknowledgment

This work was supported by Singapore NRF RSSS Grant (NRF-RSSS-002).

References

1. Liu J, Jung C, Xu J, Wang H, Deng S, Bernad L, Arenas-Huertero C, Chua NH (2012) Genome-wide analysis uncovers regulation of long intergenic noncoding RNAs in Arabidopsis. Plant Cell 24:4333–43345
2. Wang HV, Chekanova JA (2017) Long noncoding RNAs in plants. Adv Exp Med Biol 1008:133–154
3. Chekanova JA (2015) Long non-coding RNAs and their functions in plants. Curr Opin Plant Biol 27:207–216
4. Milligan JF, Groebe DR, Witherell GW, Uhlenbeck OC (1987) Oligoribonucleotide synthesis using T7 RNA polymerase and synthetic DNA templates. Nucleic Acids Res 15:8783–8798
5. Seo JS, Sun HX, Park BS, Huang CH, Yeh SD, Jung C, Chua NH (2017) ELF18-INDUCED LONG-NONCODING RNA associates with mediator to enhance expression of innate immune response genes in Arabidopsis. Plant Cell 29:1024–1038

Chapter 18

Analysis of Interaction Between Long Noncoding RNAs and Protein by RNA Immunoprecipitation in *Arabidopsis*

Jun Sung Seo and Nam-Hai Chua

Abstract

Long noncoding RNAs (lncRNAs) play important roles in several processes including control of gene expression. These RNAs function through binding to histone-modifying complexes and transcriptional machinery including transcription factor, mediator, and RNA polymerase II. We present methods for the discovery and characterization of lncRNAs. RNA immunoprecipitation (RIP) is a modified version of chromatin immunoprecipitation (ChIP), and it is now generally used in lncRNA study. The method allows for testing of lncRNA-protein interactions in vivo. RIP assay facilitates the identification of consensus sequences of preferred binding site for the RNA-binding protein under study, and identification of the binding sites can provide valuable information on the possible mechanism by which the RNA-binding protein functions.

Key words Noncoding RNA, ChIP, RNA-protein interaction, RIP, lncRNA, RNA immunoprecipitation

1 Introduction

It has recently become apparent that RNA itself, the product of transcription, is a major regulator of the transcriptional process [1]. In particular, long noncoding RNAs (lncRNAs), which are so numerous in eukaryotes, function in many cases as transcriptional regulators [2]. These RNAs function through binding to histone-modifying complexes, to DNA-binding proteins (including transcription factors), and even to RNA polymerase II [2].

A method for analyzing native RNA-protein interactions has recently been developed for yeast [3, 4] and animal cells [5, 6]. Ribonucleoprotein immunoprecipitation [5] and cross-link immunoprecipitation [6] both depend on the cross-linking of RNA-binding proteins to RNAs in vivo, followed by immunocapture of specific RNP complexes and identification of the associated RNAs. Importantly, these approaches are based on cross-

Julia A. Chekanova and Hsiao-Lin V. Wang (eds.), *Plant Long Non-Coding RNAs: Methods and Protocols*, Methods in Molecular Biology, vol. 1933, https://doi.org/10.1007/978-1-4939-9045-0_18,

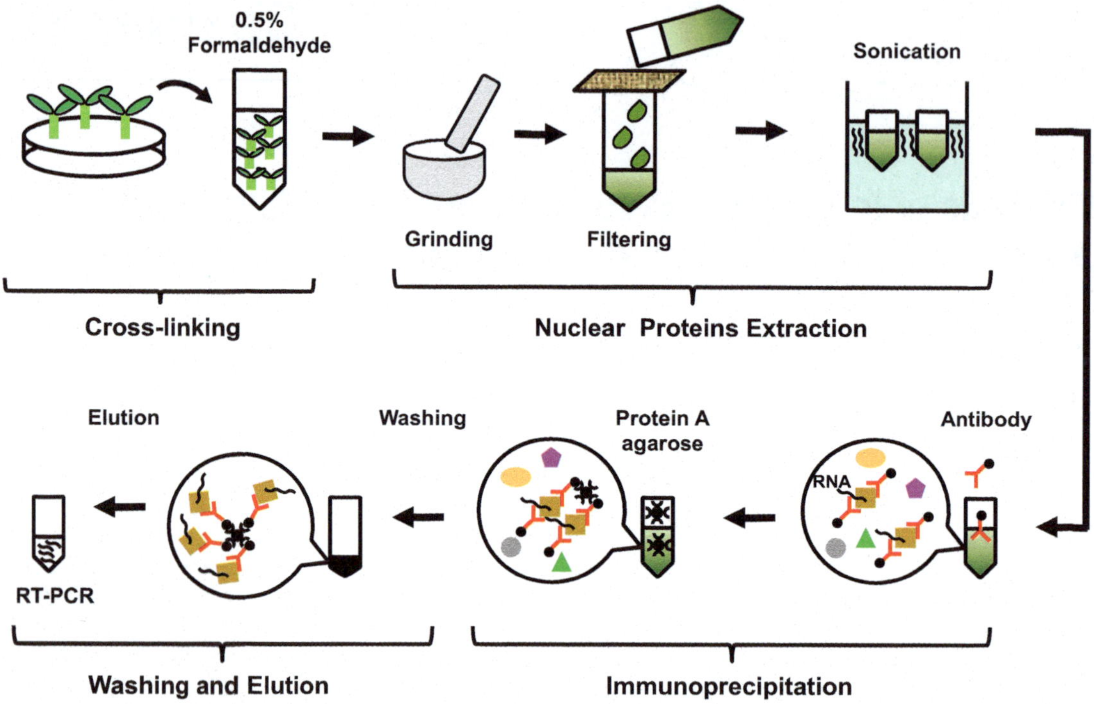

Fig. 1 Work flow of the RNA immunoprecipitation (RIP). Preparing the sample with seedlings and cross-linking them with 0.5% formaldehyde. Grinding cross-linked samples in liquid nitrogen, resuspending the powder in extraction buffer, and sonicating the samples. Immunoprecipitate with specific antibody and protein A bead. Finally, the RNAs are eluted and purified before being analyzed by real-time RT-PCR

linking to overcome a technical issue in standard immunoprecipitation: the reassociation of RNA-binding proteins with non-cognate RNAs that occur in cellular extracts [7]. Denaturing conditions during extract preparation and at the immuno-capture step prevent occurrence of such reassociation events. Native RNA-protein complexes are not affected by these conditions because they are stabilized by the cross-linking of their components. RIP has the advantage that it is a modification of the chromatin immunoprecipitation (ChIP) method that works well with *Arabidopsis* lncRNA studies (Fig. 1) [8, 9]. RIP assay facilitates the identification of consensus sequences of preferred binding site for the RNA-binding protein under study, which can provide valuable information on the mechanism by which the RNA-binding protein functions.

2 Materials

Prepare all solutions using ultrapure water, RNase-free disposal materials, and analytical grade reagents. Keep samples on ice as much as possible.

2.1 Cross-Linking Components

1. 50 mL Falcon tubes.
2. 0.5% formaldehyde.
3. 2 M glycine.
4. Paper towel and aluminum foil.
5. Liquid nitrogen.
6. Vacuum chamber.

2.2 Nuclear Protein Extraction Components

1. Honda buffer: 0.44 M sucrose, 1.25% Ficoll 400, 2.5% dextran T40, 20 mM HEPES-KOH (pH 7.4), 10 mM $MgCl_2$ (*see* **Note 1**). Add 5 mM DTT, 1 mM PMSF, 1% plant protease inhibitors (Sigma, P9599-5ml), and 8 U/mL Ribolock RNase inhibitor (Thermo Fisher) at the day of experiment.
2. Mortar and pestle.
3. Miracloth.
4. Refrigerated centrifuge (Eppendorf).
5. Nuclei lysis buffer: 50 mM Tris–HCl (pH 8.0), 10 mM EDTA, and 1% SDS. Add 5 mM DTT, 1 mM PMSF, 1% plant protease inhibitors, and 160 U/mL Ribolock RNase inhibitor at the day of experiment.
6. Bioruptor® plus sonication device (Diagenode).
7. NanoDrop spectrometer.

2.3 Immunoprecipitation Components

1. ChIP dilution buffer: 16.7 mM Tris–HCl (pH 8.0), 167 mM NaCl, 1.2 mM EDTA (pH 8.0), 1.1% Triton X-100, 350 U/mL Ribolock RNase inhibitor. Store at 4 °C. Add 1 mM PMSF and 40 U/mL Ribolock RNase Inhibitor at the day of experiment.
2. Protein A agarose/salmon sperm DNA (Upstate 16-157).
3. Binding/washing buffer (B/W): 20 mM Tris–HCl (pH 8.0), 150 mM NaCl, 2 mM EDTA (pH 8.0), 1% Triton X-100, 0.1% SDS. Add water to 500 mL and filter through 0.2 μm filter. Store at 4 °C. Add 1 mM PMSF and 40 U/mL Ribolock RNase Inhibitor at the day of experiment.
4. Specific antibody.
5. Tube rotator.
6. RIP elution buffer: 100 mM Tris–HCl (pH 8.0), 10 mM EDTA, 1% SDS. Add 400 U/mL Ribolock RNase Inhibitor at the day of the experiment.
7. Heat block.
8. RNA-grade proteinase K 20 mg/mL (Invitrogen).
9. Trizol® reagent (Ambion).
10. 3 M NaOAc (pH 5.2).

11. GlycoBlue 15 mg/mL (Ambion).
12. Absolute ethanol.

2.4 RT-PCR Components

1. RNeasy Mini Kit (Qiagen).
2. Turbo DNase 2 U/μL (Ambion).
3. SuperScript™ III First-Strand Synthesis System (Thermo Fisher).
4. SYBR Premix Ex Taq (Tli RNase H Plus) (Takara).
5. CFX Connect™ Real-Time PCR Detection System (Bio-Rad).

3 Methods

3.1 Cross-Linking

1. Harvest 3 g of 2-week-old seedlings, and place them in 50 mL Falcon tubes (*see* **Note 2**).
2. Add 0.5% formaldehyde to the Falcon tubes to a volume of 35 mL, and apply vacuum for 2 min at 85 kPa (~25 in.Hg). Release the vacuum and reapply for 8 more minutes (*see* **Note 3**).
3. Add 1.25 mL 2 M glycine and mix.
4. Apply vacuum for 1 min. Release the vacuum and reapply for 4 more minutes.
5. Rinse the plants five times with water. Remove excess water by squeezing between paper towels (*see* **Note 4**). Wrap in an aluminum foil and freeze in liquid nitrogen.

3.2 Nuclear Protein Extraction

1. Grind cross-linked samples in liquid nitrogen into a fine powder, and resuspend gently in 15 mL prepared Honda buffer with 120 U Ribolock RNase inhibitor.
2. Filter through two layers of Miracloth and squeeze the Miracloth. Then wash the Miracloth with remaining plant material in 10 mL of fresh Honda buffer with 80 U Ribolock RNase inhibitor (*see* **Note 5**).
3. Centrifuge at 3000 × *g* for 10 min at 4 °C. Resuspend the pellet in 1 mL Honda buffer with 8 U Ribolock RNase inhibitor.
4. Centrifuge at 2500 × *g* for 5 min in a micro-centrifuge at 4 °C. Repeat the resuspension and centrifuge two more times (*see* **Note 6**).
5. Resuspend the pellet in 500 μL nuclei lysis buffer with 80 U Ribolock RNase inhibitor.
6. Sonicate ten cycles (one cycle is 30 s on/30 s off with high level) in a Bioruptor plus sonication system at 4 °C.

7. Centrifuge the samples for 10 min at the maximum speed in a micro-centrifuge at 4 °C, and then transfer the supernatant to new 1.5 mL tubes.
8. Measure DNA concentration in the samples using NanoDrop. Nuclei lysis buffer with inhibitors as a blank. Ensure that chromatin concentration of all the samples is similar.
9. Prepare 100 μL aliquots of the chromatin samples, and freeze in liquid nitrogen, and also freeze one 10 μL aliquot for inputs. Store at −80 °C.

3.3 Immunoprecipitation

1. Dilute the samples tenfold with ChIP dilution buffer (100 μL chromatin sample + 900 μL ChIP dilution buffer) (*see* **Note 7**).
2. Centrifuge in a micro-centrifuge at the maximum speed for 10 min, and then transfer the supernatant to fresh 1.5 mL tubes.
3. Take 20 μL of protein A agarose/salmon sperm DNA for each sample. Wash three times with 1 mL of B/W buffer. Centrifuge at 3000 × *g* for 1 min in a micro-centrifuge after each wash. Resuspend in B/W buffer to have 100 μL of the suspension for each sample.
4. Add 2.5–5 μg of antibody and 100 μL of washed beads, and incubate on a rotator for 3–4 h at 4 °C (*see* **Note 8**). Then add protein A agarose prepared in Subheading 3.3, **step 3**, and incubate on a rotator for 2 h at 4 °C.
5. Wash the beads two times briefly with 1 mL B/W buffer (including 40 U RNase inhibitor) and then two times for 5 min with 1 mL of B/W buffer (including 40 U RNase inhibitor). After each wash, centrifuge the samples at 3000 × *g* for 1 min. During the final wash, before centrifugation, transfer the samples to fresh 1.5 mL tubes (*see* **Note 9**).
6. Add 50 μL RIP elution buffer (including 40 U RNase inhibitor), and rotate at room temperature for 10 min. Centrifuge at 3000 × *g* for 1 min and save the supernatant.
7. Prepare 1% input samples by adding 1 μL chromatin extract to 100 μL RIP elution buffer including 40 U RNase inhibitor.
8. Repeat elution at 65 °C with mixing for 10 min, and then centrifuge at 3000 × *g* for 1 min. Take the supernatant and combine it with the supernatant of Subheading 3.3, **step 6**.
9. Add 1 μL 20 mg/mL proteinase K and incubate at 65 °C for 1 h.
10. Add 100 μL Trizol (Ambion), vortex for 15 s, centrifuge at the maximum speed for 5 min, and then transfer the aqueous phase to fresh 1.5 mL tubes (*see* **Note 10**).

11. Add 10 μL of 3 M sodium acetate (pH 5.2), 60 μg GlycoBlue, and 250 μL ethanol. And leave at 80 °C for overnight.
12. Centrifuge at the maximum speed for 20 min at 4 °C.
13. Wash the pellet with 500 μL 70% ethanol and air-dry for 2 min. Then dissolve in RNase-free water (*see* **Note 11**).

3.4 RT-PCR

1. Clean up RIP samples using the Qiagen RNeasy Plant Mini Kit following the instructions with the on-column DNase digestion (*see* **Note 12**).
2. Incubate the entire RIP sample at 65 °C 10 min.
3. Transfer 1/2 RNA to a new tube for no-RT control, and transfer 1/2 RNA to a new tube for the RT reaction.
4. Add 0.4 μL 500 ng/μL random primers (Invitrogen) and 1 μL 10 mM dNTPs.
5. Incubate at 65 °C for 5 min. Keep on ice for at least 1 min.
6. Add 1 μL 0.1 M DTT, 4 μL 10 × first-strand buffer, 2 μL 25 mM $MgCl_2$ 1 μL 40 U/μL Ribolock, and 1 μL SuperScript III Reverse Transcriptase (substitute water for SuperScript III in no-RT controls). Add H_2O to a total volume of 20 μL.
7. Incubate 25 °C for 5 min, followed by 50 °C for 1 h, and then 85 °C for 10 min.
8. Add 1 μL cDNA, 0.2 μL 25 μM forward and reverse primer mix, 10 μL SYBR premix EX-Taq (Takara), and H_2O to a total volume of 20 μL.
9. Run qPCR by using CFX Connect™ Real-Time PCR Detection System including internal control (such as Actin or PP2A).

4 Notes

1. Prepare HEPES first in a volume of 350 mL, and adjust the pH before adding other components. Adjust to a final volume of 500 mL, and filter the solution through 0.2 μm filters. Store at 4 °C.
2. It is essential to have equal amounts of tissue for all samples tested.
3. Releasing the vacuum ensures that the formaldehyde has access to plant tissues during most of the incubation time because actual infiltration occurs during the release of vacuum rather than during vacuum application.
4. Leftover water makes grinding much more difficult.
5. After the first filtration, a significant amount of the nuclei is stuck in the green stuff on the Miracloth. Washing the Miracloth can increase the yield by approximately twofold.

6. After initial centrifugation the nuclear preparation is contaminated by quite a lot of chloroplasts. Repeated washes help to lyse the chloroplasts but also increase the risk of nuclei fragmentation. Some nuclear proteins may be lost in the supernatant at this step.
7. Bringing SDS concentration to 0.1% by dilution is essential as higher concentrations may denature antibodies. Centrifugation at this step is important if frozen chromatin samples are being used. The volume of diluted chromatin should not exceed 1.5 mL.
8. Antibody concentration should be optimized for each specific application.
9. Transferring the samples to new tubes is essential as even siliconized tubes bind some amount of nucleic acids which would be released during elution step.
10. Instead of Trizol, you can extract RNA sample using phenol-chloroform.
11. Centrifuge the samples briefly before air-drying to make sure all ethanol is removed. Do not dry the samples longer than 2 min to avoid overdrying.
12. It is also possible that RNA sample is cleaned up using Qiagen RNeasy Kit after DNase treatment.

Acknowledgment

This work was supported by Singapore NRF RSSS Grant (NRF-RSSS-002).

References

1. Cech TR, Steitz JA (2014) The noncoding RNA revolution—trashing old rules to forge new ones. Cell 157:77–94
2. Long Y, Wang X, Youmans DT, Cech TR (2017) How do lncRNAs regulate transcription? Sci Adv 3(9):eaao2110
3. Hieronymus H, Silver PA (2003) Genome-wide analysis of RNA-protein interactions illustrates specificity of the mRNA export machinery. Nat Genet 33(2):155–161
4. Gerber AP, Herschlag D, Brown PO (2004) Extensive association of functionally and cytotopically related mRNAs with Puf family RNA-binding proteins in yeast. PLoS Biol 2:e79
5. Niranjanakumari S, Lasda E, Brazas R, Garcia-Blanco MA (2002) Reversible cross-linking combined with immunoprecipitation to study RNA-protein interactions in vivo. Methods 26 (2):182–190
6. Ule J, Jensen KB, Ruggiu M, Mele A, Ule A, Darnell RB (2003) CLIP identifies Nova-regulated RNA networks in the brain. Science 302(5648):1212–1215
7. Mili S, Steitz JA (2004) Evidence for reassociation of RNA-binding proteins after cell lysis: implications for the interpretation of immunoprecipitation analyses. RNA 10(11):1692–1694
8. He Y, Michaels SD, Amasino RM (2003) Regulation of flowering time by histone acetylation in Arabidopsis. Science 302:1751–1754
9. Heo JB, Sung S (2011) Vernalization-mediated epigenetic silencing by a long intronic noncoding RNA. Science 331:76–79

Chapter 19

Trimolecular Fluorescence Complementation (TriFC) Assay for Visualization of RNA-Protein Interaction in Plants

Jun Sung Seo and Nam-Hai Chua

Abstract

RNA-protein interactions play important roles in various eukaryotic biological processes. Molecular imaging of subcellular localization of RNA-protein complexes in plants is critical for understanding these interactions. However, methods to image RNA-protein interactions in living plants have not yet been developed until now. Recently, we have developed a trimolecular fluorescence complementation (TriFC) system for in vivo visualization of RNA-protein interaction by transient expression in tobacco leaves. In this method, we combined conventional bimolecular fluorescence complementation (BiFC) system with the MS2 system (phage MS2 coat protein [MCP] and its binding RNA sequence [MS2 sequence]) to tag lncRNA. Target RNA is tagged with 6xMS2, and MCP and RNA-binding protein are fused with YFP fragments. DNA constructs encoding such fusion RNA and proteins are infiltrated into tobacco leaves with *Agrobacterium* suspensions. RNA-protein interaction in vivo is observed by confocal microscopy.

Key words Long noncoding RNA, RNA-protein interaction, TriFC, Tobacco transient expression, In vivo visualization

1 Introduction

Recently, a variety of types of long noncoding RNAs (lncRNAs) has been identified and shown to play important roles in transcriptional regulation and chromatin modification [1]. So far, most of the molecular mechanisms for lncRNA-mediated functions are closely related to RNA-protein interactions [1]. Therefore, characterizing RNA-protein interaction is a key step in the functional study of lncRNAs. In plants, molecular functions of lncRNAs are only beginning to be characterized, and the molecular basis of lncRNA-mediated gene regulation is still poorly understood. Although techniques for RNA visualization in plants have been well developed, a visual assay for RNA-protein interaction in the plant is still poorly developed [2]. To develop a visual assay for RNA-protein interaction in plants, we modified and combined the MS2 system for RNA imaging technique with a conventional BiFC

Julia A. Chekanova and Hsiao-Lin V. Wang (eds.), *Plant Long Non-Coding RNAs: Methods and Protocols*, Methods in Molecular Biology, vol. 1933, https://doi.org/10.1007/978-1-4939-9045-0_19,

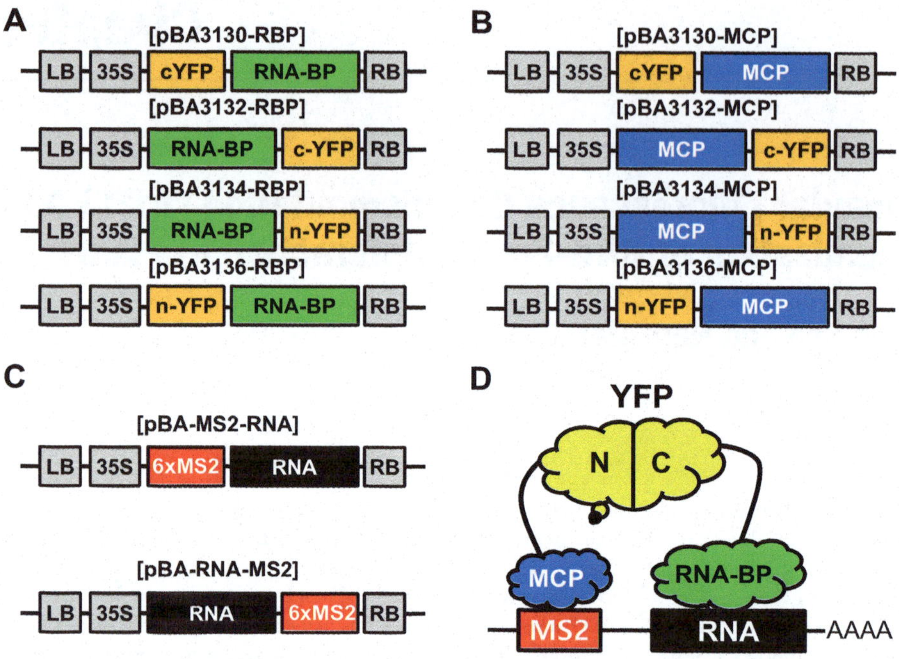

Fig. 1 Schematic representation of the TriFC system. (**a**) A schematic drawing of an expression vector encoding RNA-binding protein (RNA-BP) fused with YFP fragments. (**b**) A schematic drawing of an expression vector encoding MCP (MS2 coat protein) fused with YFP fragments. (**c**) A schematic drawing of MS2-tagged RNA expression vectors. (**d**) Illustration of the TriFC system

system for protein-protein interaction [3] (Fig. 1d). The MS2 coat protein (MCP) binds to a 19-nucleotide RNA hairpin that is recognized with high specificity [4] and with an affinity sufficient for in vivo detection [5]. Additional advantages of this RNA-binding protein are that no potential endogenous target sequences are known from eukaryotes and a fusion of MCP to GFP does not produce a specific subcellular localization by itself. For RNA imaging, the MCP-FP (fluorescence protein) fusion is co-introduced with a construct of the target RNA "tagged" by the introduction of MS2 binding sites [6, 7]. To increase the sensitivity of detection, usually, 6–24 repeats of the MS2 binding site are introduced into target RNAs, although as many as 96 repeats have been used to enable single-molecule sensitivity [8]. We generated binary Gateway vectors (pBA3130, pBA3132, pBA3134, and pBA3136) for transient BiFC assay [9] and obtained binary Gateway vectors (pBA-GW-6xMS2 and pBA-6xMS2-GW) for RNA tagging from Dr. Ulrich Z. Hammes (Fig. 1) [3]. We tested and confirmed this TriFC assay was working well in plants with lncRNA, ELENA1, and MED19a protein [9]. TriFC assay in plants will provide new insights on the interaction between lncRNAs and proteins.

2 Materials

2.1 TriFC Vector Construction Components

1. Destination vectors for TriFC: pBA3130 (35S:cYFP-DC; spectinomycin), pBA3132 (35S:DC-cYFP; spectinomycin), pBA3134 (35S:DC-nYFP; spectinomycin), pBA3136 (35S: nYFP-DC; spectinomycin) [9], pBA-6xMS2-DC (kanamycin), and pBA-DC-6xMS2 (kanamycin) [3].
2. Entry clones: pENTR-ELENA1 (kanamycin), pENTR-Med19a (kanamycin), and pENTR-MCP (kanamycin).
3. LR Clonase II (Thermo Fisher Scientific).
4. *E. coli* competent cells (strain DH5α).
5. LB agar media.
6. LB broth media.
7. Antibiotics: Spectinomycin (Sigma), Kanamycin (Sigma).
8. QIAprep Spin Miniprep Kit (Qiagen).
9. Incubator (37 °C) and shaking incubator (37 °C).

2.2 Agrobacterium Preparation Components

1. *Agrobacterium* competent cells (strain GV3101).
2. Incubator (28 °C) and shaking incubator (28 °C).
3. 10 mL culture tubes and 50 mL Falcon tubes.
4. Spectrometer.
5. Centrifuge for 50 mL tubes.
6. LB-MES (pH 5.6): LB broth with 10 mM MES (pH 5.6).
7. 100 mM acetosyringone stock: 20 mg acetosyringone (Sigma) in 1 mL ethanol or DMSO, stored at −20 °C (*see* **Note 1**).
8. Resuspension solution: 10 mM $MgCl_2$, 10 mM MES-KOH (pH 5.6), and 100 μM acetosyringone (*see* **Note 2**).

2.3 Tobacco Infiltration and Confocal Microscope Components

1. 1 mL syringe (BD).
2. *Nicotiana benthamiana* (*N. benthamiana*) plants; 2–4 weeks old (6–10 leaves stage).
3. Slide glasses and cover glasses for microscope.
4. Scissor and forceps.
5. Confocal laser scanning microscope system (ZEISS, model: LSM 780).
6. ZEN image analyzer (ZEISS).

3 Methods

3.1 TriFC Vector Construction

1. Prepare the entry and destination vectors for construction of TriFC vectors.
2. LR reaction MCP and RNA-BP entry clones with Gateway BiFC vectors (pBA3130, pBA3132, pBA3134, and pBA3136; Fig. 1a, b) following the manufacturer's instruction (*see* **Note 3**). In this study, we used pENTR-MED19a for RNA-BP entry clone and pENTR-MCP for MCP entry clone.
3. LR reaction RNA entry clones with Gateway MS2 tagging vectors (pBA-6xMS2-DC or pBA-DC-6xMS2; Fig. 1c) following the manufacturer's instruction (*see* **Note 3**). In this study, we used pENTR-ELENA1 for RNA entry clone. After LR reaction, transform competent *E. coli* cells (strain DH5α) by heat shock method.
4. Spread the cells on the LB/spectinomycin (or kanamycin) agar media, and grow overnight in an incubator at 37 °C.
5. Screen colony by PCR with specific primers.
6. Inoculate one single colony of *E. coli* in 5 mL LB with appropriate antibiotics. Grow overnight in a shaking incubator with 200 rpm at 37 °C.
7. Extract plasmid with plasmid extraction kit (Qiagen).

3.2 Agrobacterium Preparation for Tobacco Infiltration

1. Transform 50 μL *Agrobacterium* competent cells (strain GV3101) with each DNA construct (50 ng) (three different vectors; Fig. 1a–c) by heat shock method [10].
2. Spread the cells on the LB/spectinomycin (or kanamycin) agar media, and grow for 2 days in an incubator at 28–30 °C.
3. Inoculate one single colony of *Agrobacterium* in 5 mL LB with appropriate antibiotics. Grow cells overnight in a shaking incubator with 200 rpm at 28–30 °C.
4. Use 250 μL of the overnight culture to inoculate 10 mL LB-MES (pH 5.6) (with same antibiotics) plus 4 μL of 100 mM acetosyringone, and grow the culture overnight in a shaking incubator with 200 rpm at 28–30 °C.
5. Measure the A600 of overnight cultures (OD = 0.8–1.0).
6. Collect bacterial cells (4000 *g*, 10 min) at room temperature, and resuspend the pellet in resuspension solution. Adjust the final A600 to 0.4.
7. Leave on a bench (room temperature) for over 4 h (or overnight) before infiltration.

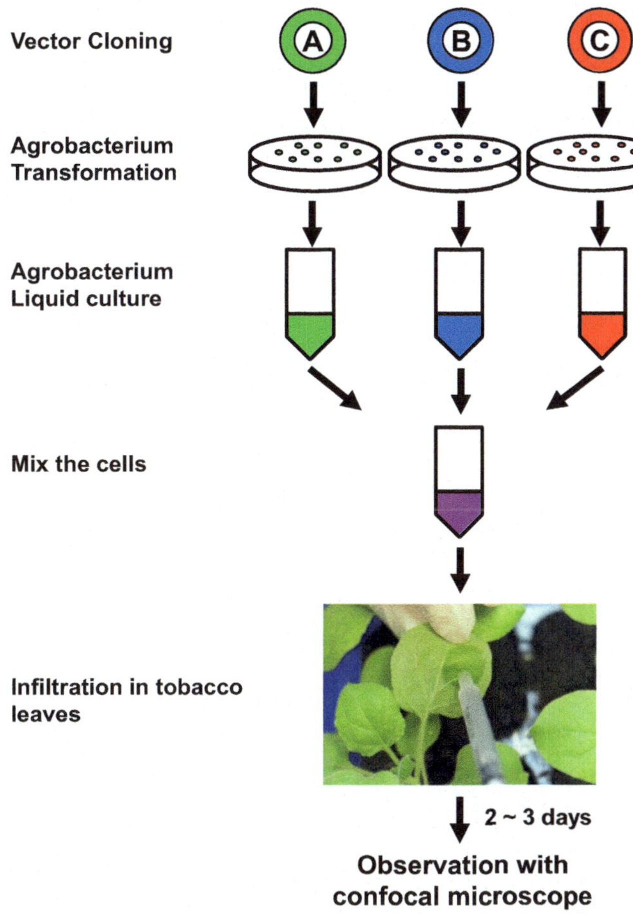

Fig. 2 Work flow of TriFC assay in *N. benthamiana*. Cloning the vectors for TriFC assay [*A* is RNA-BP construct (Fig. 1a), *B* is MCP construct (Fig. 1b), and *C* is MS2-tagged RNA construct (Fig. 1c)] and transformation of *Agrobacterium* with each vector. Selecting the colony and growing the cells. Mix three different *Agrobacterium* cells. Infiltration in tobacco leaves with syringe, and keep the plant in the growth room for 2–3 days. Observe the infiltrated region with confocal microscope

3.3 Tobacco Infiltration and Confocal Microscopy Observation

1. Transfer 1 mL of each *Agrobacterium* cell suspension into a 10 mL tube, and mix well (*see* **Note 4**).
2. Perform the infiltration with a 1 mL syringe. Simply press the syringe (no needle) on the underside of a leaf (Note: Avoid cotyledons), and exert a counter-pressure with a finger on the other side. Successful infiltration is often observed as a spreading of "wetting" area in the leaf and labels the infiltrated region (Fig. 2) (*see* **Note 5**).
3. Keep infiltrated plants in the greenhouse or growth chamber for 2–3 days.

4. Observe the YFP signals under a confocal laser scanning microscope 2–3 days after infiltration (*see* **Note 6**). The excitation wavelength is 514 nm, and detection range of emission wavelength is 520–550 nm (*see* **Note 7**).

4 Notes

1. We recommend making acetosyringone stock solution freshly before using.
2. Add acetosyringone solution after autoclaving and immediately before using.
3. We performed LR reaction through the following procedures: Add 100 ng entry vector and 100 ng destination vector in 1.5 mL tube and make total volume to 10 μL with distilled water. Then add 2 μL LR Clonase II enzyme mix, and keep the reaction mix for 1–2 h. (Optional) Add 1 μL of the proteinase K solution to each sample, and incubate samples at 37 °C for 10 min.
4. An optimal combination of 6xMS2 tag, nYFP, and cYFP orientation is sometimes critical for the TriFC assay. Therefore, other combinations should be tried if the signal is not detected.
5. Do not press the leaf strongly with a syringe. It will damage the leaf and cause cell death.
6. Posttranscriptional gene silencing (PTGS) is a major cause for the lack of efficiency in transient expression experiments [11]. Therefore, we usually co-infiltrate with *pUBQ:p19*, which suppresses PTGS, together with other constructs as it greatly improves the efficiency.
7. Adjust the detection range of emission wavelength depending on the signal intensity.

Acknowledgments

We thank Dr. Ulrich Z. Hammes for the Gateway 6xMS2 tagging vectors. This work was supported by Singapore NRF RSSS Grant (NRF-RSSS-002).

References

1. St Laurent G, Wahlestedt C, Kapranov P (2015) The Landscape of long noncoding RNA classification. Trends Genet 31 (5):239–251
2. Christensen NM, Oparka KJ, Tilsner J (2010) Advances in imaging RNA in plants. Trends Plant Sci 15(4):196–203

3. Schonberger J, Hammes UZ, Dresselhaus T (2012) In vivo visualization of RNA in plants cells using the λN22 system and a GATEWAY-compatible vector series for candidate RNAs. Plant J 71(1):173–181
4. Valegård K, Murray JB, Stockley PG, Stonehouse NJ, Liljas L (1994) Crystal structure of an RNA bacteriophage coat protein-operator complex. Nature 371(6498):623–626
5. LeCuyer KA, Behlen LS, Uhlenbeck OC (1996) Mutagenesis of a stacking contact in the MS2 coat protein-RNA complex. EMBO J 15(24):6847–6853
6. Campalans A, Kondorosi A, Crespi M (2004) Enod40, a short open reading frame-containing mRNA, induces cytoplasmic localization of a nuclear RNA binding protein in Medicago truncatula. Plant Cell 16 (4):1047–1059
7. Fujioka Y, Utsumi M, Ohba Y, Watanabe Y (2007) Location of a possible miRNA processing site in SmD3/SmB nuclear bodies in Arabidopsis. Plant Cell Physiol 48 (9):1243–1253
8. Golding I, Cox EC (2004) RNA dynamics in live Escherichia coli cells. PNAS 101 (31):11310–11315
9. Seo JS, Sun HX, Park BS, Huang CH, Yeh SD, Jung C, Chua NH (2017) ELF18-INDUCED LONG-NONCODING RNA associates with mediator to enhance expression of innate immune response genes in Arabidopsis. Plant Cell 29(5):1024–1038
10. Höfgen R, Willmitzer L (1988) Storage of competent cells for Agrobacterium transformation. Nucleic Acids Res 16(20):9877
11. Voinnet O, Rivas S, Mestre P, Baulcombe D (2003) An enhanced transient expression system in plants based on suppression of gene silencing by the p19 protein of tomato bushy stunt virus. Plant J 33:949–956

Chapter 20

In Vivo Genome-Wide RNA Structure Probing with Structure-seq

Laura E. Ritchey, Zhao Su, Sarah M. Assmann, and Philip C. Bevilacqua

Abstract

In vivo genome-wide RNA structure probing provides a global view of RNA structure as it occurs in the cell and can assist in elucidating important functional aspects of RNA structure. Structure-seq2 provides high-quality data on transcriptome-wide RNA structure in vivo but contains numerous steps that require technical precision. In this chapter we present the steps needed to produce high-quality structural data with a focus on controls and troubleshooting. Structure-seq2 can be applied to numerous organisms including plants, humans, and bacteria and is amenable to a wide variety of RNA-modifying chemicals including DMS, glyoxal, and SHAPE reagents. Notably, the data generated by the method highlighted here can be readily analyzed using our StructureFold2 computational pipeline.

Key words RNA structure probing, Structure-seq, Illumina sequencing, StructureFold2, DMS, Glyoxal, SHAPE, 1-ethyl-3-(3-dimethylaminopropyl)carbodiimide (EDC)

1 Introduction

RNA has the potential to form complex structures that increase its functionality beyond a simple intermediate molecule that transfers information between DNA and proteins. Historically, RNA structure has been probed by chemicals in a gene-specific manner, first in vitro and later in vivo [1]. Recently, various techniques have been developed to apply the same chemistry and techniques in vivo and genome-wide [2]. Among these methods, we have developed Structure-seq [3, 4], which has been optimized in Structure-seq2 [5] to produce sensitive high-quality data that identify single-stranded nucleotides. Key advantages in the preparation of Structure-seq libraries over other methods are the emphasis on single-hit kinetics [6], the conversion of RNA into cDNA at

Electronic Supplementary Material: The online version of this chapter (https://doi.org/10.1007/978-1-4939-9045-0_20) contains supplementary material, which is available to authorized users.

Julia A. Chekanova and Hsiao-Lin V. Wang (eds.), *Plant Long Non-Coding RNAs: Methods and Protocols*, Methods in Molecular Biology, vol. 1933, https://doi.org/10.1007/978-1-4939-9045-0_20,

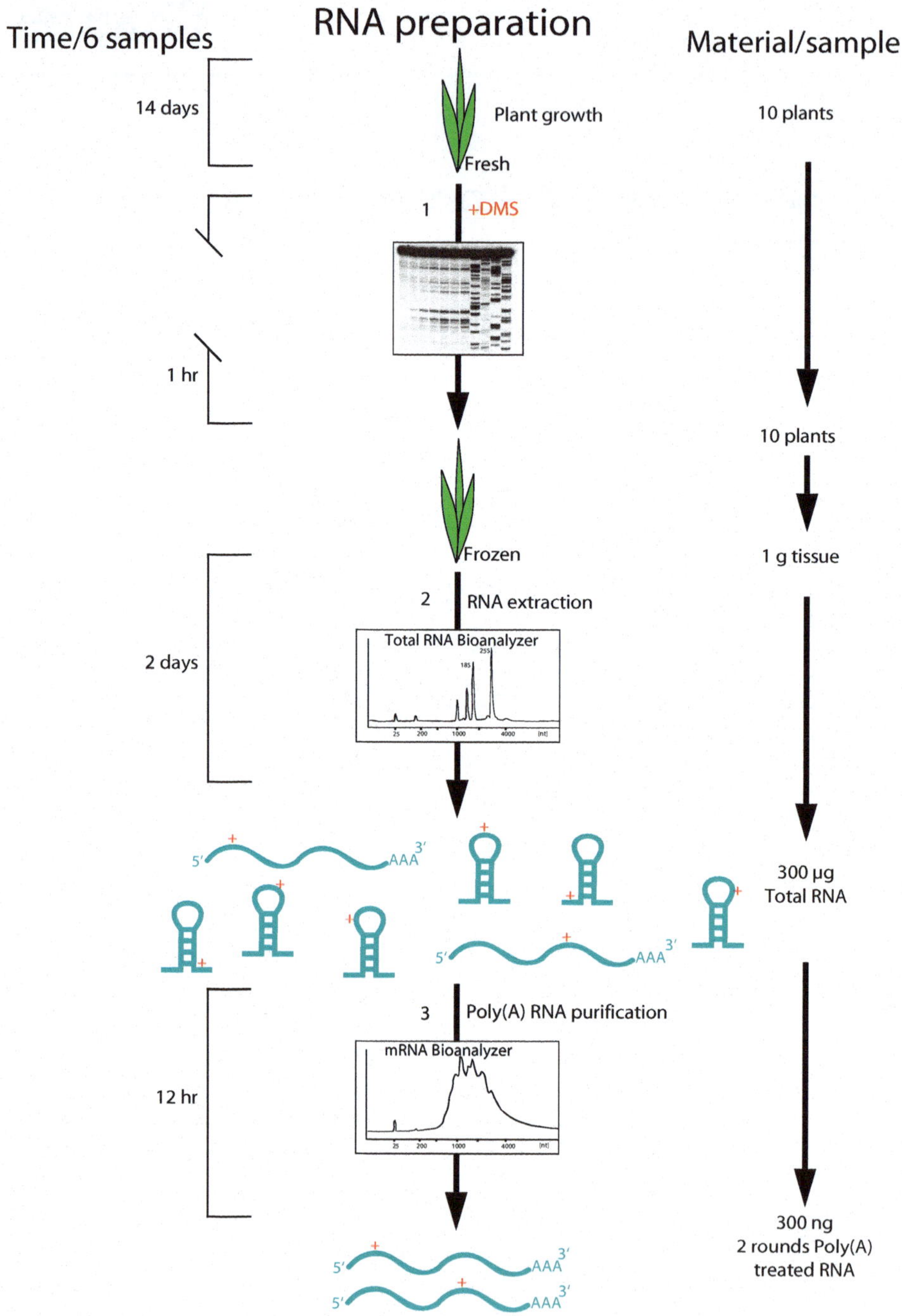

Fig. 1 RNA preparation. In order to prepare RNA for Structure-seq2 library preparation (Fig. 2), the organism must first be grown to the appropriate stage. Here, rice seedlings are grown for 14 days. After growth, the organism is treated with DMS concentrations that provide single-hit kinetics (step 1 inset) and frozen. This step should take about 1 h (step 1). The appropriate RNA extraction technique is then performed. Here the plant RNA extraction protocol takes approximately 2 days (step 2). A Bioanalyzer scan should be run as a quality check to validate RNA quality and yield (step 2 inset). For plants, 1 g of frozen tissue is sufficient for

the earliest step possible (*see* **Note 1**), a low ligation bias, and the ease of data analysis by our companion bioinformatics package, StructureFold2 [7].

Briefly, RNA is probed in vivo by a chemical in a single-hit kinetics manner (Fig. 1, step 1). We have focused to date on dimethyl sulfate (DMS) but also have obtained structural data with a suite of glyoxals [8], EDC [9], and SHAPE reagents. DMS methylates adenosine (A) and cytosine (C) residues on the Watson-Crick face. The methylated RNA is extracted (Fig. 1, step 2), enriched for the RNA type of interest (Fig. 1, step 3), for example, mRNAs, and reverse transcribed using a random hexamer with a 5′ adaptor of known sequence to obtain transcriptome-wide information (Fig. 2, step 1). Reverse transcription stops one nucleotide before the methylated base to produce truncated cDNA fragments. An adaptor is ligated to the 3′ end of these fragments (Fig. 2, step 3) to allow for amplification using Illumina sequencing primers (Fig. 2, step 5). The resulting libraries are sequenced on an Illumina sequencing platform (Fig. 2, step 6) and then analyzed using the StructureFold2 bioinformatics suite (Fig. 2, step 7) [7]. The Structure-seq2 method described herein has numerous advances over the original Structure-seq method. The specifics of these improvements, which result in higher sensitivity and sequencing quality, are highlighted elsewhere [5]. Here, we focus on the current Structure-seq2 method rather than a detailed listing of changes versus the original Structure-seq technique [3, 4].

We provide two versions of the Structure-seq2 protocol, which utilize two different types of cleanup protocols (Fig. 2, steps 2 and 4). In one version, a standard reverse transcription (RT) is performed, and then cleanup steps following RT and ligation make use of denaturing PAGE extraction (Fig. 2, left). In another version, RT with doped biotinylated dNTPs is performed, and then cleanup steps following RT and ligation make use of streptavidin pulldowns (Fig. 2, right). The former version provides several visualizations of the material and may be especially helpful to troubleshoot reverse transcription and ligation. The latter version, on the other hand, allows the libraries to be prepared on a much shorter time scale with overall times for the two versions of 9 and 3.5 days, respectively.

Fig. 1 (continued) each sample and provides at least 300 μg of total RNA (poly(A) and non-poly(A) transcripts shown in blue). Next, RNA is purified to enrich or deplete various types of RNA; typically ribosomal RNA is depleted. Here, Poly(A) purification is performed two times to enrich for mRNAs and takes about 12 h (step 3). A high-sensitivity Bioanalyzer scan should be run as a quality check to validate rRNA depletion and yield, which will be much lower than the Bioanalyzer scan run for total RNA (step 3 inset). As two rounds of poly (A) purification gives a yield of about 0.5%, 300 μg of total RNA provides over twice the amount of purified RNA after two rounds of poly(A) enrichment that is necessary for library preparation (300–500 ng)

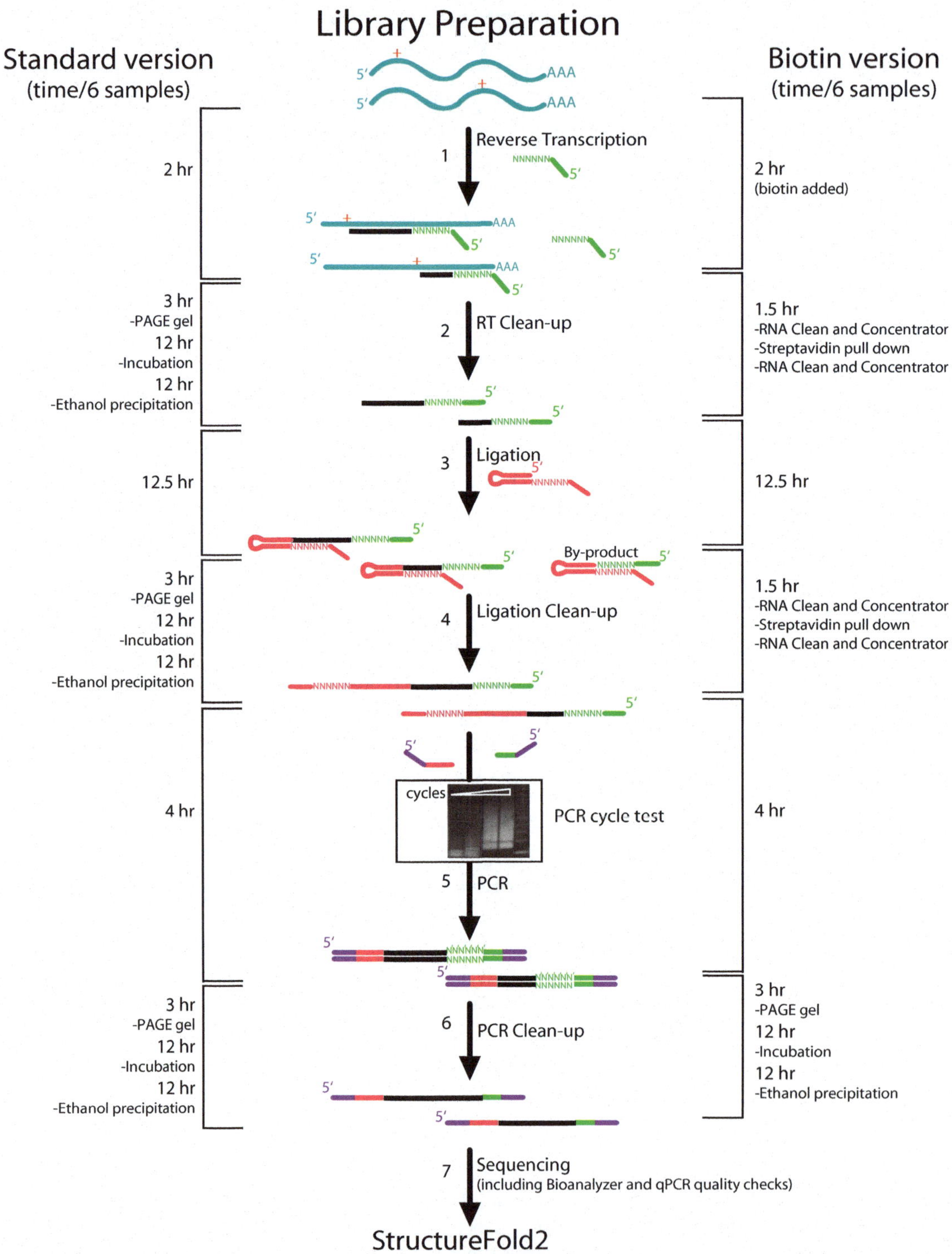

Fig. 2 Structure-seq2 library preparation. Using 300–500 ng of purified RNA (blue) from RNA preparation (Fig. 1, bottom), reverse transcription (RT) with a random hexamer fused to an Illumina adaptor (green) is performed (step 1) either with (right) or without (left) biotin incorporated. The RT reaction is then cleaned up (step 2) via streptavidin pulldown (right) or PAGE extraction (left). T4 DNA ligation is then performed (step 3)

This chapter provides details for each step necessary to produce Structure-seq2 libraries using both versions.

Structure-seq2 can easily be applied to various organisms including plants, humans, and bacteria. Herein, we describe the methods using plants as our primary example; however, we also describe in the Notes sections small changes that allow structural determination in other organisms. In particular, DMS modification, RNA extraction, and RNA purification methods can be easily varied as described in the Notes (*see* **Note 2**). Additionally, Structure-seq2 can also be used in vitro to characterize how RNA structure responds to defined test tube conditions. This is accomplished by extracting RNA prior to DMS modification. All other steps can be carried out as described.

2 Materials

RNA can readily degrade and thus all reagents and plasticware must be nuclease-free. Ultrapure water should be used to prepare all solutions and to store samples and can be tested for RNases by incubating labeled RNA overnight and analyzing for degradation by PAGE. Gloves should be worn throughout the procedure, and reagents should be high quality. Double gloves and personal protective equipment should be worn for all procedures involving DMS and other structure-probing chemicals, and those steps should be performed in a chemical fume hood. Buffers can be stored at room temperature unless otherwise noted. Periodically, the pH of solutions should be validated by spotting on pH paper.

2.1 General Supplies

1. 50 mL conical plastic centrifuge tubes that can withstand 16,000 × *g* force.
2. 1.7 mL microcentrifuge tubes.
3. 200 proof ethanol stored in a −20 °C freezer.
4. Eppendorf PCR tubes.

2.2 Instrumentation

1. −80 and −20 °C freezers.
2. Vortexer.

Fig. 2 (continued) with a hairpin adaptor (magenta), and the ligation is cleaned up (step 4) via streptavidin pulldown (right) or PAGE extraction (left). A PCR cycle test should be performed to validate library preparation (step 5 inset), followed by PCR performed with appropriate Illumina adaptors incorporated into the PCR primers (step 5). The PCR reaction is then cleaned up (step 6) via PAGE extraction (both left and right). Quality controls of a Bioanalyzer scan and Illumina-specific qPCR analysis (step 7) are performed. Following successful quality controls, the samples are sequenced, and the results are then piped into StructureFold2 [7]. The PCR cycle test (step 5, inset) is a good step to achieve successfully with test RNA samples to verify the library preparation procedure (*see* **Note 15**)

3. Microcentrifuge to 16,000 × *g*.
4. NanoDrop (Thermo Scientific).
5. Thermal cycler.
6. High-voltage power source for PAGE gels to 3000 V.
7. Gel rigs for PAGE gels of various sizes (*see* **Note 3**).
8. Large format gel dryer.
9. Incubator/shaker.
10. Centrifuge for 50 mL tubes that reaches 4 °C.
11. SpeedVac.
12. Heat blocks for 1.7 mL microcentrifuge tubes.
13. High current power source for agarose gels to 200 mA.
14. Gel rig for agarose gel (*see* **Note 3**).
15. UV light box with 254 nm wavelength.
16. Rotating/shaking tube rack for 1.7 mL microcentrifuge tubes.
17. Magnetic stand.

2.3 Necessary Oligomers

All oligomers can be ordered from IDT or a comparable source and are DNA. In order to be sure that all oligomers are sufficiently pure, they either should be PAGE purified by hand or ordered HPLC purified.

1. rRNA-specific RT primer to be varied depending on rRNA sequences of the organism of interest. In rice, the primers used are as follows:

   `5.8S primer: 5′-GCG TGA CGC CCA GGC-3′`
   `18S primer: 5′-AAC TGA TTT AAT GAG CCA TTC GCA G-3′`

2. Random hexamer with fused Illumina adaptor:

   `5′-CAG ACG TGT GCT CTT CCG ATC NNN NNN-3′`

3. Hairpin adaptor:

   `/5Phos/TGA AGA GCC TAG TCG CTG TTC ANN NNN NCT GCC CAT AGA G/3SpC3/ where /5Phos/ is a 5′ phosphate, and /3SpC3/ is a 3-carbon linker on the 3′ end`

4. Modified Illumina TruSeq universal adaptor (forward primer):

   `5′-AAT GAT ACG GCG ACC ACC GAG ATC TAC ACT CTT TCC CTA CAC GAC GCT CTT CCG ATC TTG AAC AGC GAC TAG GCT CTT CA-3′`

5. Illumina TruSeq reverse primer:

   5′-CAA GCA GAA GAC GGC ATA CGA GAT-*barcode*-GTG ACT GGA GTT CAG ACG TGT GCT CTT CCG ATC-3′, where *barcode* is a unique 6–8 nt Illumina barcode for each sample.

6. Custom sequencing primer:

   5′-TCT TCC GAT CTT GAA CAG CGA CTA GGC TCT TCA-3′

2.4 DMS Treatment

1. Organism grown to desired growth period under desired condition. For example, for experiments on rice seedlings, we grow rice for 14 days.
2. DMS reaction buffer: 40 mM HEPES (pH 7.5), 100 mM KCl, and 0.5 mM $MgCl_2$.
3. Dimethyl sulfate (DMS). DMS must be handled very carefully (*see* Subheading 3.1.1), and any material that comes into contact with DMS should be disposed of as hazardous waste (Video 1).
4. Dithiothreitol (DTT).
5. Liquid nitrogen.

2.5 RNA Extraction and Purification

1. NucleoSpin RNA plant kit (Macherey-Nagel).
2. Mortar and Pestle.
3. Poly(A) Purist MAG Kit (Thermo Fisher Scientific).
4. Liquid nitrogen.

2.6 PAGE Gels

1. 2× Formamide loading buffer: 95% formamide, 0.1× TBE, 20 mM EDTA, 0.1% xylene cyanol FF, and 0.025% bromophenol blue (1× TBE is 89 mM Tris, 89 mM boric acid, and 2 mM EDTA).

   Gel plates of various sizes (*see* **Note 3**).
2. Polyacrylamide gel electrophoresis (PAGE) reagents: 10% acrylamide (29:1) in 8.3 M urea and 1× TBE. 10% ammonium persulfate; TEMED (stored at 4 °C).
3. 10× TBE buffer.

2.7 rRNA-Specific and/or Genome-Wide Reverse Transcription

1. 150–250K cpm γ-labeled DNA primer (rRNA-specific for single-hit kinetics) or 100 μM random hexamer (genome-wide). Radiation work must be performed with proper protective equipment including lab coats, gloves, and plexiglass shields.
2. 10× RT buffer: 200 mM Tris (pH 8.4), 500 mM KCl. This buffer is supplied with the SuperScriptIII First Strand Synthesis System for RT-PCR, but can be homemade.
3. 80 mM (rRNA-specific) and 100 mM (genome-wide) $MgCl_2$.

4. 100 mM DTT.
5. 10 mM dNTP mix (each dNTP is 10 mM).
6. 10 mM ddATP, ddGTP, ddCTP, and ddTTP (rRNA-specific only; for sequencing lanes).
7. 5 mM Biotin-16-AA- 2′-dCTP (TriLink) and/or biotin-11-dUTP (Thermo Scientific) (final, combined concentration of 1 mM).
8. SuperScript III Reverse Transcriptase (200 U/μL) (Thermo Fisher).
9. 2 M NaOH (rRNA-specific only).
10. RNase H (5 U/μL) (genome-wide only).
11. GeneRuler Low Range DNA Ladder (Thermo Fisher).

2.8 Ligation

1. 100 μM hairpin adaptor.
2. 5 M betaine.
3. 50% PEG 8000.
4. 10× T4 DNA ligase buffer (NEB).
5. T4 DNA ligase (400 U/μL, NEB).
6. A custom ladder of ssDNA oligonucleotides near 67 and 91 nt is useful as 67 nt represents the undesired ligation by-product between the RT primer and hairpin adaptor and 91 nt indicates the correct location of the bottom of the cut at Subheading 3.6.1, step 5.

2.9 PCR

1. 5× Q5 reaction buffer (NEB).
2. 10 mM dNTP mix (each).
3. 10 μM forward and reverse primers.
4. Q5 DNA polymerase (NEB).
5. Agarose.
6. 1× TAE buffer: 40 mM Tris–acetate, 1 mM EDTA.
7. Blue/Orange 6× Loading Dye (Promega).
8. GeneRuler 100 bp ladder (Thermo Fisher).
9. ss100 bp ladder (Simplex Science).

2.10 PAGE Purifications

1. SYBR gold (Thermo Fisher).
2. 1× TE buffer: 10 mM Tris pH 7.5, 1 mM EDTA
3. Glass or plastic container made opaque by covering with aluminum foil on all sides, for staining agarose and PAGE gels.
4. Razor blades.
5. TEN_{250} buffer: 10 mM Tris pH 7.5, 1 mM EDTA, 250 mM NaCl.

2.11 Streptavidin Pulldown

1. RNA Clean and Concentrator-5 kit (Zymo Research).
2. Hydrophilic Streptavidin Magnetic Beads (NEB).
3. Magnetic stand.
4. Wash/binding buffer: 20 mM Tris pH 7.5, 1 mM EDTA, 500 mM NaCl.
5. Low-salt buffer: 20 mM Tris pH 7.5, 1 mM EDTA, 150 mM NaCl.
6. Elution buffer: 95% formamide, 10 mM EDTA.

2.12 Sequencing

Illumina sequencing using appropriate kits and reagents for MiSeq or HiSeq for single-end reads.

2.13 Visual Demonstrations

1. DMS treatment: some appropriate safety precautions needed for the DMS reaction were demonstrated. Also, *Arabidopsis* was used as an example to demonstrate how to perform the DMS treatment (*see* Video 1).
2. PAGE excision: how to properly dice the gel pieces at various steps throughout Structure-seq2 was demonstrated (*see* Video 2).
3. Streptavidin elution: how to elute the biotinylated products from streptavidin magnetic beads was demonstrated (*see* Video 3).

3 Methods

3.1 Single-Hit Kinetics

Single-hit kinetics (SHK) measures the extent to which each individual RNA molecule was modified by the structure-probing chemical. It is desired that, on average, each molecule is modified at most only one time per ~200 nt to minimize mapping of chemically induced structure changes and also to be able to read out each modification.

3.1.1 Chemical Modification

Although other RNA-modifying chemicals can be used, we describe DMS here. All manipulations with DMS must be conducted in a fume hood with sufficient airflow and handled with proper personal protective equipment (PPE) such as lab coats and double gloves. Also, all disposables and materials generated in the course of DMS work should be disposed of as hazardous waste.

1. Before libraries are prepared, single-hit kinetics validation must be performed (Fig. 1, step 1) [6]. This involves using varying concentrations and/or duration of DMS treatment to test for proper modification (*see* **Note 4**, Video 1).
2. Grow biological material as desired for a given experiment (*see* **Note 2**). For example, for rice seedlings, grow for 14 days, then excise 1 g of shoot tissue from the plant, and place

immediately in 20 mL of DMS reaction buffer in 50 mL conical centrifuge tubes. Add DMS to the solution to varying concentrations and/or durations, and leave one tube with no DMS as a negative control.

3. Allow the reactions to proceed for the chosen time (*see* **Note 4**) with gentle shaking to remove air bubbles surrounding the plant tissue (Video 1).
4. Quench the DMS in the reaction by adding DTT powder directly to each reaction to a final concentration that is at least five times higher than the highest DMS concentration tested. Vortex the tubes vigorously to allow the DTT to dissolve, and allow the quench to proceed for 2 min (Video 1).
5. Decant the solution into DMS liquid waste, and wash each sample twice with ~40 mL distilled deionized water. After the final wash, remove residual water by inverting the tube onto paper towels and blotting the tissue for ~2 s. Immediately freeze the tissue by wrapping dried plants in aluminum foil and then plunging into liquid nitrogen (Video 1). The frozen tissue can be stored for at least 3 months in a −80 °C freezer.

3.1.2 RNA Extraction

1. Using the relevant RNA extraction kit (*see* **Note 2**), e.g. the NucleoSpin RNA plant kit, perform RNA extraction in a chemical fume hood (Fig. 1, step 2).
2. Prepare 350 μL of buffer RA1 (from kit) in 1.7 mL microcentrifuge tubes, and add 3.5 μL β-mercaptoethanol to the tubes. (DTT or TCEP could also be used as reducing agents to a final concentration of 10–20 mM.)
3. Finely grind plant tissue under liquid nitrogen with a pre-chilled mortar and pestle (*see* **Note 5**).
4. Add no more than 100 mg of tissue to buffer RA1 (from kit) for each extraction, and vortex vigorously (*see* **Note 5**).
5. Place the NucleoSpin filter (violet ring) in the supplied 2 mL collection tube, add the lysate to the filter, and centrifuge for 1 min at 11,000 × *g*. Transfer the filtrate to a new 1.7 mL microcentrifuge tube (*see* **Note 5**), discard the NucleoSpin filter, and keep the flow-through.
6. Add 350 μL 70% ethanol to the lysate and mix by vortexing (*see* **Note 5**). Add entire 700 μL to a NucleoSpin RNA plant column (light blue ring) in a 2 mL collection tube, and centrifuge for 30 s at 11,000 × *g*. Transfer the column with bound RNA to a new 2 mL collection tube to continue (*see* **Note 5**) and discard the flow-through.
7. Add 350 μL membrane desalting buffer (MDB from kit) to the column, and centrifuge for 1 min at 11,000 × *g*. Discard the flow-through and continue with the column.

8. Add 10 μL reconstituted rDNase to 90 μL reaction buffer (from kit) in a fresh tube, mix by flicking the tube, then transfer 95 μL of this mixture directly to the center of the NucleoSpin RNA plant column, and incubate at room temperature for 15 min.
9. Add 200 μL of buffer RAW2 (from kit) to the NucleoSpin RNA plant column to inactivate the rDNase, centrifuge for 30 s at 11,000 × *g*, and then place the column into a new 2 mL collection tube to continue.
10. Add 600 μL of buffer RA3 to the NucleoSpin RNA plant column, centrifuge for 30 s at 11,000 × *g*, discard the flow-through, and place the column back into the same 2 mL collection tube to continue.
11. Add 250 μL of buffer RA3 (from kit) to the NucleoSpin RNA plant column, and centrifuge for 2 min at 11,000 × *g* to completely dry the membrane. Leave the column open at room temperature for a few minutes to ensure dryness. Place the column into a new 1.5 collection tube (supplied with kit) to continue.
12. Add 60 μL water directly to the membrane, and centrifuge for 1 min at 11,000 × *g* to elute the RNA. RNA can be kept safely frozen at −20 °C for at least 30 days or stored at −80 °C for months (*see* **Note 1**).
13. Check RNA quantity and quality by performing a NanoDrop analysis on 2 μL of the RNA. For example, 30–50 μg of RNA should be obtained from 100 mg of fresh rice shoot sample. The 260:280 ratio should be 2.0 or above.

3.1.3 Reverse Transcription of rRNA

1. Program the PCR thermal cycler to 95 °C for 1 min, followed by 35 °C for 5 min, 55 °C for 21 min, and 95 °C for 5 min, and then a hold at 4 °C.
2. Add water to a PCR tube so that the final volume of RNA and water adds up to 4.5 μL, or 3.5 μL for each of the four ddNTP lanes, and then add 1–5 μg of total RNA to each PCR tube (*see* **Note 6**). Use the negative control of no DMS for the ddNTP lanes and negative control lane, but all other reactions should be from different DMS concentrations.
3. Following all precautions for work with radioactivity, add 1 μL of radiolabeled DNA primer (*see* **Note 6**) that is ~150–250 K cpm and 1 μL of 10× RT buffer. Carefully vortex and spin down the PCR tubes.
4. Place tubes in the thermal cycler, and denature the RNA and DNA primer at 95 °C for 1 min.
5. Once the temperature has cooled to 35 °C, pause the program, and add 1 μL of 80 mM $MgCl_2$, 1 μL of 100 mM DTT, and

1 μL of 10 mM dNTPs (these components can be premade into a master mix for easier addition). In each of the ddNTP lanes, add 1 μL of a given 10 mM ddNTP. Mix by pipetting up and down with the same pipette tip used for addition of the master mix.

6. Allow the DNA primer to anneal to the RNA at 35 °C for 5 min, and then heat the reaction to 55 °C for 1 min.
7. After 1 min at 55 °C, pause the program, and then add 0.5 μL of SuperScript III reverse transcriptase; the total volume of the reaction should be 10 μL. Carefully vortex and pulse centrifuge the reactions. Return the reactions to the thermal cycler at 55 °C, and allow the reaction to proceed for 20 min.
8. After 20 min, add 0.5 μL of 2 M NaOH, carefully vortex and pulse centrifuge the reactions, then place in the prepared thermal cycler, heat to 95 °C for 5 min to cleave the RNA and denature the enzyme, and then cool the reaction to 4 °C.
9. Add 10 μL of 2× formamide loading buffer to each reaction, carefully vortex, and then pulse centrifuge.
10. Immediately before loading the samples onto a sequencing gel, denature each sample at 95 °C for 2–5 min. Load 4–5 μL of the samples onto a 10% acrylamide 8.3 M urea denaturing PAGE sequencing gel that has been pre-run for 1 h (*see* **Note 3**), and allow the gel to run at 85–90 Watts for about 1.5–2 h, or until the bottom dye reaches the bottom of the gel.
11. After running, drying, and overnight exposure of the gel, scan the gel, and analyze for single-hit kinetics (Fig. 1, step 1, inset, *see* **Note 4**).

3.2 Genome-Wide Experiments

Many of the steps in the chemical modification (Fig. 1, step 1) and RNA extraction (Fig. 1, step 2) for genome-wide experiments are similar to those for single-hit kinetics experiments. Notably, we describe the procedure that we found provides single-hit kinetics for 14-day-old rice seedlings using DMS (*see* **Note 4**).

3.2.1 Chemical Modification

1. Prepare tissue as desired for a given experiment (*see* **Note 2**). For example, for rice seedlings, grow for 14 days, then excise 1 g of tissue from the plant, and place immediately in 20 mL of DMS reaction buffer in 50 mL conical centrifuge tubes. Add DMS to the solution to the final concentration found to give single-hit kinetics, and leave one tube with no DMS as a negative control (*see* **Note 7**, Video 1).
2. Allow the reactions to proceed for 10 min with gentle shaking to promote mixing and remove any air bubbles surrounding the plant tissue (Video 1).

3. Quench the DMS in the reaction by adding DTT powder directly to each reaction (including the no DMS control) to a final concentration that is at least five times higher than the DMS concentration used. Vortex the tubes vigorously to allow the DTT to dissolve, and allow the quench to proceed for 2 min (Video 1).
4. Decant the solution into DMS liquid waste, and wash each sample twice with ~40 mL distilled deionized water. After the final wash, remove residual water by inverting the tube onto paper towels and blotting the tissue for ~2 s. Immediately freeze the tissue by placing dried plants into aluminum foil and then plunging into liquid nitrogen (Video 1). The frozen tissue can be stored for 3 months in a −80 °C freezer.

3.2.2 RNA Extraction

1. Using the NucleoSpin RNA plant kit, perform RNA extraction in a chemical fume hood (Fig. 1, step 2).
2. Prepare 350 μL of buffer RA1 (from kit) in 1.7 mL microcentrifuge tubes, and add 3.5 μL β-mercaptoethanol (DTT or TCEP could also be used as reducing agents to a final concentration of 10–20 mM) to the tubes.
3. Finely grind plant tissue under liquid nitrogen with a pre-chilled mortar and pestle (*see* **Note 5**).
4. Add 100 mg of tissue to buffer RA1 (from kit) and vortex vigorously (*see* **Note 5**).
5. Place the NucleoSpin filter (violet ring) in the supplied 2 mL collection tube, add the lysate to the filter, and centrifuge for 1 min at 11,000 × *g*. Transfer the filtrate to a new 1.7 mL microcentrifuge tube (*see* **Note 5**), and discard the NucleoSpin filter.
6. Add 350 μL room temperature 70% ethanol to the lysate, and mix by pipetting (*see* **Note 5**). Add entire 700 μL to a NucleoSpin RNA plant column (light blue ring) in a 2 mL collection tube, and centrifuge for 30 s at 11,000 × *g*. Transfer the column to a new 2 mL collection tube (*see* **Note 5**).
7. Add 350 μL membrane desalting buffer (MDB from kit) to the column, and centrifuge for 1 min at 11,000 × *g*.
8. Add 10 μL reconstituted rDNase to 90 μL reaction buffer (from kit) in a fresh tube, mix by flicking the tube, then transfer 95 μL of this mixture directly to the center of the NucleoSpin RNA plant column, and incubate at room temperature for 15 min.
9. Add 200 μL of buffer RAW2 (from kit) to inactivate the rDNase to the NucleoSpin RNA plant column, centrifuge for 30s at 11,000 × *g*, and then place into a new 2 mL collection tube.

10. Add 600 μL of buffer RA3 (from kit) to the NucleoSpin RNA plant column, centrifuge for 30 s at 11,000 × *g*, discard the flow-through, and place back into the same collection tube.
11. Add 250 μL of buffer RA3 (from kit) to the NucleoSpin RNA plant column, and centrifuge for 2 min at 11,000 × *g* to completely dry the membrane. Leave the column open at room temperature for a few minutes to ensure dryness. Place the column into a new 1.5 collection tube (supplied with kit).
12. Add 60 μL water directly to the membrane, and centrifuge for 1 min at 11,000 × *g* to elute the RNA. RNA can be kept safely frozen at −20 °C for at least 30 days or stored at −80 °C for months (*see* **Note 1**).
13. Check RNA quantity and quality by performing a NanoDrop analysis on 2 μL of the RNA. Ideally 30–60 μg of RNA should be obtained from 100 mg of sample, and the 260:280 ratio should be 2.0 or above. Additionally, a bioanalyzer scan should be performed to determine RNA quality (Fig. 1, step 2, inset, *see* **Note 8**).

3.2.3 RNA Purification

Experimenters are often interested in one particular type of RNA, i.e., mRNA. Because rRNA is very abundant, the target RNA type must be enriched or the rRNA depleted. We enrich for mRNA in plants by performing a poly(A) selection twice (Fig. 1, step 3). This could be adjusted to the interests of the experimenter.

1. Prepare a 70 °C heat block and 200 μL/sample of pre-warmed RNA Storage Solution (from Poly(A) Purist MAG Kit).
2. For each library, use 200–300 μg total RNA for the first round of poly(A) selection using the Poly(A) Purist MAG Kit (Fig. 1, step 3, *see* **Note 2**), and bring to 400 μL with water in a 1.7 mL microcentrifuge tube.
3. Add 400 μL 2× binding solution (from kit) and mix thoroughly.
4. To 20 μL Oligo(dT) MagBeads (from kit) that have been thoroughly suspended by vortexing and added to a 1.7 mL microcentrifuge tube, add 100 μL wash solution 1 (from kit), then capture the beads by placing the tube on the magnetic stand for 2 min. Remove the solution by pipetting and discard.
5. Add an additional 100 μL of wash solution 1 (from kit) to the beads, place on the magnetic stand for 2 min, and remove the solution.
6. Transfer the mixture of total RNA in binding solution (800 μL total) to the washed beads, mix by inversion to resuspend the suspension, and then incubate the RNA/Oligo(dT) mixture at 70 °C for 5 min to denature RNA secondary structure and thus promote poly(A) annealing to oligo(dT).

7. Remove from the heat block, then incubate the tubes at room temperature for 60 min with gentle agitation provided by a rotator to maximize the binding of the mRNA to the beads.
8. Place each tube in the magnetic stand for 2 min to pull the beads to the side of the tube. Then, carefully remove and discard the solution without dislodging the Oligo (dT) MagBeads.
9. Add 400 μL washing solution 1 (from kit) to remove nonspecifically bound material, mix well by inversion, then capture the beads on the magnetic stand, and discard the wash solution.
10. Repeat previous step.
11. Add an additional 400 μL washing solution 2 (from kit), mix well by inversion, then capture the beads on the magnetic stand, and discard the wash solution.
12. Repeat previous step. Be sure to remove all liquid.
13. Add 200 μL pre-warmed (70 °C) RNA storage solution (from kit) to suspend the beads. Attempt to suspend the beads by addition of this solution. Capture the beads on the magnetic stand, and transfer the solution to a new 2.0 mL microcentrifuge tube.
14. Add a second 200 μL pre-warmed RNA storage solution (from kit), capture the beads on the magnetic stand, and transfer the solution to the previous tube for a total of 400 μL.
15. Add 40 μL 5 M ammonium acetate (from kit), 1 μL glycogen (from kit), and 1.1 mL 100% ethanol, then leave the mixture in a −20 °C freezer overnight or in a −80 °C freezer for 30 min to precipitate the RNA.
16. Pellet the RNA by centrifugation at 12,000 × *g* for 30 min at 4 °C.
17. Carefully remove and discard the supernatant; add 1 mL 70% ethanol to resuspend the pellet. Centrifuge at 12,000 × *g* for 10 min at 4 °C to pellet the RNA.
18. Dissolve the poly(A) RNA pellet in 10 μL of the RNA storage solution (from kit) (*see* **Note 9**). A bioanalyzer scan should be performed to validate rRNA depletion and determine yield (Fig. 1, step 3 inset, *see* **Note 8**). RNA can be kept safely frozen at −20 °C for at least 30 days or stored at −80 °C for months (*see* **Note 1**).

3.3 Reverse Transcription (Fig. 2, Step 1)

Please note that only Subheading 3.3.1 (Fig. 2, left) or Subheading 3.3.2 (Fig. 2, right) should be followed to perform genome-wide reverse transcription, depending on cleanup method preferred.

Table 1
Reverse transcription protocol

Component/action	Volume (μL)	Temperature/ time	Final concentration
[a]Water + 250–500 ng RNA	13.5 (standard version) 11 (biotin version)		
100 μM random hexamer	0.5		2.5 μM
10× RT buffer	2		1×
1 mM biotinylated dNTP(s)	2.5 (biotin version)		125 μM
10 mM dNTP (each)	1		500 μM
Denature		90 °C/1 min	
Snap cool		Ice/1 min	
[b]100 mM $MgCl_2$	1		5 mM
[b]100 mM DTT	1		5 mM
Anneal		25 °C/10 min	
Preheat		55 °C/1 min	
SuperScript III (200 U/μL)	1		200 U
Reaction		55 °C/50 min	
Denature		85 °C/5 min to 37 °C	
RNase H (5 U/μL)	1		5 U
Reaction		37 °C/20 min	
Stop		4 °C hold	

[a]Water is added first to dilute additional components so as to not expose RNA to highly concentrated buffers.
[b]$MgCl_2$ and DTT can be combined to provide a master mix for ease of addition and mixing.

3.3.1 PAGE Version

1. Set a program on a thermal cycler to heat to 90 °C for 1 min followed by cooling to 25 °C for 10 min, then heating to 55 °C for 51 min, increasing the heat to 85 °C for 5 min, cooling to 37 °C for 20 min, and then a final hold at 4 °C (Table 1).
2. In an Eppendorf PCR tube on ice (*see* **Notes 1** and **10**), add 250–500 ng of purified mRNA per sample, then add 1 μL of 100 μM random hexamer primer to each sample, bring the volume to 28 μL with water, and mix by pipetting.
3. Pulse centrifuge and then split this sample into two PCR tubes (*see* **Note 10**) to increase the yield, and add 2 μL of 10× RT buffer and 1 μL 10 mM dNTPs to each tube, vortex briefly, and pulse centrifuge to bring the solution to the bottom.

4. Heat the reactions to 90 °C for 1 min to denature the RNA and primer, and then snap cool the reactions on ice for 1 min with the thermal cycler paused at 25 °C.
5. After 1 min, add 1 μL of 100 mM $MgCl_2$ and 1 μL of 100 mM DTT to each reaction (these components can be made into a master mix for ease of addition), and vortex briefly then pulse centrifuge.
6. Heat the reactions to 25 °C for 10 min in the thermal cycler to anneal the primer to the RNA, and then heat the reactions to 55 °C for 1 min to preheat the reactions.
7. After 1 min, add 1 μL SuperScript III (200 U/μL) to each reaction while the thermal cycler is paused, vortex briefly, pulse centrifuge, and allow the reaction to proceed for 50 min at 55 °C. The final solution will contain 1× RT buffer, 0.5 mM each dNTP, 5 mM $MgCl_2$, 5 mM DTT, and 200 U SuperScript III (Table 1).
8. Heat the reactions to 85 °C for 5 min to denature the SuperScript III, then cool the reaction to 37 °C.
9. Add 1 μL of RNase H (5 U/μL) to each reaction to degrade any RNA bound to the cDNA, and allow the reaction to proceed for 20 min at 37 °C. Finally, cool the reaction to 4 °C. The cDNA can be stored in a −20 °C freezer.

3.3.2 Biotin Version

1. Set a program on a thermal cycler to heat to 90 °C for 1 min, followed by cooling to 25 °C for 10 min and then heating to 55 °C for 51 min, increasing the heat to 85 °C for 5 min, cooling to 37 °C for 20 min, and then a final hold at 4 °C (Table 1).
2. In an Eppendorf PCR tube on ice (*see* **Notes 1** and **10**), add 250–500 ng of purified mRNA per sample, then add 1 μL of 100 μM random hexamer primer to each sample, bring the volume to 23 μL with water, and mix by pipetting.
3. Pulse centrifuge and then split this sample into two PCR tubes (*see* **Note 10**) to increase yield. To each tube, add 2 μL of 10× RT buffer, 1 μL 10 mM dNTPs, and 2.5 μL of 1 mM biotinylated dNTPs (*see* **Note 11**), vortex briefly, and pulse centrifuge.
4. In a thermal cycler, heat the reactions to 90 °C for 1 min to denature the RNA and primer, and then snap cool the reactions on ice for 1 min, while the thermal cycler is paused at 25 °C.
5. After 1 min, add 1 μL of 100 mM $MgCl_2$ and 1 μL of 100 mM DTT to each reaction (these components can be made into a master mix for ease of addition), vortex briefly, and then pulse centrifuge.

6. Heat the reactions to 25 °C for 10 min to anneal the primer to the RNA, and then heat the reactions to 55 °C for 1 min to preheat the reactions.
7. After 1 min, add 1 μL SuperScript III (200 U/μL) to each reaction while the thermal cycler is paused, vortex briefly, pulse centrifuge, and allow the reaction to proceed for 50 min at 55 °C. The final solution will contain 1× RT buffer, 0.5 mM each dNTP, 125 μM biotin-dNTP total, 5 mM $MgCl_2$, 5 mM DTT, and 200 U SuperScript III (Table 1).
8. Heat the reactions to 85 °C for 5 min to denature the SuperScript III, and then cool the reaction to 37 °C.
9. Add 1 μL of RNase H (5 U/μL) to each reaction to degrade any RNA bound to the cDNA, and allow the reaction to proceed for 20 min at 37 °C. Finally, cool the reaction to 4 °C. The cDNA can be stored in a −20 °C freezer.

3.4 Reverse Transcription Clean-up (Fig. 2, Step 2)

Please note that either Subheading 3.4.1 (Fig. 2, left) or Subheading 3.4.2 (Fig. 2, right) should be followed to clean up after reverse transcription. If biotinylated dNTPs were not included during reverse transcription, Subheading 3.4.1 must be followed.

3.4.1 PAGE Version

1. Following reverse transcription, combine both reactions of a given sample, and add 42 μL of 2× formamide loading dye to bring the total volume to 84 μL.
2. Denature the samples for 2–5 min at 95 °C, and then carefully load the entire 84 μL sample onto a 10% denaturing PAGE gel (8.3 M urea). Additionally, be sure that appropriate ladders are loaded (*see* **Note 3**).
3. Run the gel for approximately 1 h at 25–30 W (*see* **Note 3**). When the bottom dye is almost to the bottom of the gel, stop the power, remove the gel from the glass plates, and then carefully place the gel into a container with SYBR gold in 1× TE buffer. Cover the container with aluminum foil, and gently shake for 10 min (*see* **Note 12**).
4. Carefully remove the gel from the container, and place on a piece of plastic wrap that is on top of a glass plate (*see* **Note 12**).
5. Visualize the DNA using a UV light box (*see* **Note 12**), and cut out a gel strip corresponding to 50 nt and above with a clean razor blade to remove excess RT primer and retain all cDNA fragments (*see* **Note 12**, Video 2).
6. Dice each gel slice into very fine 1–2 mm^3 pieces using a razor blade, and transfer into a 50 mL centrifuge tube, or smash the gel slice against the sides of the tube to completely crush the gel piece into small fragments (*see* **Note 12**, Video 2).

7. Add 2× the gel volume of TEN_{250} buffer to the tube (*see* **Note 12**, Video 2), and place in an incubator/shaker at 37 °C overnight.
8. To remove fine gel pieces, use a 0.2 μm syringe filter attached to a 10 mL syringe. Expel the filtrate into a new 50 mL centrifuge tube, add 2× volume of 100% ethanol and 1 μL GlycoBlue to help visualize the pellet, and place into crushed dry ice for 1 h or at −20 °C overnight (recommended).
9. Spin down at 14,000 × *g* for 30 min in a centrifuge at 4 °C (*see* **Note 12**), decant the liquid, add 1–2 mL of 75% ethanol (at least 1× the original gel volume), and suspend the pellet by pipetting.
10. Spin down at 14,000 × *g* for 5 min in a centrifuge at 4 °C, decant the liquid, and spin down again briefly (<1 min). Pipette the remaining liquid off with a pipette, and then place in a 37 °C incubator (not shaker) with lids off until completely dry (*see* **Note 12**).
11. Dissolve the pellet (*see* **Note 12**) in 100 μL of water, transfer to a 1.7 mL tube (*see* **Note 12**), speed-vac to dryness, and bring up to 2 μL with water. The purified cDNA can be stored in a −20 °C freezer.

3.4.2 Streptavidin Version

1. Following reverse transcription, combine both 21 μL reactions of a given sample in a single 1.7 mL centrifuge tube, and add 8 μL of water to bring the sample to 50 μL.
2. Following the instructions for the RNA Clean and Concentrator-5 kit, add 100 μL of RNA binding buffer (from kit) to each sample, and vortex to mix. The RNA Clean and Concentrator-5 kit can be used to purify any single-stranded nucleic acids; here we use it to purify cDNA.
3. Add 150 μL of 100% ethanol to each sample and mix by pipetting. Using the same pipette, transfer to the Zymo-spin IC column in a collection tube (provided), and centrifuge at 10,000–16,000 × *g* for 30 s.
4. Discard the flow-through, then add 400 μL RNA prep buffer (from kit) to the column, and centrifuge at 10,000–16,000 × *g* for 30 s.
5. Discard the flow-through, then add 700 RNA wash buffer (from kit) (*see* **Note 13**) to the column, and centrifuge at 10,000–16,000 × *g* for 30 s.
6. Discard the flow-through, then add 400 μL RNA wash buffer (from kit) to the column, and centrifuge at 10,000–16,000 × *g* for 2 min (*see* **Note 13**).
7. Transfer the column to a new 1.7 mL centrifuge tube, being sure to not touch the column to any liquid in the original tube,

add 15 μL of water to the column, and centrifuge at 10,000–16,000 × *g* for 2 min. The flow-through contains the cDNA sample. The cDNA can be stored in a −20 °C freezer.

8. The following steps do not use any premade kits. Vortex magnetic hydrophilic streptavidin beads to ensure homogeneous suspension. Add 50 μL of beads to a 1.7 mL centrifuge tube. Wash beads by adding 50 μL 1× wash/binding buffer, vortex to mix, and then pulse centrifuge to collect droplets. Place tubes on a magnetic stand to pull beads to one side. Once the solution is clear, pipette off and discard supernatant. Repeat wash two more times for a total of three times. Each time add the buffer while the beads are still on the magnetic stand to prevent the beads from drying out (*see* **Note 13**).
9. Add 35 μL of 1× wash/binding buffer to cDNA (*see* **Note 13**) and vortex to mix.
10. Transfer the 50 μL sample to the washed beads on the magnetic stand with no buffer remaining, vortex to suspend the beads, and then incubate at room temperature for 10 min with occasional agitation by hand to keep the beads suspended (*see* **Note 13**).
11. After 10 min, briefly vortex to ensure homogeneous suspension, and then pulse centrifuge. Place the tubes on the magnetic stand, and remove the supernatant by pipetting. With the beads remaining on the stand, add 100 μL of 1× wash/binding buffer, remove from stand, vortex, and pulse centrifuge. Place the tubes on the magnetic stand, and remove the supernatant by pipetting. Repeat with fresh 100 μL wash/binding buffer.
12. With the tubes remaining on the stand after the second 1× wash/binding buffer, add 100 μL of pre-warmed (40 °C) low-salt buffer to the beads, remove from stand, vortex, and pulse centrifuge. Place the tubes on the magnetic stand and remove the supernatant by pipetting. Repeat with fresh 100 μL pre-warmed low-salt buffer (*see* **Note 13**).
13. With the tubes remaining on the stand after the second low-salt buffer, add 25 μL elution buffer to the beads, vortex to suspend the beads, wrist-flick the beads to the bottom (do not centrifuge), and incubate at 95 °C for 2 min (Video 3).
14. After 2 min (*see* **Note 13**), mix *quickly* by hand-flick, pulse centrifuge, and then immediately place the tubes on the magnetic stand and immediately remove the elution buffer to a new 1.7 mL microcentrifuge tube. Repeat the elution with 25 μL of fresh elution buffer, and add the supernatant to the first elution (Video 3).

15. Following the instructions of the RNA Clean and Concentrator-5 kit, add 100 μL of RNA binding buffer (from kit) to each sample, and vortex to mix.
16. Add 150 μL of 100% ethanol to each sample and mix by pipetting. Using the same pipette, transfer to the Zymo-Spin IC Column in a Collection Tube (provided), and centrifuge at 10,000–16,000 × *g* for 30 s.
17. Discard the flow-through, then add 400 μL RNA Prep Buffer (from kit) to the column, and centrifuge at 10,000–16,000 × *g* for 30 s.
18. Discard the flow-through, then add 700 RNA wash buffer (from kit) to the column, and centrifuge at 10,000–16,000 × *g* for 30 s.
19. Discard the flow-through, then add 400 μL RNA wash buffer (from kit) to the column, and centrifuge at 10,000–16,000 × *g* for 2 min.
20. Transfer the column to a new 1.7 mL centrifuge tube, being sure to not touch the column to any liquid in the original tube, add 15 μL of water to the column, and centrifuge at 10,000–16,000 × *g* for 2 min. The flow-through contains the cDNA sample.
21. Speed-vac the sample to dryness, and then bring up in 2 μL of water. The purified cDNA can be stored in a −20 °C freezer.

3.5 Ligation (Fig. 2, Step 3)

1. Set a program on a thermal cycler to heat to 90 °C for 90 s, followed by cooling to 16 °C for 6 h, warming to 30 °C for 6 h, heating to 65 °C for 15 min, and a final hold at 4 °C (Table 2).
2. Carefully transfer the 2 μL of purified cDNA to a PCR tube (*see* **Note 10**), and then add 1 μL of 100 μM hairpin adaptor, 1 μL of 5 M betaine, and 4 μL of 50% PEG 8000 (PEG must be pipetted slowly due to its high viscosity). Vortex the tubes briefly followed by brief centrifugation to collect droplets.
3. Place the tubes in a thermal cycler, and denature the components at 90 °C for 90 s.
4. Allow the tubes to cool to 16 °C, and then add 1 μL of 10× DNA ligase buffer and 1 μL T4 DNA ligase (400 U/μL).
5. Allow the reaction to proceed for 6 h at 16 °C and then for another 6 h at 30 °C.
6. Denature the enzyme at 65 °C for 15 min and then cool to 4 °C. The ligation product can be stored in a −20 °C freezer.

Table 2
Ligation protocol

Component/action	Volume (μL)	Temperature/time	Final concentration
cDNA	2		
100 μM Random Hexamer	1		10 μM
5 M betaine	1		0.5 M
[a]50% PEG 8000	4		20%
Denature		90 °C/90 s	
Cool		16 °C	
10× T4 ligase buffer	1		1×
T4 DNA ligase (400 U/μL)	1		400 U
Reaction		16 °C/6 h	
Reaction		30 °C/6 h	
Denature		65 °C/15 min	
Stop		4 °C hold	

[a]PEG must be pipetted and added slowly as it is a very viscous solution.

3.6 Ligation Clean-up (Fig. 2, Step 4)

Please note that either Subheading 3.6.1 (Fig. 2, left) or Subheading 3.6.2 (Fig. 2, right) should be followed to clean up after ligation. If biotinylated dNTPs were not included during reverse transcription, Subheading 3.6.1 must be followed.

3.6.1 PAGE Version

1. Following ligation, add 10 μL of 2× formamide loading dye to bring the total volume to 20 μL.
2. Denature the samples for 2–5 min at 95 °C, and then carefully load the entire 20 μL sample onto a 10% denaturing PAGE gel (8.3 M urea). Additionally, be sure that appropriate ladders are loaded (*see* **Note 3**).
3. Run the gel for approximately 2 h at 25–30 Watts (*see* **Note 3**). When the bottom dye reaches the bottom of the gel, stop the power, remove the gel from between the glass plates, and then carefully place the gel into a container with SYBR gold in 1× TE buffer. Cover the container with aluminum foil, and gently shake for 10 min (*see* **Note 12**).
4. Carefully remove the gel from the container, and place on a piece of plastic wrap that is on top of a glass plate (*see* **Note 12**).
5. Visualize the DNA using a UV light box (*see* **Note 12,** Video 2), and cut out a gel strip corresponding to 91 nt and above in each sample lane with a clean razor blade. This is to remove any by-product formed from the RT primer and ligation adaptor and to retain all sizes of ligation product.

6. Dice each gel slice into fine 1–2 mm^3 pieces using a razor blade, and transfer into a 50 mL centrifuge tube, or smash the gel slice against the sides of the tube to completely crush the gel piece into small fragments (*see* **Note 12,** Video 2).
7. Add about 2× the gel volume of TEN_{250} buffer to the tube (*see* **Note 12,** Video 2), and place in an incubator/shaker at 37 °C overnight.
8. To remove fine gel pieces, use a 0.2 μm syringe filter attached to a 10 mL syringe. Expel the filtrate into a new 50 mL centrifuge tube, add 2× volume of 100% ethanol and 1 μL GlycoBlue, and place into crushed dry ice for 1 h or at −20 °C overnight.
9. Spin down at 14,000 × *g* for 30 min in a centrifuge at 4 °C (*see* **Note 12**), decant the liquid, add 1–2 mL of 75% ethanol (at least 1× the original gel volume), and suspend the pellet by pipetting.
10. Spin down at 14,000 × *g* for 5 min in a centrifuge at 4 °C (*see* **Note 12**), decant the liquid, and spin down again briefly (<1 min). Remove the remaining liquid with a pipette, and then place in a 37 °C incubator (not shaker) until completely dry (*see* **Note 12**).
11. Dissolve the pellet in 100 μL of water (*see* **Note 12**), and transfer to a 1.7 mL tube (*see* **Note 12**), speed-vac the sample to dryness, and bring to 10 μL with water. The purified ligation product can be stored in a −20 °C freezer.

3.6.2 Streptavidin Version

1. Following ligation, add 40 μL of water to the sample to bring each sample to 50 μL, and transfer the sample to a 1.7 mL centrifuge tube.
2. Following the instructions for the RNA Clean and Concentrator-5 kit, add 100 μL of RNA binding buffer (from kit) to each sample, and vortex to mix.
3. Add 150 μL of 100% ethanol to each sample and mix by pipetting. Using the same pipette, transfer to the Zymo-spin IC column in a collection tube (provided), and centrifuge at 10,000–16,000 × *g* for 30 s.
4. Discard the flow-through, then add 400 μL RNA prep buffer (from kit) to the column, and centrifuge at 10,000–16,000 × *g* for 30 s.
5. Discard the flow-through, then add 700 RNA wash buffer (from kit) (*see* **Note 13**) to the column, and centrifuge at 10,000–16,000 × *g* for 30 s.
6. Discard the flow-through, then add 400 μL RNA wash buffer (from kit) to the column, and centrifuge at 10,000–16,000 × *g* for 2 min (*see* **Note 13**).

7. Transfer the column to a new 1.7 mL centrifuge tube, being sure to not touch the column to any liquid in the original tube, add 15 μL of water to the column, and centrifuge at 10,000–16,000 × *g* for 2 min. The flow-through contains the cDNA sample. The cDNA can be stored in a −20 °C freezer.
8. The following steps do not use any premade kits. Vortex magnetic hydrophilic streptavidin beads to ensure homogeneous suspension. Add 50 μL of beads to a 1.7 mL centrifuge tube. Wash beads by adding 50 μL 1× wash/binding buffer, vortex to mix, and then centrifuge briefly. Place tubes on a magnetic stand to pull beads to one side. Once the solution is clear, pipette off and discard supernatant. Repeat wash two more times for a total of three times. Each time add the buffer while the beads are still on the magnetic stand to prevent the beads from drying out (*see* **Note 13**).
9. Add 35 μL of 1× wash/binding buffer to cDNA (*see* **Note 13**) and vortex to mix.
10. Transfer the 50 μL sample to washed beads on the magnetic stand with no buffer remaining, vortex to suspend the beads, and then incubate at room temperature for 10 min with occasional agitation by hand to keep the beads suspended.
11. After 10 min, briefly vortex to ensure homogeneous suspension, and then pulse centrifuge to collect droplets. Place the tubes on the magnetic stand, and remove the supernatant by pipetting. With the beads remaining on the stand, add 100 μL of 1× wash/binding buffer, remove from stand, vortex, and pulse centrifuge. Place the tubes on the magnetic stand and remove the supernatant by pipetting. Repeat with fresh 100 μL wash/binding buffer.
12. With the tubes remaining on the stand after the second 1× wash/binding buffer, add 100 μL of pre-warmed (40 °C) low-salt buffer to the beads, remove from stand, vortex, and pulse centrifuge. Place the tubes on the magnetic stand and remove the supernatant by pipetting. Repeat with fresh 100 μL pre-warmed low-salt buffer (*see* **Note 13**).
13. With the tube remaining on the stand after the second low-salt buffer, add 25 μL elution buffer to the beads, vortex to suspend the beads, flick the beads to the bottom (do not centrifuge), and incubate at 95 °C for 2 min (Video 3).
14. After 2 min (*see* **Note 13**), mix *quickly* by hand, pulse centrifuge, and then immediately place the tubes on the magnetic stand and immediately remove the elution buffer to a new 1.7 mL microcentrifuge tube. Repeat the elution with 25 μL of fresh elution buffer, and add the supernatant to the first elution (Video 3).

15. Following the instructions of the RNA Clean and Concentrator-5 kit, add 100 μL of RNA binding buffer (from kit) to each sample, and vortex to mix.
16. Add 150 μL of 100% ethanol to each sample and mix by pipetting. Using the same pipette, transfer to the Zymo-spin IC column in a collection tube (provided), and centrifuge at 10,000–16,000 × *g* for 30 s.
17. Discard the flow-through, then add 400 μL RNA prep buffer (from kit) to the column, and centrifuge at 10,000–16,000 × *g* for 30 s.
18. Discard the flow-through, then add 700 RNA wash buffer (from kit) to the column, and centrifuge at 10,000–16,000 × *g* for 30 s.
19. Discard the flow-through, then add 400 μL RNA wash buffer (from kit) to the column, and centrifuge at 10,000–16,000 × *g* for 2 min.
20. Transfer the column to a new 1.7 mL centrifuge tube being sure to not touch the column to any liquid in the original tube, add 11 μL of water to the column, and centrifuge at 10,000–16,000 × *g* for 2 min.
21. Ensure that the final product contains approximately 10 μL; if not, adjust volume to 10 μL. The purified ligation product can be stored in a −20 °C freezer.

3.7 PCR Cycle Test (Fig. 2, Step 5, Inset)

The PCR cycle test is performed to determine the appropriate number of cycles needed for amplification. The results also indicate whether the libraries have been properly prepared (*see* **Note 14**).

1. Set a program on a thermal cycler to heat to 98 °C for 1 min, and then cycle through 98 °C for 8 s and 72 °C for 45 s for about 35 cycles (Table 3, *see* **Note 15**).
2. For each sample (ligation product), add 14.25 μL of water to a PCR tube (*see* **Note 10**), and then transfer 3 μL of the purified ligation product to the tube.
3. Add 1 μL of 10 μM reverse primer (*see* **Note 15**).
4. Make a master mix solution in an appropriately sized tube (0.65 mL or 1.7 mL depending on the number of samples) that contains enough of the following components for n + 1 reactions (where n is the number of samples). Each sample requires 5 μL of 5× Q5 reaction buffer, 0.5 μL 10 mM dNTP, 1 μL of 10 μM forward primer, and 0.25 μL Q5 DNA polymerase. After these components are combined, vortex briefly, and pulse centrifuge.

Table 3
PCR protocol

Component/action		Volume (μL)	Temperature/time	Final concentration
[a]Water		14.25		
Ligation product		3		
[b]5× Q5 buffer		5		1×
[b]10 mM dNTPs (each)		0.5		0.2 mM
[b]10 μM forward primer		1		0.4 μM
10 μM reverse primer		1		0.4 μM
Q5 polymerase (2 U/μL)		0.25		0.5 U
Initial denature			98 °C/1 min	
Cycle	Denature Extend		98 °C/8 s 72 °C/45 s	
Final extension			72 °C/10 min	
Cool			4 °C hold	

[a]Water is added first to dilute additional components so as to not expose RNA to highly concentrated buffers.
[b]5× Q5 buffer, dNTPs, forward primer, and Q5 polymerase can be combined to provide a master mix for ease of addition and mixing.

5. Add 6.75 μL of the master mix solution to each of the PCR tubes containing the ligation product and reverse primer. Vortex briefly, pulse centrifuge, and then place into a thermal cycler with the program ready.
6. At the appropriate cycles to test the PCR reaction, remove 5 μL of each sample, and place into new PCR tubes containing 1 μL of 6× glycerol loading dye (*see* **Note 15**). These samples can be stored in a 4 °C refrigerator.
7. Run all the collected samples on a 1.5% agarose gel in 1× TAE running buffer at 150 V for approximately 1.5 h, or until the orange dye is near the bottom of the gel.
8. Carefully place the gel into a container containing SYBR gold in 1× TE buffer, and shake at room temperature for 30 min.
9. Visualize the gel on a UV gel box or UV gel imager to determine the appropriate cycle that provides a good smear on the gel and does not produce specific bands (Fig. 2, step 5, inset, *see* **Note 15**).

3.8 PCR Library Amplification (Fig. 2, Step 5)

After one appropriate cycle number is determined for all samples, including DMS control samples, PCR can be performed to properly amplify the material needed for sequencing.

1. Set a program on a thermal cycler to heat to 98 °C for 1 min, and then cycle through 98 °C for 8 s and 72 °C for 45 s for the number of cycles determined from the PCR cycle test, followed by a 72 °C hold for 10 min (Table 3).
2. Add 14.25 μL of water to a PCR tube (*see* **Note 1**), and then transfer a fresh 3 μL (*see* **Note 15**) of the purified ligation product to the tube.
3. Add 1 μL of a ***unique*** 10 μM barcoded reverse primer per sample (*see* **Note 15**).
4. Make a master mix solution in an appropriately sized tube (0.65 mL or 1.7 mL depending on number of samples) that contains enough of the following components for n + 1 reactions (where n is the number of samples). Each sample requires 5 μL of 5× Q5 reaction buffer, 0.5 μL 10 mM dNTP, 1 μL of 10 μM forward primer, and 0.25 μL Q5 DNA polymerase. After these components are combined, vortex briefly, and pulse centrifuge.
5. Add 6.75 μL of the master mix solution to the PCR tubes containing the ligation product and reverse primer. Vortex briefly, pulse centrifuge, and then place into a thermal cycler with the program ready.
6. Run the PCR reaction following the preset program. The samples can be stored in a −20 °C freezer.

3.9 PCR Clean-Up (Fig. 2, Step 6)

Please note that PAGE extraction after PCR is performed for both versions as the amplification does not include biotinylated dNTPs due to potential interference with subsequent Illumina sequencing. Thus PAGE extraction must be performed on all samples from either version.

1. Following PCR, add 25 μL of 2× formamide loading dye to bring the total volume to 50 μL.
2. Denature the samples for 2–5 min at 95 °C, and then carefully load the entire 50 μL sample onto a 10% denaturing PAGE gel (8.3 M urea). Additionally, be sure that appropriate ladders are loaded (*see* **Note 3**).
3. Run the gel for approximately 3 h at 25–30 Watts (*see* **Note 3**). When the bottom dye is almost to the bottom of the gel, stop the power, remove the gel from between the glass plates, and then carefully place the gel into a container containing SYBR gold in 1× TE buffer. Cover the container with aluminum foil, and gently shake for 10 min (*see* **Note 12**).
4. Carefully remove the gel from the container, and place on a piece of plastic wrap that is on top of a glass plate (*see* **Note 12**).
5. Visualize the DNA using a UV light box (*see* **Note 12**, Video 2), and cut out a gel strip corresponding to 200–600 nt with a

clean razor blade to obtain inserts of appropriate size to sequence and to exclude any primer dimers or by-product from the samples.

6. Dice each gel slice into fine 1–2 mm^3 pieces using a razor blade, and transfer into a 50 mL centrifuge tube, or smash the gel slice against the sides of the tube to completely crush the gel piece into small fragments (*see* **Note 12**, Video 2).
7. Add 2× the gel volume of TEN_{250} buffer to the tube (*see* **Note 12**, Video 2), and place in an incubator/shaker at 37 °C overnight.
8. To remove fine gel pieces, use a 0.2 μm syringe filter attached to a 10 mL syringe. Expel the filtrate into a new 50 mL centrifuge tube, add 2× volume of 100% ethanol and 1 μL GlycoBlue, and place into crushed dry ice for 1 h or at −20 °C overnight.
9. Spin down at 14,000 × *g* for 30 min in a centrifuge at 4 °C (*see* **Note 12**), decant the liquid, add 1–2 mL of 75% ethanol (at least 1× the original gel volume), and suspend the pellet by pipetting.
10. Spin down at 14,000 × *g* for 5 min in a centrifuge at 4 °C (*see* **Note 12**), decant the liquid, and spin down again briefly (<1 min). Remove the remaining liquid with a pipette, and then place in a 37 °C incubator (not shaker) until completely dry (*see* **Note 12**).
11. Dissolve the pellet in 100 μL of water (*see* **Note 12**), transfer to a 1.7 mL tube (*see* **Note 12**), and speed-vac the sample to about 30 μL. Store samples in a −20 °C freezer.

3.10 Quality Checks

1. Using 2 μL of the final sample, take a NanoDrop reading (*see* **Note 8**).
2. Upon submitting the samples for sequencing, be sure that a bioanalyzer run and qPCR analysis are performed by the sequencing facility prior to sequencing to check for good quality and ascertain the amount of sample that will provide optimal Illumina clustering (*see* **Note 8**).

3.11 Sequencing

1. Sequencing is performed on a MiSeq or HiSeq instrument with the appropriate Illumina kits and reagents for single-end sequencing of 150 cycles. In order to obtain high-quality reads, it is necessary to sequence with a custom sequencing primer to increase the diversity at the first nucleotide sequenced. If genome size is small (e.g., prokaryotic genomes), MiSeq analysis alone may provide useful data. For eukaryotic samples, we typically perform MiSeq sequencing first to validate the quality of our libraries and optimize clustering parameters before obtaining HiSeq reads (Fig. 2, step 7).
2. Data can then be piped to StructureFold2 [7].

4 Notes

1. Handling RNA: RNA is subject to degradation by high pH, high temperature, high Mg^{2+} concentration, or RNases that are commonly found on skin and other surfaces. As such, it is important that all solutions containing RNA (including and up to the reverse transcription step) be kept on ice and in RNase-free surroundings and gloves be worn at all times. One must be sure that any water or other solution used is RNase-free. RNA can be kept safely frozen at −20 °C for at least 30 days or stored at −80 °C for months. Notably, unlike other structure-probing methods, Structure-seq2 converts RNA into cDNA at an early stage within library preparation in an effort to minimize RNA degradation.
2. Approaches for other organisms: Structure-seq2 is adaptable to a diversity of organisms in addition to plants. In order to apply this method to a new organism, one must determine how to perform the chemical probing treatment, including determination of single-hit kinetics conditions, RNA extraction, and RNA purification. We have successfully performed DMS modification in bacteria by directly adding DMS to the growth medium, allowing the reaction to proceed for a short time, and then quenching the reaction by directly adding DTT to the medium. After the quench is finished, the cells can be frozen and pelleted. The RNA is extracted using the RNeasy Mini Kit (Qiagen), and rRNA is removed via Ribo-Zero rRNA Removal Kit (Bacteria) (Illumina). The DMS-modified ribo-depleted RNA can then be used in Structure-seq2 starting at genome-wide reverse transcription (Subheading 3.3).

   To prepare human RNA for Structure-seq2, one should grow cells of interest to 70–80% confluency. For adherent cells, media can then be discarded and PBS added back to the cells with and without the relevant concentration of DMS, for the predetermined amount of time. The PBS/DMS solution can then be discarded into hazardous waste, and the cells washed once with PBS/DTT solution to quench the reaction. This solution can then be discarded and the cells lysed and RNA extracted using TRIzol followed by ethanol precipitation. Poly (A) RNA is then enriched using the poly(A) purist Mag kit. For nonadherent cells, a similar approach to that described above for bacteria would be appropriate. The DMS-modified poly (A) purified RNA can then be used in Structure-seq2 starting at genome-wide reverse transcription (Subheading 3.3).
3. Recommended gel sizes and ladders: All PAGE gels used in Structure-seq2 are 10% polyacrylamide 8.3 M urea denaturing gels with 1× TBE running buffer. It is important that the gels

be fully denaturing to properly visualize and separate the products. It may also be helpful to silanize the glass plates to ensure gels do not stick to the glass when removing the plates.

The width and thickness of the gel can have an effect on how the current travels through the gel and in turn affects "smiling," the sharpness of the band, and how easy the gel is to handle. The length of the gel controls how far the products travel and separate. For library preparation, it is essential that the desired product is well separated from any undesired products that may be in the reactions as well; thus the length is generally more important than the width and thickness. Additionally, the width of the RT, ligation, and PCR gels can be easily adjusted depending on the number of samples to be analyzed. It is also important that appropriate ladders be run with each experiment in order to ensure proper size selection. We use the following dimensions of gels ($L \times W \times T$) and wells ($W \times H$) and ladders for the PAGE gels in our Structure-seq2 experiments.

- Sequencing gel: 43 cm × 38.2 cm × 0.3 mm with 8 mm × 5 mm wells. The ladders used are ddNTP lanes that also serve as sequencing lanes.
- RT gel: 25 cm × 15 cm × 1.5 mm with 1.7 cm × 4.2 mm wells bridging teeth with lab tape to appropriate volume and GeneRuler low range DNA ladder (sizes 25, 50, 75, 100, 150, 200, 300, 400, 500, and 700 bp).
- Ligation gel: 25 cm × 15 cm × 1.5 mm with 1.0 cm × 4.2 mm wells bridging teeth with lab tape to appropriate volume with a handmade ladder of 67 and 91 nt DNA oligos (250 ng each) and the GeneRuler low range DNA ladder.
- PCR gel: 25 cm × 15 cm × 1.5 mm with 1.7 cm × 4.2 mm wells bridging teeth with lab tape to appropriate volume with a ss100 DNA ladder (denatured prior to addition) GeneRuler low range DNA ladder, and GeneRuler 100 bp ladder.

The PCR cycle test gel is the only agarose gel run and uses 1.5% agarose with 1× TAE running buffer. The width of the gel is dependent on the number of samples to be run, but the length used is typically 14 cm long, and 3–4 mm deep, with wells 9 mm wide and 1 mm deep. We use a 100 bp ladder.

4. Single-hit kinetics determination: To test for single-hit kinetics, we typically perform in vivo test reactions with varying DMS concentrations ranging from about 10–100 mM and keep the time constant. It is also possible to test the effect of varying time, but the assay should be performed for a maximum of 10 min to minimize the possibility of RNA degradation.

When performing these DMS concentration tests, it is helpful to stagger the reactions so that they are performed at approximately the same time and growth stage, but only one reaction at a time is handled.

To determine the appropriate DMS concentration, the full length band on the sequencing gel is analyzed. The ratio of the intensity of the full length band to the total intensity of the lane should decrease in the DMS lane 10–20% as compared to the−DMS lane for appropriate single-hit kinetics. This gel can also be used to validate DMS modification at appropriate solvent accessible A and C residues in a reference RNA such as rRNA. We typically do single-hit kinetics on rRNA since it is abundant and structures are known [9].

5. RNA extraction troubleshooting: It is vital to the experiment that the RNA extraction be done as effectively as possible. To obtain the highest-quality RNA, the mortar, pestle, and any other equipment used in handling the plant tissue should be pre-chilled under liquid nitrogen. Additionally, the plant tissue should be ground to an extremely fine powder and should be kept under liquid nitrogen for the entire grinding process, as it is essential that the tissue should never be allowed to thaw. As the Buffer RA1 contains components that protect the RNA from degradation, the RNA is safe once in the buffer. Thus, we find it easiest to add the frozen ground tissue to a pre-weighed 1.7 mL tube containing buffer RA1 and measure the amount added to the buffer. As the plant tissue being processed will likely be more than 100 mg, multiple tubes can be used per sample in order to increase the amount of RNA obtained from a given reaction.

   If the lysate solidifies with Buffer RA1, use Buffer RAP instead.

   After centrifugation of the NucleoSpin Filter, do not transfer any pellet that forms but only the supernatant.

   After addition of room temperature 70% ethanol to the supernatant, transfer everything to the NucleoSpin RNA Plant Column, even if a stringy precipitate forms.
   Only 750 μL can be placed on the NucleoSpin RNA Plant Column at a time, but the column can be loaded multiple times.

   Notably, this procedure will vary when studying different organisms. Please use the RNA extraction protocol recommended for the particular organism of interest.

6. rRNA reverse transcription troubleshooting: For rRNA-specific reverse transcription, the abundance of the RNA being targeted can change the quality of the results. Typically, we have found that using 2.5 μg of total RNA provides enough material to obtain clear results from rRNA. In addition to adjusting the amount of total RNA, one can adjust the amount

of primer or dNTPs added. To obtain clear sequencing lanes, the amount of ddNTPs may also need to be adjusted. When using radiolabeled DNA primers, we typically measure the radioactive signal rather than the concentration and add 1 μL of 150–250K cpm of purified primer. We recommend using radioactively-labeled primers but understand that this may not be possible for some users, and thus fluorescently tagged primers can also be used. All materials must be kept in the dark as much as possible, even when running the gel. With fluorescent primers, we have used 0.2–0.25 μM final concentration of primer. Additionally, fluorescently labeled primers are more expensive and do not produce as strong a signal, and the scanned image is often not as sensitive as when using radioactively-labeled primers. To visualize a gel run with fluorescent primers, one does not need to dry or expose the gel overnight but can typically scan the gel with the gel still on a borosilicate plate.

To obtain a clear full length band, primers should be designed approximately 100–200 nts from the 5′ end of a transcript and should contain a region that is predicted to be single-stranded. For rRNA, one can use the predicted secondary structures found at the Comparative RNA Web Site and Project [10] to determine this. We typically use primers that are about 20 nts in length and have a T_m of 60 °C.

7. Experimental design: In order to perform a proper Structure-seq2 experiment that can be rigorously analyzed with statistical validity, it is important to have good experimental design. We recommend performing all treatments in biological triplicates to obtain statistically significant analysis. We also recommend preparing all samples at the same time as much as possible to prevent differences in experimental material or procedure that could impact the results. For StructureFold2 analysis, it is necessary to have a negative control that has not been chemically modified, and it is best to obtain this in triplicate at the same time as the samples treated with the chemical probe.

8. Quality checks: Total RNA is considered good quality when the RIN provided by a Total RNA Bioanalyzer scan is close to 10, and there are distinct peaks for each rRNA present in the organism.

   Ribosomal RNA is considered to be depleted when the Bioanalyzer trace shows a smooth curve for the fragments analyzed. The Bioanalyzer trace should also show at best only small peaks at the sizes of rRNA species, indicating little to no rRNA remaining.

   When submitting samples for sequencing, a number of quality checks are performed. As typical, a NanoDrop analysis is run. This scan typically does not look like an ideal DNA sample with a peak at 260 and a 260:280 ratio of about 2 as there is a low concentration of nucleic acids, and there could be residual components from gel extraction.

A bioanalyzer scan of the prepared libraries is performed at the highest sensitivity available. However, this scan often suggests that there is not significant sample present. This is due to the fact that the amplified DNA was extracted from a denaturing gel and will most likely be single-stranded, while the high-sensitivity bioanalyzer analyzes double-stranded DNA. The benefit of the bioanalyzer scan is to determine if there is a high concentration of by-product, which is typically found around 172 bp. Since the bioanalyzer suggests that there is no sample, it is essential to also perform a qPCR analysis using the Illumina primers. This ensures that there is sufficient sample for sequencing, but the qPCR test cannot identify if the product is primarily by-product or real sample. One can be confident that the samples are of good quality if the bioanalyzer shows no to a minimal peak at the size of the by-product (172 bp), and the qPCR analysis indicates sufficient material that can be sequenced.

9. RNA purification troubleshooting: If very pure RNA is needed, the poly(A) procedure can be repeated a second time. Typically, in our labs, we do two rounds of poly(A) purifications, but this could be adjusted depending on the organism and system of interest.

   Notably, the RNA purification protocol will differ between different organisms or types of RNA. Use the RNA purification protocol recommended for the type of RNA to be studied in the particular organism of interest.

10. Plasticware choice: While using PCR tubes, we have noticed that certain brands did not produce reliable results. We have found that using Eppendorf brand PCR tubes provides consistent results.

11. Biotinylated reverse transcription: Depending on the organism or RNA type of interest, it can be beneficial to use different types of biotinylated dNTPs. In rice, using a final concentration of 125 μM Biotin-dCTP was successful. However, in A-/U-rich organisms or RNA type, it may be beneficial to use a combination of biotinylated dNTPs that sum to a final concentration of 125 μM.

12. PAGE extraction troubleshooting: In order to visualize the small amount of DNA present in the thin PAGE gels, there are some specialized techniques that must be performed. To do this, we prepare SYBR Gold dye using 5 μL of concentrated stain from the manufacturer in 50 mL of 1× TE buffer in a small, flat, opaque container (plastic or Pyrex with aluminum

foil surrounding it to prevent light from entering). To transfer the gel from between the glass plates into the buffer, the glass plates are separated, excess gel removed from the sides of the gel (nothing should be removed from the top or the bottom in order to aid in identification and sizing of bands), and the part of the gel to be analyzed is then carefully transferred into the buffer. If there are difficulties handling the gel, consider rolling the gel up while it is still on the glass plate, transferring it to the buffer, and then unrolling it once in the buffer. The container is covered with aluminum foil and placed on a rocker moving slowly for 10 min. During this 10 min, materials to remove the gel from the buffer and excise the proper gel region are prepared, which include a glass plate with plastic wrap on top, a glass rod that is slightly dampened with a moist paper towel, pipette tips, razor blades, labeled 50 mL conical tubes, a permanent marker, and additional plastic wrap.

After 10 min, the gel is lifted up by carefully placing the glass rod about halfway across the gel width-wise, lifting the gel out of the buffer, and unrolling it onto the plastic wrap which is on a glass plate. One can gently use a gloved hand to assist in this or to place gel pieces together that may have torn apart. A piece of plastic wrap is placed on top of the gel, and the gel is visualized on a UV light box. It is recommended that the visualization occur in a dark room to assist in identification of the bands.

The UV light exposure should be as short as possible to minimize DNA cross-linking.

It is essential to crush the excised gel strip into fine pieces, and we have found that this process is assisted if the gel has slightly dried out.

Add 2 mL TEN_{250} for every gram of gel. For instance, if the weight of the excised gel strip is determined to be 1.5 g, then 3 mL of buffer is added (2× the gel mass).

When pelleting the extracted DNA during ethanol precipitation, it is important to orient the tubes in such a way that the pellet remains at the same known position during centrifugation.

It is also helpful to dry the tube out at 37 °C while it is on its side so that the pellet remains in place. After it is dry, the pellet is often somewhat transparent and usually seems to be positioned at the bottom of the tube or at the crease in a 50 mL conical tube.

To be sure that the pellet (which could be spread out and difficult to identify) is completely dissolved, a rather large volume of 100 μL water is used. This water is continuously pipetted up and down throughout the bottom of the conical tube in any place where the pellet may be using the same pipette tip as used to dispense the 100 μL water.

After moving the 100 μL of water around, the tubes can be pulse spun to ensure all of the solution is at the bottom of the tube and then transferred to a 1.7 mL microcentrifuge tube using the same pipette tip as previously. Use a different pipette tip for each sample.

13. Streptavidin pulldown troubleshooting: The Wash Buffer of RNA Clean and Concentrator must be prepared beforehand. This involves adding ethanol to the amount indicated the first time the buffer is used.

    When using RNA Clean and Concentrator, it is important that the column is completely dry after the final wash buffer addition to ensure removal of ethanol.

    When performing streptavidin binding, one should prevent the streptavidin beads from ever drying out and ensure that the beads are thoroughly suspended when in solution. Typically, we perform streptavidin pulldown on only six samples at a time. In the preparation of the streptavidin beads, the 1× wash/binding buffer should remain on the beads until one is ready to add the cDNA.

    When adding the cDNA to the beads, the final volume should be 50 μL, which can be adjusted by adding 1× wash/binding buffer.

    Before performing streptavidin pulldown, a water bath should be set to 40 °C and a tube containing the low-salt buffer placed in this water. A heat block to accommodate 1.7 mL tubes should be set to 95 °C. The water bath and heat block should be checked periodically to ensure proper temperature.

    When performing the temperature-sensitive steps of streptavidin pulldown, timing must be precise, and solution manipulation (addition and removal) must be performed quickly to prevent the solutions from cooling; thus we typically only handle three samples at a time. Particularly during the 2 min, these three samples are staggered by 30 s to allow precise manipulation.

14. Structure-seq2 tests: Structure-seq2 contains many specialized techniques. In order to ensure proper technique, it is suggested that one prepare test libraries using total RNA from the single-hit kinetics experiments or other test experiments. After obtaining total RNA to practice Structue-seq2, one can follow the steps from reverse transcription through the PCR cycle test. If the PCR cycle test provides sufficient smears on this test RNA, one can be more confident in their technique to prepare Structure-seq2 libraries with more valuable samples.

15. PCR troubleshooting: For the PCR cycle test, one will be able to identify appropriate cycles that work for the particular experimental material. Typically we sample cycles between 10 and 35 depending on our confidence in the amount of initial

material present. We usually test every five cycles (e.g., 10, 15, 20, 25, and 30 cycles or 15, 20, 25, 30, and 35 cycles).

The reverse primer contains the barcode information for Illumina sequencing; thus a different reverse primer must be added to each sample during the preparative PCR. However, in the PCR cycle test, it is recommended that the same reverse primer be added to all samples as different sequences have the potential of running differently in the non-denaturing agarose gel, and one may misinterpret the results.

Tubes containing the 6× loading buffer can be prepared while the samples are cycling to minimize sample prep time. These tubes do not need to be Eppendorf brand as there is no thermal cycling of these tubes.

All samples must be tested during a cycle test to ensure they can all amplify, and a cycle should be chosen that works for all samples analyzed. A "good" cycle is one in which all samples show a nice smear in the appropriate range of 200 bp to >1000 bp (not too few cycles) and in which no bright specific bands are present in any sample (not too many cycles).

We note that for the PCR cycle test, 3 μL of the 10 μL purified ligation product are used per sample, and another 3 μL per sample are used in the real PCR amplification. If unforeseen problems occur, the experimenter has an extra ~3 μL of ligated sample available as backup.

Acknowledgments

We thank Dr. David Mitchell and Dr. Joseph Waldron for helpful input on the manuscript and Dr. Craig Praul (Genomics Core Facility, Penn State University) for high-throughput Illumina sequencing. This work was supported by the National Science Foundation Plant Genome Research Program (NSF-IOS-1339282), with additional support from an Innovation Award from Penn State University.

References

1. Ehresmann C, Baudin F, Mougel M, Romby P, Ebel JP, Ehresmann B (1987) Probing the structure of RNAs in solution. Nucleic Acids Res 15(22):9109–9128
2. Bevilacqua PC, Ritchey LE, Su Z, Assmann SM (2016) Genome-wide analysis of RNA secondary structure. Annu Rev Genet 50:235–266. https://doi.org/10.1146/annurev-genet-120215-035034
3. Ding Y, Tang Y, Kwok CK, Zhang Y, Bevilacqua PC, Assmann SM (2014) In vivo genome-wide profiling of RNA secondary structure reveals novel regulatory features. Nature 505 (7485):696–700. https://doi.org/10.1038/nature12756
4. Ding Y, Kwok CK, Tang Y, Bevilacqua PC, Assmann SM (2015) Genome-wide profiling of in vivo RNA structure at single-nucleotide resolution using structure-seq. Nat Protoc 10 (7):1050–1066. https://doi.org/10.1038/nprot.2015.064

5. Ritchey LE, Su Z, Tang Y, Tack DC, Assmann SM, Bevilacqua PC (2017) Structure-seq2: sensitive and accurate genome-wide profiling of RNA structure in vivo. Nucleic Acids Res 45(14):e135. https://doi.org/10.1093/nar/gkx533

6. Wan Y, Qu K, Ouyang Z, Kertesz M, Li J, Tibshirani R, Makino DL, Nutter RC, Segal E, Chang HY (2012) Genome-wide measurement of RNA folding energies. Mol Cell 48(2):169–181. https://doi.org/10.1016/j.molcel.2012.08.008

7. Tack D, Tang Y, Ritchey LE, Assmann SM, Bevilacqua PC (2018) StructureFold2: bringing chemical probing data into the computational fold of RNA structural analysis. Methods 3:12–15. https://doi.org/10.1016/j.ymeth.2018.01.018

8. Mitchell D 3rd, Ritchey LE, Park H, Babitzke P, Assmann SM, Bevilacqua PC (2017) Glyoxals as in vivo RNA structural probes of guanine base pairing. RNA. https://doi.org/10.1261/rna.064014.117

9. Mitchell, D 3rd, Renda AJ, Douds CA, Babitzke P, Assmann SM, Bevilacqua, PC (in press). In vivo RNA structural probing of uracil and guanine base pairing by 1-ethyl-3-(3-dimethylaminopropyl)carbodiimide (EDC). Nucl Acids Res

10. Cannone JJ, Subramanian S, Schnare MN, Collett JR, D'Souza LM, Du Y, Feng B, Lin N, Madabusi LV, Muller KM, Pande N, Shang Z, Yu N, Gutell RR (2002) The comparative RNA web (CRW) site: an online database of comparative sequence and structure information for ribosomal, intron, and other RNAs. BMC Bioinformatics 3:2

Chapter 21

Using Protein Interaction Profile Sequencing (PIP-seq) to Identify RNA Secondary Structure and RNA–Protein Interaction Sites of Long Noncoding RNAs in Plants

Marianne C. Kramer and Brian D. Gregory

Abstract

From the moment of transcription, RNA molecules are continuously bound by RNA-binding proteins (RBPs). While the majority of research has focused on how these RBPs regulate posttranscriptional gene regulation of messenger RNAs (mRNAs), the majority of cellular RNAs do not code for proteins, such as ribosomal RNAs, transfer RNAs, and microRNAs. Since these RNAs do not code for protein, their function is mainly determined by their interactions with RBPs as well as their intramolecular base pairing, or RNA secondary structure. One class of noncoding RNAs termed long noncoding RNAs (lncRNAs) have recently become the subject of intense research interest. To study the function of lncRNAs in eukaryotic cells, it is important to examine both their interactions with RBPs as well as their RNA secondary structure. Protein interaction profile sequencing (PIP-seq) is a genome-wide method that uses structure-specific ribonucleases (RNases) to identify regions of double-stranded and single-stranded RNA as well as regions that are protected from these RNases, which represent sites of RBP binding. This method is a very powerful way to examine RNA-protein interactions and RNA secondary structure of all lncRNAs expressed in cells and tissues and can reveal cell-type-specific or tissue-specific patterns of these RNA features. Here, we give a step-by-step account of performing this technique for comprehensively analyzing RNA–protein interactions and RNA secondary structure in plant transcriptomes.

Key words Protein interaction profiling sequencing (PIP-seq), RNA secondary structure, RNA-binding proteins (RBPs), RNA–protein interactions, Genome-wide, RNA sequencing, Structure-specific nucleases

1 Introduction

From transcription to degradation, RNA molecules undergo many posttranscriptional processes that control RNA transport, localization, and stability [1–3]. These posttranscriptional processes are largely controlled by RNA-binding proteins (RBPs) and intramolecular base pairing within RNA molecules, their so-called RNA secondary structure. Most previous research has focused on how RBPs regulate posttranscriptional processes of messenger RNAs

Julia A. Chekanova and Hsiao-Lin V. Wang (eds.), *Plant Long Non-Coding RNAs: Methods and Protocols*, Methods in Molecular Biology, vol. 1933, https://doi.org/10.1007/978-1-4939-9045-0_21,

(mRNAs), but this class of RNAs only represents a very small fraction of the total cellular RNA population. In fact, the majority of cellular RNAs do not code for proteins and are thus termed noncoding RNAs (ncRNAs) yet have essential functions at every step of an organism's life cycle [4]. Two well-studied examples are ribosomal RNAs (rRNAs) and transfer RNAs (tRNAs), which are vital for proper translation. In the past 30 years, however, it has become increasingly clear that there are many additional classes of ncRNAs such as small RNAs (smRNAs), which include microRNAs (miRNAs), small interfering RNAs (siRNAs), small nucleolar RNAs (snoRNAs), and others, as well as long noncoding RNAs (lncRNAs) [5–9].

LncRNAs are defined as having little to no protein-coding capacity and are longer than 200 nucleotides (nt) in length [9, 10]. Like mRNAs, RBPs and RNA secondary structure control the processing and function of ncRNAs. For instance, the secondary structure of rRNAs, tRNAs, and miRNA precursors is essential for proper processing, function, and binding to the correct cohort of RBPs [11, 12]. Furthermore, several examples of the function of lncRNAs interacting with RBPs have been identified in the model plant *Arabidopsis thaliana*. For example, the lncRNA *ASCO* has been found to bind and sequester the nuclear speckle RNA-binding protein NSR away from its proper splicing targets, ultimately leading to a disruption in lateral root formation [13]. Furthermore, the lncRNA *HIDDEN TREASURE 1* (*HID1*) binds and reduces the expression of *PHYTOCHROME INTERACTING FACTOR 3* (*PIF3*), leading to elongated hypocotyls under shade conditions [14].

While the function and relationship of several lncRNA-RBP interactions have been elucidated, few studies have looked globally at RBP binding to lncRNAs. Using genome-wide techniques can lead to information for all expressed lncRNAs and provide many good targets for subsequent functional analyses. Furthermore, given that the function of several classes of ncRNAs is dependent on RNA secondary structure (such as rRNAs and tRNAs), the function of many lncRNAs may also be driven by their RNA secondary structure [15, 16]. Thus, it is also very important to globally examine RNA secondary structure of lncRNAs to examine folding patterns throughout these transcripts and if their structure varies under different conditions and between different cell types.

Protein interaction profile sequencing (PIP-seq) is a method to globally examine RNA–protein interaction sites as well as RNA secondary structure [17–19]. Beginning with formaldehyde cross-linked tissue, PIP-seq can be performed on both total cell lysate and different cellular compartments, such as nuclei and cytoplasm. PIP-seq begins by splitting the cross-linked lysates into two separate groups termed the structure-only samples and the footprinting samples (Fig. 1). In the structure-only samples, the lysates are first

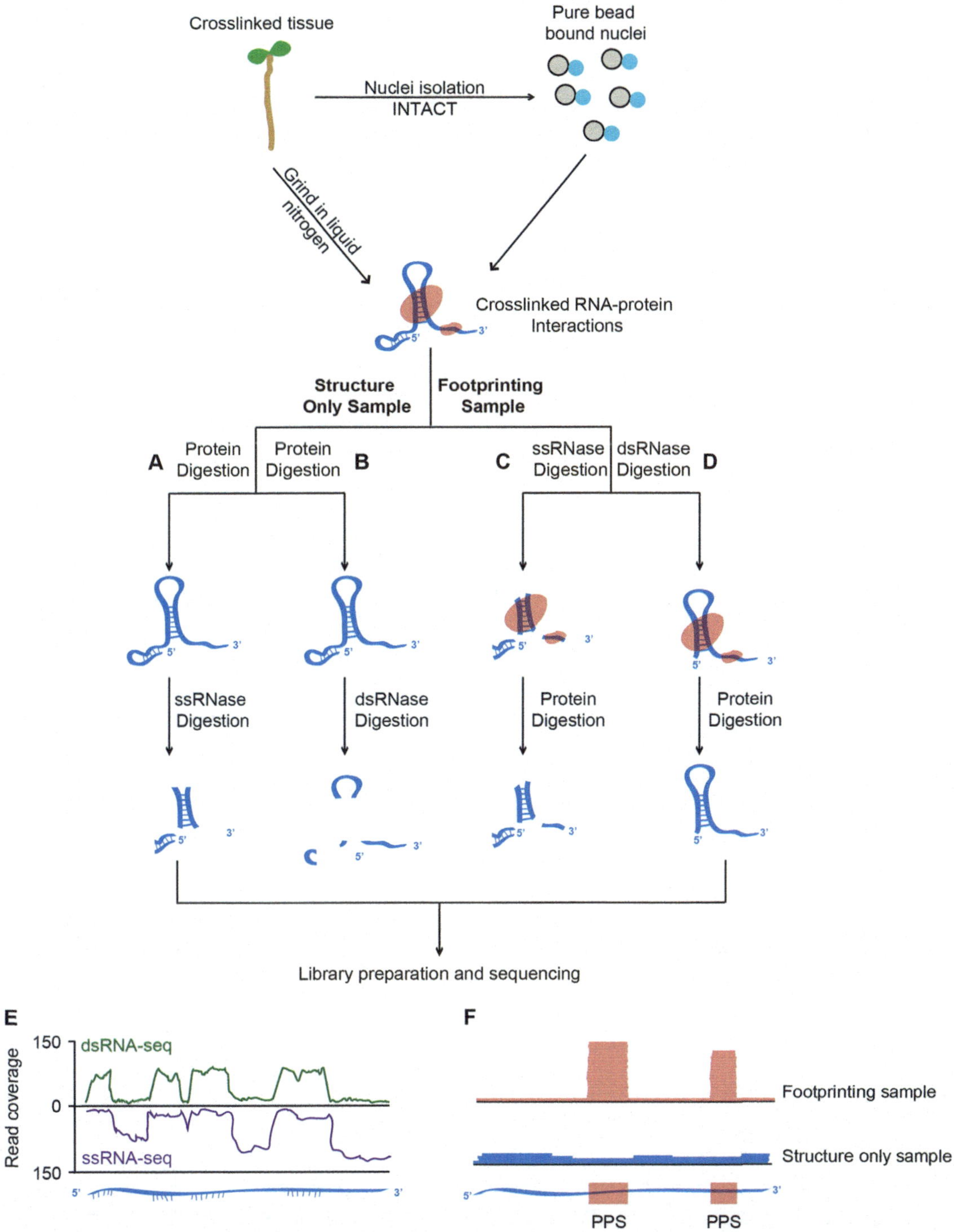

Fig. 1 Overview of protein interaction profile sequencing (PIP-seq). PIP-seq can be performed on total cell lysate or purified nuclei. The tissue is first cross-linked with 1% formaldehyde (Subheading 3.1) before being lysed and subject to PIP. (**a**, **b**) The structure-only samples are first digested with proteinase K to remove all proteins bound. The sample is then either digested with (**a**) ssRNase or (**b**) dsRNase, which will digest all ssRNA or dsRNA, respectively (Subheading 3.2.1). (**c**, **d**) The footprinting samples are first subjected to either

treated with proteinase K to remove all proteins, ultimately resulting in deproteinated RNA. These samples are then digested by structure-specific ribonucleases (RNases) that will either digest all double-stranded RNA (dsRNase) or single-stranded RNA (ssRNase) (Fig. 1a, b). In the footprinting sample, the samples are first exposed to the structure-specific nucleases to digest all accessible ssRNA or dsRNA, also leaving regions of RNAs that are bound by RBPs. The proteins are then removed by protease treatment, and all samples undergo sequencing library preparation and RNA sequencing (Fig. 1c, d) [17–19].

Through bioinformatics approaches discussed in the next chapter, this technique can allow the calculation of a structure score for each detectable nucleotide, as well as identify protein-protected sites (PPS) that are enriched in the footprinting sample compared to the structure-only sample. Using this data, the RNA secondary structure and RNA-protein interactions of lncRNAs can be observed globally. Recently, this technique was used in the nuclei of root hair and non-hair cells [19]. In this study, it was observed that there was significantly different RNA secondary structure along the entire length of lncRNAs between cell types. More specifically, there was higher structure within the 5′ region of these transcripts in hair cells compared to non-hair cells, while this trend reverses at the 3′ of these transcripts, where there is lower structure in hair cells compared to non-hair cells. Of note, however, is that while the structure varies between cell types, there was no global change in the density of protein binding, indicating that while overall these transcripts are bound by the same number of proteins in each cell type, the differences in secondary structure indicate that they are likely different cohorts of proteins that interact with these collections of molecules. Thus, confirming that PIP-seq is a useful technique to detect changes in RNA secondary structure and protein binding of lncRNAs [19]. In this chapter, we will describe the step-by-step methods for constructing PIP-seq libraries using plant tissue or subcellular compartments.

Fig. 1 (continued) digestion with (**c**) ssRNase or (**d**) dsRNase to digest all accessible RNAs in the correct structural confirmation. Following RNase digestion, the samples are removed of protein and all cross-links reversed (Subheading 3.2.2). (**e**) The structure-only PIP-seq libraries can be compared along the length of the transcript to determine regions which are more double-stranded or more single-stranded based on the amount of reads in the libraries digested with ssRNase or dsRNase, respectively. (**f**) Protein-protected sites (PPSs) can be identified by examining reads that build up in the footprinting samples compared to the structure-only samples

2 Materials

2.1 Plant Material

This technique was optimized on lysate from whole 10-day-old (10D) seedlings but has since been used on nuclei isolated from 10D seedlings as well as 4-week- and 5-week-old *Arabidopsis* leaves. Additionally, PIP-seq has been performed in human cell culture cells, allowing it to be used in a wide variety of systems [17].

2.2 PIP Buffers

1. Cross-linking buffer: 1% formaldehyde, 0.02% Silwet L-77. Add 810 μL 37% formaldehyde and 6 μL Silwet L-77 (Lehle Seeds) to 30 mL PBS. Prepare freshly before each use.
2. 1 M glycine solution in nuclease-free water.
3. RIP buffer: 25 mM Tris–HCl pH 8.6 (*see* **Note 1**), 150 mM KCl, 5 mM EDTA pH 8.0, 0.5% Igepal CA-630 in distilled, deionized water. Filter sterilized through 22 μm membrane. Store at 4 °C up to 1 year.

   Before use, to 10 mL RIP buffer, add 1 tablet protease inhibitor mini cocktail (1 tablet per 10 mL of RIP buffer) (Roche), 5 μL RNaseOUT (Fisher Scientific), and 50 μL 0.1 M DTT. Keep on ice throughout experiment.
4. 10× structure buffer: 10 mM Tris–HCl pH 7.0, 100 mM KCl, and 10 mM $MgCl_2$ in nuclease-free water. Aliquot and store at −20 °C.
5. 16× RNase stop buffer: 10% SDS, 5 mM EDTA pH 8.0, and 20 μg Proteinase K (1 μL of 20 mg/mL stock) (Denville Scientific) in nuclease-free water (*see* **Note 2**).
6. 1× DNase solution: Make resuspension solution with 550 μL DEPC-treated water and 3.85 mL RDD buffer (Qiagen). Resuspend lyophilized RNase-free DNase (Qiagen) with resuspension solution. Prepare aliquots and store at −20 °C.

2.3 PIP Materials

1. Bell-shaped vacuum container.
2. Liquid nitrogen.
3. Mortar and pestle.
4. Plastic pestles.
5. 1.7 mL tubes
6. 2.0 mL tubes
7. Centrifuge.
8. Heat blocks set to 65 and 37 °C.
9. 3 M NaOAc
10. RNase ONE (10 U/μL) (Promega).
11. RNase ONE buffer (Promega).
12. RNase V1 (Protein Labs).

13. Qiazol (Qiagen).
14. 100% ethanol
15. 70% ethanol.

2.4 Library Preparation Buffers

1. DSN hybridization buffer: 200 mM HEPES buffer (Fisher Scientific), 2 M NaCl (Fisher Scientific) in nuclease-free water.
2. DSN stop buffer: 10 mM EDTA in nuclease-free water.

2.5 Library Preparation Materials

1. 1.7 mL tubes.
2. 2.0 mL tubes.
3. 200 μL PCR tubes.
4. Thermocycler.
5. Heat blocks at 70 and 37 °C.
6. Gel box for running pre-poured gels.
7. 18-gauge needles.
8. Razor blades.
9. Chloroform/isoamyl alcohol (24:1).
10. RNeasy columns (Qiagen).
11. Buffer RWT (Qiagen).
12. Buffer RPE (Qiagen).
13. Nuclease-free water.
14. 5 mg/mL glycogen, ultrapure (Ambion).
15. 100% ethanol.
16. 80% ethanol.
17. 70% ethanol.
18. RNA fragmentation reagents (Ambion).
19. T4 polynucleotide kinase (NEB).
20. T4 DNA ligase buffer (NEB).
21. 10 mM ATP.
22. 15% Novex TBE-Urea gels (Thermo Fisher Scientific).
23. 10× TBE buffer (Bio-Rad).
24. Gel Loading Buffer II (Thermo Fisher Scientific).
25. 10 base pair (bp) ladder (Thermo Fisher Scientific).
26. 10 mg/mL ethidium bromide (Thermo Fisher Scientific).
27. Spin-X columns (Thermo Fisher Scientific).
28. Gel breaker tubes (IST Engineering Inc.).
29. 0.3 M NaCl.
30. 5 μM 3′ adapter (5′-TGGAATTCTCGGGTGCCAAGG-3′) (RA3).

31. 25 μM 5′ adapter (5′-GUUCAGAGUUCUACAGUCCGAC GAUC-3′) (RA5).
32. RNA ligase buffer (NEB).
33. RNaseOUT (Thermo Fisher Scientific).
34. RNA ligase 2, truncated (NEB).
35. T4 RNA ligase 1 (NEB).
36. 100 μM reverse transcription primer (5-′-GCCTTGGCACCCGAGAATTCCA-3′) (RTP).
37. 5× first-strand buffer (Thermo Fisher Scientific).
38. 50 mM dNTPs (Thermo Fisher Scientific).
39. 100 mM DTT.
40. Superscript II reverse transcriptase (Thermo Fisher Scientific).
41. 2× Phusion high-fidelity PCR master mix with HF buffer (NEB)
42. 5 mM betaine (MP Biomedical).
43. RNA PCR primer 1 (5′-AATGATACGGCGACCACCGAGA TCTACACGTTCAGAGTTCTACAGTCCGA-3′) (RP1).
44. RNA PCR primer indices (Illumina).
45. 6% Novex TBE gels (Thermo Fisher Scientific).
46. 25 bp ladder (Thermo Fisher Scientific).
47. 1× NEB buffer 2 (NEB).
48. 10× DSN master buffer (Evrogen).
49. Duplex-specific nuclease (DSN) enzyme (Evrogen).

3 Methods

Described below are the methods for cross-linking RNA–protein interactions with formaldehyde, protein interaction profile sequencing (PIP-seq), as well as the library preparation for PIP-seq.

3.1 Preparing the Arabidopsis Tissue for PIP

In order to identify protein-bound RNA sequences, RNA–protein interactions must be preserved through cell lysis. To do so, the tissue is cross-linked in 1% formaldehyde solution under vacuum filtration. The reaction is then quenched by vacuum infiltrating 125 mM glycine followed by several washing steps with distilled, deionized water to remove any excess buffer.

3.1.1 Preservation of RNA–Protein Interactions (Cross-Linking)

1. Prepare cross-linking buffer fresh, 30 mL per sample.
2. Weigh out 1–3 g of 10D seedlings, and add directly a 50 mL conical containing 30 mL of cross-linking solution.

3. Rotate end-over-end for 1 min to ensure all seedlings are submerged in the buffer.
4. Remove the lid, place the 50 mL conical under the bell-shaped vacuum container connected to a vacuum line, and turn on the vacuum for 10 min.
5. Detach the vacuum, and add 4.25 mL 1 M glycine (125 mM final concentration) to quench the cross-linking reaction.
6. Close the tube and turn end-over-end to mix well. Place in the bell-shaped vacuum container for another 5 min (*see* **Note 3**).
7. Remove cross-linking buffer and dispose appropriately. Wash the seedlings five times with distilled, deionized water.
8. Pat the seedlings dry on a paper towel and flash freeze in liquid nitrogen.
9. Tissue can be stored at −80 °C.

3.1.2 To Examine the RNA Secondary Structure and Protein Binding in the Total Cells

1. Crush tissue with a mortar and pestle in liquid nitrogen.
2. Resuspend tissue in 850 μL RIP buffer and transfer tissue to 1.7 mL tube.
3. Spin briefly and grind well with a plastic pestle and then pass through 1000 and 200 μL pipette tips about 20 times each.
4. Proceed to PIP protocol (Subheading 3.2).

3.1.3 To Examine the RNA Secondary Structure and Protein Binding in Cell Nuclei

1. Nuclei can be isolated using the isolation of nuclei in specific cell types (INTACT) procedure [20, 21]. If using this method, it is important to note that the *Arabidopsis* tissue to be cross-linked and used in the PIP-seq protocol should express the INTACT construct. Additionally, two million nuclei are needed for each replicate to ensure enough RNA for proper PIP-seq library construction.
2. After two million nuclei per replicate have been isolated using INTACT, resuspend the nuclei in 850 μL RIP buffer.
3. Pass through 1000 and 200 μL pipette tips about 20 times each.
4. Proceed to PIP protocol (Subheading 3.2).

3.2 Protein Interaction Profiling (PIP)

Protein interaction profiling begins by digesting any DNA present, to ensure there will be no contaminating genomic DNA in the sequencing libraries, and thus all reads will be from RNA.

3.2.1 DNA Digestion Prior to Protein Interaction Profiling

1. Prepare RIP buffer (5 mL per replicate) and 16× RNase stop buffer (450 μL per replicate) freshly before each experiment.
2. Add 160 μL DNase solution to the whole tissue or nuclei sample from Subheading 3.1. Mix well and incubate at room temperature for 30 min.

3. Split the sample into four aliquots by pipetting 250 μL of lysate into two 2.0 mL tubes (structure-only samples; Fig. 1a, b) and two 1.7 mL tubes (footprinting samples; Fig. 1c, d) (*see* **Note 4**).
4. Add 600 μL of RIP buffer to each tube and mix well.

3.2.2 Structure-Only Samples (Fig. 1a, b)

For these samples, all proteins are first removed using proteinase K, resulting in RNA that is absent of bound proteins. Following a brief precipitation, the samples are digested with structure-specific ribonucleases (RNases) that will either digest all single-stranded RNA (ssRNase) or all double-stranded RNA (dsRNase). Following library preparation and sequencing, these libraries can be compared to calculate RNA secondary structure throughout the transcriptome by looking at read coverage in the ssRNase libraries (representing dsRNA) and the dsRNase libraries (representing ssRNA) (Fig. 1e).

Proteinase K Treatment Followed by RNA Precipitation

1. To the samples in the 2 mL tubes, add 75 μL 16× RNase stop buffer. Incubate at room temperature for 15 min.
2. Precipitate RNA by adding 90 μL 3 M NaOAc and 1 mL 100% ethanol. Mix well and store at −80 °C for at least 1 h.
3. After the 1 h incubation, spin samples for 45 min at 20,000 *g* (max speed) at 4 °C.
4. Remove the supernatant, and wash the pellet with 700 μL 70% ethanol by spinning for 5 min at 20,000 *g* at 4 °C.
5. Remove the supernatant and any excess ethanol, and allow the pellet to dry for 10 min. Resuspend the pellet in 850 μL RIP buffer (*see* **Note 5**).

Digestion with ssRNase

To one of the tubes, digest with the ssRNase (Fig. 1a):

1. Add 100 μL RNase ONE buffer and 10 μL RNase ONE.
2. Mix well and incubate at 37 °C for 1 h and 15 min, mixing the samples every 15 min.
3. Add 75 μL 16× RNase stop buffer, and incubate at room temperature for 10 min.
4. To reverse the cross-links, transfer tubes to 65 °C for 2 h, mixing every 30 min.
5. Split the sample in half by pipetting ~500 μL to two new 2 mL tubes.
6. Add 700 μL Qiazol to each tube and store at −20 °C. Samples can be stored until the library preparation.

Digestion with dsRNase

To the other 2 mL tube, digest with the dsRNase (Fig. 1b):

1. Add 100 μL 10× RNA structure buffer and 25 μL RNase V1.
2. Mix well, and incubate at room temperature for 1 h and 15 min, mixing the samples every 15 min.
3. Add 75 μL 16× RNase stop buffer, and incubate at room temperature for 10 min (*see* **Note 6**).
4. To reverse the cross-links, transfer tubes to 65 °C for 2 h, mixing every 30 min.
5. Split the sample in half by pipetting ~500 μL to two new 2 mL tubes.
6. Add 700 μL Qiazol to each tube and store at −20 °C. Samples can be stored until the library preparation.

3.2.3 Footprinting Samples (Fig. 1c, d)

For these samples, the samples are first digested with the structure-specific nucleases that will digest all single-stranded RNA (ssRNase) or double-stranded RNA (dsRNase). These RNases will digest any accessible RNA in the correct structural configuration that is not protected by proteins. Following RNase digestion, proteins are removed using proteinase K, and cross-links are reversed. Following library preparation and sequencing, these libraries along with the structure-only samples (Subheading 3.2.2) are used to identify regions of RNA that are bound by proteins, so-called protein-protected sites (PPSs). A PPS is called as enrichment in reads in the footprinting sample compared to the structure-only sample (Fig. 1f).

Digestion with ssRNase

To one of the 1.7 mL tubes, first digest with the ssRNase (Fig. 1c):

1. Add 100 μL RNase ONE buffer and 10 μL RNase ONE.
2. Mix well and incubate at 37 °C for 1 h, mixing the samples every 15 min.

Digestion with dsRNase

To the other 1.7 mL tube, digest with the dsRNase (Fig. 1d):

1. Add 100 μL 10× RNA structure buffer and 25 μL RNase V1.
2. Mix well and incubate at room temperature for 1 h, mixing the samples every 15 min.

Proteinase K Treatment and Reversing of the Cross-linking

1. To each tube add 75 μL 16× RNase stop buffer, and incubate at room temperature for 10 min (*see* **Note 6**).
2. To reverse the cross-links, transfer tubes to 65 °C for 2 h, mixing every 30 min.
3. Split the sample in half by pipetting ~500 μL to two new 2 mL tubes.

4. Add 700 μL Qiazol to each tube and store at −20 °C. Samples can be stored until the library preparation.

3.3 Sequencing Library Preparation

All precipitations during the library prep are done by adding 3 μL glycogen, 1/10 the volume of 3 M NaOAc (final concentration 0.3 M), and three times the volume of 100% ethanol.

For all precipitations before the reverse transcription Subheading 3.3, **step** 7, samples are stored at −80 °C for at least 2 h, followed by centrifugation at 20,000 × *g* for at least 80 min at 4 °C. Pellets are then washed with 700 μL 80% ethanol for 5 min and allowed to dry for 10 min before resuspension. Samples are kept on ice for 25 min before proceeding to the next step.

Following and including Subheading 3.3, **step** 7, samples are stored at −80 °C for at least 2 h, followed by centrifugation at 20,000 × *g* for at least 45 min at 4 °C. Pellets are then washed with 700 μL 70% ethanol for 5 min and allowed to dry for 10 min before resuspension. Samples are kept on ice for 25 min before proceeding to the next step.

3.3.1 RNA Isolation

For each sample, there will be eight tubes: two tubes for the ssRNase structure-only samples, two tubes for the dsRNase structure-only samples, two tubes for the ssRNase footprinting samples, and two tubes for the dsRNase footprinting samples.

1. Remove structure-only and footprinting samples from −20 °C, and thaw at room temperature.
2. Add 280 μL chloroform/isoamyl alcohol (24:1).
3. Vortex thoroughly and allow to sit at room temperature for 5 min.
4. Spin for 15 min at 12,000 *g* at 4 °C.
5. Transfer aqueous layer to new 2 mL tube, and fill the tube to capacity with 100% ethanol. Mix well.
6. Transfer 700 μL to a miRNeasy column. Spin at 20,000 *g* for 30 s at 4 °C.
7. Remove the flow-through, and repeat **step 6** until the entire sample has been applied to the column. Both tubes for each sample can be put over the same column.
8. Wash with 700 μL Buffer RWT. Spin at 20,000 *g* for 30 s at 4 °C, and remove the flow-through.
9. Wash with 500 μL Buffer RPE. Spin at 20,000 *g* for 30 s at 4 °C, and remove the flow-through.
10. Wash once more with 500 μL Buffer RPE. Spin at 20,000 *g* for 2 min at 4 °C and remove the flow-through.
11. Transfer tubes to new 1.7 mL tubes for elution.

12. Add 100 μL nuclease-free water to the center of the column. Allow the samples to sit at room temperature for 5 min. Spin 1 min at 20,000 *g* at 4 °C.
13. Repeat **step 12** once more.
14. Precipitate samples and resuspend in 9 μL nuclease-free water.

3.3.2 RNA Fragmentation

1. To each 9 μL sample, add 1 μL 10× fragmentation reagent, and immediately incubate samples at 70 °C for 3 min.
2. Immediately add 1 μL of 10× Stop solution.
3. Raise volume to 100 μL and precipitate RNA.
4. Resuspend in 16 μL nuclease-free water.

3.3.3 T4 Polynucleotide Kinase (T4 PNK) Treatment

1. To 16 μL of sample, add 2 μL NEB DNA ligase buffer (*see* **Note** 7), 1 μL T4 PNK, and 1 μL 10 mM ATP.
2. Incubate at 37 °C for 1 h.
3. Raise volume to 100 μL and precipitate.
4. Resuspend in 10 μL.

3.3.4 Size Selection 1 (Fig. 2a)

The purpose of this size selection is to eliminate sizes that are too small or too large to represent protein-bound sites.

1. Assemble 15% Novex TBE-Urea gel in a gel box with ample amounts of 1× TBE (~500 mL).
2. Preclear each well using a syringe with 18-gauge needle.
3. Pre-run the gel at 155 V for 25 min.
4. To the 10 μL sample, add 10 μL Gel Loading Buffer II.
5. Mix well and incubate at 70 °C for 5 min to denature any secondary structure, and then immediately place on ice for 3 min.
6. Prepare 1.5 μg 10 bp ladder in 10 μL nuclease-free water and 10 μL Gel Loading Buffer II.
7. Preclear each well again using a syringe with an 18-gauge needle.
8. Load samples onto gel, keeping one lane in between each sample to avoid sample-to-sample contamination, and run at 155 V for ~90 min or until the bromophenol blue band has traveled 80% of the gel.
9. Stain gel in 100 mL 1× TBE with 140 μg ethidium bromide (14 μL of 10 mg/mL ethidium bromide) for 10 min.
10. Using ethanol-cleaned razor blades, excise gel regions 15–150 nt for each sample.
11. Place gel slices in gel breaker tubes in new 2 mL tubes (*see* **Note 8**). Spin 20,000 *g* for 2 min.

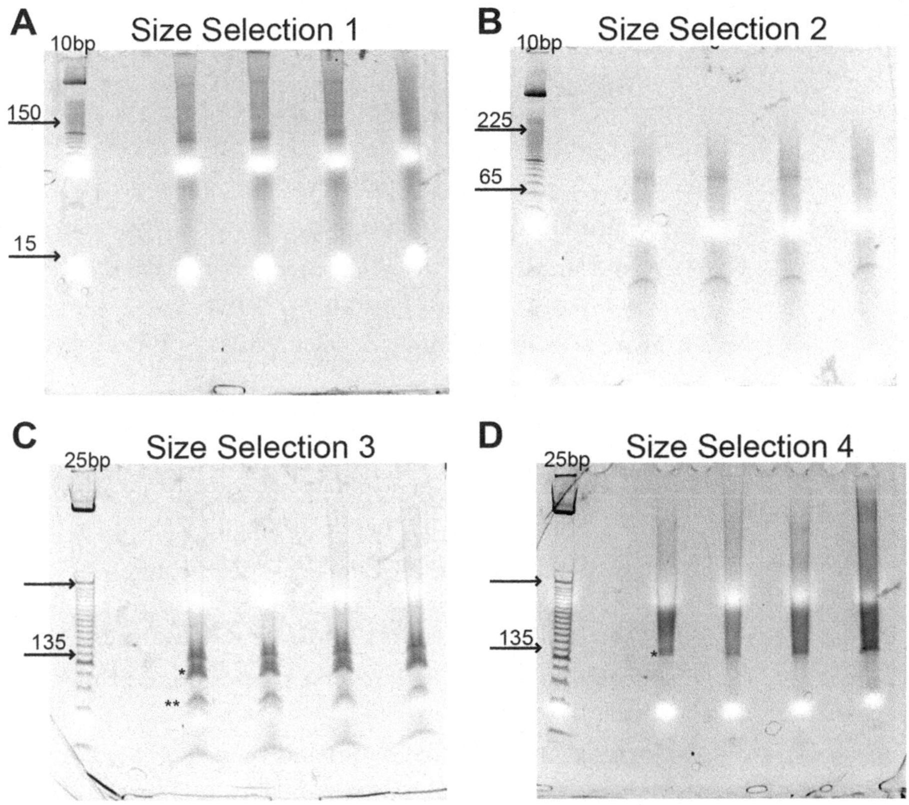

Fig. 2 Examples of gels for each size selection. (**a**) Size selection 1: samples are run on a 15% TBE-Urea gel for ~90 min, and RNA from 15 nt to 150 nt are selected from the gel (Subheading 3.3.4). (**b**) Size selection 2: samples are run on a 15% TBE-Urea gel for ~90 min, and RNA from 65 nt to 225 nt are selected from the gel (Subheading 3.3.6). (**c**) Size selection 3: samples are run on a 6% TBE gel for ~30 min and DNA from 135 nt to the end of the signal are selected from the gel (Subheading 3.3.6). (**d**) Size selection 4: samples are run on a 6% TBE gel for ~30 min, and DNA from 135 nt to the end of the signal are selected from the gel (Subheading 3.3.6)

12. Add 300 μL 0.3 M NaCl, and rotate end-over-end for 4 h at room temperature (*see* **Note 9**).
13. Transfer entire sample to Spin-X column (*see* **Note 10**). Spin for 3 min, 20,000 *g*, at 4 °C.
14. Transfer the flow-through to a new tube and precipitate.
15. Resuspend in 5 μL nuclease-free water.

3.3.5 Adapter Ligation (See Note 11)

1. Transfer 5 μL sample to PCR tube and add 1 μL 3′ RNA adapter (RA3).
2. Mix well and incubate at 70 °C for 2 min and then 4 °C for 2 min.
3. To each sample add 2 μL RNA ligase buffer, 1 μL RNaseOUT, and 1 μL T4 RNA ligase 2, truncated.

4. Incubate at 28 °C for 1 h.
5. Near the end of the hour incubation, add 5′ RNA adapter (RA5) into a new tube. Heat at 70 °C for 2 min and then on ice for 2 min. RA5 for all samples can be denatured in a single tube.
6. Add 1 μL 10 mM ATP and 1 μL T4 RNA ligase 2 to the denatured 5′ RNA adapter.
7. Add 3 μL of the 5′ RNA adapter mix to each sample. Mix well and incubate for 1 h 15 min at 28 °C.
8. Store at −20 °C until size selection 2 (Subheading 3.3.6).

3.3.6 Size Selection 2 (Fig. 2b)

This size selection step is to remove any excess adapters and any adapter-adapter ligation products.

1. Run samples on 15% TBE-Urea gel as was done in Subheading 3.3.4, excising the gel between 65 and 225 nt (*see* **Note 12**).
2. Resuspend pellet in 6 μL nuclease-free water.

3.3.7 Reverse Transcription and PCR

1. Transfer the 6 μL sample into a new PCR tube.
2. To each tube, add 1 μL reverse transcription primer (RTP) and mix well.
3. Incubate at 70 °C for 2 min and then 4 °C for 2 min.
4. To each sample, add 2 μL 5× first-strand buffer, 0.5 μL 50 mM dNTPs, 1 μL 0.1 M DTT, 1 μL RNaseOUT, and 1 μL Superscript II reverse transcriptase.
5. Incubate at 50 °C for 1 h.
6. Make a PCR master mix of 50 μL Phusion high-fidelity PCR master mix with HF buffer, 33.5 μL 5 mM betaine, and 2 μL RNA PCR primer 1 (RP1).
7. Add 85.5 μL master mix to each RT reaction.
8. Add 2 μL RNA PCR Primer Index to each sample (*see* **Note 13**).
9. Divide each 100 μL PCR reaction into four tubes, each with 25 μL each.
10. Incubate samples on a thermocycler using the following setting: 1 cycle of 30 s at 95 °C; 12 cycles of 10 s 95 °C, 30 s 60 °C, and 15 s 72 °C; and 1 cycle of 10 min at 72 °C. Hold at 4 °C.
11. Combine all four PCR tubes and precipitate.
12. Resuspend pellet in 10 μL nuclease-free water.

3.3.8 Size Selection 3 (Fig. 2c)

This size selection step is to remove any adapter-adapter ligation products amplified by PCR and PCR primer dimer.

1. Assemble 6% TBE gel in gel box in ample amounts of 1× TBE (~500 mL).
2. Preclear each well using a syringe with an 18-gauge needle.
3. Add 10 μL Gel Loading Buffer II to each sample.
4. Prepare 1.5 μg 25 bp ladder in 10 μL water and 10 μL Gel Loading Buffer II.
5. Mix well and load gel, leaving a blank well in between each sample to avoid sample-to-sample contamination.
6. Run the gel at 155 V for 30 min or until bromophenol blue has traveled ~80% of the gel.
7. Stain gel in 100 mL 1× TBE with 140 μg ethidium bromide (14 μL of 10 mg/mL ethidium bromide) for 10 min.
8. Using ethanol-cleaned razor blades, excise gel regions 135 bp and above for each sample. Avoid band at 120 bp, which represents adapter-adapter artifacts.
9. Place gel slices in gel breaker tubes in new 2 mL tubes. Spin 20,000 *g* for 2 min.
10. Add 300 μL 1× NEB buffer 2, and rotate end-over-end for 2 h at room temperature.
11. Transfer entire sample to Spin-X column (*see* **Note 10**). Spin for 3 min, 20,000 *g*, at 4 °C.
12. Transfer the flow-through to a new tube and precipitate.
13. Resuspend in 15.5 μL nuclease-free water.

3.3.9 Treatment with Duplex-Specific Nuclease (DSN)

The treatment with duplex-specific nuclease (DSN) is to remove the highly abundant fraction of our sample that may represent ribosomal RNA reads. Here, we first denature the DNA at a high temperature and slowly allow it to renature. After 5 h, we add the DSN enzyme, to digest double-stranded DNA that has already renatured, representing the most abundant fraction in the sample, thus removing ribosomal RNA contamination.

1. Transfer a total of 100 ng of library product from Subheading 3.3.8 into 13.5 μL nuclease-free water in a 200 μL PCR tube (*see* **Note 14**). The remainder of the library can be stored at −80 °C.
2. Add 4.5 μL DSN hybridization buffer to each tube. Mix thoroughly, and incubate at 98 °C for 2 min and then 65 °C for 5 h (*see* **Note 15**).
3. With about 4.5 h remaining in the incubation, dilute 4 μL 10× DSN master buffer in 16 μL nuclease-free water per sample to make 2× DSN master buffer.

4. Incubate the 2× DSN master buffer at 68 °C for the remainder of the incubation.
5. Add 20 μL 2× DSN to each sample and mix thoroughly by pipette. Do not allow the samples to cool down.
6. Incubate at 68 °C for 10 min.
7. Add 2 μL DSN enzyme to each sample. Mix by pipetting and spin down briefly. Do not allow the samples to cool down.
8. Incubate at 68 °C for 25 min.
9. Add 40 μL DSN Stop solution. Mix thoroughly.
10. Transfer samples to new 1.7 mL tubes, and raise volume to 100 μL and precipitate.
11. Resuspend pellet in 6 μL.

3.3.10 PCR Reaction 2

1. Make a PCR master mix of 50 μL Phusion high-fidelity PCR master mix with HF buffer, 40 μL 5 mM betaine, and 2 μL RNA PCR primer 1 (RP1).
2. Add 85.5 μL master mix to each RT reaction.
3. Add 2 μL RNA PCR Primer Index to each sample (*see* **Note 16**).
4. Divide each 100 μL PCR reaction into four tubes, each with 25 μL each.
5. Incubate samples on a thermocycler using the following setting: 1 cycle of 30 s at 95 °C; 12 cycles of 10 s 95 °C, 30 s 60 °C, and 30 s 72 °C; and 1 cycle of 5 min at 72 °C.
6. Combine all four PCR tubes and precipitate.
7. Resuspend pellet in 10 μL nuclease-free water.

3.3.11 Size Selection 4 (Fig. 2d)

This size selection step is to remove excess primer dimer and adapter-adapter products.

1. Run samples on 6% TBE-Urea gel as was done in Subheading 3.3.8, excising the gel from 135 bp to the end of the signal nucleotides.
2. Resuspend pellet in 30 μL nuclease-free water. This is the final library.

3.4 PIP Library Quantitation and Quality Control

1. To determine the concentration in the final PIP libraries, 1.5 μL of the library can be read by NanoDrop. The 260/280 nm ration should be within 1.5–2.5.
2. Additionally, 1 μL of the library can be run on an Agilent Technologies 2100 Bioanalyzer to determine a more accurate concentration and size distribution (Fig. 3).

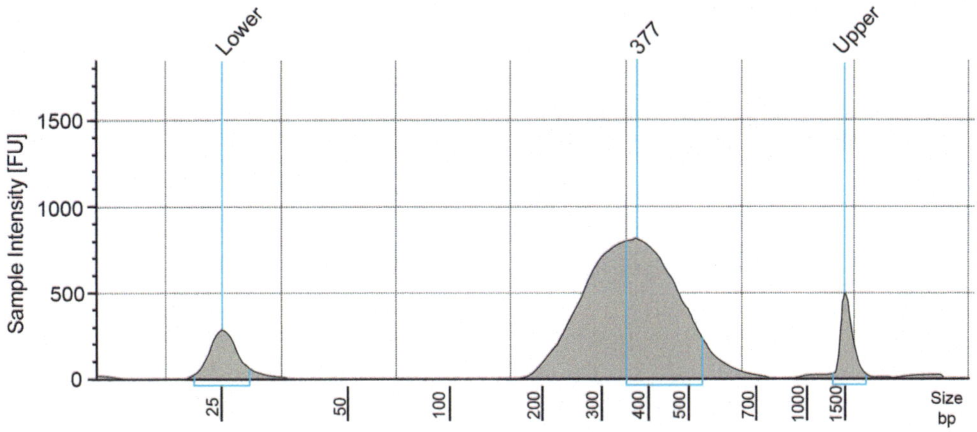

Fig. 3 Example of bioanalyzer results for a PIP-seq library. PIP libraries generally range in size from 200 bp to 700 bp. No adapter band around 170 bp representing adapter contamination is present, indicating the library is free of contamination (Subheading 3.4)

3. PIP libraries generally range in size from 200 bp to 700 bp.
4. Strong bands around 170 bp are likely a result of adapter contamination.

4 Notes

1. This pH has been optimized for plant tissue due to the presence of the vacuole. The pH of Tris–HCl can be modified for mammalian samples. In the past, pH of 7.4 has been used for mammalian cell culture.
2. We find it best to prepare the RNase stop buffer freshly before each experiment. Do not keep on ice, as SDS will precipitate out of solution at 4 °C.
3. After cross-linking and quenching, the *Arabidopsis* seedling cotyledons and leaves will be slightly translucent, indicating that the cross-linking has worked.
4. To avoid pipetting errors and ensure equal amounts of lysate in each tube, aliquot slightly less than 250 μL to each tube. If there is extra, try to split evenly among the four tubes.
5. It is essential that the pellet doesn't fully dry. Allowing the pellet to fully dry can affect the RNA secondary structure. The pellet will also be very hard to resuspend; pipette and vortex thoroughly to ensure the entire pellet is fully resuspended. The resulting solution may be somewhat soapy looking due to the presence of SDS.

6. When 10× RNase stop buffer is added to samples containing structure buffer, a precipitate will form. This is normal and will not affect the enzyme activity.
7. 10× NEB DNA ligase buffer is not provided with T4 PNK enzyme and must be purchased separately. This buffer is used because it contains ATP.
8. For the high percentage gels (15% TBE-Urea), we recommend widening the holes of the gel breaker tubes with a 21-gauge needle.
9. The goal is to have the resuspended gel pieces easily turn end-over-end. If the gel piece is too large and the sample is not moving fluidly, more 0.3 M NaCl should be added.
10. It may be difficult to pipette all of the gel slurry into the Spin-X column. Pipette as much of the liquid as possible. The liquid contains the RNA samples; thus the gel pieces can be left behind.
11. Here we use two different RNA ligases to obtain a strand-specific library. T4 RNA ligase 2, truncated, will ligate the 3' RNA adapter in the absence of ATP, while T4 RNA ligase 1 requires ATP to ligate the 5′ RNA adapter.
12. The goal is to purify the RNA molecules that were successfully ligated to 5′ and 3′ adapters. Each adapter is 25 nt at this point in the protocol; therefore the size selection is 50 nt larger than the first size selection.
13. Each sample should have a different RNA PCR Primer Index. The index is used after pooling and sequencing to pull out reads mapping to each sample. If more than one sample has the same index, they will not be able to be differentiated after sequencing. Take note of the index for each sample.
14. 100 ng is the ideal amount of DNA to use in DSN reactions, but if one or more sample does not have enough, this amount can vary so that all samples have the same amount of starting DNA input into the reactions.
15. This time has been optimized for *Arabidopsis* tissue. For mammalian samples, we have found 7 h to be the best length of time.
16. This RNA PCR Primer Index should be the same as used in Subheading 3.3.7.

Acknowledgments

The authors would like to thank the members of the Gregory Laboratory both past and present for helpful discussions. This work was funded by NSF grants MCB-1243947, MCB-1623887, and IOS-1444490 to B.D.G.

References

1. Izaurralde E, Jarmolowski A, Beisel C et al (1997) A role for the M9 transport signal of hnRNP A1 in mRNA nuclear export. J Cell Biol 137:27–35
2. Garneau NL, Wilusz J, Wilusz CJ (2007) The highways and byways of mRNA decay. Nat Rev Mol Cell Biol 8:113–126
3. Lunde BM, Moore C, Varani G (2007) RNA-binding proteins: modular design for efficient function. Nat Rev Mol Cell Biol 8:479–490
4. Mattick JS (2004) RNA regulation: a new genetics? Nat Rev Genet 5:316–323
5. Borges F, Martienssen RA (2015) The expanding world of small RNAs in plants. Nat Rev Mol Cell Biol 16:727–741
6. Jones-Rhoades MW, Bartel DP, Bartel B (2006) MicroRNAS and their regulatory roles in plants. Annu Rev Plant Biol 57:19–53
7. Axtell MJ (2013) Classification and comparison of small RNAs from plants. Annu Rev Plant Biol 64:137–159
8. Rodor J, Letelier I, Holuigue L et al (2010) Nucleolar RNPs: from genes to functional snoRNAs in plants. Biochem Soc Trans 38:672–676
9. Kapranov P, Cheng J, Dike S et al (2007) RNA maps reveal new RNA classes and a possible function for pervasive transcription. Science 316:1484–1488
10. Wang H, Chung PJ, Liu J et al (2014) Genome-wide identification of long noncoding natural antisense transcripts and their responses to light in Arabidopsis. Genome Res 24:444–453
11. Noller HF, Kop J, Wheaton V et al (1981) Secondary structure model for 23S ribosomal RNA. Nucleic Acids Res 9:6167–6189
12. Sprinzl M, Horn C, Brown M et al (1998) Compilation of tRNA sequences and sequences of tRNA genes. Nucleic Acids Res 26:148–153
13. Bardou F, Ariel F, Simpson CG et al (2014) Long noncoding RNA modulates alternative splicing regulators in Arabidopsis. Dev Cell 30:166–176
14. Wang Y, Fan X, Lin F et al (2014) Arabidopsis noncoding RNA mediates control of photomorphogenesis by red light. Proc Natl Acad Sci 111:10359–10364
15. Li R, Zhu H, Luo Y (2016) Understanding the functions of long non-coding RNAs through their higher-order structures. Int J Mol Sci 17 (5):702
16. Wan Y, Kertesz M, Spitale RC et al (2011) Understanding the transcriptome through RNA structure. Nat Rev Genet 12:641–655
17. Silverman IM, Li F, Alexander A et al (2014) RNase-mediated protein footprint sequencing reveals protein-binding sites throughout the human transcriptome. Genome Biol 15:R3
18. Gosai SJ, Foley SW, Wang D et al (2015) Global analysis of the RNA-Protein interaction and RNA secondary structure landscapes of the Arabidopsis nucleus. Mol Cell 57:376–388
19. Foley SW, Gosai SJ, Wang D et al (2017) A global view of RNA-Protein interactions identifies post-transcriptional regulators of root hair cell fate. Dev Cell 41:204–220.e5
20. Deal RB, Henikoff S (2010) A simple method for gene expression and chromatin profiling of individual cell types within a tissue. Dev Cell 18:1030–1040
21. Deal RB, Henikoff S (2011) The INTACT method for cell type-specific gene expression and chromatin profiling in Arabidopsis thaliana. Nat Protoc 6:56–68

Chapter 22

Computationally Characterizing Protein-Bound Long Noncoding RNAs and Their Secondary Structure Using Protein Interaction Profile Sequencing (PIP-Seq) in Plants

Mengge Shan, Zachary D. Anderson, and Brian D. Gregory

Abstract

Two major components of posttranscriptional regulation are RNA–protein interactions and RNA secondary structure. While noncoding RNAs are far more abundant than messenger RNAs in eukaryotic systems, their functions remain largely unstudied. Evidence suggests that RNA–protein interactions and RNA secondary structure also regulate the function of long noncoding RNAs (lncRNAs), which are noncoding RNAs over 200 nucleotides (nt) in length. Protein interaction profile sequencing (PIP-seq) allows researchers to perform an unbiased screen of protein-bound regions and secondary structure of RNAs throughout a transcriptome of interest. Using a peak calling approach, our pipeline is able to identify protein-protected sites (PPSs), which are putative RNA–protein interaction sites. Additionally, by taking the ratio of read coverages in double-stranded RNA (dsRNA)-seq compared to single-stranded RNA (ssRNA)-seq libraries, our analysis can also calculate an RNA secondary structure score that reflects the likelihood of a region being comprised of double- or single-stranded ribonucleotides. Researchers can also use this pipeline to look at specific regions of interest, such as known lncRNAs, and determine their protein-bound status as well as elucidate their secondary structure.

Key words Long noncoding RNA (lncRNA), RNA secondary structure, RNA–protein interaction, RNA-binding proteins (RBPs), Messenger RNAs, RNA sequencing

1 Introduction

RNA-binding proteins (RBPs) and RNA secondary structure play integral roles in all stages of the RNA life cycle and help determine correct posttranscriptional processing of transcripts, as well as their storage, export, and translation [1]. Studies have shown that RBPs interact with mRNAs as well as long noncoding RNAs (lncRNAs) to regulate a broad spectrum of biological processes such as plant development and stress response. With the ability to sequence the transcriptome to greater and greater depth more cost effectively, it

Julia A. Chekanova and Hsiao-Lin V. Wang (eds.), *Plant Long Non-Coding RNAs: Methods and Protocols*, Methods in Molecular Biology, vol. 1933, https://doi.org/10.1007/978-1-4939-9045-0_22,

is imperative that researchers have the tools to characterize the interplay between lncRNAs, RBPs, and RNA secondary structure. The need is made timelier by evidence suggesting that RNA structure may be more important in the function and conservation of lncRNAs than their primary nucleotide sequences [2, 3]. However, the majority of approaches exploring lncRNA-protein interactions of these RNAs rely on either the a priori identification of the RBP or the ability to pull down the RNA of interest [4]. Protein interaction profile sequencing (PIP-seq) globally captures the transcriptome-wide patterns of RNA–protein interactions and RNA secondary structure in transcriptomes of interests without previous knowledge of any RBP or RNA of interest. To date, the technique has already successfully identified sites of RNA–protein interactions and uncovered RNA secondary structure in *Arabidopsis* [5, 6] and human HeLa cells [7].

PIP-seq analysis takes advantage of the peak calling package ChIP-seq analysis in R (CSAR) to detect significantly enriched regions in footprinting samples as compared to structure-only samples, which are indicative of regions likely to be protected by the binding of RBPs. Furthermore, the structure-only samples from PIP-seq can be used to calculate a generalized log ratio of dsRNA-seq (paired bases) and ssRNA-seq (unpaired bases) read coverages, which allows users to determine whether the region is more likely to be double- or single-stranded. Here, we outline the basic PIP-seq analysis pipeline and offer advice on how to efficiently take sequenced PIP-seq libraries and detect statistically significant regions of protein binding and their corresponding RNA secondary structure. The resulting BED files can be used for a variety of downstream analyses such as using a list of regions of interest (i.e., list of potential lncRNA sites) to parse out protein binding status and/or RNA secondary structure in this emerging and important class of RNAs.

2 Materials

2.1 Hardware Requirements

Desktop computer or server (multicore server with 32 GB RAM is recommended when starting with raw sequenced reads).

2.2 Software Requirements

1. Linux-based operating system (CentOS, Ubuntu, Debian, etc.) (*see* **Note 1**).
2. C compiler (g++).
3. Standard POSIX programs (awk, grep, bash) (*see* **Note 2**).

Packages #4–8 can be installed manually if necessary (*see* **Note 3**). The external programs are available from Linux repositories and can be installed using package managers such as apt-get in Ubuntu. Sample commands for downloading and installing SAMtools and R are provided below:

sudo apt-get install samtools

sudo apt-get install r-base

4. SAMtools suite [8, 9].
5. Python 2 (version v2.7.x or above) (*see* **Note 4**).
6. Perl (version v.5.8.8 or above).
7. R program (version 3.1.x or above)
8. BEDtools suite [10].

2.3 Tools Required for Processing Raw Reads

The following tools are required for processing raw reads:

1. SRA toolkit (https://trace.ncbi.nlm.nih.gov/Traces/sra/sra.cgi?cmd=show&f=software&m=software&s=software); needed when analyzing data downloaded from the Sequencing Read Archive
2. Cutadapt (http://cutadapt.readthedocs.io/en/stable/guide.html) is used to trim the adapters from the reads [11].
3. TopHat2 (https://ccb.jhu.edu/software/tophat/manual.shtml) and Bowtie2 (http://bowtie-bio.sourceforge.net/bowtie2/manual.shtml) are recommended for mapping the raw reads to the transcriptome [12]. These tools are required for users planning on using the wrapper scripts provided. Please note that TopHat2 relies on Bowtie2 to function, so users will need to install both.

2.4 Python Wrapper Scripts Developed by the Gregory Lab

In order to streamline the analysis, using the following scripts is recommended:

1. `PIP_align_wrapper.py`
2. `PIP_PPS_calling_bambased.py`
3. Various other scripts as listed in Table 1.

The scripts are packaged in an archive (`PIPseq_scripts.tar.gz`) and can be downloaded from GitHub (http://github.com/GregoryLab/pipSeq) to a script directory the user makes (i.e. <ProjectDirectory/scripts>).

Table 1
List of scripts available on GitHub for the analysis of protein-protected sites and RNA secondary structure using PIP-seq data

Script name	Description
PIP_align_wrapper.py	Python wrapper that aligns *fastq files to the transcriptome and produces *bam files, one for each library
PIP_PPS_calling_bambased.py	Python wrapper that takes aligned *bam files and produces *bed files containing PPSs
shuffle_reads_BAM.pl	Perl script that randomly shuffles reads between control and sample (to generate a background for CSAR peaks)
split_coverage_by_chrom.pl	Perl script to split the coverage calculations into each chromosome
generate_empty_coverage_files.rb	Ruby script that generates coverage files for any chromosomes/scaffolds that do not have coverages
make_CSAR_files.R	Converts coverage TXT files into CSAR-compatible input files
run_CSAR_shuffled.R	Calculates statistical thresholds based upon the shuffled reads
run_CSAR_saturation.R	Determines statistically significant peaks (i.e., PPSs) using the threshold established in run_CSAR_shuffled.R
filter_bed_vNOsplit.py	Filters reference BED12 files based upon read coverage in UTRs
calc_bw_strucScores.py	Calculates the structure score and outputs in bigWig format
standardize_structScores.py	Standardizes raw structure score by read coverage
extract_bigwig_to_bed12.py	Converts the bigWig format to user-readable BED12+1 format
extract_structure_score_at_point.pl	Extracts the secondary structure score user-specified window around a nt

3 Methods

The protocol assumes that users have generated the appropriate dsRNAse and ssRNAse digested libraries as recommended in the associated chapter (*see* Chap. 21). For each sample, the user should have four libraries: dsRNAse-treated footprinting sample, ssRNAse-treated footprinting sample, dsRNAse-treated structure-only sample, and ssRNAse-treated structure-only samples. A general overview of the workflow is outlined in Fig. 1a. Since the pipeline requires mapped files as input, the first step is to process and align raw reads to the corresponding transcriptome. The

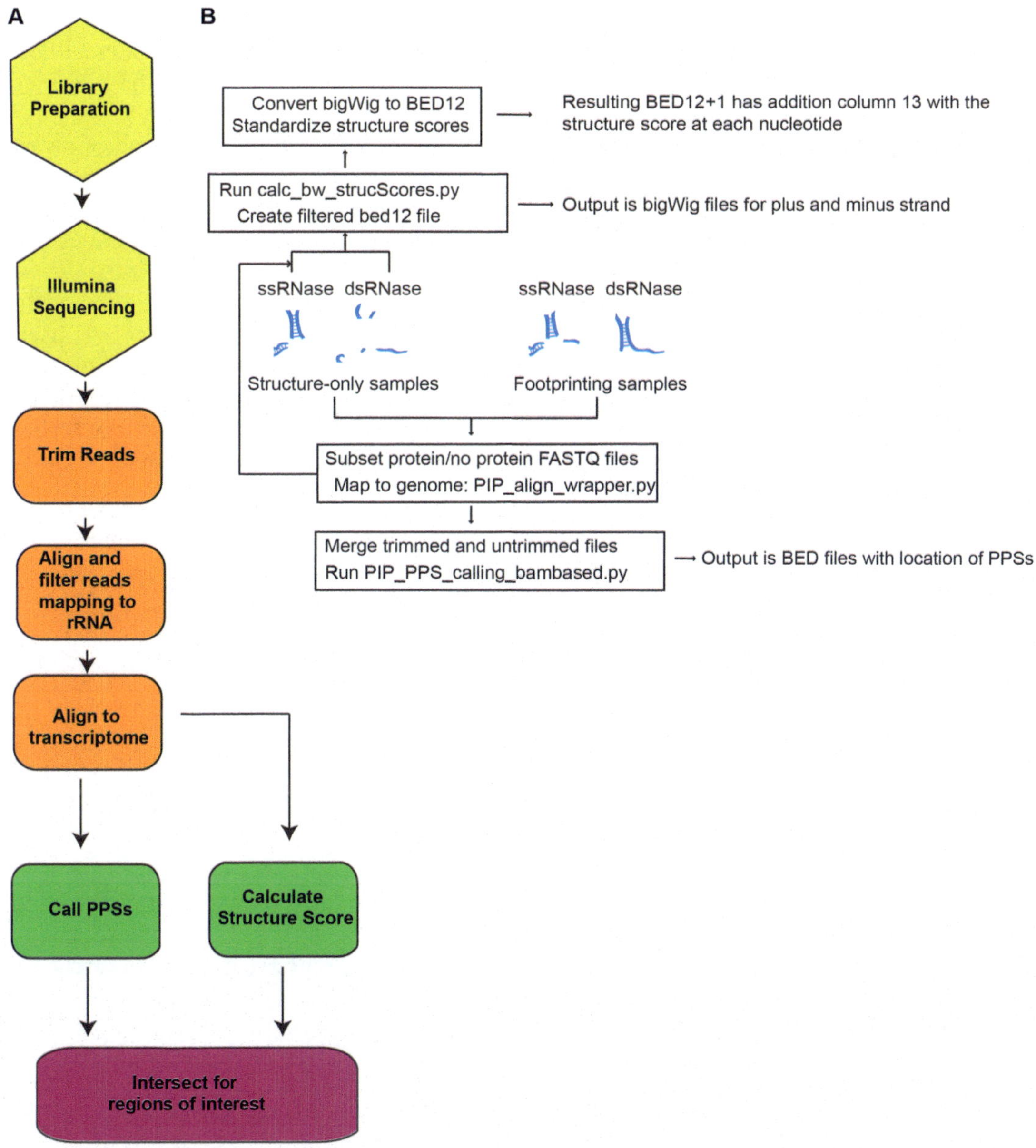

Fig. 1 Overview of the steps involved in PIP-seq analysis. (**a**) A general overview of the workflow that takes users from library preparation to the detection of protein-protected sites (PPSs) and calculating RNA secondary structure score. (**b**) The steps and libraries involved in detecting PPSs and calculating RNA secondary structure score are outlined, with the specific scripts involved in each step noted below the pertinent analysis step

mapped files are then run through the PIP-seq processing pipeline, which results in a transcriptome-wide list of potentially protein-bound sites that can be further filtered for those that are statistically significant (default: FDR < 0.05). Concurrently, the user can also use the structure libraries to determine the RNA secondary structure of detectable transcripts. Assuming the user has a file containing regions of interest (i.e., lncRNA-producing regions), the user can then identify any overlapping protein-protected sites (PPSs) and the RNA secondary structure information in the specified regions of interest.

The following instructions are current as of the time of publication, but subsequent updates may add or remove functionality. In all instances, users are encouraged to double check any accompanying README files for the most up-to-date installation and execution instructions.

3.1 Downloading and Installing the PIP-Seq Processing Pipeline

Upon successful completion of the PIP-seq analysis pipeline, users should have BED6 files containing the genomic locations of PPSs as well as a BED12+1 file with the standardized RNA secondary structure score for each nucleotide in detectable transcripts (Fig. 1b). The steps of this analysis pipeline are described below.

1. Create a unique directory for your project. In the project directory, make sure to create two subdirectories, one for data and one for the scripts you need to analyze the data.

```
mkdir <ProjectDirectory>
mkdir <ProjectDirectory/data>
mkdir <ProjectDirectory/scripts>
cd <ProjectDirectory>
```

2. Download the scripts from GitHub (https://github.com/GregoryLab/pipSeq) to the script directory <ProjectDirectory/scripts>.

   Example command to download the scripts from the command line:
   `wget` https://github.com/GregoryLab/pipSeq/archive/PIPseq_scripts.tar.gz `-O PIPseq_scripts.tar.gz`
3. Extract any archived files using

```
tar -xzvf PIPseq_scripts.tar.gz
```

4. Make sure that you have the following scripts in the script directory.

```
cd <ProjectDirectory/scripts>
ls
```

Among the collection of scripts necessary for analyzing PIP-seq data (*see* Table 1 for a complete list), users should see the following Python scripts:

```
PIP_align_wrapper.py
PIP_PPS_calling_bambased.py
```

Users interested in running the alignment component separately (i.e., replacing PIP_align_wrapper.py with their own alignment pipeline) may refer to Subheading 3.3.2 for a more detailed breakdown of each step in the alignment pipeline.

5. Downloading and installing the custom CSAR package [13]. The modified package can be found in the GitHub repository as <CSAR_PPS>. The package should be downloaded to a directory where the user has read and write permissions. Sample command to install the custom R library:

```
R CMD INSTALL CSAR_PPS -- library=[PATH_TO_DIRECTORY]
```

3.2 Preparing the Reference Genome

3.2.1 Download the Reference Genome

The genomes of many model organisms can be found at the UCSC Genome Browser (http://hgdownload.soe.ucsc.edu/downloads.html). Additionally, some plant genomes can be found from other sources such as Phytozome (https://phytozome.jgi.doe.gov/pz/portal.html) (*see* **Note 5**) or Ensembl Plant Project (*see* **Note 6**).

```
mkdir <ProjectDirectory/annotations>

cd <ProjectDirectory/annotations>
```

Sample command for downloading the Arabidopsis TAIR10 genome:

```
wget ftp://ftp.ensemblgenomes.org/pub/release-25/plants/fasta/arabidopsis_thaliana/dna/Arabidopsis_thaliana.TAIR10.25.dna.genome.fa.gz
```

3.2.2 Build the Bowtie Index for the Organism

Bowtie, which is part of the TopHat package, has pre-built indices for human, mouse, fruit fly, and other model organisms available here: http://bowtie-bio.sourceforge.net/bowtie2/index.shtml. For other organisms, users will need to build the Bowtie index manually (*see* **Note 7**).

```
bowtie2-build [options]* [REFERENCE_FASTA_FILE] [PREFIX_-
FOR_OUTPUT_FILES]
```

For example:

```
bowtie2-build -f Arabidopsis.genome.fa At_TAIR10
```

3.2.3 Preparing rRNA Genome Index

Our analysis first maps reads to ribosomal RNA (rRNA) to filter out reads from these highly abundant RNA species, which can decrease PPS calls due to their highly protein-bound nature. All unmapped reads from that analysis are then mapped to the genome of the organism of interest. If users do not already have an rRNA genome index, they can generate one using a BED file with the rRNA locations by using the following commands:

```
bedtools -getfasta -s -name -i [INPUT_FASTA_FILE] - bed
[INPUT_BED_FILE]
```

The above command will create the FASTA file. The resulting FASTA can be used as in Subheading 3.2.2 to build the rRNA Bowtie genome index.

Users can download https://bitbucket.org/galaxy/galaxy-central/src/61b09dc1dff2/tools/filters/bed_to_gff_converter.py and run

```
python bed_to_gff_converter.py [INPUT_BED_FILE] [OUT-
PUT_GFF_FILE]
```

to convert the BED file to GFF, which can be used for aligning reads with TopHat.

3.3 Processing the PIP-Seq Raw Reads

The FASTQ file obtained from an Illumina sequencer needs to be mapped to the transcriptome, producing a mapped file format (BAM). Before mapping, there are some preprocessing steps that will ensure the resulting mapped file is compatible with downstream tools. The exact steps are outlined in Fig. 2 and include trimming away the sequencing adapters and removing reads that are PCR amplification duplicates. Unless specified, all sorting is done with respect to the genomic coordinates (i.e., chromosomal location). Please note that all steps outlined in the analysis have been optimized to work in Linux-based operating systems.

3.3.1 Aligning FASTQ Files to the Transcriptome

The “PIP_align_wrapper.py” script will take raw FASTQ files, trim away adapters, and map reads to the transcriptome, filtering out any reads that map to rRNA. All mapping is done using TopHat. Experienced users who are comfortable exploring other transcriptome alignment tools may do so, following the general steps outlined in Fig. 3 and substituting tools as desired.

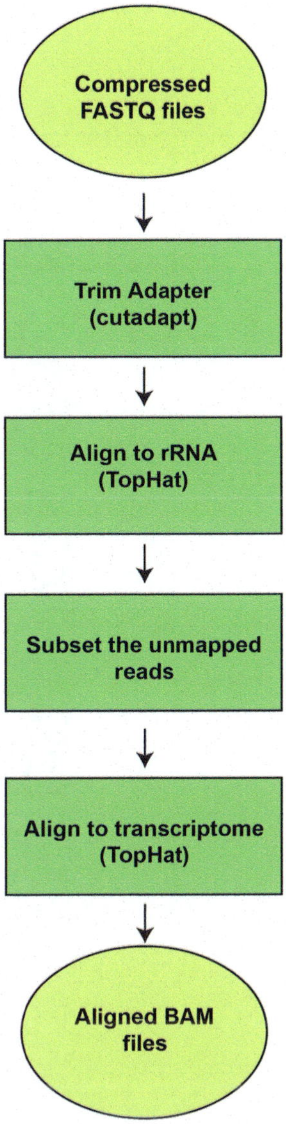

Fig. 2 Steps in the PIP_align_wrapper.py for aligning raw FASTQ files to the genome of interest. First the sequencing adapters are trimmed away, and then the reads are aligned to rRNAs to remove these highly abundant RNA sequences. Next, the sequences that do not map to rRNAs are then aligned to the transcriptome in general, creating the aligned BAM file that can be used in downstream analyses

For users who choose to align their reads independently of PIP_align_wrapper.py, please note that each library should be aligned separately and the resulting BAM file should follow the naming convention below for use in downstream analysis.

```
'[unique_tag]_(d/s)sRNase_(no/)protein_((un/)trimmed/
merged)/NR_hits.bam'
```

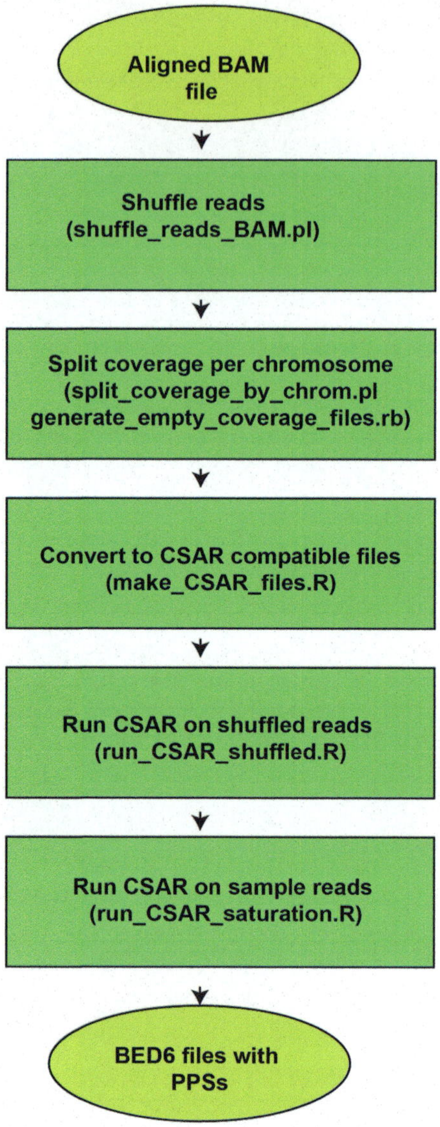

Fig. 3 Steps used by PIP_PPS_calling_bambased.py. Users start off with paired aligned BAM files that are then randomly shuffled to create a background control dataset. After splitting the coverage into chromosomes for easier processing, the random background is used to determine the statistical threshold for calling protein-protected sites (PPSs). Lastly, the actual data is run through the CSAR analysis pipeline to determine the actual protein-protected sites, which are written into a BED6 file for downstream analyses

where:

unique_tag = experiment specific descriptor (i.e. At_nuc or Pp_light)

(d/s)sRNAse = samples treated with ds- or ssRNAse

(no/)protein = footprinting (protein) or structure control (noprotein) samples

((un/)trimmed/merged) = whether the reads had their adapters trimmed away

PIP_align_wrapper.py has several requirements: the FASTQ files should be in <ProjectDirectory/data/fastqFiles>, scripts should be kept in <ProjectDirectory/scripts>, and the "PIP_align_wrapper.py" should be called directly in the directory <ProjectDirectory/data/fastqFiles>.

Upon successful completion, the mapped BAM files are found in two directories. <ProjectDirectory/data/topOut_rRNA_free> contains the reads that should be used for the downstream PPS calling algorithm, while <ProjectDirectory/data/mapped_rRNA> contains reads that mapped to rRNA.

Command:

```
python2.7 PIP_align_wrapper.py [DIRECTORY_WITH_BOWTIE_INDEX_FILES] [DIRECTORY_WITH_rRNA_FILES] -G [GENOME_ANNOTATION.gtf] -ti [BOWTIE_TRANSCRIPTOME_INDEX_FILE] -p [NUMBER_OF_PROCESSORS] -in [SPACE_SEPARATED_LIST_OF_FASTQ_FILES]
```

Please note that PIP_align_wrapper.py assumes the libraries were created with Illumina sequencing technology and therefore has hardcoded the adapter sequence into its script. If your adapter sequence varies from TGGAATTCTCGGGTGCCAAGGAACTCCAGTCACnnnnnnATCTCGTATGCCGTCTTCTGCTTG, you can modify the appropriate line in

```
PIP_align_wrapper.py
```

for your adapter sequence. The script has also hardcoded the optimal parameter values from our analyses, though experienced users are welcome to change the parameters for their own needs. The adapter sequence and parameter options can be found on line 105 of the

```
PIP_align_wrapper.py
```

file.

3.3.2 Individual Components of PIP_align_wrapper.py

Below is a brief description of each separate component of the wrapper script described in Subheading 3.3.1. More experienced users may refer to the header section in each script for a description of the relevant parameters in their own analysis.

1. gunzip: decompresses the compressed FASTQ file.
2. cutadapt: trims away the sequencing adapters.
3. TopHat (mapping to rRNA): We first use TopHat to map the trimmed reads to rRNA sequences to detect any possible rRNA contamination. We take the unmapped reads from this portion of the analysis and map them to the entire transcriptome.
4. TopHat (mapping to the wider transcriptome): This step takes any reads that did not map to rRNA and maps them to the wider transcriptome.

3.4 Calling Statistically Significant Protein-Protected Sites (PPSs)

The provided “PIP_PPS_calling_bambased.py” script is a one-step solution that will take aligned BAM files from the “PIP_align_wrapper.py” and use the R statistical package CSAR to determine statistically significant protein-bound sites. CSAR calculates read enrichment on a single nucleotide basis and uses a permutation strategy to establish a threshold for determining significant sites. Before running the command, users should prepare two reference files: (1) a chromosome length file in TXT format and (2) a BED6 file with insignificant locations that should not be considered during CSAR analysis (e.g., rRNAs and tRNAs). The chromosome length file should have two columns: (1) chromosome name followed by (2) the number of nucleotides in the chromosome. This information can be easily calculated from reference genome FASTA files. Please note that the chromosome naming convention should exactly match the naming convention of the reference files (i.e., if the reference FASTA denotes chromosome 1 as chr1, then the chromosome length file should do the same). The BED6 file with locations to be masked provides the peak calling tool with a set of genomic locations that the user does not want to use in subsequent PPS (peak) calling (i.e., tRNA, rRNA, etc.).

The provided python wrapper, “PIP_PPS_calling_bambased.py”, will take the aligned BAM files and generate BED6 output files with the PPS locations. While individual components of the wrapper are outlined in Fig. 3, we recommend utilizing the wrapper to ensure uniform processing of the samples. Please note that if users chose to align FASTQ files without using “PIP_align_wrapper.py” due to preference for another alignment tool, the naming convention as outlined in Subheading 3.3.1 should be followed exactly. In addition, please note that “PIP_PPS_calling_bambased.py” is run twice for each condition, once for the dsRNAse-treated samples and once for the ssRNAse-treated samples (*see* **Note 10** or Subheading 3.4.1).

3.4.1 Calling Statistically Significant PPSs

Run the following commands:

1. `python2.7 PIP_PPS_calling_bambased.py [ALIGNMENT_TAG] [TRIM_STATUS] [CHROMOSOME_LENGTH_FILE] -do -mrg -filter_-bam [BED_FILE_WITH_REGIONS_TO_MASK]`

2. `python2.7 PIP_PPS_calling_bambased.py [ALIGNMENT_TAG] [TRIM_STATUS] [CHROMOSOME_LENGTH_FILE] -so -mrg -filter_-bam [BED_FILE_WITH_REGIONS_TO_MASK]`

where the ALIGNMENT_TAG refers to the unique name for the experimental condition (e.g., At_nuc_rep1 or leaf_cyto_rep2) and TRIM_STATUS refers to whether the reads were trimmed or untrimmed.

The script outputs several BED6 files (*see* **Note 9**), one of which contains the list of putative RNA–protein interaction sites that have passed a statistical threshold of FDR < 0.05. This file will have the naming convention [file_prefix].CSAR_PPS_FDR05.bed where file_prefix is derived from the input BAM file names. The BED6 files are tab-delimited and organized in the following manner: chromosome (column 1), PPS start site (column 2), PPS stop site (column 3), peak name (column 4), score (column 5), and strand (column 6).

3.4.2 Merging Replicates

Users can then merge or intersect sites at will using BEDtools commands. In general, we recommend using mergeBed to merge overlapping PPSs in biological replicates at this step.

1. Run the following command to merge all replicates into a single file.

```
awk 'BEGIN {OFS="\t"; FS="\t";} {print $1,$2,$3, "[REPLICA-
TE_IDENTIFIER]_"$4,$5,$6}' [INPUT_BED] >>TEMP_OUTPUT_BED
```

2. After all replicates have been put into the single TEMP_INPUT_BED, use BEDtools's mergeBed function to collapse overlapping PPSs into one PPS for that particular biological condition/experiment. We recommend having three separate BED6 files for each experimental condition: ssRNAse PPSs, dsRNAse PPSs, and a merged file containing all PPSs found in the PIP-seq experiment. The same mergeBed parameters can then be used to create the final BED6 files, swapping out the TEMP_INPUT_BED file as needed.

```
mergeBed -c 4,5,6 -o collapse,collapse,distinct -s -i
[TEMP_OUTPUT_BED]>[OUTPUT_BED]
```

3.5 Determining the RNA Secondary Structure of the Transcript

RNA secondary structure is another feature of RNAs that affects posttranscriptional regulatory processes in eukaryotic systems. PIP-seq libraries not only contain information on potential RBP–RNA interactions but also capture a transcriptome-wide snapshot of RNA secondary structure for the transcriptome of interest. The structure score for a particular nucleotide is calculated as follows:

$$S_i = \text{glog}(ds_i) - \text{glog}(ss_i)$$
$$= \log_2\left(ds_i + \sqrt{1 + ds_i^2}\right) - \log_2\left(ss_i + \sqrt{1 + ss_i^2}\right)$$

$$ds_i = n_{ds}\frac{\max(L_{ds}, L_{ss})}{L_{ds}}, \quad ss_i = n_{ss}\frac{\max(L_{ds}, L_{ss})}{L_{ss.}}$$

where S_i is the structure score, ds_i and ss_i are the normalized read coverages, and L_{ds} and L_{ss} are the total covered length by mapped dsRNA-seq and ssRNA-seq reads, respectively. By taking a generalized log ratio of single-stranded RNAs (ssRNAs) and double-stranded RNAs (dsRNAs) for a given region, PIP-seq analysis can estimate the likelihood of the region being single- or double-stranded. The higher the score, the more likely the region is to be double-stranded (i.e., more structured) at the time of sample cross-linking. The script will calculate a generalized log ratio score for ssRNA and dsRNA libraries in the structure-only samples (samples A and B in Figs. 1a, b in the associated chapter (*see* Chap. 21)). Please note that the pipeline requires a BED12 file, which contains additional information regarding relative exon start sites and sizes (*see* **Note 10**).

3.5.1 Filter BED12 File

Since structure scores are calculated based upon read coverage, we suggest first creating a subset of transcripts that have enough read depth across regions of interest annotated in the REFERENCE_BED12 file. We recommend a minimum of 50 reads in each of these regions. In order to accurately compare transcripts, the final BED12 file should contain transcripts that have enough read coverage in ALL conditions. If there are biological replicates, we recommend merging the aligned BAM files first and then calculating the read coverage, keeping ssRNAse- and dsRNAse-treated samples separately.

1. To merge BAM files.

```
samtools merge [FIRST_INPUT_BAM] ... [Nth_INPUT_BAM] [OUTPUT_BAM_FILE_NAME]
```

2. To create a BAM file index (needed to filter the BED12 file).

```
samtools index [MERGED_BAM_FILE]
```

3. To filter the BED12 file based on read coverage in the UTRs, one for each experimental condition.

```
python2.7 filter_bed_vNOsplit.py -str yes -cov [COVERAGE] -UTR3 [MIN_LENGTH_UTR3] -UTR5 [MIN_LENGTH_UTR5] -ncPass -read [MERGED_BAM_FILE_FOR_COND1] [REFERENCE_BED12] [OUT-PUT_FILTERED_BED12]
```

4. If you have multiple experimental conditions, it is necessary to identify the subset of transcripts that has sufficient read coverage for ALL conditions. The final filtered BED12 file can be found by running the following command.

```
intersectBed -s -a [FILTERED_BED12_FIRST_EXPERIMENTAL_CON-DITION] -b [FILTERED_BED12_SECOND_EXPERIMENTAL_CONDITION] ... [FILTERED_BED12_LAST_EXPERIMENTAL_CONDITION] > [FI-NAL_FILTERED_BED12]
```

3.5.2 Calculate Structure Score

The script should be run once for every pair of ssRNA and dsRNA libraries, and each pair of files should be given a unique output file prefix (i.e., At_control_rep1, At_light_rep1, At_dark_rep1, etc.). The script will produce several bigWig files as well as a BED12+1 file where the unstandardized structure score for each nucleotide is listed in a comma-delimited format in the 13th column (*see* **Note 11**). Run the following command:

```
python2.7 calc_bw_strucScores.py [FILTERED_REFEREN-CE_BED12] [CHR_LEN_FILE] [OUTPUT_DIRECTORY] [OUTPUT_PRE-FIX] -ds
[BAM_FILE_OF_STRUCTURE_SAMPLE_TREATED_WITH_ssRNASE] -ss
[BAM_FILE_OF_STRUCTURE_SAMPLE_TREATED_WITH_dsRNASE]
```

3.5.3 To Obtain Standardized Structure Scores

Run the following command:

```
python2.7 <ProjectDirectory/Scripts/>standardize_struct-Scores.py [OUTPUT_DIRECTORY] [CHROMOSOME_LENGTH_INFO] [RAW_-PLUS_STRAND_BIGWIG] [RAW_MINUS_STRAND_BIGWIG]
```

The script produces two bigWig files, one for each strand, that are converted in the next step to the appropriate BED12+1 file.

3.5.4 Converting bigWig Format to BED12+1 Format

Run the following command:

```
python2.7 <ProjectDirectory/Scripts/>extract_bigwig_to_bed12.
py -plus [PLUS_STRAND_BIGWIG] -minus [MINUS_STRAND_BIGWIG]
-NAval 0 [FILTERED_REFERENCE_BED12] [OUTPUT_FILE_NAME]
```

3.6 Identifying Protein-Bound lncRNAs and Their Secondary Structure

PIP-seq can be used to determine RNA secondary structure information for regions hypothesized to produce lncRNAs. Additionally, PIP-seq data can also be used to identify RBP occupancy within lncRNAs. If the user has a list of lncRNAs in BED format, PPSs that overlap lncRNAs can be identified using commands available in the BEDtools suite. Output files from Subheading 3.5.4. contain the structure score for each nucleotide in the last column of the BED12+1 file, separated by commas. Users can parse out the structure score for a particular region of interest including all or specific lncRNAs. The following section will give examples of the types of possible downstream analyses.

3.6.1 Identifying Protein-Protected Regions that Intersect lncRNAs

Since the PPSs are identified in a traditional BED6 file format, users can use the intersectBed tool from BEDtools to identify PPSs that overlap lcRNAs.

```
intersectBed -a [PPS_BED] -b [LNCRNA_BED] -s -u>[OUTPUT_BED]
```

3.6.2 RNA Secondary Structure of Transcripts Containing lncRNAs

The BED12+1 file produced in Subheading 3.5.4 contains a comma-separated list of RNA secondary structure scores in the last column. For users specifically interested in lncRNA-producing regions, the following script will extract a user-specified number of bases up- and downstream of a user-chosen position, and the scores can then be visualized using Excel or other graphing tools. The script requires the following input files: (1) tab-delimited file with the transcript name in the first column and the genomic position of interest in the second column and (2) the BED12+3 file created as an output in Subheading 3.5.4.

Run the following command:

```
perl extract_structure_score_at_point.pl [STD_STRUCTURE_-
SCORE_BED13] [POSITION_TO_EXTRACT_TXT_FILE]
[NUM_OF_BASES_UP_DOWNSTREAM]>[OUTPUT_FILE]
```

Please note the transcript names in the POSITION_TO_EXTRACT_TXT_FILE must exactly match with the corresponding transcript name in the FILTERED_REFERENCE_BED12 file that was used to generate the structure scores. Transcript names can be repeated as long as there is only one transcript per line in the file. The Perl script will extract a user-specified number of nucleotides around the position specified in the second column of the

POSITION_TO_EXTRACT_TXT_FILE (i.e., if the lncRNA starts at 30100 and the script is told to extract 100 nucleotides around the start site, the final output will contain scores from 30,000 to 30,200). The output file will be a tab-delimited TXT file with the transcript name in the first column and the extracted structure scores in the second column.

4 Notes

1. Please ensure that the user has read and write privileges on the computer and/or server. If your server is administered by professional staff, please contact them for help installing server-compatible versions of the programs. The protocol also assumes a familiarity with UNIX. Beginners can refer to online tutorial manuals such as "Unix Tutorial for Beginners" (http://www.ee.surrey.ac.uk/Teaching/Unix/) for an introduction to the basic commands.
2. The programs required for the protocol are all freely and readily available for popular Linux distributions.
3. Another way to install programs is to build them in your own directory. Download the required packages, paying attention to the version required for the pipeline, and follow the distributor's installation directions. The instructions are usually provided in *readme.txt* files or on the package's website. A good practice is to check for dependencies (i.e., packages/tools a program needs to function properly) and make sure those are available on your computer or server.
4. Use --version to check the version of most programs (e.g., python --version should show you the version of python that is available).
5. Phytozome requires users to create a free account before accessing the downloads page and provide information on the institution you are affiliated with.
6. The *Arabidopsis* genome can be found at ftp://ftp.ensemblgenomes.org/pub/release-25/plants/fasta/arabidopsis_thaliana/dna/Arabidopsis_thaliana.TAIR10.25.dna.genome.fa.gz.
7. The current version of TopHat will build genome indices on the fly using the Bowtie2 package. However, pre-indexing the genome FASTA file will allow for faster execution of the mapping step.
8. The dsRNAse and ssRNAse samples should be run separately, using the –do and –so flags, respectively. The rest of the parameters should stay the same.

9. BED6, BED12, or BED12 + 1 files are best viewed directly on the terminal using commands such as cat, cut, or less. They are tab-delimited by default, and relevant columns can be extracted using a combination of cut and awk. Structure scores are captured in the 13th column of the output file(s) produced in Subheading 3.5.2 and with nucleotides separated by commas.
10. Information on BED12 format can be found at https://genome.ucsc.edu/FAQ/FAQformat.html.
11. Information on bigWig file format can be found at https://genome.ucsc.edu/goldenpath/help/bigWig.html.

Acknowledgments

The authors would like to thank the members of the Gregory lab both past and present for helpful discussions. This work was funded by NSF grants MCB-1243947, MCB-1623887, and IOS-1444490 to B.D.G.

References

1. Vandivier LE, Anderson SJ, Foley SW et al (2016) The conservation and function of RNA secondary structure in plants. Annu Rev Plant Biol 67:463–488. https://doi.org/10.1146/annurev-arplant-043015-111754
2. Ponjavic J, Pontig CP, Lunter G (2007) Functionality or transcriptional noise? Evidence for selection within long noncoding RNAs. Genome Res 17:556–565
3. Chodroff RA, Goodstadt L, Sirey TM, Oliver PL et al (2010) Long noncoding RNA genes: conservation of sequence and brain expression among diverse amniotes. Genome Biol 11: R72. https://doi.org/10.1186/gb-2010-11-7-r72
4. Barra J, Leucci E (2017) Probing long non-coding RNA-protein interactions. Front Mol Biosci 4:45. https://doi.org/10.3389/fmolb.2017.00045
5. Foley SW, Gosai SJ, Wang D et al (2017) A global view of RNA-protein interactions identifies post-transcriptional regulators of root hair cell fate. Dev Cell 41:204–220. https://doi.org/10.1016/j.devcel.2017.03.018
6. Gosai SJ, Foley SW, Wang D et al (2015) Global analysis of the RNA-protein interaction and RNA secondary structure landscapes of the *Arabidopsis* nucleus. Mol Cell 57:376–388. https://doi.org/10.1016/j.molcel.2014
7. Silverman IM, Li F, Alexander A et al (2014) RNase-mediated protein footprint sequencing reveals protein-binding sites throughout the human transcriptome. Genome Biol 15:R3. https://doi.org/10.1186/gb-2014-15-1-r3
8. Li H, Handsaker B, Wysoker A et al (2009) The Sequence alignment/map (SAM) format and SAMtools. Bioinformatics 25:2078–2019. https://doi.org/10.1093/bioinformatics/btp352
9. Li H (2011) A statistical framework for SNP calling, mutation discovery, association mapping and population genetical parameter estimation from sequencing data. Bioinformatics 27:2987–2993. https://doi.org/10.1093/bioinformatics/btr509
10. Quinlan AR, Hall IM (2010) BEDtools: a flexible suite of utilities for comparing genomic features. Bioinformatics 26:841–842. https://doi.org/10.1093/bioinformatics/btq033
11. Martin M (2011) Cutadapt removes adapter sequences from high-throughput sequencing reads. EMBnet J 17(1):10–12. https://doi.org/10.14806/ej.17.1.200
12. Trapnell C, Pachter L, Salzberg S (2009) TopHat: discovering splice junctions with RNA-Seq. Bioinformatics 25:1105–1111. https://doi.org/10.1093/bioinformatics/btp120
13. Muino J, Kaufmann K, van Ham R et al (2011) ChIP-seq Analysis in R (CSAR): an R package for the statistical detection of protein-bound genomic regions. Plant Methods 7:11. https://doi.org/10.1186/1746-4811-7-11

Chapter 23

Stalking Structure in Plant Long Noncoding RNAs

Karissa Y. Sanbonmatsu

Abstract

Long noncoding RNAs play important roles in plant epigenetic processes. While many extensive studies have delineated the range of their functions in plants, few detailed studies of the structure of plant long noncoding RNAs have been performed. Here, we review genome-wide and system-specific structural studies and describe methodology for structure determination.

Key words Long noncoding RNA, Plants, Evolution, Selection, Secondary structure, COOLAIR

1 Introduction

Epigenetics has altered our understanding of gene regulation, development, and inheritance. Epigenetic effects often involve changes in phenotype without corresponding changes in genotype. These effects are typically caused by (1) DNA methylation, (2) chemical modifications of histone tails, and (3) noncoding RNAs. One aspect of epigenetics that has generated excitement in pharmaceutical and bioengineering communities is the high specificity of epigenetic effects with respect to tissue type and developmental stage. While many chromatin-modifying proteins and corresponding epigenetic marks have been identified, the mechanism of specificity remains poorly understood in many cases. Since long noncoding RNAs are highly specific, these molecules have been proposed as a solution to the specificity problem: by recruiting chromatin-modifying factors to appropriate gene loci with high spatiotemporal precision, long noncoding RNAs could in principle result in tissue-specific and developmental stage-specific epigenetic gene regulation.

Long noncoding RNAs range between ~200 nt and ~100 kB in length and are typically polyadenylated, capped, and spliced. They are often associated with epigenetic histone marks such as H3K27me3 and play important roles in development. Solving the

Julia A. Chekanova and Hsiao-Lin V. Wang (eds.), *Plant Long Non-Coding RNAs: Methods and Protocols*, Methods in Molecular Biology, vol. 1933, https://doi.org/10.1007/978-1-4939-9045-0_23,

detailed mechanism of individual long noncoding RNA molecules requires (1) extensive functional studies in realistic environments and (2) biochemical, biophysical, and atomistic studies of the RNA alone and with protein partners to uncover the physical mechanism of action. While several functional studies have been performed on individual long noncoding RNAs to determine phenotypes, there is a dearth of structural studies on individual long noncoding RNAs and even fewer structure-function relationship studies. Regarding (2), very few large RNA systems have been studied in atomistic detail. Historical examples that have been studied include the ribosome complex and the group II intron. In the case of the ribosome, atomistic understanding is still ongoing after 50 years of research. After the sequence of the ribosome was determined, a critical step leading to the 3-D structures of the ribosome was the in vitro mapping of the 2-D structures of the RNA alone. The 2-D maps provided an essential framework for biochemical studies of mechanism as well as for the determination of the 3-D structure via cryo-EM and X-ray crystallography. In the case of long noncoding RNAs, determination of the in vitro 2-D structure is also critical for biochemical studies and for 3-D studies. Challenges in interpreting biochemical studies and 3-D studies are compounded by the enormous size of long noncoding RNAs, which are typically similar in size to the ribosome but extend up to 100 kB (about 20 times the size of a eukaryotic ribosome). Here, we describe methods capable of determining 2-D in vitro structures of long noncoding RNAs, with a particular emphasis on plant long noncoding RNAs, noting that these maps are critical for interpreting in vivo secondary structures, structure-function relationships, and 3-D studies.

Regarding model organisms, plant systems are exciting models for epigenetic studies in light of their relatively slow regulatory time scales, clearly demarcated cell types, and simpler developmental plans. Plant epigenetic systems contain analogous pathways and regulatory mechanisms while offering the advantages of slower, more easily observable phenomena. These systems have enabled quantitative, highly reproducible studies and testable hypotheses. A prime example is vernalization, where plants use epigenetic noncoding RNAs to sense long-term exposure to cold temperatures.

Long noncoding RNAs play important roles in plant epigenetics and gene regulation [1]. Genome-wide studies of long noncoding RNAs in plants report differential expression in conditions of drought and depleted heat have suggested a role for long noncoding RNAs in critical processes such as wood formation [2, 3]. Several databases have been constructed, allowing researchers to more easily search known information about plant long noncoding RNAs [4–6]. Interestingly, long noncoding RNAs have also been shown to be a source for miRNAs in plants [7] and to be responsive to hypoxic stress [8]. Key individual miRNAs regulate plant development.

A major epigenetic mechanism is the regulation of flowering in *A. thaliana* and other species, where FLC downregulates flowering in the autumn and the COOLAIR long noncoding RNA helps upregulate flowering in the spring by suppressing FLC [9–14]. Here, the PHD complex is analogous to the human epigenetic complex polycomb 2 (PRC2). Chemical probing studies demonstrate the secondary structure of COOLAIR to be conserved from *A. thaliana* to *B. rapa*, suggesting that the secondary structure may be important in COOLAIR mechanism. [15] An RNA motif common to several other long noncoding RNAs (the right hand turn motif) was observed to occur twice in COOLAIR [16, 17]. SNPs in one of the extended helices show this helix to be conserved despite significant sequence variation. In addition to COOLAIR, the COLDAIR long noncoding RNA plays a role in vernalization. In an additional study, Lu and co-workers used a pipeline including RNA-seq, and infernal was used to identify conserved structural motifs in plants [2].

2 Materials

2.1 RNA Synthesis

1. Use dsDNA templates in runoff transcription with the T7-scribe Standard RNA IVT kit from CELLSCRIPT.
2. Extract RNA products with phenol-chloroform, precipitating with addition of 1 volume of 5 M ammonium acetate and 2.5 volumes of ethanol.
3. Check the RNA on agarose and polyacrylamide gels.

2.2 Design of RNA Fragments for Shotgun Analysis

1. The full-length transcript should be analyzed. In addition the full-length transcript should be bisected once, producing fragments 1 and 2.
2. The resulting two fragments should then be bisected again. Fragment 3 runs from the 5′-end of the full transcript to the midpoint of fragment 1. Fragment 5 runs from the midpoint of fragment 2 to the 3′-end of the full transcript.
3. Fragment 4 runs from the midpoint of fragment 1 to the midpoint of fragment 2. We note that the fragmentation approach has been successful for RNAs of lengths ~1 kB. For significantly longer RNAs, additional rounds of fragmentation can be performed.

2.3 Primer Design

1. Design the primers with desired T_M. Primers should target regions of the lncRNA separated by approximately 200 nts.
2. Primers can be ordered directly with attached fluorophore at the 5′-end. For in-house labeling with Alexa 488, use DNA oligos synthesized with an amino moiety on their 5′-end

(IDTDNA) and an Alexa Fluor 488 amine reactive ester (Invitrogen).

3. Purify the fluorophore-labeled primers on reverse phase HPLC.

3 Methods

Secondary structure determination of long noncoding RNAs is challenging but possible when combining a variety of experimental strategies. Chemical probing techniques have been applied extensively. These techniques are akin to foot printing of immobile regions of RNA, allowing one to identify specific nucleotides involved in base pairs, which are typically much less mobile that nucleotides in single-stranded regions of the RNA. In inline probing, base-paired nucleotides are significantly less reactive to cleavage than single-stranded regions. In selective 2′-hydroxyl acylation by primer extension (SHAPE), 2′-hydroxyl groups of nucleotides in base pairs are much less reactive to particular SHAPE reagents (e.g., 1M7) in comparison with single-stranded regions. The situation is similar for the classical dimethyl sulfide (DMS) probing reaction, where adenines and cytosines in base pairs are less reactive than adenines and cytosines in single-stranded regions [17]. In each of these and related methods, a foot printing-like scenario ensues, whereby base paired nucleotides are less susceptible to a diagnostic reaction. Thus, to obtain an accurate picture of the RNA secondary structure (i.e., the map of base pairs in the RNA), other occluding molecules, such as proteins, cannot be present (Fig. 1). These molecules, by stabilizing more mobile single-stranded regions, would have the effect of artificially lowering the reactivity of single-stranded regions of the RNA, identifying single-stranded nucleotides as base paired, thereby creating false positives.

While several excellent chemical probing studies of long noncoding RNA secondary structures have been performed in vivo [18], these RNAs exist in an environment of a multitude of proteins that, when bound specifically or non-specifically to the long noncoding RNA in question, will likely convolute the chemical probing signal by decreasing the mobility of single-stranded regions of RNA, identifying them as double stranded [19]. As the length of the RNA increases, the potential for binding of multiple proteins increases, exponentiating the number of false positively identified base pairs. In vitro chemical probing prior to in vivo chemical probing allows one to establish a secondary structure framework in vitro for the naked RNA (*see* **Note 1**).

The problem of solving the secondary structure of long noncoding RNAs is challenging, even in vitro. We devised a technique to identify modularly folding sub-domains of the long noncoding

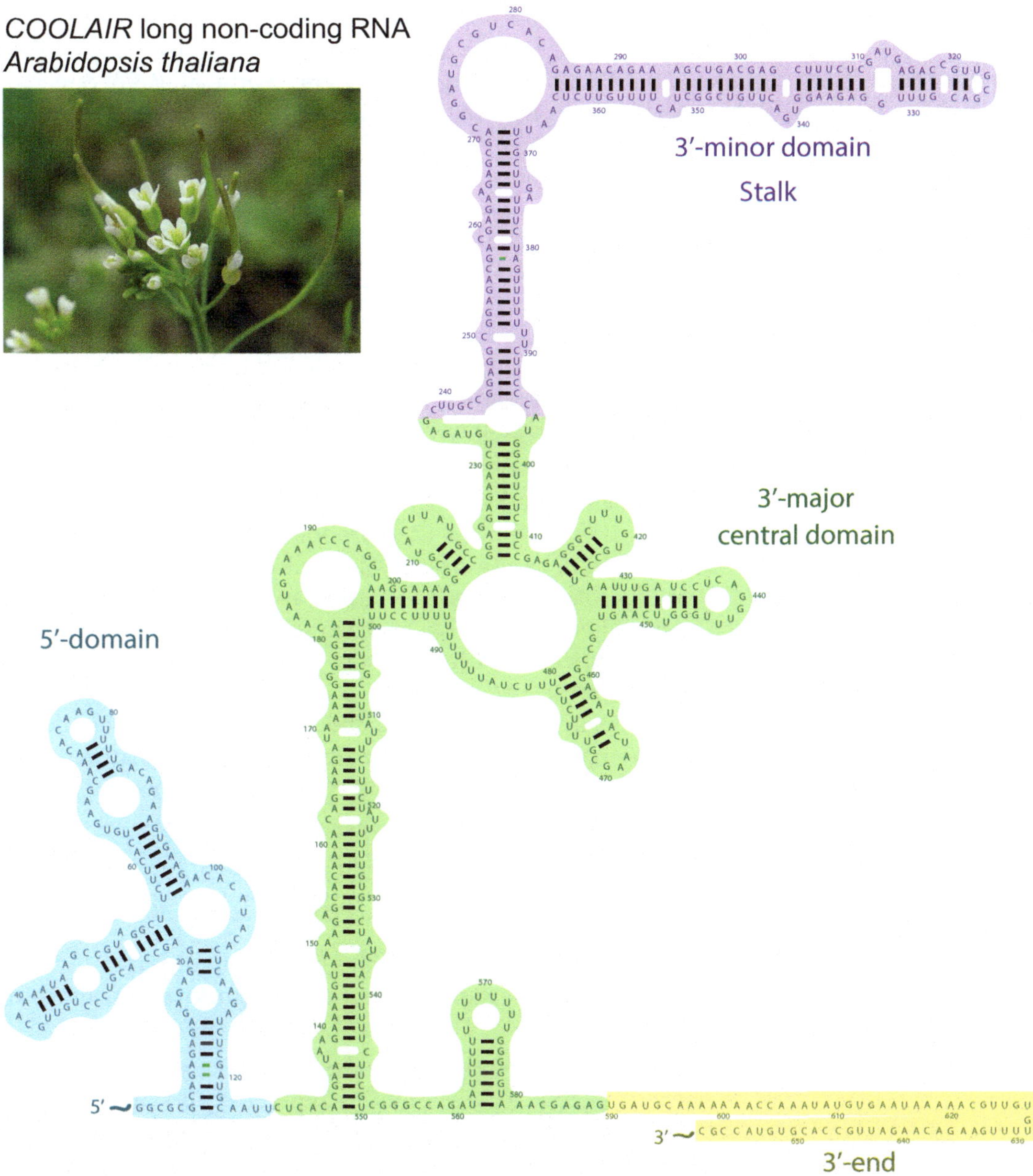

Fig. 1 Secondary structure of the COOLAIR long noncoding RNA in *A. thaliana* as determined by shotgun secondary structure chemical probing (3S). 3S produces strong evidence for three modular sub-domains: 5′-domain, 3′-major domain ("central"), and 3′-minor domain ("stalk")

RNA, dramatically reducing the number of possible secondary structures (*see* **Note 2**). The technique has been applied to SRA-1, HOTAIR, Braveheart, RepA of Xist, P21-lncRNA, and NEAT1 [16, 17, 20, 21]. In plants, the technique has been applied to COOLAIR [15]. Xist has also been studied extensively [22, 23]. Also computational techniques trained on chemical probing data for long noncoding RNAs have been implemented [24]. Recent evidence has suggested that long noncoding RNAs guide ARGONAUTE4 (AGO4) to chromatin in RNA-directed DNA methylation (RdDM) in plants [25].

3.1 Shotgun Secondary Structure Determination (3S) of Plant Long Noncoding RNAs

1. To determine the secondary structure of lncRNA molecules, we follow studies used for the 16S rRNA secondary structure and the riboswitches (Fig. 1). To determine nucleotides that are highly mobile and likely to reside in looping regions, we perform chemical probing experiments, adding reagents that are highly reactive to these nucleotides, but not nucleotides sequestered in base pairs (e.g., nucleotides inside RNA double helices). Nucleotides with low mobility are likely to participate in Watson-Crick base pairs.
2. For SHAPE probing, folded RNA is probed using 1M7. Parallel RNA samples are treated with DMSO as a blank. For CMCT analysis, 1-cyclohexyl-(2-morpholinoethyl) carbodiimide metho-p-toluenesulfonate (Sigma-Aldrich) is added to 50 mM. Both are reacted for 5 min at 22 °C and precipitated. The modified sites of RNA are analyzed by reverse transcription using site-specific 50-fluorophore-labeled primers and SuperScript III reverse transcriptase (Life Technologies). The samples, supplemented with the dideoxy terminate sequencing products of Cy3-labeled primer extension, are denatured and loaded on an ABI PRISM 3100-Avant genetic analyzer.
3. To manage the large RNA size, we employ 3S (shotgun secondary structure) [17, 26]. Here, the entire RNA is probed. This is followed by subsequent rounds of probing on smaller regions of the RNA.
4. We identify modular sub-domains by identifying regions where the signals of short segments match the signals of the full RNA experiments.
5. The resulting secondary structure can be used to improve existing phylogenetic sequence alignments and, in principle, can be used to find instances of the lncRNA not previously found in other species. In our studies, we begin with alignments generated by genome browser, or alignments using synteny. Then we use the initial secondary structure to improve these sequence alignments, focusing on alignment of helical regions. Covariance analysis helps to validate each helix. Next, we use the helices with the most covariant base pairs to further improve the sequence alignment. This process is performed iteratively, with improved or validated helices enabling improved sequence alignments and improved sequence alignments enabling more accurate covariant measures.

4 Notes

1. Because many proteins may bind to the RNA in vivo and possibly obfuscate the chemical probing signal, it is important to first perform probing experiments in vitro, establishing the naked RNA secondary structure as a baseline.
2. Various software packages can be used to fold each sub-domain, using chemical reactivity data as constraints on the RNA fold. Phylogenetic data can be used to improve the secondary structure.

Acknowledgment

This work was supported by the National Institutes of Health Grant.

References

1. Chekanova JA (2015) Long non-coding RNAs and their functions in plants. Curr Opin Plant Biol 27:207–216
2. Di C, Yuan J, Wu Y, Li J, Lin H, Hu L, Zhang T, Qi Y, Gerstein MB, Guo Y, Lu ZJ (2014) Characterization of stress-responsive lncRNAs in Arabidopsis thaliana by integrating expression, epigenetic and structural features. Plant J 80:848–861
3. Chen J, Quan M, Zhang D (2015) Genome-wide identification of novel long non-coding RNAs in Populus tomentosa tension wood, opposite wood and normal wood xylem by RNA-seq. Planta 241:125–143
4. Jin J, Liu J, Wang H, Wong L, Chua NH (2013) PLncDB: plant long non-coding RNA database. Bioinformatics 29:1068–1071
5. Xuan H, Zhang L, Liu X, Han G, Li J, Li X, Liu A, Liao M, Zhang S (2015) PLNlncRbase: a resource for experimentally identified lncRNAs in plants. Gene 573:328–332
6. Szczesniak MW, Rosikiewicz W, Makalowska I (2016) CANTATAdb: a collection of plant long non-coding RNAs. Plant Cell Physiol 57:e8
7. Ma X, Shao C, Jin Y, Wang H, Meng Y (2014) Long non-coding RNAs: a novel endogenous source for the generation of Dicer-like 1-dependent small RNAs in Arabidopsis thaliana. RNA Biol 11:373–390
8. Wu J, Okada T, Fukushima T, Tsudzuki T, Sugiura M, Yukawa Y (2012) A novel hypoxic stress-responsive long non-coding RNA transcribed by RNA polymerase III in Arabidopsis. RNA Biol 9:302–313
9. Yang H, Howard M, Dean C (2014) Antagonistic roles for H3K36me3 and H3K27me3 in the cold-induced epigenetic switch at Arabidopsis FLC. Curr Biol 24:1793–1797
10. Bastow R, Mylne JS, Lister C, Lippman Z, Martienssen RA, Dean C (2004) Vernalization requires epigenetic silencing of FLC by histone methylation. Nature 427:164–167
11. Swiezewski S, Liu F, Magusin A, Dean C (2009) Cold-induced silencing by long antisense transcripts of an Arabidopsis Polycomb target. Nature 462:799–802
12. Angel A, Song J, Dean C, Howard M (2011) A Polycomb-based switch underlying quantitative epigenetic memory. Nature 476:105–108
13. Coustham V, Li P, Strange A, Lister C, Song J, Dean C (2012) Quantitative modulation of polycomb silencing underlies natural variation in vernalization. Science 337:584–587
14. Ietswaart R, Wu Z, Dean C (2012) Flowering time control: another window to the connection between antisense RNA and chromatin. Trends Genet 28:445–453
15. Hawkes EJ, Hennelly SP, Novikova IV, Irwin JA, Dean C, Sanbonmatsu KY (2016) COOLAIR antisense RNAs form evolutionarily conserved elaborate secondary structures. Cell Rep 16:3087–3096
16. Xue Z, Hennelly S, Doyle B, Gulati AA, Novikova IV, Sanbonmatsu KY, Boyer LA (2016) A G-Rich motif in the lncRNA braveheart

interacts with a zinc-finger transcription factor to specify the cardiovascular lineage. Mol Cell 64:37–50

17. Novikova IV, Hennelly SP, Sanbonmatsu KY (2012) Structural architecture of the human long non-coding RNA, steroid receptor RNA activator. Nucleic Acids Res 40:5034–5051
18. Wan Y, Qu K, Zhang QC, Flynn RA, Manor O, Ouyang Z, Zhang J, Spitale RC, Snyder MP, Segal E, Chang HY (2014) Landscape and variation of RNA secondary structure across the human transcriptome. Nature 505:706–709
19. Sanbonmatsu KY (2016) Towards structural classification of long non-coding RNAs. Biochim Biophys Acta 1859:41–45
20. Lin Y, Schmidt BF, Bruchez MP, McManus CJ (2018) Structural analyses of NEAT1 lncRNAs suggest long-range RNA interactions that may contribute to paraspeckle architecture. Nucleic Acids Res 46(7):3742–3752
21. Somarowthu S, Legiewicz M, Chillon I, Marcia M, Liu F, Pyle AM (2015) HOTAIR forms an intricate and modular secondary structure. Mol Cell 58:353–361
22. Lu Z, Zhang QC, Lee B, Flynn RA, Smith MA, Robinson JT, Davidovich C, Gooding AR, Goodrich KJ, Mattick JS, Mesirov JP, Cech TR, Chang HY (2016) RNA duplex map in living cells reveals higher-order transcriptome structure. Cell 165:1267–1279
23. Chu C, Zhang QC, da Rocha ST, Flynn RA, Bharadwaj M, Calabrese JM, Magnuson T, Heard E, Chang HY (2015) Systematic discovery of Xist RNA binding proteins. Cell 161:404–416
24. Delli Ponti R, Marti S, Armaos A, Tartaglia GG (2017) A high-throughput approach to profile RNA structure. Nucleic Acids Res 45:e35
25. Lahmy S, Pontier D, Bies-Etheve N, Laudie M, Feng S, Jobet E, Hale CJ, Cooke R, Hakimi MA, Angelov D, Jacobsen SE, Lagrange T (2016) Evidence for ARGONAUTE4-DNA interactions in RNA-directed DNA methylation in plants. Genes Dev 30:2565–2570
26. Novikova IV, Dharap A, Hennelly SP, Sanbonmatsu KY (2013) 3S: shotgun secondary structure determination of long non-coding RNAs. Methods 63:170–177

Chapter 24

Transcriptome-Wide Mapping 5-Methylcytosine by m^5C RNA Immunoprecipitation Followed by Deep Sequencing in Plant

Xiaofeng Gu and Zhe Liang

Abstract

Transcriptome-wide mapping RNA modification is crucial to understand the distribution and function of RNA modifications. Here, we describe a protocol to transcriptome-wide mapping 5-methylcytosine (m^5C) in plant, by a RNA immunoprecipitation followed by deep sequencing (m^5C-RIP-seq) approach. The procedure includes RNA extraction, fragmentation, RNA immunoprecipitation, and library construction.

Key words RNA methylation, 5-Methylcytosine, m^5C, m^5C-RIP-seq, RNA immunoprecipitation, Plant

1 Introduction

Among over 100 distinct chemically modifications found in RNA, many of them are identified in eukaryotic messenger RNAs (mRNAs) [1] and include N^6-methyladenosine (m^6A), 5-methylcytosine (m^5C), N^1-methyladenosine (m^1A), 5-hydroxymethylcytosine (hm^5C), inosine, and pseudouridine [2, 3]. The recent advances in high-throughput sequencing techniques have provided potent tools for profiling of RNA modification in transcriptome-wide. Functional studies of m^6A, the most abundant internal modification of eukaryotic mRNA, in multicellular animals and plants have shed light on regulatory roles of mRNA modifications in eukaryotes [2–4]. We adapted the method of mapping m^6A in mammals [5], to transcriptome-wide profiling of m^5C sites in *Arabidopsis*, through m^5C-RIP-seq using an m^5C-specific antibody [6]. The high sensitivity and specificity of the m^5C antibody facilitate effective immunoprecipitation, thus allowing detection of low-abundance methylated RNAs without the requirement of extremely deep sequencing. Using this approach, we identified 6045 m^5C peaks in 4465 expressed genes in young seedlings of *Arabidopsis* and found that m^5C is enriched in coding sequences

Julia A. Chekanova and Hsiao-Lin V. Wang (eds.), *Plant Long Non-Coding RNAs: Methods and Protocols*,
Methods in Molecular Biology, vol. 1933, https://doi.org/10.1007/978-1-4939-9045-0_24,

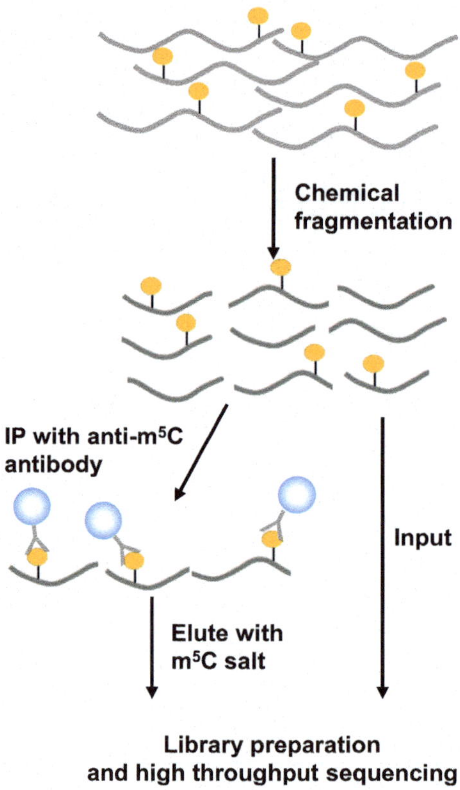

Fig. 1 Schematic diagram of the m^5C-RIP-seq procedure, modified from previous publication [6]

with two peaks located immediately after start codons and before stop codons; subsequently analysis revealed several novel features of m^5C in *Arabidopsis* mRNA [6]. The protocol could be used for mapping other RNA modifications in plant as well. The basic workflow of m^5C-RIP-seq is shown in Fig. 1. First, the RNA was chemically fragmented. Second, the fragmented RNA was subjected to RNA immunoprecipitation using a m^5C antibody. Third, sequence the RNA fragments enriched by an Illumina platform. Forth, identify the m^5C peaks and other downstream analysis. In this chapter, we will describe the protocol of m^5C-RIP-seq used for transcriptome-wide mapping m^5C RNA modification in plant.

2 Materials

All solutions and equipment used in this protocol should be RNase-free. The plant material must be harvested directly into liquid nitrogen (*see* **Note 1**).

2.1 Equipment

1. Eppendorf tubes (1.5 mL) and PCR tubes (200 μL).
2. Mortar and pestle.
3. Pipettes.
4. Pipette tips.
5. Head-over-tail rotator.
6. NanoDrop spectrophotometer.
7. Thermocycler machine.
8. Gel electrophoresis system.
9. Centrifuge.
10. Shaker.

2.2 Solutions and Chemicals

1. Liquid nitrogen.
2. RNeasy Plus Mini kit (Qiagen).
3. Dynabeads® mRNA Purification Kit (Ambion).
4. Terminator™ 5′-Phosphate-Dependent Exonuclease (Epicenter).
5. RNA Nano 6000 Assay Kit.
6. 0.5 M EDTA.
7. Anti-m^{5}C antibody (1 μg/μL, Diagenode) (*see* **Note 2**).
8. RNasin Plus RNase inhibitor (Promega).
9. Protein A/G Plus-Agarose (Santa Cruz).
10. Bovine serum albumin (BSA).
11. 5-Methylcytosine (Sigma-Aldrich).
12. Ethanol.
13. Sodium acetate (pH 5.2).
14. Tris–HCl.
15. $ZnCl_2$.
16. NaCl.
17. Igepal CA-630.
18. Glycogen.
19. Qubit RNA Assay Kit.
20. NEB next Ultra Directional RNA Library Prep Kit for Illumina (NEB).

2.3 Buffers

1. 10× RNA fragmentation buffer: 100 mM Tris–HCl, pH 7.0, and 100 mM $ZnCl_2$.
2. 10× IP buffer: 7.5 M NaCl, 5% Igepal CA-630, and 500 mM Tris–HCl, pH 7.4.
3. Elution buffer: 6.7 mM m^{5}C and 0.4 U/μL RNasin® plus RNase inhibitor in 1× IP buffer.

3 Methods

3.1 RNA Isolation and Fragmentation

1. Ground the plant material to a fine powder using mortar and pestle with sufficient liquid nitrogen. Extract total RNA using RNeasy Plus Mini kit according to manufacturer's instructions (*see* **Note 3**). We recommend to isolate more than 500 μg total RNA. The RNA concentration can be measured with Nano-Drop spectrophotometer (*see* **Note 4**).
2. Validate RNA integrity by agarose gel electrophoresis or analysis on an Agilent 2100 Bioanalyzer (*see* **Note 5**).
3. Adjust the RNA concentration to 1–1.5 μg/μL. Distribute 500 μL RNA into 20 PCR tubes (25 μL each tube), and add 2.78 μL 10× RNA fragmentation buffer to each tube. Incubate the PCR tubes for 5 min at 94 °C in a preheated thermocycler machine.
4. Take out the tubes and immediately add 2.78 μL of 0.5 M EDTA for each tube. Vortex and spin down the tubes and place them on ice (*see* **Note 6**).
5. Gather all the fragmented RNA into 2 Eppendorf tubes. Check the RNA size distribution by an agarose gel (*see* **Note** 7).
6. Add 1/10 volumes of 3 M sodium acetate (pH 5.2), glycogen (100 μg/mL final), and 2.5 volumes of 100% ethanol. Vortex and incubate 2 h or overnight at −80 °C.
7. Centrifuge the tubes at 12,000 for 30 min at 4 °C. Discard the supernatant (*see* **Note 8**), and wash the pellet with 1 mL of 75% ethanol and centrifuge at 13,523 × *g* for 15 min at 4 °C.
8. Air-dry the pellet. Elute the pellet in 150 μL of RNase-free water, and combine the RNA from two tubes into one (*see* **Note 4**).

3.2 RNA Immunoprecipitation

1. Save several microgram RNA from Subheading 3.1, **step 8** as input control. Dilute the restfragmented RNA with RNase-free water to 880 μL, and add 10 μL anti-m^5C monoclonal antibody and 100 μL 10× IP buffer supplemented with 10 μL RNasin® Plus RNase inhibitor, the total volume is 1 mL, mix and incubate with head-over-tail rotation for 2 h at 4 °C.
2. While the samples are incubating, wash 200 μL of protein A/G plus-agarose beads twice with 1 mL 1× IP buffer. Resuspend the beads with 1 mL 1× IP buffer supplemented with BSA (0.5 mg/ mL), and incubate with head-over-tail rotation for 2 h at 4 °C.
3. Spin down, discard the supernatant, and wash the beads twice with 1 mL 1× IP buffer.
4. Spin down and discard the supernatant.

5. Mix the IP reactions from Subheading 3.2, **step 1** and beads from Subheading 3.2, **step 4**, and incubate with head-over-tail rotation for 2 h at 4 °C.
6. Spin down the beads, and carefully remove and retain the supernatant (*see* **Note 9**). Wash the beads with 1 mL 1× IP buffer.
7. Repeat Subheading 3.2, **step 6** three times, and then spin down the beads and carefully remove and retain the supernatant.
8. Add 100 μL of elution buffer to the beads. Incubate the mixture for 1 h with continuous shaking at 4 °C.
9. Spin down the beads and carefully remove and retain the supernatant.
10. Add 100 μL of 1× IP buffer to the beads, and incubate the mixture for 5 min with continuous shaking at 4 °C. Spin down the beads and carefully remove and retain the supernatant.
11. Combine the supernatant from Subheadings 3.2, **steps 8** and **9**, and add 1/10 volumes of 3 M sodium acetate (pH 5.2) and 2.5 volumes of 100% ethanol. Mix and incubate overnight at −80 °C.
12. Centrifuge the tubes at 12,000 for 30 min at 4 °C. Discard the supernatant, wash the pellet with 500 μL of 75% ethanol, and centrifuge at 13,523 × *g* for 15 min at 4 °C.
13. Air-dry the pellet. Elute the pellet in 20 μL of RNase-free water (*see* **Note 4**).
14. Measure the RNA concentration by a Qubit RNA Assay Kit.

3.3 Library Construction

1. NEB Next Ultra Directional RNA Library Prep Kit for Illumina (NEB) was used to construct the libraries from 100 ng IP RNA and 100 ng input RNA, respectively, according to manufacturer's instructions (*see* **Note 10**).
2. Submit the libraries for high-throughput sequencing on the Illumina platform.

4 Notes

1. Plant materials can be stored at −80 °C until further use for up to 1 year.
2. The specificity and sensitivity of the antibody from Diagenode for RNA m^{5}C methylation was tested.
3. One option is to purify mRNA from total RNA using Dynabeads® mRNA Purification Kit according to the manufacturer's instructions. The remaining rRNA could be further digest by

Terminator™ 5′-Phosphate-Dependent Exonuclease according to the manufacturer's instructions.

4. RNA can be stored at −80 °C until further use for up to 1 year.
5. RNA quality and integrity is very important for the downstream experiment.
6. Work quickly at this stage.
7. The RNA fragments size should be ~100 nt. Repeat Subheadings 3.1, **steps 3** to **5** if the RNA was not sufficiently fragmented.
8. Be careful not to disrupt the RNA pellet.
9. The supernatant could be used for quality control.
10. Skip the fragmentation step because the RNA was already fragmented.

Acknowledgments

This work was supported by Ministry of Science and Technology of the People's Republic of China (2016YFD0101001) to X.G., National Natural Science Foundation of China (31671670) to X. G., and Recruitment program of Global Youth Expert of China to X.G.

References

1. Machnicka MA, Milanowska K, Osman O et al (2013) MODOMICS: a database of RNA modification pathways—2013 update. Nucleic Acids Res 41:262–267
2. Zhao BS, Roundtree IA, He C (2017) Post-transcriptional gene regulation by mRNA modifications. Nat Rev Mol Cell Biol 18(1):31–42
3. Gilbert WV, Bell TA, Schaening C (2016) Messenger RNA modifications: form, distribution, and function. Science 352(6292):1408–1412
4. Shen LS, Liang Z, Gu XF et al (2016) N^6-Methyladenosine RNA modification regulates shoot stem cell fate in *Arabidopsis*. Dev Cell 38 (2):186–200
5. Dominissini D, Moshitch-Moshkovitz S, Salmon-Divon M, Amariglio N, Rechavi G (2013) Transcriptome-wide mapping of N^6-methyladenosine by m^6A-seq based on immunocapturing and massively parallel sequencing. Nat Protoc 8(1):176–189
6. Cui XA, Liang Z, Shen LS et al (2017) 5-Methylcytosine RNA methylation in *Arabidopsis thaliana*. Mol Plant 10(11):1387–1399

Part VI

Databases of Plant lncRNAs and How to Use Them

Chapter 25

A Walkthrough to the Use of GreeNC: The Plant lncRNA Database

Andreu Paytuvi-Gallart, Walter Sanseverino, and Riccardo Aiese Cigliano

Abstract

Experimentally validated plant lncRNAs have been shown to regulate important agronomic traits such as phosphate starvation response, flowering time, and interaction with symbiotic organisms, making them of great interest in plant biology and in breeding. We developed a pipeline to annotate lncRNAs and applied it to 37 plant species and 6 algae, resulting in the annotation of more than 120,000 lncRNAs. To facilitate the study of lncRNAs for the plant research community, the information gathered is organized in the Green Non-Coding Database (GreeNC, http://greenc.sciencedesigners.com/). This chapter contains a detailed explanation of the content of GreeNC and how to access both programmatically and with a web browser.

Key words lncRNAs, Plants, Annotation, Pre-miRNA, Folding energy, lncRNAs database, Plant lncRNAs database, Database, Repository

1 Introduction

The Encyclopedia of DNA Elements (ENCODE) was launched by the US National Human Genome Research Institute (NHGRI) in September 2003. The aim was to uncover the role of the noncoding regions of the human genome, concluding that 80.4% of the human genome participated in at least one biochemical RNA or chromatin-associated event [1]. Other publications reinforce the idea that a great part of the genome has at least some kind of function; for instance, it has been reported that about 80% of *Homo sapiens*' genome is transcribed [2]. Today, long noncoding RNAs (lncRNAs) are being increasingly studied as they might have an important contribution to this amount of expression.

In plants, very few lncRNAs have been functionally characterized with some exceptions. For instance, IPS1 is a lncRNA expressed in *Arabidopsis thaliana* upon phosphate starvation, and it is thought to counteract the activity of miR399 on PHO2, which in turn regulates the expression of phosphate transporter genes

Julia A. Chekanova and Hsiao-Lin V. Wang (eds.), *Plant Long Non-Coding RNAs: Methods and Protocols*,
Methods in Molecular Biology, vol. 1933, https://doi.org/10.1007/978-1-4939-9045-0_25,

[3]. It has been shown that the lncRNA COLDAIR recruits the histone methylase PRC2 to interact with the PRC2 complex, so maintaining a stable silenced state of FLC to repress flowering during vernalization [4]. COOLAIR and ASL are other *Arabidopsis* lncRNAs that regulate the FLC expression [5, 6]. In rice, the lncRNA LDMAR has been found to control photosensitive male sterility by regulating DNA methylation levels in the promoter region of LDMAR [7]. Finally, in *Medicago truncatula*, the lncRNA Enod40 has been shown to participate in establishing symbiotic interactions with soil bacteria by affecting nodule formation [8]. These findings highlight the potential interest of lncRNAs in plant biology and in regulating important agronomic traits.

To further our knowledge of lncRNAs in plant biology, their comprehensive annotation is very important. Recently, several genome-wide studies have been performed in several plant species [9–17], so there is a need to store this information and provide it to the researchers. Many lncRNA databases exist, but most of them are focused on few specific species or with few number of annotated lncRNAs [18–23].

We created the Green Non-Coding Database (GreeNC), which is a database that provides information on the sequence, genome position, coding potential, and folding energies of >200,000 lncRNAs. Its aim is to be the most comprehensive database of lncRNAs thus becoming a meeting point for the plant lncRNA research. The lncRNAs annotated in GreeNC come from an in-house made pipeline from official genome annotations. Below in Subheading 3, we outlined the database of GreeNC and guide the readers how to navigate through it.

2 Materials

2.1 Data Source

The transcriptomes of the analyzed species in FASTA format were downloaded from Phytozome [24]. Only the genomes available for genomic studies according to the restriction of data usage were used [25–65].

2.2 Identification of lncRNAs

Two bash scripts were written to identify lncRNAs (*see* **Note 1**), followed by dividing the lncRNAs into high- and low-confidence groups (*see* Subheadings 2.2.1, 2.2.2, and 2.2.3).

2.2.1 First Script: Identification of Putative lncRNA

The first script followed the approach developed at the McGinnis lab [9] to identify lncRNAs in transcriptomes and it is based on identifying the coding potential of each transcript and on similarity with known proteins [11] (Fig. 1).

The script retains transcripts longer than 200 nt and with an ORF shorter than 120 aa by using Ugene (1.13) (http://ugene.unipro.ru/). Sequences were then blasted (blastx, 2.2.28+) (ftp://

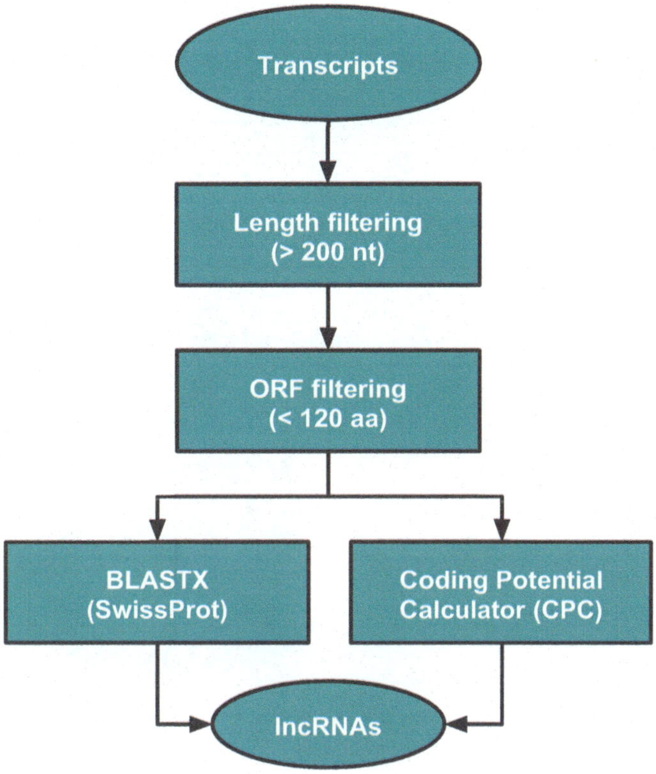

Fig. 1 Overview of the pipeline that identifies lncRNAs

ftp.ncbi.nlm.nih.gov/blast/executables/LATEST) against SwissProt (2013/11) [66]. CPC (0.9-r2) (http://cpc.cbi.pku.edu.cn/) [67] was also used, with the FrameFinder parameter -r set to "True" or "False" and the BLASTX parameter -S set to "3" or "1," depending on the group of transcripts being analyzed.

2.2.2 Second Script: Discrimination of Other Noncoding Transcripts from lncRNAs

The second script was written to discriminate other noncoding transcripts from lncRNAs and to identify possible miRNA precursors (Fig. 2).

Transcripts were analyzed by cmscan (Infernal 1.1rc4) against the RFAM database (release 11) [68]. In addition, BLASTn (2.2.28+) was used against a database of mature plant miRNAs from miRBase (release 20) [69], and the putative miRNA coordinates were validated by MIReNA (v2.0) (http://www.lcqb.upmc.fr/mirena/). Finally, MIReNA was called again, using the parameters –valid, –x, –mfei –0.69, –amfe –32, –ratiomin 0.83, and –ratiomax 1.17.

2.2.3 Benchmarks

The first script for the annotation of lncRNAs was tested with 480 lncRNAs and 1268 coding genes annotated in *Arabidopsis thaliana* (TAIR10) resulting in a sensitivity of 92% and a specificity

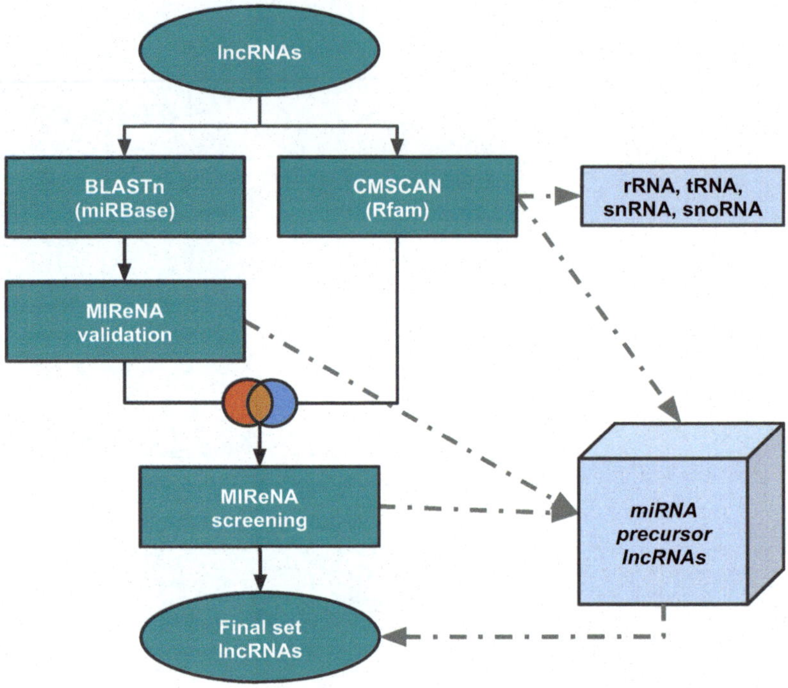

Fig. 2 Overview of the pipeline that discriminates other noncoding transcripts from long noncoding transcripts

of 94.95%. The second script was tested with 480 lncRNAs annotated in *Arabidopsis thaliana* (TAIR10) resulting in a sensitivity of 93% and a specificity of 97.6%.

2.2.4 Classification of lncRNAs into High- and Low-Confidence Groups

The final set of lncRNAs was divided into high-confidence and low-confidence. The transcripts without hits in BLASTX and described as noncoding by CPC, and considered non-precursors of miRNA, were classified as high-confidence lncRNAs. Transcripts without hits in the BLASTx step but described as coding by CPC, and transcripts with hits in the BLASTx step but described as noncoding by CPC, were considered low-confidence lncRNAs, as well as the transcripts identified as putative precursors of miRNAs. Transcripts having predicted repetitive regions by RepeatMasker (http://www.repeatmasker.org/) were also classified as low-confidence in order to exclude putative transposons.

2.3 Annotation of Repetitive Elements

RepeatMasker (open-4.0.5) (http://www.repeatmasker.org/) was used for repetitive element identification with the parameters: -species Viridiplantae, −no_is, −gff, and -nolow. The search engine used was RMBLAST (2.2.23+) against the RepBase database (released: 31 January 2014) [70].

2.4 Database Structure

Data was imported into a MySQL (5.5)-based relational database stored in an Ubuntu server (14.04). This database was then integrated into a MediaWiki by mapping relational data fields against wiki predefined templates via Semantic MediaWiki. Using templates makes it easy to print information and style it for different page types (e.g., genes and species). The template approach exposes the fields which may be queried, enhancing the search possibilities of the site. All transcript sequences were kept in a FASTA file with the same IDs as in the MySQL and then formatted using NCBI makeblastdb. In this way, sequences can be retrieved using their ID with blastdbcmd, and, at the same time, other BLAST programs can be run against the resulting BLAST database. Taking advantage of this, an Express Node.js API web service was created to expose both sequence retrieval and BLAST searches via client JavaScript from the MediaWiki interface.

3 Methods

GreeNC includes approximately 170,000 gene pages with information on more than 200,000 transcripts (63% classified as high-confidence lncRNA) from 39 plants and 6 algae. All information can be accessed through a graphical interface using any web browser or can be programmatically accessed via RESTful API.

3.1 Graphical Interface

3.1.1 Main Page

GreeNC is available at the following link: http://greenc.sciencedesigners.com. This address brings the user to the main page of the database, which can be divided into four different parts (*see* below **steps 1–4**):

1. Navigation bar: There is a black bar at the top of the web page with two drop-down menus and a search box (*see* **Note 2**). The first drop-down menu is called navigation and allows the access to other sections of the database. The second drop-down menu lists every available species in GreeNC. Finally, the search box allows the search of any species or gene (Fig. 3).
2. Species panel: This panel contains a picture for each available species in order to access to any of them in a fast and visual way (Fig. 4).
3. Miscellaneous panel: This panel stores a general description of the database. Below this description there are four buttons that allow fast access to other sections of the database, such as BLAST, advanced search, frequently asked questions (FAQ), or a page to contact the authors. To the right part of this panel, there is a list of news created by the maintainers (Fig. 5).

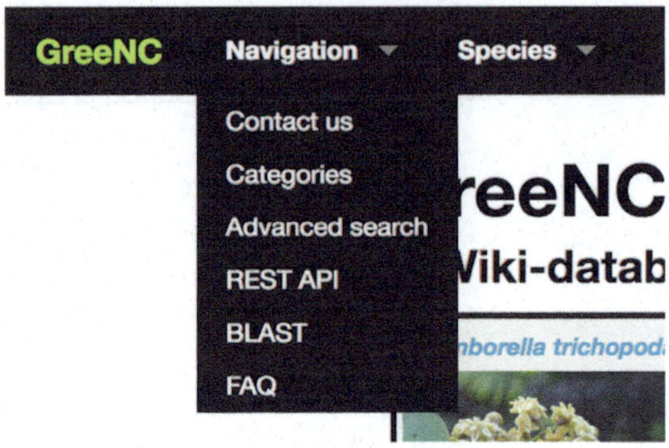

Fig. 3 Main navigation bar and its content

Fig. 4 Species navigation panel

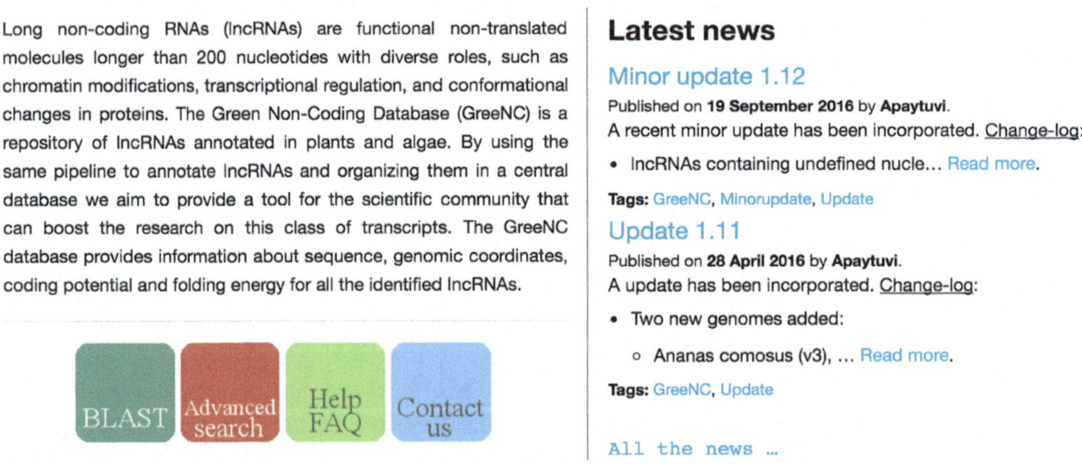

Fig. 5 Miscellaneous navigation panel to access to BLAST, Advanced search, Help, and Contact pages

4. Statistics panel: This panel shows a table with statistics about the genome assembly version, the number of genes, and of lncRNAs per species (Fig. 6).

Statistics

Species	Assembly	Gene number	lncRNAs	High confidence	Low confidence	Repetitive elements	miRNA precursors
Amborella_trichopoda	v1.0	5698	5698	4156	1542	103	347
Ananas_comosus	v3	3376	3376	1674	1702	163	935
Arabidopsis_lyrata	v1.0	4363	4363	2622	1741	158	60

Fig. 6 Preview of the species general statistics including genome assembly version, number of lncRNA genes, lncRNA transcripts, high- and low-confidence lncRNA transcripts, repetitive elements, and pre-miRNAs

Species

Arabidopsis thaliana

This release is based on the genome version **TAIR10** (Phytozome internal code **167**)
Arabidopsis thaliana, Arabidopsis thaliana (L.) Heynh., Arabidopsis thaliana (thale cress), Arabidopsis_thaliana, Arbisopsis thaliana, mouse-ear cress, thale cress, thale-cress

- NCBI Taxonomy
- Download lncRNAs into a FASTA file

Fig. 7 Screenshot of the species description panel

Genes

Gene alias	Chrom.	Start	End	Num. of lncRNAs
Athaliana AT1G01046	Chr1	28500	28706	1
Athaliana AT1G01170	Chr1	73931	74737	2
Athaliana AT1G01448	Chr1	163376	166344	3
Athaliana AT1G01471	Chr1	173251	173466	1
Athaliana AT1G01725	Chr1	269792	270775	1

Fig. 8 Preview of the species gene content showing coordinates of lncRNA loci and the number of transcribed lncRNAs

3.1.2 Species Page

From the navigation bar (second drop-down menu) or from the species panel, it is possible to access to a species page. GreeNC is hierarchically organized into species pages, and under the species page, anyone can access to gene pages from the corresponding species. The species page contains two different sections (*see* below **steps 1** and **2**).

1. Species title and description: This section contains the name of the species with its associated picture, synonyms of the species name, information about the used genome version, and links that point to the corresponding NCBI taxonomy page and to the FASTA file of the lncRNAs for that species (Fig. 7) (*see* **Note 3**).
2. Gene list: This section contains a table showing genes that transcribe lncRNAs. This table also contains the chromosome, start and end positions of the gene and the number of lncRNAs it transcribes (Fig. 8).

3.1.3 Gene Page

The gene page contains four sections (*see* below **steps 1–4**):

1. Gene information: A table shows the gene name and alias, its coordinates, the database source and assembly, the species it comes from, and whether this gene also transcribes coding transcripts or not (Fig. 9).
2. Transcript features: A table shows all lncRNAs the gene transcribes. Each row is relative to a lncRNA and displays its confidence (high or low), whether the lncRNA might be a miRNA precursor or not, its length, its sequence, and other features such as its ORF, coding potential, folding energies, or GC content (Fig. 10).
3. Matches to external databases: This table displays whether the lncRNAs have matches to miRBase, Rfam, Swissprot, RepBase, or NONCODE. It includes the database version, the hit name linked to the corresponding web page in the reference database, and the e-value (Fig. 11).

Gene information

Gene alias / *name*	Chrom.	Start	End	Source	Assembly	Species	Details
Athaliana_AT1G01046 *AT1G01046*	Chr1	28500	28706	Phytozome	TAIR10	*Arabidopsis_thaliana*	Coding?

Fig. 9 Preview of the Gene Content table, showing locus coordinates, genome assembly source and version, the species, and additional details

Transcript features

Transcript alias / *name*	Confidence	Pre-miRNA	Length	Sequence	Details
Athaliana_AT1G01046.1 AT1G01046.1	Low	Yes	207	TACTTTTCTAATATCACGAGGACTTACATGGCCTCAAGTCACCTGTGGTGTTGTGCAAGAAGGAGAAGCAAAGTCTGTCT ATGTATTATGAGATAGCTACTTCTATGGCTAGGATATATGTTGTACAAGACCGGCTTTTCTTCTACTTCTTGCACAACCT GAGGTTATTGAGGCTATACAAGTCTTCTTCTATAATGTTATTTATTA	Get ORF Coding potential... Folding energies... GC content

Fig. 10 Preview of the Transcripts Feature table, showing the type of lncRNA (high-/low-confidence), whether it is a pre-miRNA, transcript length, sequence, and additional details

Matches to external databases

Transcript alias	Database	Version	Hit	Evalue
Athaliana_AT1G01046.1	miRBase	20.0	ath-miR838	1.0e-5
Athaliana_AT1G01046.1	miRBase	20.0	aly-miR838-3p	5.0e-4
Athaliana_AT1G01046.1	miRBase	20.0	aly-miR838-5p	0.002
Athaliana_AT1G01046.1	Rfam	11.0	MIR164	1.3e-6

Fig. 11 Preview of the Matches to External Databases table showing any match to external databases such as miRBase or RFAM

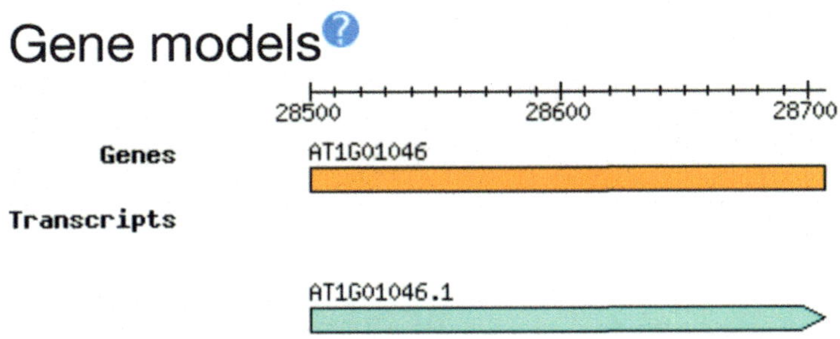

Fig. 12 Gene model representation containing from top to bottom: genomic coordinates, gene structure (in orange), and transcripts. Transcripts can be colored as magenta (coding transcripts), green (high-confidence lncRNAs), cyan (low-confidence lncRNAs), or salmon (other ncRNAs)

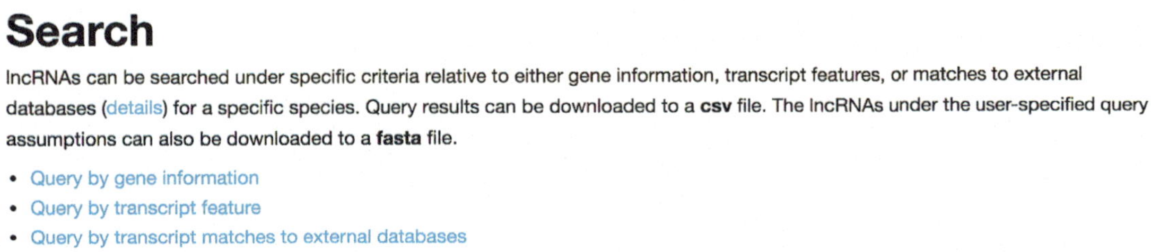

Fig. 13 Advanced search page. This page displays the three query options available: (1) query by gene information, (2) query by transcript feature, and (3) query by transcript matches to external databases

4. Gene model: This section shows a picture of the gene model – The axis that shows the coordinates of the gene, the gene feature, and all transcripts being associated with it. Coding transcripts are drawn in magenta, while high-confidence lncRNAs are drawn in green and low-confidence lncRNAs are drawn in cyan (Fig. 12).

3.1.4 Advanced Search Page

The *Advanced search* section can be accessed via the *Miscellaneous panel* from the main page of from the first drop-down menu in the navigation bar (*see* **Note 4**). There are three different query options to choose in this page (*see* below **steps 1–3**) (Fig. 13).

1. Query by gene information: This query option allows the user to filter the lncRNAs by species, coordinates (*see* **Notes 5** and **6**), and whether the genes transcribing lncRNAs also transcribe coding transcripts. The data can be also downloaded in a CSV table by selecting csv in the output option (Fig. 14).
2. Query by transcript feature: This query option allows the user to filter the lncRNAs by species, confidence, length, coding potential, folding energies (AMFE and MFEI), or GC content. The data can be also downloaded in a CSV table by selecting csv in the output option (Fig. 15) (*see* **Note 7**).

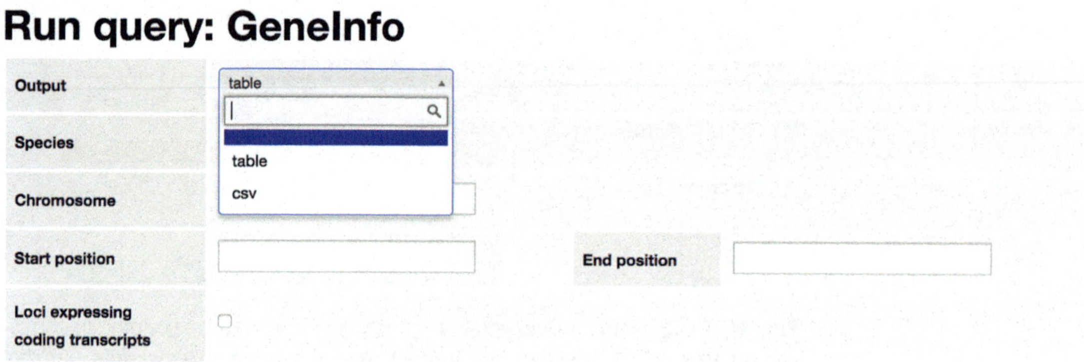

Fig. 14 Advanced Gene search page. Output format can be selected, and then genes can be queried according to coordinates or for the ability to express protein coding transcripts

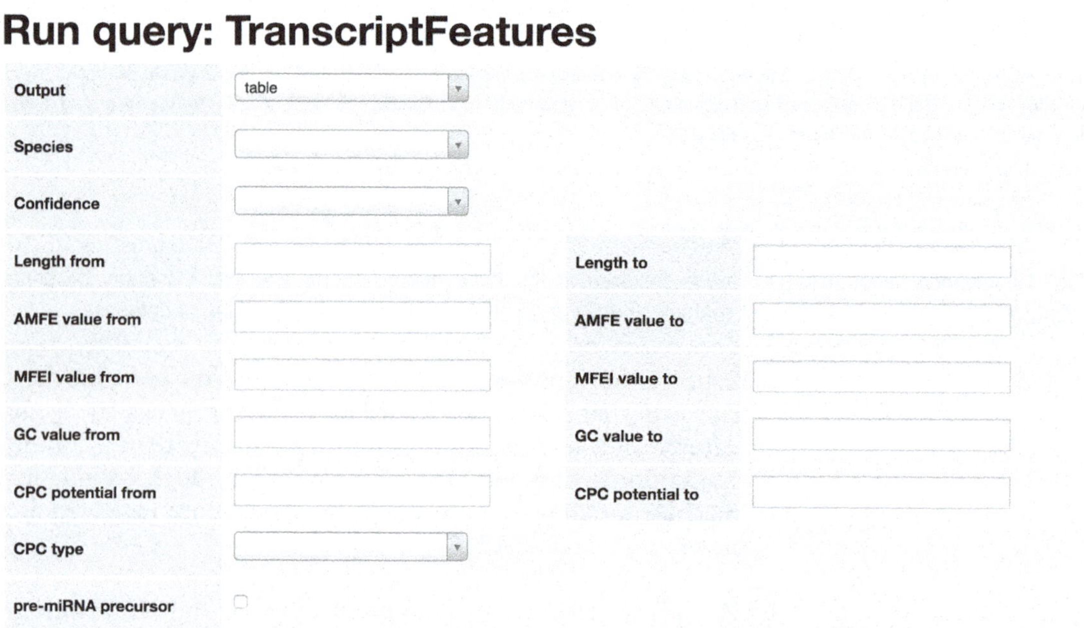

Fig. 15 Advanced Transcript search page. Output format can be selected, and then transcripts can be queried according to several parameters including species, confidence (high- or low-), transcript length, physical/chemical parameters, or coding potential

3. Query by transcript matches to external databases: This query option allows the user to filter the lncRNAs by species, hit to database, hit name, and e-value. The data can be also downloaded in a CSV table by selecting csv in the output option (Fig. 16) (*see* **Note 7**).

3.1.5 Miscellaneous Panel

The user might also be interested in searching lncRNAs based on the similarity of some sequences. For this reason, we also made available a BLAST page where the user is able to BLASTn some

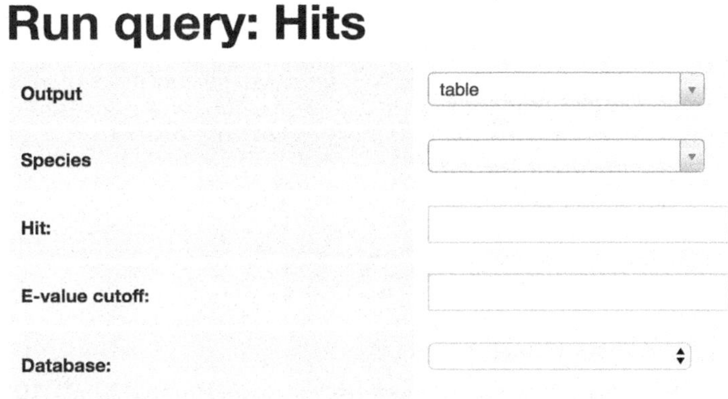

Fig. 16 Advanced Transcript search page based on matches to external databases. Output format can be selected, and then transcripts can be queried for keywords, e-value cutoff, and database

sequence against the whole GreeNC database. The BLAST section can be accessed via the *Miscellaneous panel* from the main page of from the first drop-down menu in the navigation bar (Fig. 17) (*see* **Note 8**).

3.2 Programmatic Access

The graphical interface is always a nice way to see the results. However, it fails when there is a need to retrieve the information through programming scripts. GreeNC incorporates a RESTful API that provides the information via HTTP GET requests (Fig. 18) (*see* **Note 9**).

It can be accessed under /api/ location. It contains three different resources (*see* Subheadings 3.2.1, 3.2.2, and 3.2.3).

3.2.1 /db/: Access to the GreeNC lncRNA Sequence Database

The "db" function shows the list of the available BLAST databases in GreeNC:

1. We only have a unique database currently containing all lncRNAs and its name is *greenc*.

```
$ curl http://greenc.sciencedesigners.com/api/db/
{"nucl":{"greenc":{"path":"/home/ubuntu/db/lncrna/
lncRNAall.fa"}}}
```

2. Once we have chosen the database to use, the sequences of specific lncRNAs can be retrieved by adding the database name after /db/ followed by /entry/, the transcript alias, and /fasta (for instance /db/greenc/entry/Athaliana_AT1G01170.1/fasta).

```
$ curl http://greenc.sciencedesigners.com/api/db/greenc/
entry/Athaliana_AT1G01170.1/fasta
```

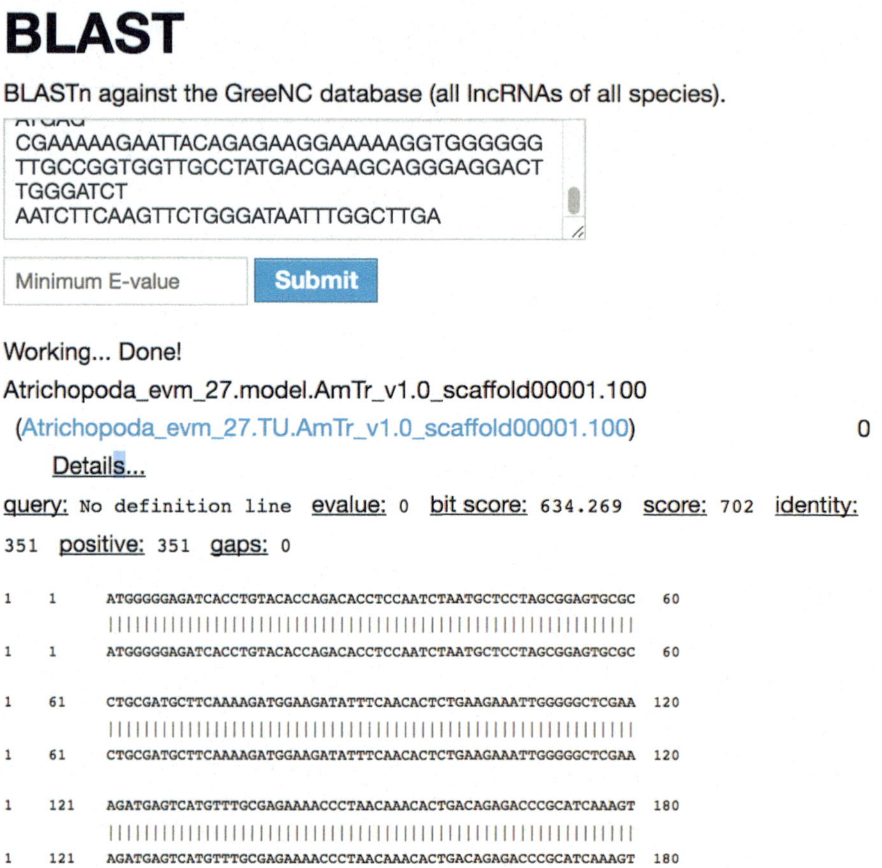

Fig. 17 BLAST page with an example of results. It displays the basic information about the alignments (hit linked with its corresponding page, e-value, bit score, matches, gaps, and the corresponding alignment)

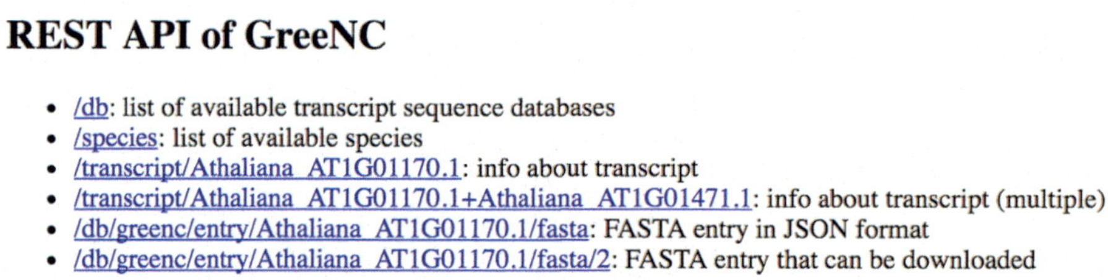

Fig. 18 REST API main page with the available resources to retrieve information

{"def":">lcl|Athaliana_AT1G01170.1:1-515","seq":"AC-
GACCGTCTTCCACCGTTGAATTCTTCTGGAACTGGAGTCCACTGTTTAAGCTT-
CACTGTCTCTGAATCGGCAAAGCTT\nTAGAAGAAAATGGCATCAGGAGGTAAAGC-
CAAGTACATAATCGGTGCTCTCATCGGTTCTTTCGGAATCTCATACATCTT\nCGA-
CAAAGTTATCTCTGATAATAAGATCTTTGGAGGGACTACTCCAGGAACTGTCTC-
TAACAAAGAATGGTGGGCAGCAA\nCGGATGAGAAATTCCAAGCATGGCCAA-
GAACCGCTGGTCCTCCCGTTGTTATGAATCCCATTAGCCGTCAGAATTTCATC
\nGTCAAGACTCGTCCGGAATGAGAAAATAATAAGTTCAATGCTTTGATTTTCA-

GAATAAGATGAACGATGACGATGTTTTC\nTAAATCCGAGCTTGTACTAAATAA-
CAATACATTACAACACGGTTTGCGGAACTACTCCACAGTCTATCTTCTGTTAAAAA
\nACTCAAACAAGCTATTGCAAAAAGCCCTTACGAGA"}

3. The output is in JSON format and the sequence contains breaklines every 80 bases. FASTA format can be also retrieved by adding /2 at the end of the URL.

```
$ curl http://greenc.sciencedesigners.com/api/db/greenc/
entry/Athaliana_AT1G01170.1/fasta/2
>lcl|Athaliana_AT1G01170.1:1-515
ACGACCGTCTTCCACCGTTGAATTCTTCTGGAACTGGAGTCCACTGTTTAAGCTT-
CACTGTCTCTGAATCGGCAAAGCTT
TAGAAGAAAATGGCATCAGGAGGTAAAGCCAAGTACATAATCGGTGCTCT-
CATCGGTTCTTTCGGAATCTCATACATCTT
CGACAAAGTTATCTCTGATAATAAGATCTTTGGAGGGACTACTCCAG-
GAACTGTCTCTAACAAAGAATGGTGGGCAGCAA
CGGATGAGAAATTCCAAGCATGGCCAAGAACCGCTGGTCCTCCCGTTGTTAT-
GAATCCCATTAGCCGTCAGAATTTCATC
GTCAAGACTCGTCCGGAATGAGAAAATAATAAGTTCAATGCTTTGATTTTCAGAA-
TAAGATGAACGATGACGATGTTTTC
TAAATCCGAGCTTGTACTAAATAACAATACATTACAACACGGTTTGCGGAAC-
TACTCCACAGTCTATCTTCTGTTAAAAA
ACTCAAACAAGCTATTGCAAAAAGCCCTTACGAGA
```

3.2.2 /species/: Available Species

This function provides all species that GreeNC stores in a list/array.

```
$ curl http://greenc.sciencedesigners.com/api/species/
["Amborella_trichopoda","Ananas_comosus","Arabidopsis_lyra-
ta","Arabidopsis_thaliana","Brachypodium_distachyon","Capsel-
la_grandiflora","Capsella_rubella","Carica_papaya","Chlamydo-
monas_reinhardtii","Citrus_clementina","Citrus_sinensis","-
Coccomyxa_subellipsoidea_C-169","Cucumis_sativus","Eucalyp-
tus_grandis","Eutrema_salsugineum","Fragaria_vesca","Glyci-
ne_max","Gossypium_raimondii","Linum_usitatissimum","Malus_-
domestica","Manihot_esculenta","Medicago_truncatula","Micro-
monas_pusilla_CCMP1545","Micromonas_pusilla_RCC299","Mimu-
lus_guttatus","Musa_acuminata","Oryza_sativa_Japonica_Grou-
p","Ostreococcus_lucimarinus","Phaseolus_vulgaris","Physcomi-
trella_patens","Populus_trichocarpa","Prunus_persica","Rici-
nus_communis","Selaginella_moellendorffii","Setaria_itali-
ca","Solanum_lycopersicum","Solanum_tuberosum","Sorghum_bico-
lor","Spirodela_polyrhiza","Theobroma_cacao","Triticum_aesti-
vum","Vitis_vinifera","Volvox_carteri","Zea_mays","Zostera_-
marina"]
```

3.2.3 /transcript/: Transcript Information

This function shows the transcript information for a transcript in JSON format. It is necessary to specify the transcript alias at the end or the URL. If information about more than one transcript needs to be retrieved, the transcript aliases need to be concatenated separated by "+" and placed at the end of the URL (for instance, Athaliana_AT1G01170.1 + Athaliana_AT1G01471.1).

```
$ curl http://greenc.sciencedesigners.com/api/transcript/
Athaliana_AT1G01170.1
```

[{"Athaliana_AT1G01170.1":{"transcript_name":"AT1G01170.1","-features":{"length":515,"cpc_type":"noncoding","cpc_potential":-0.756,"mfei":-21.573,"amfe":-0.519,"gc_content":41.553},"swissprot":{},"rfam":{},"repbase":{},"confidence":"High","gene_alias":"Athaliana_AT1G01170","gene_name":"AT1G01170","coord":{"chromosome":"Chr1","start":73931,"end":74737,"strand":"-","species":"Athaliana"}}}]

3.3 Future Updates

GreeNC will be evenly updated annually in order to add new sequences from other species and to update existing genome annotations. New information will also be made available, such as tissue-specific expression levels obtained from publicly available RNA-seq data, conservation across different species, and phylogeny. Also, the gene models in the gene pages will be redesigned by adding an HTML iframe that will show the gene model through a genome browser such as JBrowse.

4 Notes

1. All lncRNAs from GreeNC have been annotated in silico from reference transcripts using highly specific and sensitive in-house bioinformatics pipelines (*see* Subheading 2.2). Several methods exist to annotate lncRNAs both in plant and animal sequences; however, there is not a consensus method on which the scientific community agrees. Hence different genomes are often analyzed with different pipelines thus making difficult to perform comparisons across species. GreeNC tries to solve this problem by showing lncRNAs annotated with the same pipeline on many different species thus allowing to reduce the noise generated by the use of different pipelines.
2. The search box at the navigation bar (*see* Subheading 3.1.1, **step 1**) can be used to search and locate gene aliases and species names only.
3. All lncRNA sequences from a species can be downloaded in FASTA format: the user should go to the corresponding species page and click "Download lncRNAs into a FASTA file" (*see* Subheading 3.1.2).

4. All information about the lncRNAs from a species can be exported into a file. Whatever the query type selected in the Advanced search page (*see* Subheading 3.1.4), the user must select (1) the corresponding species in the species field and (2) the *csv* option in the output field. Afterward, click to "Run query" and "More results."
5. The "Query by gene information" page allows to select all the lncRNA genes within specific genomic coordinates, as indicated by the Chromosome, Start and End position. This function can be useful for those working on QTLs or which are focusing on broad genomic areas (*see* Subheading 3.1.4, **step 1**).
6. Gene aliases (*see* Subheading 3.1.4, **step 1**) are made of the abbreviated species name and the gene name in the assembly, separated by an underscore. Therefore, taking gene AT1G01046 as an example in *A. thaliana*, it has the GreeNC alias **Athaliana_AT1G01046**, which can be directly accessed via /wiki/Gene:**Athaliana_AT1G01046**. This gene alias can also be searched in the search box AT the navigation bar.
7. Querying by "transcript feature" and by "transcript matches to external databases" in the Advanced search page (*see* Subheading 3.1.4, **steps 2** and **3**) allows downloading the lncRNA sequences from those matching the query criteria. After clicking "Run query," click to "Download the lncRNAs under these query assumptions."
8. The default e-value in the BLAST section (*see* Subheading 3.1.5) is 10. We recommend using more stringent thresholds such as an e-value of 0.01 or lower to start considering the results significant.
9. The RESTful API is only suitable to get specific information (including sequences) about one or few lncRNAs (*see* Subheading 3.2).

Acknowledgments

We thank Dr. Antonio Hermoso Pulido from Centre for Genomic Regulation (CRG), who was also involved in the database development.

References

1. Encode Consortium, Carolina N, Hill C (2013) For junk DNA. Nature 489:57–74
2. Djebali S, Davis CA, Merkel A et al (2012) Landscape of transcription in human cells. Nature 489:101–108
3. Franco-Zorrilla JM, Valli A, Todesco M, Mateos I, Puga MI, Rubio-Somoza I, Leyva A, Weigel D, García JA, Paz-Ares J (2007) Target mimicry provides a new mechanism for regulation of microRNA activity. Nat Genet 39:1033–1037

4. Heo JB, Sung S (2011) Vernalization-mediated epigenetic silencing by a long intronic noncoding RNA. Science 331:76–79
5. Swiezewski S, Liu F, Magusin A, Dean C (2009) Cold-induced silencing by long antisense transcripts of an Arabidopsis Polycomb target. Nature 462:799–802
6. Shin JH, Chekanova JA (2014) Arabidopsis RRP6L1 and RRP6L2 function in FLOWERING LOCUS C silencing via regulation of antisense RNA synthesis. PLoS Genet 10: e1004612. https://doi.org/10.1371/journal.pgen.1004612
7. Ding J, Lu Q, Ouyang Y, Mao H, Zhang P, Yao J, Xu C, Li X, Xiao J, Zhang Q (2012) A long noncoding RNA regulates photoperiod-sensitive male sterility, an essential component of hybrid rice. Proc Natl Acad Sci 109:2654–2659
8. Campalans A (2004) Enod40, a short open Reading frame-containing mRNA, induces cytoplasmic localization of a nuclear RNA binding protein in Medicago truncatula. Plant Cell Online 16:1047–1059
9. Boerner S, McGinnis KM (2012) Computational identification and functional predictions of long noncoding RNA in Zea mays. PLoS One 7:e43047. https://doi.org/10.1371/journal.pone.0043047
10. Li L, Eichten SR, Shimizu R et al (2014) Genome-wide discovery and characterization of maize long non-coding RNAs. Genome Biol 15:R40. https://doi.org/10.1186/gb-2014-15-2-r40
11. Lu T, Zhu C, Lu G et al (2012) Strand-specific RNA-seq reveals widespread occurrence of novel cis-natural antisense transcripts in rice. BMC Genomics 13:721. https://doi.org/10.1186/1471-2164-13-721
12. Shuai P, Liang D, Tang S, Zhang Z, Ye CY, Su Y, Xia X, Yin W (2014) Genome-wide identification and functional prediction of novel and drought-responsive lincRNAs in Populus trichocarpa. J Exp Bot 65:4975–4983
13. Wen J, Parker BJ, Weiller GF (2007) In Silico identification and characterization of mRNA-like noncoding transcripts in Medicago truncatula. In: In Silico Biol, vol 7, pp 485–505
14. Xin M, Wang Y, Yao Y, Song N, Hu Z, Qin D, Xie C, Peng H, Ni Z, Sun Q (2011) Identification and characterization of wheat long non-protein coding RNAs responsive to powdery mildew infection and heat stress by using microarray analysis and SBS sequencing. BMC Plant Biol 11:61. https://doi.org/10.1186/1471-2229-11-61
15. Flórez-Zapata NMV, Reyes-Valdés MH, Martínez O (2016) Long non-coding RNAs are major contributors to transcriptome changes in sunflower meiocytes with different recombination rates. BMC Genomics 17:490. https://doi.org/10.1186/s12864-016-2776-1
16. Joshi RK, Megha S, Basu U, Rahman MH, Kav NNV (2016) Genome wide identification and functional prediction of long non-coding RNAs responsive to Sclerotinia sclerotiorum infection in Brassica napus. PLoS One 11: e0158784. https://doi.org/10.1371/journal.pone.0158784
17. Jain P, Sharma V, Dubey H, Singh PK, Kapoor R, Kumari M, Singh J (2017) Open access Identification of long non-coding RNA in rice lines resistant to Rice blast pathogen Maganaporthe oryzae. Bioinformation 13:249–255
18. Xie C, Yuan J, Li H, Li M, Zhao G, Bu D, Zhu W, Wu W, Chen R, Zhao Y (2014) NONCODEv4: exploring the world of long non-coding RNA genes. Nucleic Acids Res 42:D98. https://doi.org/10.1093/nar/gkt1222
19. Yi X, Zhang Z, Ling Y, Xu W, Su Z (2015) PNRD: a plant non-coding RNA database. Nucleic Acids Res 43:D982–D989
20. Jin J, Liu J, Wang H, Wong L, Chua NH (2013) PLncDB: plant long non-coding RNA database. Bioinformatics 29:1068–1071
21. Xuan H, Zhang L, Liu X, Han G, Li J, Li X, Liu A, Liao M, Zhang S (2015) PLNlncRbase: a resource for experimentally identified lncRNAs in plants. Gene 573:328–332
22. Quek XC, Thomson DW, Maag JLV, Bartonicek N, Signal B, Clark MB, Gloss BS, Dinger ME (2015) lncRNAdb v2.0: expanding the reference database for functional long non-coding RNAs. Nucleic Acids Res 43: D168–D173
23. Szcześniak MW, Rosikiewicz W, Makałowska I (2016) CANTATAdb: a collection of plant long non-coding RNAs. Plant Cell Physiol 57:e8
24. Goodstein DM, Shu S, Howson R et al (2012) Phytozome: a comparative platform for green plant genomics. Nucleic Acids Res 40:D1178. https://doi.org/10.1093/nar/gkr944
25. DePamphilis CW, Palmer JD, Rounsley S et al (2013) The Amborella genome and the evolution of flowering plants. Science 342:1241089
26. Hu TT, Pattyn P, Bakker EG et al (2011) The Arabidopsis lyrata genome sequence and the basis of rapid genome size change. Nat Genet 43:476–483

27. Lamesch P, Berardini TZ, Li D et al (2012) The Arabidopsis Information Resource (TAIR): improved gene annotation and new tools. Nucleic Acids Res 40:D1202. https://doi.org/10.1093/nar/gkr1090
28. Vogel JP, Garvin DF, Mockler TC et al (2010) Genome sequencing and analysis of the model grass Brachypodium distachyon. Nature 463:763–768
29. Slotte T, Hazzouri KM, Ågren JA et al (2013) The Capsella rubella genome and the genomic consequences of rapid mating system evolution. Nat Genet 45:831–835
30. Ming R, Hou S, Feng Y et al (2008) The draft genome of the transgenic tropical fruit tree papaya (Carica papaya Linnaeus). Nature 452:991–996
31. Merchant SS, Prochnik SE, Vallon O et al (2007) The Chlamydomonas genome reveals the evolution of key animal and plant functions. Science 318:245–251
32. Wu GA, Prochnik S, Jenkins J et al (2014) Sequencing of diverse mandarin, pummelo and orange genomes reveals complex history of admixture during citrus domestication. Nat Biotechnol 32:656–662
33. Blanc G, Agarkova I, Grimwood J et al (2012) The genome of the polar eukaryotic microalga Coccomyxa subellipsoidea reveals traits of cold adaptation. Genome Biol 13:R39. https://doi.org/10.1186/gb-2012-13-5-r39
34. Bartholomé J, Mandrou E, Mabiala A, Jenkins J, Nabihoudine I, Klopp C, Schmutz J, Plomion C, Gion JM (2015) High-resolution genetic maps of Eucalyptus improve Eucalyptus grandis genome assembly. New Phytol 206:1283–1296
35. Yang R, Jarvis DE, Chen H et al (2013) The reference genome of the halophytic plant Eutrema salsugineum. Front Plant Sci 4. https://doi.org/10.3389/fpls.2013.00046
36. Shulaev V, Sargent DJ, Crowhurst RN et al (2011) The genome of woodland strawberry (Fragaria vesca). Nat Genet 43:109–116
37. Schmutz J, McClean PE, Mamidi S et al (2014) A reference genome for common bean and genome-wide analysis of dual domestications. Nat Genet 46:707–713
38. Schmutz J, Cannon SB, Schlueter J et al (2010) Genome sequence of the palaeopolyploid soybean. Nature 463:178–183
39. Wang Z, Hobson N, Galindo L et al (2012) The genome of flax (Linum usitatissimum) assembled de novo from short shotgun sequence reads. Plant J 72:461–473
40. Velasco R, Zharkikh A, Affourtit J et al (2010) The genome of the domesticated apple (Malus × domestica Borkh.). Nat Genet 42:833–839
41. Prochnik S, Marri PR, Desany B et al (2012) The cassava genome: current Progress, future directions. Trop Plant Biol 5:88–94
42. Young ND, Debellé F, Oldroyd GED et al (2011) The Medicago genome provides insight into the evolution of rhizobial symbioses. Nature 480:520–524
43. Worden AZ, Lee JH, Mock T et al (2009) Green evolution and dynamic adaptations revealed by genomes of the marine picoeukaryotes micromonas. Science 324:268–272
44. Droc G, Larivière D, Guignon V et al (2013) The banana genome hub. Database 2013. https://doi.org/10.1093/database/bat035
45. Ouyang S, Zhu W, Hamilton J et al (2007) The TIGR rice genome annotation resource: improvements and new features. Nucleic Acids Res 35:D883. https://doi.org/10.1093/nar/gkl976
46. Palenik B, Grimwood J, Aerts A et al (2007) The tiny eukaryote Ostreococcus provides genomic insights into the paradox of plankton speciation. Proc Natl Acad Sci 104:7705–7710
47. Tuskan GA, DiFazio S, Jansson S et al (2006) The genome of black cottonwood, Populus trichocarpa (Torr. & Gray). Science 313:1596–1604
48. Paterson AH, Wendel JF, Gundlach H et al (2012) Repeated polyploidization of Gossypium genomes and the evolution of spinnable cotton fibres. Nature 492:423–427
49. Verde I, Abbott AG, Scalabrin S et al (2013) The high-quality draft genome of peach (Prunus persica) identifies unique patterns of genetic diversity, domestication and genome evolution. Nat Genet 45:487–494
50. Chan AP, Crabtree J, Zhao Q et al (2010) Draft genome sequence of the oilseed species Ricinus communis. Nat Biotechnol 28:951–956
51. Banks JA, Nishiyama T, Hasebe M et al (2011) The selaginella genome identifies genetic changes associated with the evolution of vascular plants. Science 332:960–963
52. Bennetzen JL, Schmutz J, Wang H et al (2012) Reference genome sequence of the model plant Setaria. Nat Biotechnol 30:555–561
53. Sato S, Tabata S, Hirakawa H et al (2012) The tomato genome sequence provides insights into fleshy fruit evolution. Nature 485:635–641

54. Xu X, Pan S, Cheng S et al (2011) Genome sequence and analysis of the tuber crop potato. Nature 475:189–195
55. Paterson AH, Bowers JE, Bruggmann R et al (2009) The Sorghum bicolor genome and the diversification of grasses. Nature 457:551–556
56. Wang W, Haberer G, Gundlach H et al (2014) The Spirodela polyrhiza genome reveals insights into its neotenous reduction fast growth and aquatic lifestyle. Nat Commun 5:3311. https://doi.org/10.1038/ncomms4311
57. Motamayor JC, Mockaitis K, Schmutz J et al (2013) The genome sequence of the most widely cultivated cacao type and its use to identify candidate genes regulating pod color. Genome Biol 14:r53. https://doi.org/10.1186/gb-2013-14-6-r53
58. Jaillon O, Aury JM, Noel B et al (2007) The grapevine genome sequence suggests ancestral hexaploidization in major angiosperm phyla. Nature 449:463–467
59. Prochnik SE, Umen J, Nedelcu AM et al (2010) Genomic analysis of organismal complexity in the multicellular green alga volvox carteri. Science 329:223–226
60. Schnable PS, Ware D, Fulton RS et al (2009) The B73 maize genome: complexity, diversity, and dynamics. Science 326:1112–1115
61. Huang S, Li R, Zhang Z et al (2009) The genome of the cucumber, Cucumis sativus L. Nat Genet 41:1275–1281
62. International Wheat Genome Sequencing Consortium (IWGSC) (2014) A chromosome-based draft sequence of the hexaploid bread wheat (Triticum aestivum) genome. Science 345:1251788
63. Hellsten U, Wright KM, Jenkins J, Shu S, Yuan Y, Wessler SR, Schmutz J, Willis JH, Rokhsar DS (2013) Fine-scale variation in meiotic recombination in Mimulus inferred from population shotgun sequencing. Proc Natl Acad Sci 110:19478–19482
64. Rensing SA, Lang D, Zimmer AD et al (2008) The Physcomitrella genome reveals evolutionary insights into the conquest of land by plants. Science 319:64–69
65. Zimmer AD, Lang D, Buchta K, Rombauts S, Nishiyama T, Hasebe M, Van de Peer Y, Rensing SA, Reski R (2013) Reannotation and extended community resources for the genome of the non-seed plant Physcomitrella patens provide insights into the evolution of plant gene structures and functions. BMC Genomics 14:498. https://doi.org/10.1186/1471-2164-14-498
66. Bateman A, Martin MJ, O'Donovan C et al (2015) UniProt: a hub for protein information. Nucleic Acids Res 43:D204–D212
67. Kong L, Zhang Y, Ye ZQ, Liu XQ, Zhao SQ, Wei L, Gao G (2007) CPC: assess the protein-coding potential of transcripts using sequence features and support vector machine. Nucleic Acids Res 35:W345. https://doi.org/10.1093/nar/gkm391
68. Nawrocki EP, Burge SW, Bateman A et al (2015) Rfam 12.0: updates to the RNA families database. Nucleic Acids Res 43:D130–D137
69. Griffiths-Jones S (2010) MiRBase: MicroRNA sequences and annotation. Curr Protoc Bioinformatics 34:1291–12910
70. Bao W, Kojima KK, Kohany O (2015) Repbase update, a database of repetitive elements in eukaryotic genomes. Mob DNA 6:11. https://doi.org/10.1186/s13100-015-0041-9

Chapter 26

CANTATAdb 2.0: Expanding the Collection of Plant Long Noncoding RNAs

Michał Wojciech Szcześniak, Oleksii Bryzghalov, Joanna Ciomborowska-Basheer, and Izabela Makałowska

Abstract

Long non-coding RNAs (lncRNAs) are a class of potent regulators of gene expression that are found in a wide array of eukaryotes; however, our knowledge about these molecules in plants is very limited. In particular, a number of plant species with important roles in biotechnology, agriculture and basic research still lack comprehensively identified and annotated sets of lncRNAs. To address these shortcomings, we previously created a database of lncRNAs in 10 model species, called CANTATAdb, and now we are expanding this online resource to encompass 39 species, including three algae. The lncRNAs were identified computationally using publicly available RNA sequencing (RNA-Seq) data. Expression values, coding potential calculations and other types of information were used to provide annotations for the identified lncRNAs. The data are freely available for searching, browsing and downloading from an online database called CANTATAdb 2.0 (http://cantata.amu.edu.pl, http://yeti.amu.edu.pl/CANTATA/).

Key words Database, Long noncoding RNAs, Plant RNAs, lncRNA identification

1 Introduction

lncRNAs are defined as transcripts longer than 200 bases that do not encode proteins. They represent a large portion of transcriptomes; for instance, there are 172,216 lncRNAs annotated in the human transcriptome at NONCODE 2016 [1], compared with 89,041 protein-coding transcripts at ENSEMBL 91 [2]. In plants, the numbers of known lncRNAs are at least an order of magnitude lower than those in animals, but lncRNAs still constitute an important component of plant transcriptomes. For example, in *Arabidopsis thaliana*, there are 3008 lncRNAs in GreeNC [3], 4761 in CantataDB 1.0 [4] and 3763 in NONCODE 2016 [1]. Plant lncRNAs have been associated with a variety of biological phenomena, such as vernalization [5], fertility [6], photomorphogenic processes [7], phosphate homeostasis [8], and nodule

Julia A. Chekanova and Hsiao-Lin V. Wang (eds.), *Plant Long Non-Coding RNAs: Methods and Protocols*, Methods in Molecular Biology, vol. 1933, https://doi.org/10.1007/978-1-4939-9045-0_26,

organogenesis [9]. They play these roles by modulating different steps of gene expression. First, some lncRNAs function as primary transcripts for small regulatory RNAs, such as microRNAs (miRNAs) and small interfering RNAs (siRNAs) [10]. For instance, in *Arabidopsis thaliana*, 24 nt siRNAs were shown to originate from at least five long noncoding transcripts: npc34, npc351, npc375, npc520, and npc523. Another example is target mimicry, during which interactions between miRNAs and their targets are hampered by a lncRNA molecule [11]. Plant lncRNAs are also involved in alternative splicing regulation, e.g., ASCO-lncRNA and the nuclear speckle RNA-binding (NSR) protein form an alternative splicing regulatory module, which has been linked to the development of lateral roots [12]. Moreover, lncRNAs participate in transcriptional regulation through chromatin remodeling [12] and mRNA translation modulation [13].

The abovementioned important roles played by plant lncRNAs contrast with their relatively poor annotation, which motivated us to perform a large-scale prediction of lncRNAs. To this end, we took advantage of publicly available sets of RNA sequencing (RNA-Seq) data, and using a custom computational pipeline for their prediction, we found 239,631 lncRNAs in 39 plant and algae species (Table 1). A search of the Ensembl Plants database indicated that only 9% of the identified lncRNAs corresponded to already known transcripts, as most of the species have no lncRNAs annotated at all. On the other hand, in a search of *Arabidopsis thaliana*, which has a relatively well-annotated genome, more than half of the candidates were identical to already known noncoding transcripts. The identified lncRNAs have been subject to basic annotation, which involved, among other steps, estimating their expression values and comparing them against databases of coding and noncoding sequences. The computational results are made available through an online database that we called CANTATAdb 2.0 (http://cantata.amu.edu.pl, http://yeti.amu.edu.pl/CANTATA/).

2 Materials

For most species, genome sequences as well as reference annotation data in GTF or GFF format (*see* **Note 1**) were retrieved from Ensembl Plants [14] using BioMart and a download page; for the remaining six species (*Cucumis sativus*, *Ananas comosus*, *Chenopodium quinoa*, *Malus domestica*, *Oryza sativa (Japonica group)*, and *Manihot esculenta*), the data came from NCBI. Details about the used genome and annotation versions are provided in Table 2 and at http://cantata.amu.edu.pl/download.php. Paired-end RNA--Seq reads were downloaded from the Sequence Read Archive [15], totaling 328 experiments (*see* Table 3 and http://cantata.amu.edu.pl/download.php).

Table 1
A summary of long non-coding RNAs and transcripts identified in 39 plant and algae species using RNA-Seq data and reference annotations from Ensembl Plants and NCBI

Species	No. of lncRNAs	No. of transcripts
Amborella trichopoda	5511	56,249
Ananas comosus	10,404	59,924
Arabidopsis lyrata	7593	60,381
Arabidopsis thaliana	4373	55,199
Brachypodium distachyon	4945	52,646
Brassica napus	12,010	148,644
Brassica oleracea	7338	82,803
Brassica rapa	8501	91,190
Chenopodium quinoa	17,526	220,408
Chlamydomonas reinhardtii	3425	37,010
Chondrus crispus	224	9623
Corchorus capsularis	6459	60,629
Cucumis sativus	7348	45,343
Galdieria sulphuraria	1917	12,135
Glycine max	3096	94,609
Hordeum vulgare	7970	146,853
Leersia perrieri	6402	68,790
Malus domestica	10,924	64,730
Manihot esculenta	9504	84,947
Medicago truncatula	3590	70,547
Musa acuminata	3001	61,213
Oryza barthii	7062	70,683
Oryza brachyantha	6004	61,514
Oryza nivara	8955	81,534
Oryza punctata	3459	58,395
Oryza rufipogon	10,261	86,459
Oryza sativa	2788	91,585
Physcomitrella patens	1498	53,265
Populus trichocarpa	4322	65,690
Prunus persica	2902	42,061
Selaginella moellendorffii	2267	50,231

(continued)

Table 1
(continued)

Species	No. of lncRNAs	No. of transcripts
Setaria italica	4208	61,521
Solanum lycopersicum	4716	53,395
Solanum tuberosum	5790	72,129
Sorghum bicolor	2600	53,317
Theobroma cacao	5256	56,895
Trifolium pratense	10,179	76,080
Vitis vinifera	4542	52,921
Zea mays	10,761	116,922

3 Methods

3.1 Data Population and Construction of the Database

3.1.1 Ab Initio Transcriptome Assembly

1. The RNA-Seq reads in FASTQ format (*see* **Note 1**) were subjected to quality filtering and adapter trimming with BBDuk v37.02 (https://jgi.doe.gov) with the following settings (*see* **Note 2**):

```
qtrim=w, trimq=20, maq=10, rref=bbmap/resources/adapters.fa, k=23, mink=11, hdist=1, tbo, tpe, minlength=50, removeifeitherbad=t.
```

2. The reads from a given species were then mapped against the corresponding plant genome with STAR v2.5.3a [16]. STAR settings were as follows (*see* **Note 2**):

```
--outSAMattributes All, --outSAMattrIHstart 0, --outSAMtype BAM Unsorted, --outSAMunmapped Within, --outSAMstrandField intronMotif, --outFilterIntronMotifs RemoveNoncanonical, --outFilterType BySJout, --outFilterMultimapNmax 20, --alignSJoverhangMin 8, --alignSJDBoverhangMin 1, --outFilterMismatchNmax 999, --outFilterMismatchNoverLmax 0.04, --alignIntronMin 20, --alignIntronMax 1000000, --alignMatesGapMax 1000000, --twopassMode Basic, --chimSegmentMin 12, --chimJunctionOverhangMin 12, --chimSegmentReadGapMax 3.
```

3. The resulting BAM file was used for ab initio transcriptome assembly with StringTie [17], where downloaded annotations in GTF or GFF format served as a reference. This produced a new GTF file with a custom transcriptome, one per species.

Table 2
Sources of genome sequences and genome annotations

Species	Genome file	Annotations file	Source
Amborella trichopoda	Amborella_trichopoda.AMTR1.0.dna.toplevel.fa	Amborella_trichopoda.AMTR1.0.37.gtf	Ensembl
Ananas comosus	ftp://ftp.ncbi.nih.gov/genomes/Ananas_comosus/Assembled_chromosomes/seq/	ref_ASM154086v1_top_level.gff3	NCBI
Arabidopsis lyrata	Arabidopsis_lyrata.v.1.0.dna.toplevel.fa	Arabidopsis_lyrata.v.1.0.37.gtf	Ensembl
Arabidopsis thaliana	Arabidopsis_thaliana.TAIR10.dna.toplevel.fa	Arabidopsis_thaliana.TAIR10.37.gtf	Ensembl
Brachypodium distachyon	Brachypodium_distachyon.v1.0.dna.toplevel.fa	Brachypodium_distachyon.v1.0.37.gtf	Ensembl
Brassica napus	Brassica_napus.AST_PRJEB5043_v1.dna.toplevel.fa	Brassica_napus.AST_PRJEB5043_v1.37.gtf	Ensembl
Brassica oleracea	Brassica_oleracea.v2.1.dna.toplevel.fa	Brassica_oleracea.v2.1.37.gtf	Ensembl
Brassica rapa	Brassica_rapa.IVFCAASv1.dna.toplevel.fa	Brassica_rapa.IVFCAASv1.37.gtf	Ensembl
Chenopodium quinoa	ftp://ftp.ncbi.nih.gov/genomes/Chenopodium_quinoa/CHR_Un/	ref_ASM168347v1_top_level.gff3	NCBI
Chlamydomonas reinhardtii	Chlamydomonas_reinhardtii.v3.1.dna.toplevel.fa	Chlamydomonas_reinhardtii.v3.1.37.gtf	Ensembl
Chondrus crispus	Chondrus_crispus.ASM35022v2.dna.toplevel.fa	Chondrus_crispus.ASM35022v2.37.gtf	Ensembl
Corchorus capsularis	Corchorus_capsularis.CCACVL1_1.0.dna.toplevel.fa	Corchorus_capsularis.CCACVL1_1.0.37.gtf	Ensembl
Cucumis sativus	ftp://ftp.ncbi.nih.gov/genomes/Cucumis_sativus/Assembled_chromosomes/seq/	ref_ASM407v2_top_level.gff3	NCBI
Galdieria sulphuraria	Galdieria_sulphuraria.ASM34128v1.dna.toplevel.fa	Galdieria_sulphuraria.ASM34128v1.37.gtf	Ensembl
Glycine max	Glycine_max.V1.0.dna.toplevel.fa	Glycine_max.V1.0.37.gtf	Ensembl
Hordeum vulgare	Hordeum_vulgare.Hv_IBSC_PGSB_v2.dna.toplevel.fa	Hordeum_vulgare.Hv_IBSC_PGSB_v2.37.gtf	Ensembl
Leersia perrieri	Leersia_perrieri.Lperr_V1.4.dna.toplevel.fa	Leersia_perrieri.Lperr_V1.4.37.gtf	Ensembl

(continued)

Table 2 (continued)

Species	Genome file	Annotations file	Source
Malus domestica	ftp://ftp.ncbi.nih.gov/genomes/ Malus_domestica/Assembled_ chromosomes/seq/	ref_MalDomGD1.0_top_level.gff3	NCBI
Manihot esculenta	ftp://ftp.ncbi.nih.gov/genomes/ Manihot_esculenta/Assembled_ chromosomes/seq/	ref_Manihot_esculenta_v6_top_level. gff3	NCBI
Medicago truncatula	Medicago_truncatula. MedtrA17_4.0.dna.toplevel.fa	Medicago_truncatula. MedtrA17_4.0.37.gtf	Ensembl
Musa acuminata	Musa_acuminata.MA1.dna. toplevel.fa	Musa_acuminata.MA1.37.gtf	Ensembl
Oryza barthii	Oryza_barthii.O.barthii_v1.dna. toplevel.fa	Oryza_barthii.O.barthii_v1.37.gtf	Ensembl
Oryza brachyantha	Oryza_brachyantha. Oryza_brachyantha.v1.4b.dna. toplevel.fa	Oryza_brachyantha. Oryza_brachyantha.v1.4b.37.gtf	Ensembl
Oryza nivara	Oryza_nivara.AWHD00000000. dna.toplevel.fa	Oryza_nivara.AWHD00000000.37. gtf	Ensembl
Oryza punctata	Oryza_punctata.AVCL00000000. dna.toplevel.fa	Oryza_punctata. AVCL00000000.37.gtf	Ensembl
Oryza rufipogon	Oryza_rufipogon.OR_W1943.dna. toplevel.fa	Oryza_rufipogon.OR_W1943.37.gtf	Ensembl
Oryza sativa	ftp://ftp.ncbi.nih.gov/genomes/ Oryza_sativa_Japonica_Group/ Assembled_chromosomes/seq/	ref_IRGSP-1.0_top_level.gff3	NCBI
Physcomitrella patens	Physcomitrella_patens.ASM242v1. dna.toplevel.fa	Physcomitrella_patens. ASM242v1.37.gtf	Ensembl
Populus trichocarpa	Populus_trichocarpa.JGI2.0.dna. toplevel.fa	Populus_trichocarpa.JGI2.0.37.gtf	Ensembl
Prunus persica	Prunus_persica.Prupe1_0.dna. toplevel.fa	Prunus_persica.Prupe1_0.37.gtf	Ensembl
Selaginella moellendorffii	Selaginella_moellendorffii.v1.0. dna.toplevel.fa	Selaginella_moellendorffii.v1.0.37. gtf	Ensembl
Setaria italica	Setaria_italica.JGIv2.0.dna. toplevel.fa	Setaria_italica.JGIv2.0.37.gtf	Ensembl
Solanum lycopersicum	Solanum_lycopersicum.SL2.50. dna.toplevel.fa	Solanum_lycopersicum.SL2.50.37. gtf	Ensembl
Solanum tuberosum	Solanum_tuberosum.SolTub_3.0. dna.toplevel.fa	Solanum_tuberosum. SolTub_3.0.37.gtf	Ensembl

(continued)

Table 2 (continued)

Species	Genome file	Annotations file	Source
Sorghum bicolor	Sorghum_bicolor. Sorghum_bicolor_v2.dna. toplevel.fa	Sorghum_bicolor. Sorghum_bicolor_v2.37.gtf	Ensembl
Theobroma cacao	Theobroma_cacao. Theobroma_cacao_20110822. dna.toplevel.fa	Theobroma_cacao. Theobroma_cacao_20110822.37. gtf	Ensembl
Trifolium pratense	Trifolium_pratense.Trpr.dna. toplevel.fa	Trifolium_pratense.Trpr.37.gtf	Ensembl
Vitis vinifera	Vitis_vinifera.IGGP_12x.dna. toplevel.fa	Vitis_vinifera.IGGP_12x.37.gtf	Ensembl
Zea mays	Zea_mays.AGPv4.dna.toplevel.fa	Zea_mays.AGPv4.37.gtf	Ensembl

3.1.2 Identification of lncRNAs

Transcript sequences in FASTA format were extracted from the 39 plant and algae genomes (*see* **Note 3**) taking advantage of GTF files (*see* **Note 1**) from ab initio assembly using the gffread utility from Cufflinks package v2.2.1 [18]. Additionally, annotations from the GTF file were compared against reference annotations using Cuffcompare from the Cufflinks package, with -R (*consider only the reference transcripts that overlap any of the input transfrags*) and -C (*include the "contained" transcripts in the combined gtf file*) options. Then, identification of lncRNAs was performed in the following steps, as implemented in in-house Python scripts (Fig. 1):

1. Discarding transcripts with Cuffcompare class codes =, *j*, *o* if reference transcripts were not classified as lncRNAs at Ensembl (e.g., rRNAs, protein-coding transcripts).
2. Transcript length >= 200 bases.
3. Discarding transcripts containing open reading frames (ORFs) as identified using TransDecoder 5.0.2 [19] with -m **100** (*minimum protein length*; default value) and -S (*strand-specific*) options (*see* **Note 2**).
4. Discarding transcripts classified as *coding* by Coding Potential Calculator (CPC) v0.9-r2 [20] with default settings.
5. A BLASTN v2.2.26 search against sequences from the Rfam database [21] was performed, and all sequences with hits of E-value <1e-5 were removed.
6. Transcripts originating from plastids or mitochondria were discarded.

Regardless of the TransDecoder and CPC results, all RNAs classified into the following lncRNA-related Ensembl transcript

Table 3
A list of Sequence Read Archive runs used when building the database

Species	Sequence Read Archive run ID
Solanum lycopersicum	SRR5932812, SRR5933152, SRR4423523, SRR5933175, SRR5932970, SRR5932905, SRR5932800, SRR5933037, SRR5031855
Cucumis sativus	SRR5273771, SRR6122525, SRR3156544, SRR3156511, SRR3310311, SRR3156529, SRR3207673, SRR3156524, SRR3156522, SRR3156508
Amborella trichopoda	SRR5293276, SRR3309446, SRR5293272, SRR5293275, SRR5293261, SRR3309445, SRR5293264, SRR5293277, SRR5293268, SRR5293262
Brachypodium distachyon	SRR3691905, SRR3691904, SRR3691908, SRR5011899, SRR5515075, SRR6127925, SRR3691903, SRR3691906, SRR1635409, SRR1635430
Manihot esculenta	SRR3629832, SRR1538491, SRR2041050, SRR5886936, SRR5310831, SRR3629822, SRR1538829, SRR1539558, SRR4019432, SRR1539560
Glycine max	SRR3732343, SRR3732348, SRR3680391, SRR830186, SRR830207, SRR925840, SRR3680387, SRR3475904
Malus domestica	ERR313227, SRR5132038, SRR2906811, SRR2940115, SRR1564463, ERR313236, SRR1660058, ERR313237, SRR2906848, SRR3036781
Chenopodium quinoa	SRR5572149, SRR4300167, SRR5974432, SRR5572173, SRR5572167, SRR5572152, SRR4300199, SRR4300183, SRR4300203, SRR4300188
Physcomitrella patens	SRR787295, SRR3350450, SRR787291, SRR790623, SRR3350451, SRR790660, SRR3350453, SRR1707403, SRR787292, SRR3350458
Corchorus capsularis	SRR2089352, SRR5040292, SRR1867786, SRR5046774, SRR5046775, SRR5046773, SRR5040295
Populus trichocarpa	SRR3019764, SRR3019824, SRR3019266, SRR3019883, SRR3019616, SRR3019929, SRR3019621, SRR3019318, SRR3019785, SRR3019588
Oryza nivara	SRR1722159, SRR1722160, SRR1722161, SRR1722162, SRR1171629, SRR1171630, SRR1171631
Arabidopsis thaliana	ERR1588811, SRR1734321, SRR5205776, SRR1773584, SRR5478940, SRR3664421, SRR3999707, ERR1295174, SRR4113367, SRR2177547
Solanum tuberosum	SRR2989645, SRR4032341, SRR2989629, SRR3371838, SRR2932322, SRR6281220
Chondrus crispus	ERR2041136
Chlamydomonas reinhardtii	SRR5058879, SRR5058880, ERR1000646, ERR1000649, ERR1000645, SRR5058877, SRR402023, ERR1000648, SRR5058876, ERR1000644
Setaria italica	SRR3536745, ERR2003564, SRR3536617, SRR3536772, ERR2003561, SRR5748805, ERR2003563, ERR2003556, SRR3536722, ERR2003565
Oryza brachyantha	SRR1174794
Musa acuminata	SRR4450939, SRR3732000, SRR5877238, SRR5877231, SRR5877244, SRR5877227, SRR5877209, SRR5877228, SRR5877230
Brassica rapa	SRR3189868, SRR1238072, SRR3280638, SRR3280774, SRR3280503, SRR3280621, SRR3280851, SRR1238066, SRR3280502, SRR6158359

(continued)

Table 3 (continued)

Species	Sequence Read Archive run ID
Oryza sativa Japonica Group	DRR036954, SRR1564283, SRR5925437, SRR5925537, SRR5925633, SRR5925491, SRR5925506, ERR2061597, SRR1985015, SRR3718442
Trifolium pratense	SRR6251246, SRR5312540, SRR5312541, SRR5312542, SRR5312543, SRR5312544, SRR5312545, ERR351508
Oryza rufipogon	DRR001371, DRR001374, SRR3264876, SRR3264877, SRR3180884, SRR3182795, SRR1220645, SRR1220646, SRR1220647
Sorghum bicolor	SRR3938636, SRR3452726, DRR059883, SRR5005259, SRR3087372, SRR3707947, SRR2924913, SRR2924918, SRR1822329, ERR1109876
Brassica napus	SRR5905533, SRR2016207, SRR2155562, SRR3355338, SRR1781543, SRR5905541, SRR1793860, SRR1793856, SRR3357562, SRR4242079
Selaginella moellendorffii	SRR5499402
Oryza punctata	SRR1171005, SRR1171006, SRR1171007
Hordeum vulgare	ERR1744570, SRR3305232, SRR5452122, SRR5452103, SRR3298414, SRR5452100, SRR3137027, SRR1577845, SRR5452119
Galdieria sulphuraria	SRR4450072, SRR4450073, SRR4450074, SRR4450075, SRR4450076, SRR4450077, SRR4450078, SRR4450079, SRR4450080
Brassica oleracea	SRR2296884, SRR4333195, SRR3473270, SRR2177446, SRR2296894, SRR2296910, SRR2296886, SRR3473214, SRR5643652
Arabidopsis lyrata	SRR3020584, SRR1868052, SRR1868063, SRR3020589, SRR3993767, SRR3020599, SRR1868044, SRR1868059, SRR1868042, SRR3993764
Oryza sativa	SRR1005276, SRR2931146, SRR1117255, SRR3124595, SRR2541230, SRR2072479, SRR2931502, SRR2931284
Prunus persica	SRR2086172, SRR2290949, SRR1662173, SRR5925640, SRR2290951, SRR2085964, SRR5405089, SRR3658831
Ananas comosus	SRR2983907, SRR5633635, SRR5633648, SRR2983912, SRR2983888, SRR5633613, SRR2976829, SRR2976850
Leersia perrieri	SRR1171869, SRR1171870, SRR1171884
Medicago truncatula	ERR2193167, SRR1523070, ERR2193169, SRR3735573, SRR949269, SRR5217271, SRR2060977, SRR1523075, SRR1979260, SRR5217285
Zea mays	SRR2558684, SRR3085202, ERR965605, SRR2626637, SRR5985056, ERR965650, SRR1207503, ERR1805322
Theobroma cacao	SRR747765, SRR3217281, SRR747762, SRR747776, SRR3217279, SRR747779, SRR3217277, SRR747774, SRR3217282, SRR3217319
Oryza barthii	SRR1722231, SRR1722290, SRR1722291, SRR1722292, SRR1170762, SRR1170742, SRR1170744
Vitis vinifera	SRR3069676, SRR3288880, SRR3106046, SRR5944323, SRR3046824, SRR3288867, SRR3056886, SRR2081021, SRR5984214, SRR3938631

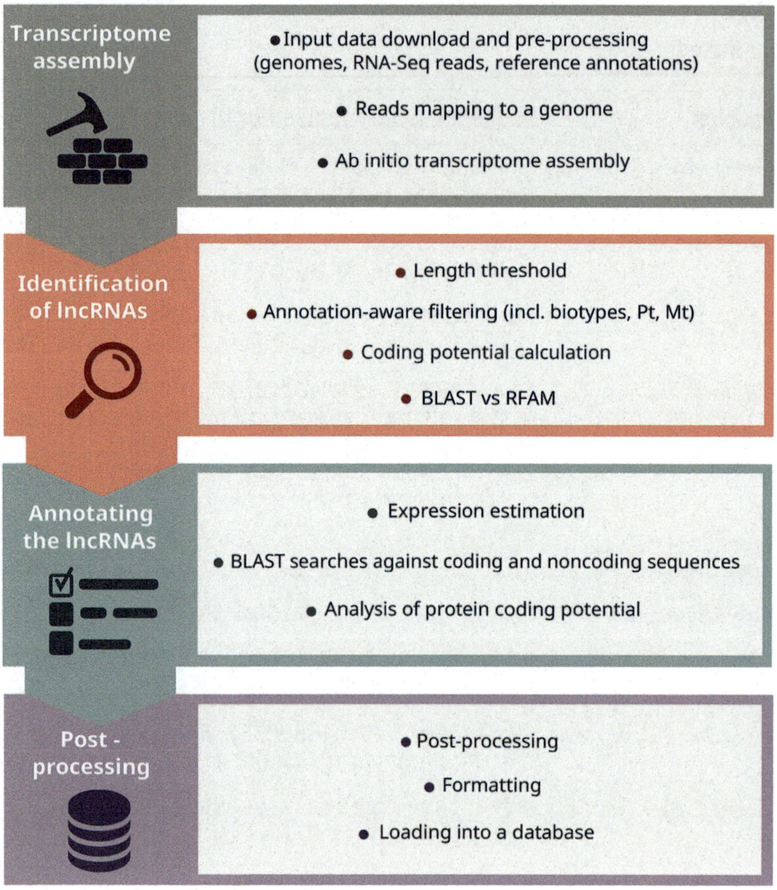

Fig. 1 A computational pipeline used for ab initio transcriptome assembly and identification of lncRNAs

types were retained: *retained_intron*, *sense_intronic*, *sense_overlapping*, *antisense_RNA*, *antisense*, *lincRNA*, *macro_lncRNA*, *bidirectional_promoter_lncRNA*, and *lncRNA* (*see* **Note 2**).

3.1.3 Annotation of Identified lncRNAs

To provide more information on the identified lncRNAs, we performed a number of additional computations. First, we used CNCI [22] in a plant mode (*model* parameter set to *pl*) to calculate the coding potential for all lncRNAs, independent of CPC. With TransDecoder [19], we identified putative peptides and proteins that could originate from the sense strand of lncRNAs. To this point, we used a –m 30 parameter, which enables identification of open reading frames (ORFs) that are at least 30 amino acids long (the default length threshold is 100 amino acids). In the search procedure, ORFs were not required to possess start or stop codons. Having identified ORFs, we then ran a BLASTP v2.2.26 search against a non-redundant and manually curated set of proteins from the Swiss-Prot database [23]. If a complete ORF showed significant similarity to an annotated protein (defined as a hit with an E-value <1e-20), the candidate was designated a *low-confidence lncRNA*; there are only 64 such records in CANTATAdb 2.0. We also ran a BLASTX search against Swiss-Prot proteins, with an E-value threshold set to 1e-5, to identify any lncRNA sequence similarity to known proteins, including regions outside identified ORFs.

Expression estimation was performed with SALMON [24] in a quasi-mapping mode, with *seqBias* and *gcBias* parameters enabled, to account for biases in nucleotide composition of transcript sequences.

3.1.4 Database Implementation and Testing

The database was constructed using Hypertext Markup Language (HTML), Cascading Style Sheets (CSS), PHP 5.4 (http://www.php.net/), MySQL 5.5 (http://www.mysql.com/), and a Bootstrap 3 (http://getbootstrap.com/) framework. The database layout was also enhanced with the JavaScript Highcharts library (http://www.highcharts.com/) for interactive data plotting.

The web interface and functionalities of the database were tested on Windows (XP, 7, 8, 8.1, and 10) and Mac OS X (Snow Leopard, Lion, Mountain Lion, Mavericks, and Yosemite) operating systems using Internet Explorer (versions 6, 7, 8, 9, 10, and 11), Mozilla Firefox (versions 37, 40, and 58), Chrome (versions 42 and 65), Opera (versions 12.15 and 12.16), and Safari (versions 5.1, 6, 6.1, 7, and 8) web browsers. Only in the case of Internet Explorer versions 6 and 7 was the web interface visualized improperly, but this did not affect the functionality of CANTATAdb 2.0.

3.2 Database Composition and Usage

To access the data stored at CANTATAdb 2.0, the user is required to select a species from the drop-down list in the main menu. A search page will appear with the following components:

3.2.1 Search Page

1. Search options. Here, the user can select criteria for data search and filtering: CANTATAdb lncRNA id, CNCI status (coding or noncoding), length of the longest peptide found in a lncRNA, maximal expression value, number of exons, lncRNA confidence, and options for ordering of the results.
2. Search summary. In this section, user-defined search parameters are provided, and the number of records found is reported. To download the filtered data into a tab-delimited text file, one needs to press the *Download current results* button. Additionally, a pie chart summarizes the CNCI status of filtered lncRNAs.
3. Search results. The search results are displayed in a table with one row per lncRNA. The presented data include lncRNA id, species, genomic location, best hit against Swiss-Prot proteins (if any), maximal length of all peptides found in a lncRNA

(if any), maximal expression value across used RNA-Seq libraries (in TPM), number of exons, and confidence. Low-confidence records are highlighted with a pale red color.

3.2.2 Search/Detail Page

There is also a *Details* button—upon clicking this; the user is able to access more information on the selected lncRNA, which includes the following:

1. The lncRNA sequence.
2. BLAST search results against Swiss-Prot and *A. thaliana* lncRNAs from NONCODE 2016.
3. The lncRNA expression profile across analyzed RNA-Seq libraries presented as a bar chart.
4. Information about peptides predicted for the lncRNA.

From this details page, the user can return to the search page without losing the search parameters by clicking a button on top of the page.

3.2.3 BLAST Page

Here, one can perform a sequence-based search of data stored in CANTATAdb 2.0 using BLAST version 2.2.26. A user-submitted nucleotide sequence should be in FASTA format (*see* **Note 1**). It is possible to select between two tools, BLASTN and MEGABLAST. Except for the expectation value (E-value) and maximal number of found hits, the search parameters are left as the default, and they are provided for user reference on the same page.

3.2.4 Download Page

Although the search results can be easily downloaded from the search page, we also enabled a bulk data download option. All lncRNAs can be downloaded in FASTA and GTF format (*see* **Note 1**) for each species separately.

4 Notes

1. **Data Formats and Sources**

   As mentioned earlier (*see* **Materials**), genomic sequences and annotation for all organism were downloaded from Ensembl Plants [14] or NCBI. In Ensembl Plants there is a single file that stores the genomic sequences, which is named using the following pattern: *Latin_name_of_species.assembly_version.dna.toplevel.fa*. In NCBI the genomic sequences are divided into chromosomes or scaffolds in separated FASTA files which can be described as: *Short_name_or_TaxonomyID.ref.assembly_version.name_of_chromosome/scaffold.fa*. That is why, in this case we provide direct links to FTP server with particular datasets.

Details of our set can be found in Materials, Table 2 and at http://cantata.amu.edu.pl/download.php.

The genome annotations files used as a reference in our computational pipeline were either in GTF (Gene Transfer Format) or GFF3 (General Feature Format) formats. Also the output predictions with lncRNAs are provided in a GTF format (*see* Subheadings 3.1.1 and 3.2.4). These tab-delimited files contain nine columns: (1) name of chromosome or scaffold, (2) source of the feature (program or database), (3) feature type, (4) feature start position, (5) feature end position, (6) score (a floating point value if exists), (7) strand (+/−), (8) frame (if exists), (9) attribute—a set of additional information on the given feature, semicolon-separated.

More details about these formats can be found at https://www.ensembl.org/info/website/upload/gff.html.

Data coming from RNA-Seq experiments were downloaded from Sequence Read Archive database [15] in a FASTQ format. FASTQ is a text-based format that stores nucleotide sequences together with a corresponding quality score for each position. Normally one sequence is described in four lines: (1) begins with '@' and followed by identifier or optional description, (2) raw sequence letters, (3) begins with '+' and optionally followed by identifier or description, (4) quality values for the sequence encoded in ASCII code (https://www.asciitable.com/). We are providing accession numbers for the used RNA-Seq data (*see* **Materials**, Table 3 and http://cantata.amu.edu.pl/download.php), which pinpoint the source of sequences (SR—SRA NCBI, ER—ENA EBI, DR—DDBJ) and indicate particular runs of sequencing machines.

2. **Improvements in CANTATAdb 2.0.**

Here, we are using an updated pipeline for both ab initio transcriptome assembly and **identification of lncRNAs, with a number of bona fide improvements**. First, the input RNA-Seq reads were paired-end only and were subject to quality filtering with BBDuk, a step omitted in CANTATAdb 1.0 (*see* Subheading 3.1.1). Second, read mapping against reference plant genomes was performed with STAR, one of the leading splice-aware mappers, which enables a 2-pass mapping strategy for a more accurate spliced alignment to novel junctions (for detailed settings, *see* Subheading 3.1.1 of Methods). Third, for 33 plant species with genomes and annotations downloaded from Ensembl Plants, handling of the transcript type information was improved. For instance, all expressed transcripts classified as lncRNAs at Ensembl are kept in CANTATAdb 2.0 (*see* Subheading 3.1.2). Fourth, to minimize the chances of classifying protein-coding transcripts as lncRNAs, candidates with open reading frames of at least 100 amino acids were discarded (*see* Subheading 3.1.2).

There are also essential differences in lncRNA annotation available in CANTATAdb 2.0. First, we replaced our custom method for expression estimation with Salmon, one of the leading tools in the area. However, we refrained from assigning functions for the identified lncRNAs due to limitations in data availability. Last but not least, the current version of the database encompasses lncRNAs for 39 species (36 plants and 3 algae) compared with 10 species in the previous version of CANTATAdb.

3. Comparison with other databases: Considering the number of analyzed species makes CANTATAdb 2.0 one of the largest currently available resources of plant lncRNAs (*see* Subheading 3.1.2). For comparison, NONCODE 2016 [1] stores lncRNAs from 17 organisms, including a single plant species, A. thaliana, with 3763 lncRNA transcripts. The Plant Non-coding RNA Database [25] includes lncRNAs from 21 plants, while PLNlncRbase [26] contains manually collected data from nearly 200 publications, covering a total of 1,187 plant lncRNAs in 43 plant species. To the best of our knowledge, the most comprehensive database of plant lncRNAs at the moment is GreeNC [3], containing lncRNAs identified in 37 plant species and six algae. However, 15 species included in CANTATAdb 2.0 are missing at GreeNC, including such model organisms as Brassica rapa, Brassica napus and Hordeum vulgare.

Acknowledgments

This work was supported by the National Science Centre (grant No. 2014/15/D/NZ2/00525 to M.W.S); the KNOW Poznan RNA Centre (grant No. 01/KNOW2/2014); the Polish Ministry of Science and Higher Education (decision No. 1268/MOB/IV/2015/0 to M.W.S - Mobility Plus project).

References

1. Zhao Y, Li H, Fang S, Kang Y, Wu W, Hao Y et al (2016) NONCODE 2016: an informative and valuable data source of long non-coding RNAs. Nucleic Acids Res 44:D203–D208
2. Zerbino DR, Achuthan P, Akanni W, Amode MR, Barrell D, Bhai J et al (2018) Ensembl 2018. Nucleic Acids Res 46:D754–D761
3. Paytuvi Gallart A, Hermoso Pulido A, Anzar Martinez de Lagran I, Sanseverino W, Aiese Cigliano R (2016) GREENC: a Wiki-based database of plant lncRNAs. Nucleic Acids Res 44:D1161–D1166
4. Szczesniak MW, Rosikiewicz W, Makalowska I (2016) CANTATAdb: a collection of plant long non-coding RNAs. Plant Cell Physiol 57:e8
5. Kim DH, Xi Y, Sung S (2017) Modular function of long noncoding RNA, COLDAIR, in the vernalization response. PLoS Genet 13:e1006939

6. Zhang YC, Liao JY, Li ZY, Yu Y, Zhang JP, Li QF et al (2014) Genome-wide screening and functional analysis identify a large number of long noncoding RNAs involved in the sexual reproduction of rice. Genome Biol 15:512
7. Wang Y, Li J, Deng XW, Zhu D (2017) Arabidopsis noncoding RNA modulates seedling greening during deetiolation. Sci China Life Sci 61:199. https://doi.org/10.1007/s11427-017-9187-9
8. Yuan J, Zhang Y, Dong J, Sun Y, Lim BL, Liu D et al (2016) Systematic characterization of novel lncRNAs responding to phosphate starvation in Arabidopsis thaliana. BMC Genomics 17:655
9. Zhang J, Wang C, Ke N, Bliesath J, Chionis J, He QS et al (2007) A more efficient RNAi inducible system for tight regulation of gene expression in mammalian cells and xenograft animals. RNA 13:1375–1383
10. Ben Amor B, Wirth S, Merchan F, Laporte P, d'Aubenton-Carafa Y, Hirsch J et al (2009) Novel long non-protein coding RNAs involved in Arabidopsis differentiation and stress responses. Genome Res 19:57–69
11. Deng F, Zhang X, Wang W, Yuan R, Shen F (2018) Identification of Gossypium hirsutum long non-coding RNAs (lncRNAs) under salt stress. BMC Plant Biol 18:23
12. Bardou F, Ariel F, Simpson CG, Romero-Barrios N, Laporte P, Balzergue S et al (2014) Long noncoding RNA modulates alternative splicing regulators in Arabidopsis. Dev Cell 30:166–176
13. Bazin J, Baerenfaller K, Gosai SJ, Gregory BD, Crespi M, Bailey-Serres J (2017) Global analysis of ribosome-associated noncoding RNAs unveils new modes of translational regulation. Proc Natl Acad Sci U S A 114:E10018–E10027
14. Bolser DM, Staines DM, Perry E, Kersey PJ (2017) Ensembl plants: integrating tools for visualizing, mining, and analyzing plant genomic data. Methods Mol Biol 1533:1–31
15. Kodama Y, Shumway M, Leinonen R (2012) The Sequence Read Archive: explosive growth of sequencing data. Nucleic Acids Res 40: D54–D56
16. Dobin A, Davis CA, Schlesinger F, Drenkow J, Zaleski C, Jha S et al (2013) STAR: ultrafast universal RNA-seq aligner. Bioinformatics 29:15–21
17. Pertea M, Pertea GM, Antonescu CM, Chang TC, Mendell JT, Salzberg SL (2015) StringTie enables improved reconstruction of a transcriptome from RNA-seq reads. Nat Biotechnol 33:290–295
18. Trapnell C, Williams BA, Pertea G, Mortazavi A, Kwan G, van Baren MJ et al (2010) Transcript assembly and quantification by RNA-Seq reveals unannotated transcripts and isoform switching during cell differentiation. Nat Biotechnol 28:511–515
19. Haas BJ, Papanicolaou A, Yassour M, Grabherr M, Blood PD, Bowden J et al (2013) De novo transcript sequence reconstruction from RNA-seq using the Trinity platform for reference generation and analysis. Nat Protoc 8:1494–1512
20. Kong L, Zhang Y, Ye Z, Liu X, Zhao S, Wei L et al (2007) CPC: assess the protein-coding potential of transcripts using sequence features and support vector machine. Nucleic Acids Res 35:W345–W349
21. Kalvari I, Argasinska J, Quinones-Olvera N, Nawrocki EP, Rivas E, Eddy SR, Bateman A et al (2018) Rfam 13.0: shifting to a genome-centric resource for non-coding RNA families. Nucleic Acids Res 46:D335–D342
22. Sun L, Luo H, Bu D, Zhao G, Yu K, Zhang C et al (2013) Utilizing sequence intrinsic composition to classify protein-coding and long non-coding transcripts. Nucleic Acids Res 41:e166
23. The UniProt Consortium (2017) UniProt: the universal protein knowledgebase. Nucleic Acids Res 45:D158–D169
24. Patro R, Duggal G, Love MI, Irizarry RA, Kingsford C (2017) Salmon provides fast and bias-aware quantification of transcript expression. Nat Methods 14:417–419
25. Yi X, Zhang Z, Ling Y, Xu W, Su Z (2015) PNRD: a plant non-coding RNA database. Nucleic Acids Res 43:D982–D989
26. Xuan H, Zhang L, Liu X, Han G, Li J, Li X et al (2015) PLNlncRbase: a resource for experimentally identified lncRNAs in plants. Gene 573:328–332

Chapter 27

Experimentally Validated Plant lncRNAs in EVLncRNAs Database

Bailing Zhou, Huiying Zhao, Jiafeng Yu, Chengang Guo, Xianghua Dou, Feng Song, Guodong Hu, Zanxia Cao, Yuanxu Qu, Yuedong Yang, Yaoqi Zhou, and Jihua Wang

Abstract

Plant long noncoding RNAs (lncRNAs) play important functional roles in various biological processes. Most databases deposit all plant lncRNA candidates produced by high-throughput experimental and/or computational techniques. There are several databases for experimentally validated lncRNAs. However, these databases are small in scale (with a few hundreds of lncRNAs only) and specific in their focuses (plants, diseases, or interactions). Thus, we established EVLncRNAs by curating lncRNAs validated by low-throughput experiments (up to May 1, 2016) and integrating specific databases (lncRNAdb, LncRAN-Disease, Lnc2Cancer, and PLNIncRBase) with additional functional and disease-specific information not covered previously. The current version of EVLncRNAs contains 1543 lncRNAs from 77 species, including 428 plant lncRNAs from 44 plant species. Compared to PLNIncRBase, our dataset does not contain any lncRNAs from microarray and deep sequencing. Moreover, 40% of entries contain new information (interaction and additional information from NCBI and Ensembl). The database allows users to browse, search, and download as well as to submit experimentally validated lncRNAs. The database is available at http://biophy.dzu.edu.cn/EVLncRNAs.

Key words Database, LncRNA, Plant, Experimentally validated, Low-throughput, Manually curated

1 Introduction

Long noncoding RNAs (lncRNAs) are generally longer than 200 nt and could not code protein. Mammalian lncRNAs could regulate gene expression and other cellular processes and are involved in disease etiology [1–3] including cancers [2, 4–6]. Plant lncRNAs are also shown to have function as molecular sponges and could regulate transcription, silencing and flowering [7]. Along with an

Bailing Zhou and Huiying Zhao contributed equally to this work.

Julia A. Chekanova and Hsiao-Lin V. Wang (eds.), *Plant Long Non-Coding RNAs: Methods and Protocols*, Methods in Molecular Biology, vol. 1933, https://doi.org/10.1007/978-1-4939-9045-0_27,

increasing number of discovered plant lncRNAs, some databases [7, 8] have been established for facilitating plant research.

Most plant lncRNA databases deposit lncRNA candidates from high-throughput techniques. Examples include GREENC with >120,000 lncRNA from 37 plants and 6 algae [9], PNRD with 25,739 ncRNAs from 150 plants [10], and PlantNATsDB with 2,146,803 natural antisense transcripts from 70 plant species [11]. PLNlncRbase [12] collected 1187 experimentally identified plant lncRNAs from 43 plant species. However, they also included the lncRNAs from high-throughput microarray and deep sequencing. Recognizing potential false positives in high-throughput techniques, several databases were built to collect lncRNAs validated by low-throughput experiments only. However, they were designed for respective particular purposes, for example, lncRNAdb [13] only for eukaryotic lncRNAs, Lnc2Cancer [14] and LncRNADisease [15] only for cancer/disease-related lncRNAs, and lncRInter [16] only for interaction of lncRNAs. Thus, we established EVLncRNAs [17] by manually curating all lncRNAs validated by low-throughput experiments (prior to May 1, 2016) and collected detailed information of sequence, structure, function, and related disease. The current dataset contains 1543 lncRNAs from 77 species including 44 plant species. The database is available at http://biophy.dzu.edu.cn/EVLncRNAs. Here we will focus on plant lncRNAs collected in this database.

2 Materials

1. Excel: Excel software was employed to collect the information manually extracted from publication and used the Excel-formatted file prior to inputting the information to the database.
2. Python: Some python scripts were written for downloading annotation information of lncRNAs from NCBI and Ensembl databases, for example, the alias, position at chromosome, and sequence accession.
3. Software for database construction: All the data of the database were organized using MySQL. The web interface was designed by PHP. The web services were built on Apache Tomcat.

3 Methods

3.1 Database Description

In this database, there are 428 plant lncRNAs from 44 plant species. The top 20 most studied plant species on lncRNA were shown in Fig. 1. We manually examined and integrated the entries in the PLNlncRbase [12] and just kept those validated by

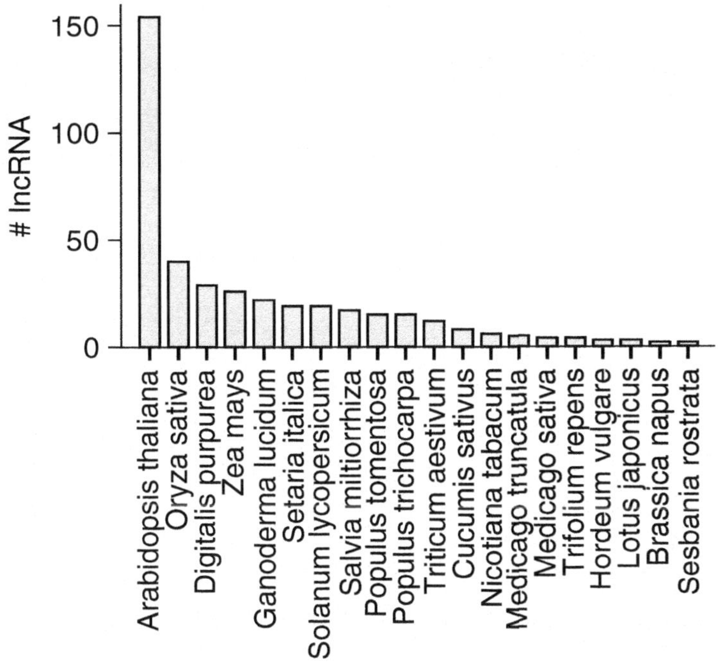

Fig. 1 Top 20 plant species for the number of experimentally validated lncRNAs

Table 1
Statistics of EVLncRNAs with lncRNAdb and PLNlncRbase in plant lncRNAs

Database	# LncRNAs	# Species	# Diseases	# Disease associations	# Functions (excl. interactions)	# Interactions
lncRNAdb-plant lncRNAs	17	10	N/A	N/A	17	9
PLNlncRbase	420[a]	43	N/A	N/A	420[a]	N/A
EVLncRNAs-plant lncRNAs	428	44	2	2	429	80
EVLncRNAs-total lncRNAs	1543	77	338	2324	793	1163

[a]After excluding 767 lncRNA from high-throughput microarray and RNA-seq

low-throughput experiments. For these validated entries, we curated interaction information not covered previously from literature and additional annotation information from NCBI [18] and Ensembl [19] database. Forty percent of entries contain new information. Furthermore, we manually integrated plant lncRNAs in lncRNAdb [13] and extracted information on newly validated plant lncRNAs from new publications. Only seven additional plant lncRNAs were found from new publications when the data was collected. Table 1 compares the statistics of plant lncRNAs collected by PLNlncRbase, lncRNAdb, and EVLncRNAs.

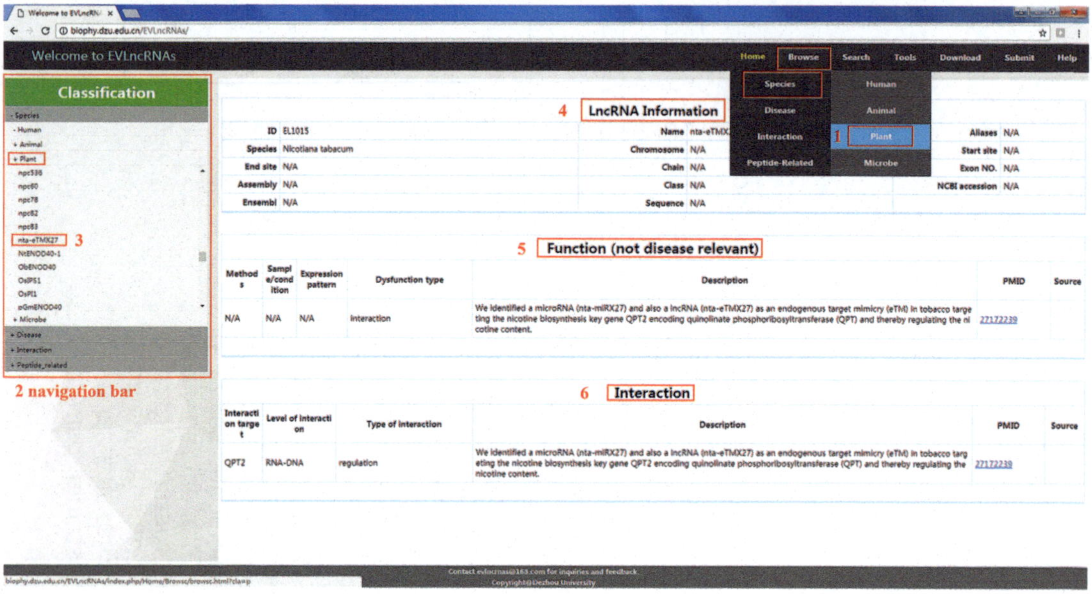

Fig. 2 Browse page of EVLncRNAs

The interface of EVLncRNAs allows users to browse, search, and download. As shown in Fig. 2, users can browse plant entries by clicking sub-menu "Plant" of "Species" in "Browse" (located in labeled position 1). Alternatively, users can click navigation bar (located in position 2) on the left to view the corresponding page or entry. Using "nta-eTMX27" as an example (position 3), the page of this entry displays the lncRNA information (position 4), its related disease (if exists), function (position 5), and interaction information (if exists, position 6). In the "Search" page, EVLncRNAs enables users to search by any keywords, such as lncRNA name, alias, species, and samples. EVLncRNAs supports users to download the database data by category in the "Download" page. Plant researchers could just download information of plant lncRNAs. In addition, EVLncRNAs provides links to the tools for lncRNA prediction in the "Tools" page. One is CPC [20] and another is COME [21]. Users could skip to their online servers by clicking the links and calculating coding potential of transcripts accordingly. EVLncRNAs also permits users to submit novel experimentally validated lncRNAs and related function in the "Submit" page. The submitted record will be included in the next release after manually checked by our curators. A detailed user tutorial is available in the "Help" page.

By browsing our database, we could find some lncRNAs have orthologs in different plant species. For example, ENOD40 has been reported in 16 plants, and it could bind protein MtRBP1 and is involved in MtRBP1 re-localization into cytoplasmic granules in

Medicago truncatula [22]. Furthermore, we included information on the differentially expressed lncRNAs during a certain stage of development such as flower development, fruit ripening, reproductive stage, and wood formation stage and under various external conditions such as cold, heat, drought, and salt. We listed the peptide-coding lncRNAs separately, including one soybean lncRNA ENOD40. ENOD40 encodes two peptides that bind to sucrose synthase, and ENOD40 peptides are involved in the control of sucrose use in nitrogen-fixing nodules [23]. EVLncRNAs provides some basic and functional information for plant lncRNAs, which will be useful for future studies.

3.2 Collection of Publications

We collected the publications by searching the PubMed with keywords matching "lncRNA," "long noncoding RNA," "long ncRNA," or "long noncode RNA" along with their plural forms.

3.3 Manual Extraction of Information from Publications

The information required to be collected was listed as column names in an Excel file, including lncRNA name, species, related disease, methods, samples, interaction target, and descriptions. The information was extracted from publications and entered in relevant columns. For this database, we only kept lncRNAs confirmed by low-throughput experiments such as quantitative reverse transcription-polymerase chain reaction (qRT-PCR), knockdown, western blot, northern blot, and luciferase reporter assays.

3.4 Integration of Specific Databases

The data from some experimentally validated databases (lncRNAdb, LncRANDisease, Lnc2Cancer, and PLNIncRBase) were also downloaded from their web interface. Each entry of their databases was manually checked, and we only retained the entries validated by low-throughput experiments. The additional information for our database (such as related-disease and interaction information) was manually curated from original publications. Entries from these databases with nonofficial names were manually corrected and merged with corresponding entries in our database.

3.5 Supplement with Annotation Information

The annotation information of lncRNAs, for example, the alias, position at chromosome, and sequence accession, were batch downloaded from NCBI and Ensembl database by python scripts and were added to corresponding lncRNA entries.

3.6 Construction of Web Interface

After cross-checking, all data were imported to a database organized by MySQL. A web interface designed by PHP was built to access the database. The website includes the home page and pages of browse, search, tools, download, submit, and help. The web services were built on Apache Tomcat.

4 Notes

1. Some lncRNAs are reported with unofficial names. This increased the difficulties of data collection and data integration. We minimized this problem by combining manual and automated computational checks.
2. Plant lncRNAs in the EVLncRNAs database were collected until May 1, 2016. Many new plant lncRNAs were discovered since then. An updated version is in progress.

Acknowledgments

The National Natural Science Foundation of China [61671107, 61271378], Taishan Scholars Program of Shandong province of China [Tshw201502045], National Health and Medical Research Council [1059775, 1083450 to Y.Z.], National Health and Medical Research Council Early Career Fellowship [1091816 to H.Z.], Projects of Science and Technology for Colleges and Universities in Shandong Province [J15LE09 to F.S.], Taishan Young Scholars Program of Shandong province of China [tsqn20161049 to F.S.], and Talent Introduction Project of Dezhou University of China [320111 to B.Z.].

References

1. Mercer TR, Dinger ME, Mattick JS (2009) Long non-coding RNAs: insights into functions. Nat Rev Genet 10(3):155–159
2. Hon C-C, Ramilowski JA, Harshbarger J, Bertin N, Rackham OJ, Gough J, Denisenko E, Schmeier S, Poulsen TM, Severin J (2017) An atlas of human long non-coding RNAs with accurate 5′ ends. Nature 543 (7644):199–204
3. Toiber D, Leprivier G, Rotblat B (2017) Long noncoding RNA: noncoding and not coded. Cell Death Discov 3:16104
4. Gibb EA, Brown CJ, Lam WL (2011) The functional role of long non-coding RNA in human carcinomas. Mol Cancer 10:38
5. Van Roosbroeck K, Pollet J, Calin GA (2013) miRNAs and long noncoding RNAs as biomarkers in human diseases. Expert Rev Mol Diagn 13(2):183–204
6. Yang Y, Chen L, Gu J, Zhang H, Yuan J, Lian Q, Lv G, Wang S, Wu Y, Yang YCT, Wang D, Liu Y, Tang J, Luo G, Li Y, Hu L, Sun X, Wang D, Guo M, Xi Q, Xi J, Wang H, Zhang MQ, Lu ZJ (2017) Recurrently deregulated lncRNAs in hepatocellular carcinoma. Nat Commun 8:14421
7. Wang HV, Chekanova JA (2017) Long noncoding RNAs in plants. Adv Exp Med Biol 1008:133–154
8. Fritah S, Niclou SP, Azuaje F (2014) Databases for lncRNAs: a comparative evaluation of emerging tools. RNA 20(11):1655–1665
9. Paytuví Gallart A, Hermoso Pulido A, Anzar Martínez de Lagrán I, Sanseverino W, Aiese Cigliano R (2016) GREENC: a Wiki-based database of plant lncRNAs. Nucleic Acids Res 44(D1):D1161–D1166
10. Yi X, Zhang Z, Ling Y, Xu W, Su Z (2015) PNRD: a plant non-coding RNA database. Nucleic Acids Res 43(Database issue): D982–D989
11. Chen D, Yuan C, Zhang J, Zhang Z, Bai L, Meng Y, Chen LL, Chen M (2012) PlantNATsDB: a comprehensive database of plant natural antisense transcripts. Nucleic Acids Res 40(Database issue):D1187–D1193
12. Xuan H, Zhang L, Liu X, Han G, Li J, Li X, Liu A, Liao M, Zhang S (2015) PLNlncRbase:

a resource for experimentally identified lncRNAs in plants. Gene 573(2):328–332

13. Quek XC, Thomson DW, Maag Jesper LV, Bartonicek N, Signal B, Clark MB, Gloss BS, Dinger ME (2015) lncRNAdb v2.0: expanding the reference database for functional long non-coding RNAs. Nucleic Acids Res 43(D1): D168–D173
14. Ning S, Zhang J, Wang P, Zhi H, Wang J, Liu Y, Gao Y, Guo M, Yue M, Wang L (2016) Lnc2Cancer: a manually curated database of experimentally supported lncRNAs associated with various human cancers. Nucleic Acids Res 44(D1):D980–D985
15. Chen G, Wang Z, Wang D, Qiu C, Liu M, Chen X, Zhang Q, Yan G, Cui Q (2013) LncRNADisease: a database for long-non-coding RNA-associated diseases. Nucleic Acids Res 41(D1):D983–D986
16. Liu C-J, Gao C, Ma Z, Cong R, Zhang Q, Guo A-Y (2017) lncRInter: a database of experimentally validated long non-coding RNA interaction. J Genet Genomics 44(5):265–268
17. Zhou B, Zhao H, Yu J, Guo C, Dou X, Song F, Hu G, Cao Z, Qu Y, Yang Y, Zhou Y, Wang J (2018) EVLncRNAs: a manually curated database for long non-coding RNAs validated by low-throughput experiments. Nucleic Acids Res 46(D1):D100–D105
18. Ncbi Resource Coordinators (2017) Database resources of the National Center for Biotechnology Information. Nucleic Acids Res 45 (Database issue):D12–D17
19. Aken BL, Achuthan P, Akanni W, Amode MR, Bernsdorff F, Bhai J, Billis K, Carvalho-Silva D, Cummins C, Clapham P, Gil L, Girón CG, Gordon L, Hourlier T, Hunt SE, Janacek SH, Juettemann T, Keenan S, Laird MR, Lavidas I, Maurel T, McLaren W, Moore B, Murphy DN, Nag R, Newman V, Nuhn M, Ong CK, Parker A, Patricio M, Riat HS, Sheppard D, Sparrow H, Taylor K, Thormann A, Vullo A, Walts B, Wilder SP, Zadissa A, Kostadima M, Martin FJ, Muffato M, Perry E, Ruffier M, Staines DM, Trevanion SJ, Cunningham F, Yates A, Zerbino DR, Flicek P (2017) Ensembl 2017. Nucleic Acids Res 45(Database issue): D635–D642
20. Kong L, Zhang Y, Ye Z-Q, Liu X-Q, Zhao S-Q, Wei L, Gao G (2007) CPC: assess the protein-coding potential of transcripts using sequence features and support vector machine. Nucleic Acids Res 35(Suppl 2):W345–W349
21. Hu L, Xu Z, Hu B, Lu ZJ (2017) COME: a robust coding potential calculation tool for lncRNA identification and characterization based on multiple features. Nucleic Acids Res 45(1):e2–e2
22. Campalans A, Kondorosi A, Crespi M (2004) Enod40, a short open reading frame-containing mRNA, induces cytoplasmic localization of a nuclear RNA binding protein in Medicago truncatula. Plant Cell 16 (4):1047–1059
23. Rohrig H, Schmidt J, Miklashevichs E, Schell J, John M (2002) Soybean ENOD40 encodes two peptides that bind to sucrose synthase. Proc Natl Acad Sci U S A 99(4):1915–1920

Part VII

Long-Range Chromatin Interactions and Chromosome Conformation Capture

Chapter 28

In Situ Hi-C for Plants: An Improved Method to Detect Long-Range Chromatin Interactions

Sudharsan Padmarasu, Axel Himmelbach, Martin Mascher, and Nils Stein

Abstract

Recently, long noncoding RNAs (lncRNAs) are shown to be implicating nuclear domain organization and gene regulation by mediating long-range chromatin interactions. Chromosome conformation capture (3C) is a method used to study such long-range interaction between two different loci in the 3D nuclear space. Through successive improvement in resolution and throughput, 3C, chromosome conformation capture on chip (4C), and chromosome conformation capture carbon copy (5C) to Hi-C methods were developed to study interactions between loci from one versus one scale to an unprecedented genome-wide resolution. In situ Hi-C is a variant of Hi-C in which proximity ligation is performed at the intact nuclei to improve the signal-to-noise ratio and throughput of the experiment to provide useful genome-wide contact frequency matrix/maps. The contact frequency maps obtained could be used for physical ordering of scaffolds in complex genome assembly projects, in deducing the nuclear domain organization in high resolution and in identifying specific long-range interactions between genomic regions of interest. In this chapter, we describe in detail a protocol for in situ Hi-C used on crops like barley, wheat, rye, oat, and evening primrose.

Key words In situ Hi-C, Chromatin interaction, Genome assembly, Scaffolding, Long noncoding RNA, Nuclear domain organization, Triticeae

1 Introduction

Recent transcriptomic studies have revealed that >90% of the genome is transcribed but only a fraction of it has the protein-coding potential [1, 2]. Transcripts longer than 200 nucleotides that do not translate to proteins are called long noncoding RNAs (lncRNAs, [3, 4]). LncRNAs are shown to be involved in gene expression regulation, genome stability maintenance, and nuclear domain organization by bringing together distant genomic regions by targeting specific proteins involved in epigenetic processes to a specific locus or by acting as decoy for miRNA targets [5]. Chromosome conformation capture methods are used to decipher these long-range genomic interactions.

Julia A. Chekanova and Hsiao-Lin V. Wang (eds.), *Plant Long Non-Coding RNAs: Methods and Protocols*, Methods in Molecular Biology, vol. 1933, https://doi.org/10.1007/978-1-4939-9045-0_28,

Chromosome conformation capture (3C) was first developed by Dekker and colleagues and used to study the yeast chromosome III organization at G1 phase of the cell cycle. In the 3C method (1), the intact nuclei were isolated, and formaldehyde was used to cross-link the proteins against protein and DNA in the nuclei, respectively. This step fixes DNA fragments in close proximity/contact as they occur in the interphase nucleus through the protein–protein cross-links (2). The cross-linked nuclei are then fragmented by its restriction digest (3). Resulting sticky end fragments are then religated (proximity ligation) in a dilute solution in order to favor ligation between cross-linked (intramolecular) fragments over random (intermolecular) fragments (4). This is followed by reverse cross-linking of the ligated fragments (5). Control template DNA is prepared from simply digesting and ligating the purified DNA where the interactions between loci are random and uniform (6). Then the ligation junction-containing fragments are amplified through semiquantitative PCR reactions using the reverse cross-linked DNA or control DNA as template and by using the primers specific for the loci/genomic region of interest (7). Interaction between two loci could be inferred from the ratio of amount of ligation frequency observed between loci of interest in the cross-linked nuclei to ligation frequency observed between loci of interest in control template DNA [6]. However, this method could only be used for determining the one-to-one locus interactions and required use of specific primers for loci of interests. The 3C method was adapted to plants in 2009 and used to study the tissue and expression level-specific chromatin looping between a hepta-repeat and transcription start site of b1 locus of maize [7]. The methodological improvement led to chromosome conformation capture on chip (4C) and chromosome conformation capture carbon copy (5C) methods to study one-to-many and many-to-many interaction pairs in parallel [8, 9]. Hi-C is a recent version of chromatin conformation capture which combines 3C with next-generation sequencing. In this procedure (1), the digested cross-linked DNA fragments from 3C experiments are biotin-labeled during the end fill-in reaction followed by (2) the blunt-end ligation (3). Biotin incorporated ligation fragments are enriched by using streptavidin C1 beads (4). Then the sequencing libraries are prepared from biotin-enriched fragments and sequenced (5). Bioinformatic analysis of sequencing data will then generate a contact matrix at the genome-wide scale [10].

Multiple variations of Hi-C methods were developed to improve the signal-to-noise ratio and throughput. Some of the variations include tethered chromosome conformation capture (TCC), in situ Hi-C, single-cell Hi-C, and capture Hi-C. In case of TCC, the chromatin is biotinylated and tethered to the streptavidin T1 beads, and the intramolecular ligation is performed on the beads to reduce spurious interactions. Due to this variation, signal-to-noise ratios are improved [11]. In case of in situ Hi-C, the restriction digestion and ligation are performed on intact nuclei in a small volume of solution compared to the ligation in diluted solution for Hi-C. Due to this, the signal-to-noise ratio and throughput of the reaction are much improved [12, 13]. Capture Hi-C is a method in which the Hi-C libraries are enriched for specific regions such as promoters by using in-solution hybridization against biotinylated RNA baits specific for genomic regions of interest. This method helps with identification of long-range interactions between the genomic region of interest (promoter) and other genomic regions without the need for very high-coverage deep sequencing [14]. By the use of DNA tagging methods and by successive rounds of tagging (combinatorial indexing) of intact nuclei, individual nuclei are labeled and single-cell Hi-C can be performed by using these labeled nuclei. It helps with the identification of chromatin conformation at single-cell level in contrast to the population-wide average that is typically sampled in other Hi-C variants [15]. Hi-C methods even use *micrococcal* nuclease (MNase) and deoxy ribonuclease (DNase) for digestion of DNA to improve resolution and to study interactions specifically at active chromatin regions [16]. Methods such as CHIA-PET and CHIP-loop were developed by combining the protocols of chromatin immunoprecipitation and Hi-C to study the chromatin interactions mediated by specific protein or transcription factor of interest [17].

Hi-C works based on the principle of distance-dependent decay of contact probabilities in the interphase nucleus, i.e., the two loci, which are closer together in the 3D-nuclei space, would have higher contact probability than the two loci that are more distant to each other. By using this principle and from the genome-wide contact matrix generated, nuclear organization of any organism could be elucidated, and it could be used for physical ordering of scaffolds from de novo assemblies of complex genomes [18–20]. This approach was initially used to study human nuclear domain organization, which was followed by research in other animals and on plants nuclear domain organization. The Hi-C method detailed in this chapter was successfully used for generating physical mapping data for genome assembly projects of large and complex genomes such as barley and wheat in our laboratory.

2 Materials

All solutions are prepared using purified water (GenPure Pro UV/UF; Thermo Fisher Scientific, Waltham, MA, USA) or double-distilled water and stored at room temperature unless otherwise specified. The solutions were sterilized by either autoclaving or by filter sterilization based on the reagents used. Use of freshly prepared solutions and handling of the sample during different steps need to be followed accurately as described. Improper handling and use of old stock solutions at certain steps lead to suboptimal experimental results. Good laboratory practice has to be applied throughout the experiment. We recommend trying out this protocol on one sample initially. Once comfortable, this protocol can be scaled up to perform Hi-C on to eight samples in parallel.

2.1 Plant Material, Growth Conditions, and Tissue Harvesting

1. Seeds of the plant material (25 seeds for barley/wheat/rye are sufficient—*see* **Note 1**).
2. Greenhouse equipped with automatic shading and supplementary light. For barley, the plants were grown at the temperature of 18 °C (night) and 21 °C (day).
3. Compost soil.
4. Pots (16 cm diameter).
5. Scissors.
6. Ice bucket filled with ice.
7. Aluminum foil.

2.2 Cross-Linking of Leaf Tissue

1. Nuclei Isolation Buffer (NIBF) with formaldehyde: 20 mM HEPES, pH 8.0, 250 mM sucrose, 1 mM $MgCl_2$, 5 mM KCl, 40% (v/v) glycerol, 0.25% (v/v) Triton X-100, 2% (w/v) formaldehyde, 0.1 mM PMSF, and 0.1% (v/v) β-mercaptoethanol [21]. Add PMSF, β-mercaptoethanol, and formaldehyde under the fume hood just before use (*see* **Note 2**).
2. 2 M glycine: Weigh 6.0 g glycine (Sigma-Aldrich, St. Louis, MO, USA), dissolve it, and adjust the volume to 40 mL with water. Store in a 50 mL plastic tube (−20 °C). Thaw in a warm water bath prior to use.
3. Desiccator connected to a vacuum pump with manometer and condensation trap.
4. 50 mL plastic tubes (e.g., Falcon; Thermo Fisher Scientific, Waltham, MA, USA) and appropriate racks fitting into the desiccator.
5. Self-made Polystyrene plugs of 0.5–1.0 cm thickness fitting into the 50 mL plastic tubes.
6. Disposable 25 mL plastic pipettes.
7. Sieve.
8. Paper towels.
9. Fume hood: All steps involving the NIBF buffer must be performed under a fume hood. Follow the safety regulations of your laboratory during the manipulations and for the waste disposal.

2.3 Nuclei Isolation

1. Liquid nitrogen and appropriate protective gear.
2. Dewar vessel for transport and storage of liquid nitrogen.
3. Mortar with pestle.
4. Metal spoon.
5. Deep freezer (−80 °C).
6. Funnels fitting into 50 mL tubes.
7. Miracloth (Merck Millipore, Darmstadt, Hessen, Germany).
8. Sefar Nitex 03-55/32 (55 μm mesh opening, 32% open area) (Sefar AG, Heiden, Switzerland).
9. Disposable 10 and 20 mL plastic pipettes.
10. Protease inhibitors: Aprotinin, leupeptin, and pepstatin (Sigma-Aldrich, St. Louis, MO, USA). Aprotinin and leupeptin were dissolved in water (1 mg/mL), and pepstatin was dissolved in ethanol (1 mg/mL) and stored in aliquots at −20 °C [22].
11. Nuclei isolation buffer with protease inhibitors (NIB-P): 20 mM HEPES, pH 8.0, 250 mM sucrose, 1 mM $MgCl_2$, 5 mM KCl, 40% (v/v) glycerol, 0.25% (v/v) triton X-100, 0.1 mM PMSF, 0.1% (v/v) β-mercaptoethanol, 1 μg/mL aprotinin, 1 μg/mL leupeptin, and 1 μg/mL pepstatin [21] (*see* **Note 3**). Add PMSF and β-mercaptoethanol under the fume hood freshly before use.
12. Sucrose cushion (NIB-PD cushion): 20 mM HEPES, pH 8.0, 1.7 M sucrose, 1 mM $MgCl_2$, 5 mM KCl, 0.25% (v/v) triton X-100, 0.1 mM PMSF, 0.1% (v/v) β-mercaptoethanol, 1 μg/mL aprotinin, 1 μg/mL leupeptin, and 1 μg/mL pepstatin (*see* **Note 3**). Add PMSF and β-mercaptoethanol under the fume hood freshly before use.
13. Nuclei isolation buffer with 1.5 M sucrose (NIB-PD): 20 mM HEPES, pH 8.0, 1.5 M sucrose, 1 mM $MgCl_2$, 5 mM KCl, 0.25% (v/v) triton X-100, 0.1 mM PMSF, 0.1% (v/v) β-mercaptoethanol, 1 μg/mL aprotinin, 1 μg/mL leupeptin, and 1 μg/mL pepstatin (*see* **Note 3**). Add PMSF and β-mercaptoethanol under the fume hood freshly before use.
14. Cooling swing-out centrifuge for 50 mL tubes (3000 × *g* required).
15. Tubes (1.5 and 2.0 mL; Eppendorf, Hamburg, Germany).
16. Cooling centrifuge for Eppendorf tubes (1.5 mL, 2.0 mL) (16,000 × *g* required).

2.4 DpnII Digestion, End Fill-in by Biotinylated Nucleotide, and Ligation

1. *Dpn*II, 10 U/μL (New England BioLabs, Ipswich, MA, USA). Store at −20 °C.
2. RE buffer (10×): 1 M NaCl, 500 mM Tris–HCl (pH 7.9), 100 mM $MgCl_2$, 10 mM 1,4-dithiothreitol (DTT).
3. Rocking platform at room temperature.
4. Incubator cabinet (37 °C) with rocking platform.
5. Deoxynucleotides: 10 mM dATP, 10 mM dTTP, and 10 mM dGTP. Diluted from 100 mM stocks and stored at −20 °C.
6. 0.4 mM biotin-14-dCTP (Thermo Fisher Scientific, Waltham, MA, USA). Store at −20 °C.
7. Klenow DNA polymerase, large fragment (10 U/μL) (Thermo Fisher Scientific, Waltham, MA, USA). Store at −20 °C.
8. Tube rotator (multiple vendors).
9. 2% SDS solution.
10. 10% triton X-100.
11. Blunt-end ligation buffer (10×): 300 mM Tris–HCl (pH 7.8), 100 mM $MgCl_2$, 100 mM DTT, and 1 mM ATP.
12. T4 DNA ligase (5 U/μL) (Thermo Fisher Scientific, Waltham, MA, USA). Store at −20 °C.
13. Incubator cabinet (16 °C) with rocking platform.

2.5 Reversion of Cross-Link and DNA Extraction

1. SDS lysis buffer: 50 mM Tris–HCl, pH 8.0, 10 mM EDTA, pH 8.0, and 1% (v/v) SDS.
2. 4 M NaCl: Dissolve 23.34 g NaCl (Sigma-Aldrich, St. Louis, MO, USA) in 80 mL water. Then make up the volume to 100 mL using water.
3. RNase A: Dissolve RNase A (DNase-free; Macherey-Nagel, Düren, Germany) in water to 25 mg/mL, and store at 4 °C.
4. Heating block (37 °C).
5. Water bath (65 °C).
6. Proteinase K: To make a 20 mg/mL solution, dissolve 3 mg proteinase K (Thermo Fisher Scientific, Waltham, MA, USA) in 150 μL 10 mM Tris–HCl, pH 8.0. Prepare solution freshly just before use. Store proteinase K stock powder at 4 °C.
7. Phenol-chloroform-isoamyl alcohol (25:24:1) (Carl Roth, Karlsruhe, Germany) is supplemented with 0.1% hydroxyquinoline as described [23] and stored at 4 °C.
8. Chloroform (Carl Roth, Karlsruhe, Germany).
9. Glycogen (5 mg/mL): Dilute 10 μL glycogen (20 mg/mL) (Thermo Fisher Scientific, Waltham, MA, USA) with 30 μL water, and store at −20 °C.

10. 100% ethanol (Carl Roth, Karlsruhe, Germany): Precool on ice.
11. 80% ethanol (v/v): Mix 80 mL 100% ethanol and 20 mL water. Store in a tightly closed bottle. Precool on ice.
12. 3 M sodium acetate, pH 5.2: Prepare 100 mL as described [24].
13. EB: 10 mM Tris–HCl, pH 8.0. To make 50 mL of the solution, mix 500 μL 1 M Tris–HCl, pH 8.0, with 49.5 mL water.
14. Qubit 2.0 fluorometer with assay tubes, ds DNA HS assay and ds DNA BR assay reagents (Thermo Fisher Scientific, Waltham, MA, USA).
15. 6× loading dye (Thermo Fisher Scientific, Waltham, MA, USA).
16. Agarose gel electrophoresis equipment and accessories [microwave, tray (15 × 15 cm), combs, power supply, electrophoresis buffer, agarose, etc.].

2.6 Removal of Biotin from Non-ligated DNA Ends and Covarisation

1. 10× NEBuffer 2.1 (New England BioLabs, Ipswich, MA, USA) stored at −20 °C.
2. T4 DNA polymerase (5 U/μl) (New England BioLabs, Ipswich, MA, USA) stored at −20 °C.
3. 10 mM dATP and 10 mM dTTP: Dilute 10 μL of 100 mM stock with 90 μL water (Thermo Fisher Scientific, Waltham, MA, USA).
4. 0.5 M EDTA, pH 8.0, prepared as described [25].
5. Covaris S220 AFA Ultrasonicator (Covaris Ltd., Brighton, UK) and associated equipment such as Snap-Cap microTUBEs (with AFA fiber and presplit septum), chiller, software, computer.
6. Heat blocks set to 37 and 70 °C.
7. AMPure XP beads (Beckman Coulter Inc., Brea, CA, USA) stored at 4 °C. Equilibrate to room temperature and mix prior to use.
8. Magnetic particle concentrator (MPC) for 1.5 mL tubes (DynaMag-2 Magnet; Thermo Fisher Scientific, Waltham, MA, USA).
9. 80% (v/v) ethanol: Mix 80 mL 100% ethanol and 20 mL water. Prepare freshly.
10. EBT: Add 500 μL 1 M Tris–HCl, pH 8.0, and 250 μL 10% (v/v) tween 20 to 49.25 mL water.

2.7 End Repair and A-Tailing

1. 10× Tango buffer (Thermo Fisher Scientific, Waltham, MA, USA) stored at −20 °C.
2. 2.5 mM dNTP mix: Add 100 μL dNTP mix (25 mM each) (Thermo Fisher Scientific, Waltham, MA, USA) to 900 μL water, and store at −20 °C.
3. 100 mM ATP (Thermo Fisher Scientific, Waltham, MA, USA) stored at −20 °C.
4. T4 DNA polymerase (5 U/μL) (New England BioLabs, Ipswich, MA, USA) stored at −20 °C.
5. T4 polynucleotide kinase (10 U/μL) (New England BioLabs, Ipswich, MA, USA) stored at −20 °C.
6. Klenow DNA polymerase large fragment (10 U/μL) (Thermo Fisher Scientific, Waltham, MA, USA) stored at −20 °C.
7. Klenow fragment 3′→5′ exo- (5 U/μL) (New England BioLabs, Ipswich, MA, USA) stored at −20 °C.
8. Heat blocks set to 20, 37 and 65 °C.
9. AMPure XP beads (Beckman Coulter Inc., Brea, CA, USA) stored at −20 °C. Equilibrate to room temperature and mix prior to use.
10. Magnetic particle concentrator (MPC) for 1.5 mL tubes (DynaMag-2 magnet; Thermo Fisher Scientific, Waltham, MA, USA).
11. 80% ethanol: Mix 80 mL 100% ethanol and 20 mL water. Prepare freshly.
12. TLE: Add 500 μL 1 M Tris–HCl, pH 8.0, and 10 μL 0.5 M EDTA. Adjust with water to a final volume of 50 mL.

2.8 Biotin Pulldown

1. Dynabeads MyOne streptavidin C1 (Thermo Fisher Scientific, Waltham, MA, USA) stored at 4 °C.
2. DNA LoBind tubes (1.5 mL) (Eppendorf, Hamburg, Germany).
3. Rocking platform (multiple vendors).
4. Tube rotator (multiple vendors).
5. Tween wash buffer (TWB): Add 250 μL 1 M Tris–HCl, pH 8.0, 50 μL 0.5 M EDTA, pH 8.0, 250 μL 10% tween 20, and 12.5 mL 4 M NaCl. Adjust with water to a final volume of 50 mL [10].
6. 2× binding buffer (2× BB): Add 500 μL 1 M Tris–HCl, pH 8.0, 100 μL 0.5 M EDTA, pH 8.0, and 25 mL 4 M NaCl. Adjust with water to a final volume of 50 mL [10].
7. 1× binding buffer (1× BB): Dilute 10 mL 2× BB with 10 mL water.

8. $1\times$ ligation buffer: Add 90 μL $10\times$ T4 DNA ligation buffer (Thermo Fisher Scientific, Waltham, MA, USA), 90 μL 50% PEG 4000 (Thermo Fisher Scientific, Waltham, MA, USA), 720 μl water, and store at −20 °C.

2.9 Illumina Paired-End Adapter Ligation and PCR

1. Paired-end DNA adapter (AD-series) (Illumina, San Diego, CA, USA) stored at −20 °C.
2. $10\times$ T4 DNA ligation buffer (Thermo Fisher Scientific, Waltham, MA, USA) stored at −20 °C.
3. 50% PEG 4000 (Thermo Fisher Scientific, Waltham, MA, USA) stored at −20 °C.
4. T4 DNA ligase (5 U/μL) (Thermo Fisher Scientific, Waltham, MA, USA) stored at −20 °C.
5. Heat block set to 22 °C.
6. Thermocycler (multiple vendors) and appropriate PCR tubes.
7. Q5 hot start high-fidelity DNA polymerase (2 U/μL) and $5\times$ Q5 hot start buffer (New England BioLabs, Ipswich, MA, USA) stored at −20 °C.
8. 2.5 mM dNTP mix.
9. Primer 1 (100 μM, 5′-AATGATACGGCGACCACCGAGAT-3′) and primer 2 (100 μM, 5′-CAAGCAGAAGACGGCATA CGA-3′). Prepare a 10 μM mixture of primers 1 and 2 by combining 10 μL 100 μM primer 1, 10 μL 100 μM primer 2, and 80 μL water.
10. Tween wash buffer (TWB).
11. $1\times$ binding buffer (BB).
12. $10\times$ NEBuffer 2.1 (New England BioLabs, Ipswich, MA, USA) stored at −20 °C.

2.10 Cleanup and Size Selection

1. EBT: Add 500 μL 1 M Tris–HCl, pH 8.0, and 250 μL 10% tween 20–49.25 mL water.
2. AMPure XP beads (Beckman coulter Inc., Brea, CA, USA) stored at 4 °C. equilibrate to room temperature and mix prior to use.
3. Magnetic particle concentrator (MPC) for 1.5 mL tubes (DynaMag-2 magnet; Thermo Fisher Scientific, Waltham, MA, USA).
4. 80% (v/v) ethanol: Mix 80 mL 100% ethanol and 20 mL water. Prepare freshly.
5. SYBR gold (Thermo Fisher Scientific, Waltham, MA, USA) stored at 4 °C.

6. UltraPure agarose (Thermo Fisher Scientific, Waltham, MA, USA).
7. GeneRuler 50 bp DNA ladder (Thermo Fisher Scientific, Waltham, MA, USA) stored at 4 °C.
8. 1.5 mL DNA LoBind tubes (Eppendorf, Hamburg, Germany).
9. 6× loading dye (Thermo Fisher Scientific, Waltham, MA, USA).
10. 50× TAE: Dissolve 242 g Tris base in deionized water, and add 57.1 mL glacial acetic acid and 100 mL 0.5 M EDTA, pH 8.0. Adjust the volume to 1 L and mix thoroughly [26]. The 50× TAE stock is diluted with water to 1× TAE working solution (mix thoroughly).
11. Agarose gel electrophoresis equipment and accessories [microwave, tray (15 × 15 cm), combs, power supply, etc.].
12. Disposable scalpels (multiple vendors).
13. Dark reader blue light transilluminator (Clare Chemical Research, Dolores, CO, USA).
14. QIAquick Gel Extraction Kit (Qiagen GmbH, Hilden, Germany).
15. 100% Isopropanol.

2.11 Quality Controls, Quantification, and Sequencing

1. Agilent TapeStation (Agilent, Santa Clara, CA, USA) and associated material (e.g., computer, Agilent D1000 high sensitivity tape, loading buffer and ladder).
2. *Cla*I (10 U/μL) with 10× Tango reaction buffer (Thermo Fisher Scientific, Waltham, MA, USA) stored at −20 °C.
3. Qubit 2.0 fluorometer with assay tubes, ds DNA HS Assay (Thermo Fisher Scientific, Waltham, MA, USA).
4. Real-time PCR system (e.g., 7900 HT Fast Real-Time PCR system from Applied Biosystems) and associated material (e.g., optical clear adhesive, 384-well plates).
5. SYBR Green PCR Master Mix (Qiagen GmbH, Hilden, Germany).
6. Illumina sequencing device (e.g., HiSeq2500 instrument; Illumina, San Diego, CA, USA) and associated material (e.g., cBot).
7. 6× loading dye (Thermo Fisher Scientific, Waltham, MA, USA).
8. Agarose gel electrophoresis equipment and accessories [microwave, tray (15 × 15 cm), combs, power supply, electrophoresis buffer, agarose, etc.].

3 Methods

3.1 Plant Material Harvesting and Tissue Fixation

Prepare NIBF without FA, PMSF, and β-mercaptoethanol and store it on ice.

Precool four 50 mL tubes.

1. Place 25 seeds on a wet filter paper containing gibberellic acid (GA3) solution (10^{-5} M), and store the setup at 4 °C for one overnight. (This is done to break any dormancy and improve germination of barley/wheat/rye seeds.)
2. Plant the GA3-treated seeds in compost soil, and grow them under greenhouse conditions for 1 week.
3. Collect 1.5 g of leaves from 1-week-old seedlings from greenhouse, store it on ice, and bring it to lab.
4. Add formaldehyde, PMSF, and β-mercaptoethanol to the previously prepared NIB-FA without FA, and mix it by keeping on magnetic stirrer for 5 min, and store it on ice.
5. While mixing NIB-FA, cut the leaves into 0.5–1 cm pieces, transfer the leaf segments into precooled 50 mL tube, and store it on ice.
6. Add 15 mL ice-cold NIBF to each tube. Mix the contents gently with the swirling of the pipette tip.
7. Plug the tube containing leaf segments in NIBF solution with self-made polystyrene plugs as shown in Fig. 1.
8. Vacuum infiltrate the leaf tissue for 1 h in vacuum desiccator (RT; 150–165 mbar). Cut the vacuum every 15 min to facilitate the entering of fixative into the leaf tissue.
9. During the first 15 min of vacuum infiltration, take out glycine 2 M stock from −20 °C for thawing; once thawed, store it on ice.
10. At the end of 1 h vacuum infiltration, remove polystyrene plug. Add 2 mL glycine (2 M) to stop the cross-linking reaction. Mix carefully by pipetting up and down few times with 10 mL pipette.
11. Insert polystyrene plugs and vacuum infiltrate for 5 min at room temperature. Check for change of leaf color from light green to dark green. This is a control step to check for efficient cross-linking (Fig. 1).
12. Decant liquid and wash the leaves three times with plenty of double-distilled water (Fig. 2).
13. Dry leaves well with a paper towel. Transfer the dried leaves into a fresh 50 mL tube. Close the tube with a punctured cap. Freeze the tube in liquid nitrogen, and store it in −80 °C freezer or continue to nuclei isolation (*see* **Note 4**).

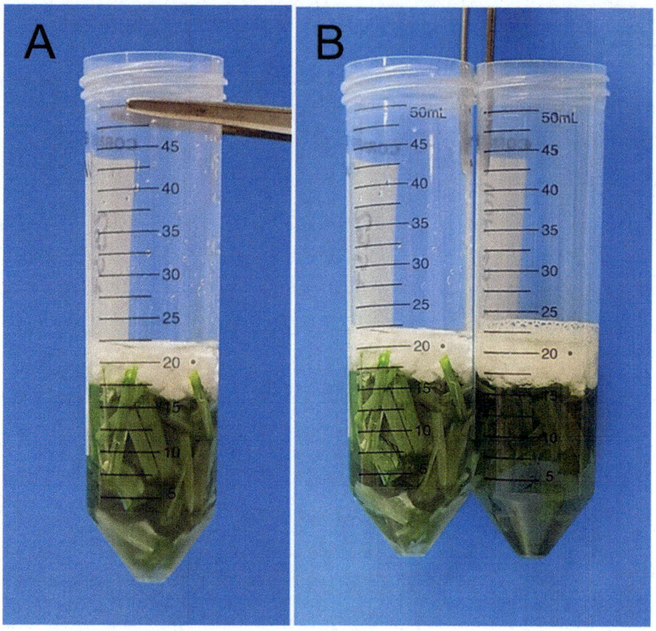

Fig. 1 Tube setup for cross-linking of leaves. (**a**) Setup of a 50 mL tube containing 0.5–1 cm leaf sections covered with polystyrene foam ready for cross-linking. (**b**) Comparison of change of leaf color from light green to dark green following 1 h of cross-linking (right)

Fig. 2 Washing setup of cross-linked leaves. Washing of cross-linked leaves using a small pore sieve and large volume of distilled water

3.2 Nuclei Isolation

Performing the steps of nuclei isolation at 4 °C yields better results.

Prepare NIB-P, NIB-PD, and NIB-PD Cushion.
Precool mortar and pestle using liquid nitrogen.
Set a water bath to 62 °C and an incubator to 37 °C.

1. Grind 1.5 g of crosslinked leaves to fine powder using pre-cooled pestle and mortar (Fig. 3). Transfer the powder into a 50 mL tube.

Fig. 3 Grinding of plant material. Finely ground powder from cross-linked leaf tissue for nuclei isolation

This material can be stored in −80 °C in precooled tubes or used immediately.

2. Add 15 mL NIB-P slowly to the ground leaf powder and thaw the powder.
3. Mix carefully by swirling the contents with the pipette tip until no clogs are visible (Fig. 4), and keep it on ice.
4. Filter the contents through one layer of Miracloth and one layer of Sefar Nitex (50 μm) placed in a funnel at cold room (4 °C). (Miracloth faces the plant material and Sefar Nitex faces the funnel) [21]. Collect the filtrate in a fresh precooled 50 mL tubes. It is important not to squeeze the filter to accelerate the process as it leads to contamination with cell debris [21] (Fig. 5).
5. Spin filtrate at 3000 × *g* for 15 min at 4 °C.
6. Carefully remove and discard the supernatant using a 10 mL pipette.
7. Suspend the pellet in 1 mL ice-cold NIB-PD by gentle mixing with pipette tip. Transfer the supernatant by pipetting slowly using wide-bore tips into a precooled 1.5 mL tube.
8. Spin the tube at 1900 × *g* for 5 min at 4 °C.
9. Discard the supernatant and resuspend the pellet in 300 μL ice-cold NIB-PD. Mix gently by swirling the solution with pipette tip. This suspension contains the nuclei.
10. Pipette 1.5 mL NIB-PD cushion (ice-cold) into a 2 mL precooled Eppendorf tube. Overlay it slowly with 300 μL ice-cold nuclei suspension from **step 9** using wide-bore tips.
11. Spin the contents at 16,000 × *g* for 1 h at 4 °C.

    During this centrifugation time, prepare RE buffer and store it on ice, and 0.5% SDS solution, and store it at room temperature.

Fig. 4 Ground plant material in nuclei isolation buffer. Ground material from cross-linked leaf tissue suspended in NIBP buffer for nuclei isolation

Fig. 5 Filtration setup of the ground material for nuclei isolation. Ground material from cross-linked leaf tissue suspended in NIBP buffer filtered through one layer of Miracloth and one layer of Sefar Nitex cloth for removal of cell debris

12. Discard the supernatant and suspend the pellet in 100 μL ice-cold NIB-PD. Mix gently by swirling the contents gently with pipette tip (nuclei suspension) (*see* **Note 5**).
13. Collect 5 μL of nuclei suspension as an intermediate nuclei preparation control before proceeding to next step, and store it at −20 °C.
14. Spin the nuclei suspension at 1900 × *g* for 5 min at 4 °C.
15. Discard the supernatant.
16. Gently and slowly suspend the pellet in 300 μL of RE buffer using wide-bore tips.
17. Centrifuge the contents at 3000 × *g* for 5 min at 4 °C. Discard the supernatant.
18. Gently and slowly suspend the pellet in 150 μL of 0.5% SDS using wide-bore tips. Transfer the resuspension in 50 μL aliquots into three 2.0 mL tubes.
19. Incubate the nuclei suspension in 0.5% SDS solution at 62 °C for 15 min (*see* **Note 6**).

    Prepare fresh 10% triton X-100 during this incubation.
20. To each tube, add 145 μL of water and 25 μL of freshly prepared 10% (v/v) Triton X-100. Mix the contents gently by inverting the tube, and incubate at 37 °C for 15 min with 350 rpm shaking (*see* **Note** 7).

3.3 DpnII Digestion, End Fill-in with Biotinylated Nucleotide, and Blunt-End Ligation

Set a water bath to 62 °C.

Set an incubator to 16 °C.

Prepare fresh 10× blunt-end ligation buffer.

1. To each tube containing permeabilized nuclei, add 25 μL of 10× RE buffer and 60 U of *Dpn*II. Mix gently by inverting the tube 8–10 times, and incubate overnight at 37 °C with 250 rpm shaking.
2. Incubate the tubes at 62 °C for 20 min. Cool the tubes to room temperature (~5 min). Transfer the tubes on ice (*see* **Note 8**).
3. Take out 5 μL of digested DNA from each tube before proceeding to next step as an intermediate digested DNA control, and store it at −20 °C.
4. To each tube containing the restriction digested DNA, add 1 μL of 10 mM dTTP, 1 μL of 10 mM dATP, 1 μL of 10 mM dGTP, 25 μl of 0.4 mM biotin-14-dCTP, 14 μL of water, and 4 μL of Klenow fragment (40 U). Mix gently by inverting the tube, and incubate at 37 °C for 2 h at 350 rpm (*see* **Note 9**).
5. To each tube containing end filled-in products, add 663 μL of water, 120 μL of 10× blunt-end ligation buffer, 100 μL of 10% (v/v) triton X-100, and 50 U of T4 DNA ligase. Mix gently and incubate at 16 °C overnight with 65 rpm shaking.

3.4 Reversion of Cross-Link

Prepare SDS lysis buffer fresh and store it at room temperature.

Prepare proteinase K solution in SDS lysis buffer and store it at room temperature.

Set up a thermomixer to 55 °C and a water bath to 65 °C.

1. Centrifuge the blunt-end ligation products at 3000 × *g* for 10 min at room temperature. Discard the supernatant carefully, and suspend the pellet with 750 μL of SDS lysis buffer.
2. To each tube, add 10 μL of proteinase K (18 mg/mL) solution prepared in SDS lysis buffer. Mix gently by inverting the tube and incubate at 55 °C for 30 min at 350 rpm.
3. To each tube, add 37.5 μL of 4 M NaCl. Mix gently by inverting the tube and incubate at 65 °C for overnight.

3.5 DNA Extraction

Precool the centrifuge at 4 °C; precool 12 × 2,0 mL Eppendorf tubes.

Switch on thermomixer and set it at 37 °C.

Thaw glycogen and store it on ice.

1. Thaw nuclei preparation control and digested DNA control, and make up the volume to 100 μL using EB buffer.
2. To each tube containing reverse cross-linked DNA from Subheading 3.4, add 750 μL (100 μL each for nuclei preparation control and digested DNA control) of phenol-chloroform-IAA (25:24:1) solution. Mix well vigorously for 2 min. Centrifuge the tubes at 13,000 × *g* for 5 min at room temperature. Transfer the upper aqueous phase to a new 2.0 mL tube.
3. Add 75 μL (10 μL for nuclei preparation control and digested DNA control) of 3 M sodium acetate (pH 5.2); add 3.75 μL (0.5 μL each for nuclei preparation control and digested DNA control) of glycogen (5 mg/μL) and 750 μL (100 μL for nuclei preparation control and digested DNA control) of ice-cold isopropanol.
4. Mix by inverting tubes 10 times. Incubate the tubes on ice for 20 min.
5. Centrifuge the tubes at 13,000 × *g* for 30 min at 4 °C. Carefully remove the supernatant. It is better to store the supernatant if unsure about the presence of the pellet.
6. Wash the pellet with 500 μL 80% (v/v) ethanol. Centrifuge the tubes at 13,000 × *g* for 20 min at 4 °C. carefully remove the supernatant.
7. Air-dry the pellets (5–10 min), and dissolve it in 100 μL (20 μL for nuclei preparation control and digested DNA control, and store it at 4 °C) of EB buffer.
8. Combine the dissolved DNA from three tubes to one tube. Add 2 μL of RNase A (10 mg/mL). Mix by pipetting and incubate at 37 °C for 30 min at 350 rpm.

9. Based on the sample volume (~300 μL), add 1/10 volume of 3 M sodium acetate (~30 μL) and an equal volume of ice-cold isopropanol (~300 μL). Mix by inverting the tube.
10. Incubate on ice for 20 min.
11. Centrifuge the tube at 13,000 × *g* for 30 min at 4 °C. Carefully remove the supernatant.
12. Wash the pellet with 500 μL 80% (v/v) ethanol. Centrifuge at 13,000 × *g* for 20 min at 4 °C. Carefully remove the supernatant completely.
13. Air-dry the pellet, and dissolve it in 100 μL of EB buffer.
14. Measure DNA concentration using Qubit HS.
15. Run a 0.8% agarose gel to check the nuclei preparation control (15 μL + 3 μL 6× loading dye), digested DNA control (15 μL + 3 μL 6× loading dye), and blunt-end ligated DNA (at least 50 ng in 15 μL + 3 μL 6× loading dye) for 30 min at 150 V (Fig. 6) (*see* **Note 10**).

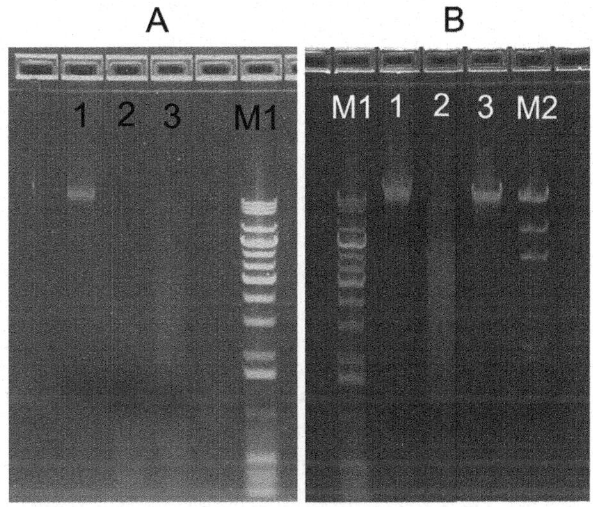

Fig. 6 Agarose gel image of intermediate control for nuclei isolation, digested DNA, and ligated DNA. (**a**) 0.8% agarose gel image of nuclei isolation control (1), *Dpn*II-digested DNA control (2), and blunt-end ligation products (3). (**b**) 0.8% agarose gel image of nuclei isolation control (1), *Hind*III-digested DNA control (2), and blunt-end ligation products (3). M1 corresponds to the banding pattern from 1 kb extension ladder from Invitrogen Corp., Carlsbad, CA, USA. M2 corresponds to 100 bp ladder from Thermo Fisher Scientific, Waltham, MA, USA. The size profile of nuclei isolation control shows fragments over 40 kb in size. *Hind*III-digested DNA runs as a smear from low molecular DNA fragments to fragments close 40 kb in size. *Dpn*II-digested DNA runs as a smear from 100 bp to 1 kb in size. Blunt-end ligation products run as a smear from low molecular weight fragments to fragments close to 40 kb in size

3.6 Removal of Biotin from Non-ligated DNA Ends and Covarisation

Set up thermomixer at 20 °C.

Bring AMPure XP beads to RT.

1. Start with 5 μg of blunt-end ligated DNA. Add 10 μL of 10× NEB2.1 buffer, 2 μL of 10 mM dTTP, 2 μL of 10 mM dATP, 20 U (4 μL of 5 U/μL) of T4 DNA polymerase, and water to bring the volume up to 100 μL. Incubate at 20 °C for 4 h at 350 rpm (*see* **Note 11**).
2. To each tube, add 5 μL of 0.5 M EDTA solution, and mix well by pipetting up and down to stop the reaction. Store it at −20 °C or proceed to the next step.
3. Adjust the volume of the solution to 130 μL using water. Transfer the contents to microtube Covaris tubes without any bubbles.
4. Shear DNA to 100–500 bp using a Covaris S220 using the following protocol (*see* **Note 12**).

Duty factor [%]	10
PIP	175 W
Cycles per burst	200
Set mode	Frequency sweeping
Time	30 s

5. Number of cycles (repetitions): 2 × 30 s.
6. Add 1.8× volume of AMPure XP beads (equilibrated to room temperature). Mix well by pipetting up and down.
7. Incubate at room temperature for 10 min. Reclaim beads in MPC for 3 min. Discard the supernatant.
8. Wash beads two times with 400 μL 80% (v/v) ethanol while placed in the MPC.
9. Air-dry beads completely. Add 52 μL EB, resuspend beads, and incubate for 10 min at room temperature. Tap the tubes every 2 min. Place the tube in MPC for 3 min, and transfer 51.1 μL of the supernatant to a fresh Eppendorf tube. Sample can be stored at −20 °C at this step.

3.7 End Repair and A-Tailing

Set a water bath to 65 °C.

Take out AMPure XP beads for equilibrating to room temperature.

Set a thermomixer to 20 °C.
Set a thermomixer to 37 °C.
Set a thermomixer to 22 °C.

1. Take out the AMPure XP bead purified covaris fragments from −20 °C freezer. Thaw it by keeping on ice. Add the following reagents in the order indicated:

10× Tango (Fermentas)	7.0 μL
2.5 mM dNTP Mix	7.0 μL
100 mM ATP	0.7 μL
T4 DNA polymerase (5 U/μL)	1.5 μL
T4 polynucleotide kinase (10 U/μL)	2.5 μL
Klenow DNA polymerase, large fragment (10 U/μL)	0.25 μL (total volume 70 μL)

2. Mix by pipetting up and down and incubate for 30 min at 20 °C.
3. Add 1.8 times the volume of AMPure XP beads (126 μL) to the end-repaired DNA (70 μL). Mix by pipetting.
4. Incubate at room temperature for 10 min. Reclaim the beads by placing it on MPC for 3 min. Discard the supernatant.
5. Wash beads two times with 400 μL 80% (v/v) ethanol while placed in the MPC.
6. Air-dry beads completely. Add 42 μL TLE, and resuspend beads by mixing the contents with the pipette.
7. Incubate at room temperature for 10 min. Mix the contents by tapping the tube every 2 min. Place the tube in MPC for 1–3 min. Transfer 38 μL of eluate to a new 1.5 mL tube.
8. Perform A-tailing on each eluate. Add the following in the order indicated (total volume 50 μL):

Purified DNA (one eluate, 5 μg)	38 μL
10× NEB2	5 μL
10 mM dATP	1 μL
Klenow fragment (3 → 5 exo-) (5 U/μL)	3 μL

9. Incubate in the heat block at 37 °C for 30 min 350 rpm.
10. Add 1 μL of 0.5 M EDTA to stop the reaction. Mix well by pipetting and incubate at 65 °C for 20 min.
11. Place samples on ice immediately after the incubation at 65 °C for 20 min.

3.8 Biotin Pulldown

All subsequent steps are performed in LoBind tubes with low retention pipette tips.

Prepare 1× ligation buffer.

1. Vortex the C1 Dynabeads, and transfer 10 μL of beads to 1.5 mL LoBind Eppendorf tube.
2. Add 400 μL of TWB and pipette beads up and down. Incubate on a rocking platform for 3 min.

3. Reclaim beads by placing the tube in MPC for 1 min. Discard supernatant.
4. Resuspend beads in 400 μL TWB, reclaim beads in MPC (1 min), and discard supernatant.
5. Resuspend beads in 50 μL 2× BB, and add the 50 μL of end-repaired DNA from **step 11** of end repair and A-tailing (3.7).
6. Incubate the tubes at room temperature for 30 min with gentle rotation.
7. At the end of 30 min incubation, place the tube in MPC for 1 min. Discard supernatant.
8. Resuspend beads in 800 μL 1× BB. Transfer suspension into a new LoBind tube. Incubate at room temperature for 5 min. Reclaim beads by placing it in MPC for 5 min.
9. Repeat **step 8** for a total of two times.
10. Wash beads with 100 μL 1× ligation buffer.
11. Place the tube in MPC for 1 min. Discard supernatant.
12. Resuspend beads in 38.75 μL 1× ligation buffer.

3.9 Illumina Adapter Ligation and Paired-End PCR

Prepare 1× NEB2.1.

1. Set up the adapter ligation reaction by adding the components in the order indicated.
2. Illumina PE adapter (*see* **Note 13**) 6 μL.

TTC DNA bound to Dynabeads in 1× ligation buffer	38.75 μL
10× T4 DNA ligation buffer (Fermentas)	1.13 μL
PEG 4000 (50%)	1.13 μL
Water	1.0 μL
T4 DNA ligase (5 U/μL) Fermentas	2.0 μL

Mix well by pipetting up and down five times. Short spin to bring the contents to the bottom of the tube.

3. Incubate the ligation reaction tube at 22 °C for 1 h.
4. Reclaim beads by placing in MPC for 3 min. Discard supernatant.
5. Resuspend beads in 400 μL TWB. Incubate the tube at room temperature for 5 min with gentle rotation. Place the tube in MPC for 1 min. Discard supernatant.
6. Repeat **steps 4** and **5** twice.
7. Resuspend beads in 200 μL 1× BB.

8. Reclaim beads by placing the tube in MPC for 1 min. Discard supernatant.
9. Resuspend beads in 200 μL 1× NEB2.1.
10. Place the tube in MPC for 1 min. Discard supernatant.
11. Repeat **steps 10** and **11**.
12. After the last wash, resuspend the beads in 20 μL 1× NEB2.1. Transfer the contents to a new tube and store on ice.

    Library preparation can be stopped at this step. Sample can be stored at −20 °C or continued to the next step immediately.
13. Setup a PCR reaction as given below to determine the optimal number of cycles.

    PCR to titrate number of cycles (one reaction/sample; volume 25 μL):

Bead-bound Hi-C DNA	1.5 μL
Illumina forward and reverse primers	1.25 μL
Q5 Hot Start (5×)	5.0 μL
2.5 mM dNTP	2.0 μL
Q5 Hot Start Pol (2 U/μL)	0.25 μL
Wate	15.0 μL

Program

Step 1: 98 °C	30 s	
2: 98 °C	10 s	
3: 68 °C	30 s	for Q5 Hot Start Pol
4: 72 °C	30 s	
5: Go to **step 2** and repeat 8×, 11×, and 14×		
6: 72 °C	3 min	
Keep at 8 °C		

Take 5 μL aliquots after 9, 12, and 15 cycles by pausing the PCR reaction.

14. Mix 1 μL of 6× loading dye to the PCR product aliquots, and load it on 2% agarose gel along with 50 bp ladder from Thermo Scientific. Run the gel for 30 min at 150 V (Fig. 7).
15. Build a calibration curve by plotting the DNA quantity against the number of cycles, and find the suitable cycle number to have complex sequencing libraries with minimum duplicates. Increasing the number of cycles can lead to changes in library size profile and complexity. So it is advised to use less number

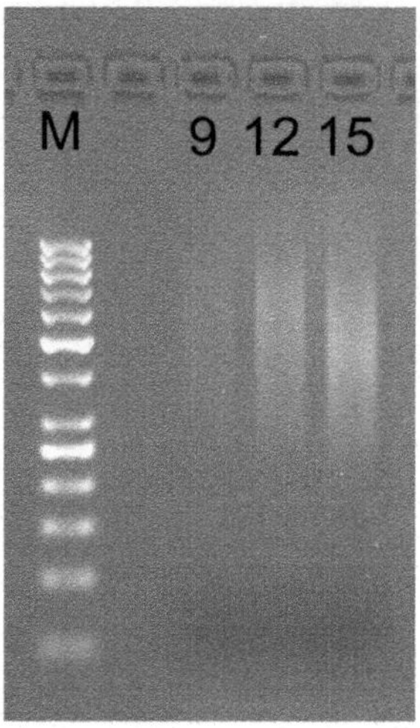

Fig. 7 Agarose gel image of titration PCR products. Typical 2% agarose gel showing concentration and size profile of titration PCR products after 9, 12, and 15 cycles of PCR. M corresponds to 100 bp ladder from Thermo Fisher Scientific, Waltham, MA, USA. Optimal number of cycle is 12 cycles in this case as it yields sufficient amplification and maintains the original library size profile

of cycles and perform more number of PCR reactions (*see* **Note 14**).

16. Perform eight PCR reactions as given above with chosen cycle number.
17. Pool the PCR products from all eight reactions into a new Eppendorf tube. Place the tube in MPC for 1 min, and transfer supernatant into a new 1.5 mL Eppendorf tube.
18. Determine the volume of the PCR product (expected volume is around 180–200).
19. Use AMPure XP beads for reaction cleanup and primer dimer removal.

    Equilibrate AMPure XP beads to room temperature. Vortex well and add 1.8 × volume of AMPure XP beads to the PCR product.
20. Mix well by pipetting and spin down briefly.
21. Incubate the tube for 10 min at room temperature. Place the tube on MPC for 5 min. Discard supernatant.
22. Wash beads two times with 400 μL 80% ethanol while in the MPC.
23. Air-dry completely.
24. Suspend the beads in 27 μL EB buffer. Incubate the tube for 10 min at room temperature. Mix the contents of the tube by tapping the tube every 2 min.

25. Reclaim beads in MPC (5 min), and transfer 25 μl containing the hi-C library into a new 1.5 mL Eppendorf tube.
26. Measure the concentration using Qubit BR reagent kit.

3.10 Size Selection

1. Add 3 g agarose in 150 mL TAE to prepare 2% agarose gel. Boil in 500 mL Erlenmeyer flask.
2. Cool to 60 °C and add 15 μL SYBR gold. Cover gel because of light-sensitive nature of SYBR gold.
3. Add 4 μL 6× loading dye (Fermentas) to 24 μL of the Hi-C library, and load 14 μL into two adjacent lanes.
4. Load 1 μL of 50 bp ladder from Thermo Scientific as a marker for size selection.
5. Run the electrophoresis for 60 min at 150 V. Cover the gel as SYBR gold is light sensitive.
6. Place gel on Dark Reader Transilluminator (Fig. 8) (*see* **Note 15**).
7. Cut the gel containing fragment sizes in the region of 250–500 bp with a clean scalpel, and place the gel piece in a 5 mL Eppendorf tube.

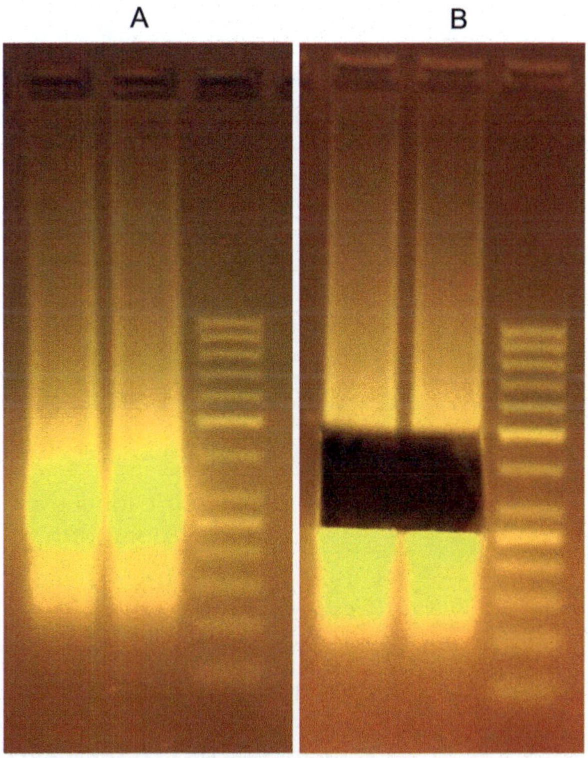

Fig. 8 Agarose gel stained with SYBR Gold for size selection. 2% agarose gels stained with SYBR Gold dye containing the PCR amplified in situ Hi-C libraries. (**a**) The left image contains the typical Hi-C library size profile, and (**b**) the image on right shows the gel size selected for 250–500 bp fragments

8. Determine the weight of the gel piece.
9. Add six times the volume of QG buffer compared to the weight of gel slice. Rotate the tube gently using the rotator for 20 min at room temperature or till the gel is dissolved completely.
10. Add one volume of isopropanol to the dissolved gel in QG buffer. Rotate for 5 min.
11. Apply 720 μL of the contents to MinElute columns, and spin for 1 min at 13,000 × *g* at room temperature. Discard flow-through. Repeat this step until all the solution is finished.
12. Add 500 μL QG to the columns, and spin for 1 min at 13,000 × *g* at room temperature. Discard flow-through.
13. Add 740 μL PE solution. Incubate for 1 min at room temperature, and spin for 1 min at 13,000 × *g*. Discard flow-through.
14. Rotate column 180° and spin for another 1 min at 13,000 × *g*.
15. Place column in a fresh 1.5 mL lo bind Eppendorf tube.
16. Add 50 μL pre-warm EB to the column kept inside an Eppendorf tube. Incubate for 5 min at room temperature, and spin for 1 min at 13,000 × *g* at room temperature.
17. Collect the eluate (size selected Hi-C library) and store it at −20 °C freezer.

3.11 Quality Control of the Library and Sequencing

1. Measure the concentration using Qubit HS reagent kit.
2. Set up a digestion reaction with *Cla*I by adding the following reagents (Fig. 9) (*see* **Note 16**):

Negative control: no *Cla*I	
Hi-C	50 ng
10× Tango buffer	2 μL
*Cla*I	±1 μL
Water	to 20 μL

3. Incubate the tubes at 37 °C overnight with 350 rpm shaking.
4. Add 3 μL of 6× loading dye to 20 μl reaction products. Run on 2% agarose gel along with 50 bp ladder at 150 V for 30 min.
5. Run 2 μL of Hi-C library on Agilent D1000 high sensitivity tape to check for the average insert size of the library (Fig. 10).
6. Perform qPCR along with known concentration standards to identify the library concentration containing the adapter-ligated fragments. Based on the concentration, dilute the libraries further to the appropriate loading concentration, and sequence them on Illumina sequencers as per the manufacturer's instructions [27].

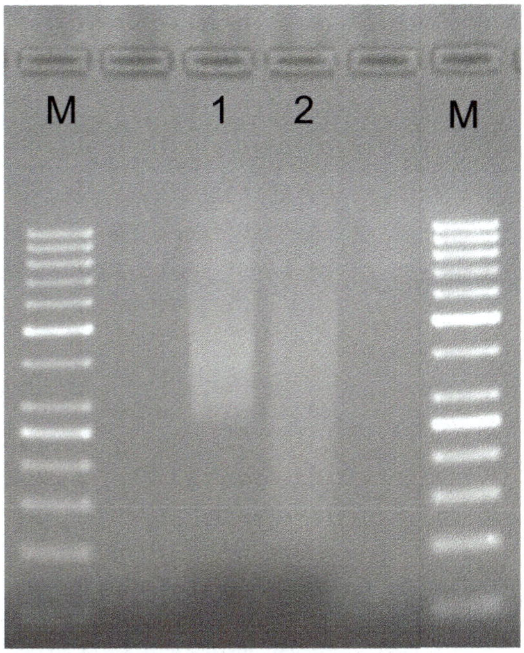

Fig. 9 Agarose gel image of in situ Hi-C library quality control. 2% agarose gel image showing (1) undigested library control and (2) *Cla*I digestion of in situ Hi-C library compared to (M) the 100 bp ladder

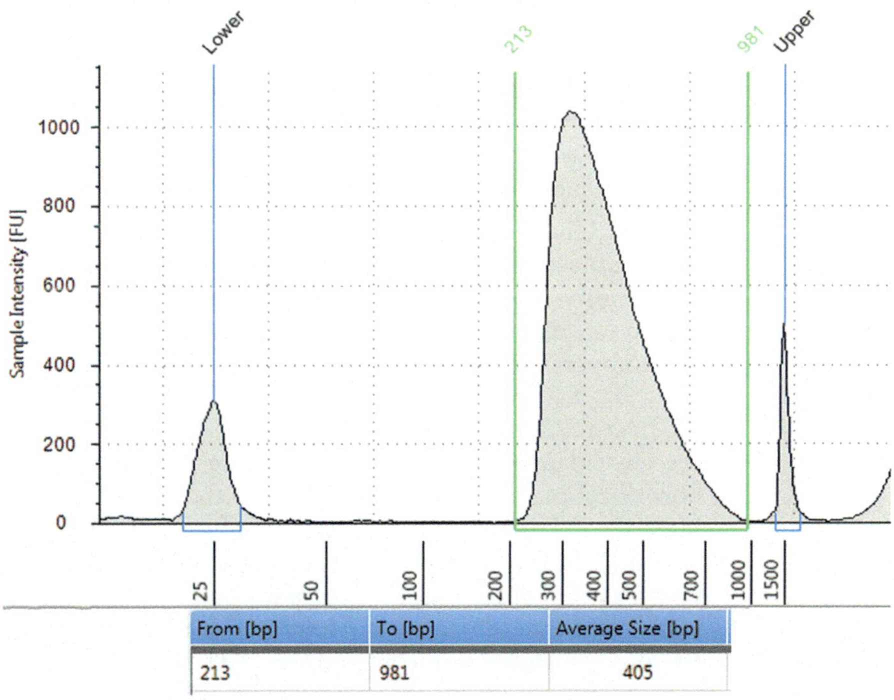

From [bp]	To [bp]	Average Size [bp]
213	981	405

Fig. 10 Agilent TapeStation size profile of the in situ Hi-C library. Agilent TapeStation D1000 High Sensitivity profile of a typical in situ Hi-C library along with the average size

3.12 Primary Sequence Data Analysis

1. Run the Illumina CASAVA pipeline (http://support.illumina.com/sequencing/sequencing_software/casava.html) to obtain deconvoluted read files in FASTQ format.
2. Trim the reads at the junction site with cutadapt [28] using the adapter sequence "GATCGATC".
3. Align the trimmed read pairs to an appropriate reference genome with BWA mem [29]. The two reads should be mapped as single ends by specifying the parameters "-S" and "-P". the parameter "-M" should be used to mark shorter split hits as secondary.
4. Sort BAM files by reference position for duplicates removal. Then sort the BAM files again by read name to have the two ends of a pair at adjacent rows of the BAM file. One of the tools SAMtools [30], Picard (http://broadinstitute.github.io/picard/), or Novosort (http://www.novocraft.com/products/novosort/) can be used.
5. Remove reads with nonunique alignments, and discard secondary alignments, unmapped reads, and duplicated reads with command "samtools view –q 10 –F1284".
6. Assign reads to restriction fragments with BEDTools [31]. You will need a BED file (https://genome.ucsc.edu/FAQ/FAQformat.html#format1) with the positions of restriction fragments (i.e., regions between two *Dpn*II sites) obtained from a *Dpn*II in silico digest of your reference genome sequence. Use the command "bedtools pairtobed –bedpe –f 1 –type both -abam". The resulting BEDPE file is a tab-separated file that can be processed with standard UNIX text processing tools such as AWK or Perl. The assignments of the two ends of a pair to restriction fragments are on adjacent lines.
7. Use UNIX scripts (e.g., in the AWK language) to calculate the insert size based on the distance of alignment start positions to neighboring *Dpn*II sites. Discard fragments with insert sizes above 500 bp, plot the insert size distribution (Fig. 11), and compare it to that obtained with the Agilent Bioanalyzer.

3.13 Hi-C Data Analysis

We refer to the review article of Lajoie et al. [32] for a summary of the Hi-C data analysis methods including primary read alignment, normalization, and biological questions that can be addressed with Hi-C data. These include the analysis of genome compartmentalization [33], topologically associating domains [34] and structural variants [35]. Moreover, Hi-C data has been used for chromosome-scale scaffolding of genome sequence assemblies [36, 37] and species delineation in metagenomics [38].

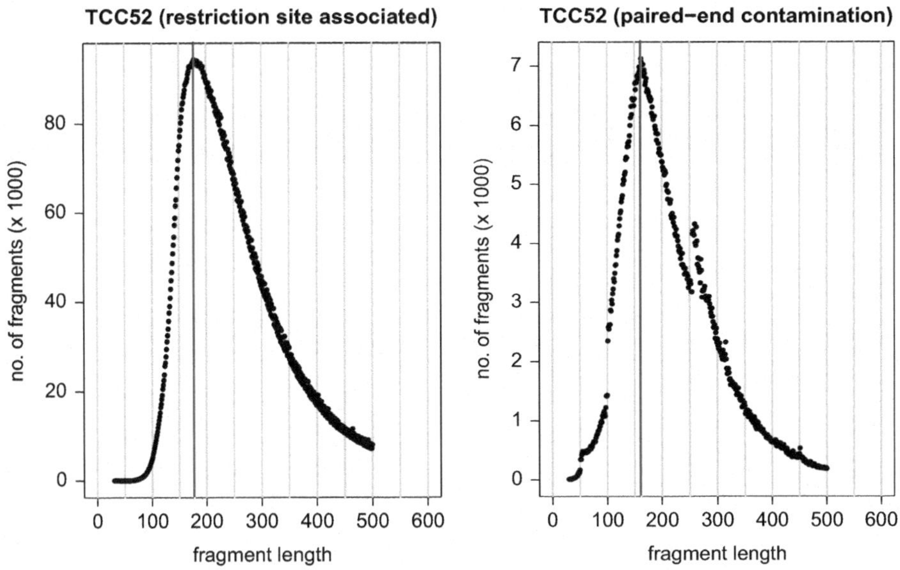

Fig. 11 Size profile of real interaction pairs and paired-end contamination reads. Distribution of real intramolecular interaction pairs from in situ Hi-C library sequencing data (left) along with the paired-end contaminant read distribution (right)

4 Notes

1. In case of barley/wheat/rye, 25 1-week-old seedlings provide about 1.2–1.5 g of fresh leaf material, which is sufficient for this method. For any other species, the amount of seeds to be sown and leaf material to be harvest may vary. We recommend growing enough plant material to obtain at least 10^6–10^7 nuclei.
2. During the preparation of NIBF, once all the components are added, stir thoroughly using a magnetic stirrer, and precool the solution on ice. PMSF is unstable in water. Due to the short half-life of 35 min at pH 8.0 [30], PMSF must be added just prior to use [39]. Formaldehyde slowly oxidizes to formic acid under normal atmospheric oxygen concentrations. Therefore, buy only small quantities and use them shortly after purchase. Poor-quality formaldehyde will adversely affect the experiment. Store formaldehyde at room temperature, because a precipitate of trioxymethylene may be formed during storage at cold conditions.
3. For the preparation of NIB-P, NIB-PD, and NIB-PD cushion, add the reagents in order provided, and add PMSF, β-mercaptoethanol, aprotinin, leupeptin, and pepstatin just before use. Mix the solution thoroughly by using a magnetic stirrer, and precool on ice before using it. Glycerol is viscous (density 1.26 g/cm^3); thus it sticks to the pipette tips leading

to inaccurate measurement of quantities. So 75.6 g of glycerol (60 mL × 1.26 g/cm^3) can be weighed using a balance instead of measuring 60 mL glycerol.

4. Fixed leaf tissues appear dark green in color compared to the initial leaf tissue used for fixation (Fig. 1). If darkening of the leaves is not observed, the fixative might not have entered completely, thus indicating incomplete cross-linking. Optimize the infiltration conditions, for example, by increasing the vacuum or cutting shorter leaf segments. Properly fixed leaf tissues are either processed immediately or can be stored at −80 °C for 2–3 days before proceeding to nuclei isolation.

5. To check for quality of nuclei, stain 7 μL nuclei suspension with 7 μL Vectashield containing DAPI for the integrity examination [21]. Mount the stained sample onto a microscopic slide and add the cover slip. Analyze the nuclei using the epifluorescence microscope (×100 objective, ×1000 magnification, DAPI-filter, absorption 358 nm, emission 461 nm). Intact nuclei appear round or oval and show sharp contours as shown in Hovel et al. [21]. To check for quantity of nuclei, stain another aliquot of the nuclei suspension with Vectashield containing DAPI. Pipette mixture onto the counting chamber, and count it using the epifluorescence microscope using a ×20 objective lens at ×200 magnification using the DAPI-filter. Determine the concentration of the nuclei, and continue the Hi-C library preparation with approximately 10^7 nuclei.

6. Addition of SDS and incubation at 62 °C for 15 min help with accessibility of chromatin for restriction digestion, inactivation of endogenous nucleases, and removal of non-cross-linked proteins from DNA. The duration of incubation at this step has to be optimized in a sample-dependent manner. Shorter incubation time leads to inefficient or partial digestion due to inaccessibility of chromatin to restriction enzyme. Longer incubation time leads to destruction of clear chromatin territories and may even lead to reversion of cross-links [40].

7. Triton X-100 is used at this step to quench the SDS from solution, which inhibits the enzymatic activities of restriction enzymes, Klenow fragment, and T4 DNA ligase. Based on our experience, use of freshly prepared Triton X-100 yields better results.

8. This step is used for inactivation of the restriction enzyme used. The conditions given here are for inactivation of *Dpn*II. Inactivation of *Eco*RI or *Xho*I can be performed by addition of 1.6% SDS and heating at 65 °C for 20 min [1]. So appropriate conditions for inactivation of restriction enzymes used need to be selected.

9. During this reaction, conduct a 3C control reaction using no Klenow fragment. Then perform a PCR reaction from both 3C control and Hi-C sample using the primers designed from two DNA fragments, which are in closer proximity in 3D space of the nuclei. For barley, primers 3 and 4 with sequences 10 μM, 5′-ATCTTCATGCGAGGCAGAGT-3′ and 10 μM, 5′-ACCG TTGAACCATCTTCAGG-3′, respectively, can be used. Upon digestion of PCR products using *Dpn*II and *Cla*I and double digestion of 3C control and hi-C products and checking the fragments on agarose gel, biotin incorporation and blunt-end ligation efficiency can be determined. Complete digestion of Hi-C products only with *Cla*I and not with *Dpn*II shows good end filling and ligation efficiency.
10. DNA from nuclear preparation runs as a band of >40 kb, digested DNA runs as a smear of low molecular weight fragments, and the ligated DNA runs as a smear to band of intermediate sizes between digested DNA and DNA from nuclear preparation. This shows the efficient digestion and blunt-end ligation [41].
11. This reaction is performed to remove the biotinylated nucleotides from un-ligated dangling-end products. Suboptimal performance of this reaction can lead to higher amounts of sequencing reads from unwanted dangling-end products and not from real interaction pairs. Exonuclease III-based biotinylated nucleotide removal from non-ligated products may be used as an alternative for this step. Exonuclease III from *E. coli* catalyzes the stepwise removal of mononucleotides from 3-′-hydroxyl termini of duplex DNA. Thus, exonuclease III allows for the removal of biotinylated residues from the non-ligated DNA ends. Because phosphorothioate linkages located 5′ to the biotinylated cytosine residue are not cleaved by exonuclease III, true ligation junctions are preserved [42].
12. Covaris fragmentation conditions given were tested for barley and wheat. We recommend checking the size profile of Covaris fragments from your sample. Modify the Covaris conditions, if the fragment size profile is not in the expected 100–500 bp interval.
13. Adapters for Hi-C library construction should be chosen by the use of Illumina Experiment Manager (IEM). Different compatible combinations of adapters must be used, if the sequencing of two or more Hi-C libraries is planned to be performed on a single lane of an Illumina flow cell.
14. In routine experiments, 11–12 amplification cycles are sufficient. Eight PCR reactions (25 μL volume per reaction) should yield sufficient DNA for qPCR-based quantification of sequencing libraries and sequencing. To sustain the complexity

of libraries, keep the number of cycles low, and increase the number of PCR reactions only to obtain sufficient library for sequencing.

15. Do not use ethidium bromide-stained gels and prolonged exposure of gels to ultraviolet radiation that will damage the DNA. So SYBR Gold dye and visible blue light emitted from a "Dark Reader" transilluminator are recommended to be used as the source of excitation.
16. Ligation of filled-in *Hind*III sites (AAGCTT) constitutes novel sites for the restriction enzyme *Nhe*I (GCTAGC). Ligation of filled-in *Dpn*II sites (GATC) generates sites for the restriction enzyme *Cla*I (ATCGAT). To control for effectiveness of fill-in reaction and blunt-end ligation, each Hi-C library is digested with the specific restriction enzyme, and the products are checked on 2% agarose gel under UV light. The shift in size profile of the digested library toward low molecular weight fragments shows the efficient end fill-in and blunt-end ligation and presence of true interaction pairs [41].

Acknowledgments

This work was financially supported by core funding of the Leibniz Institute of Plant Genetics and Crop Plant Research (IPK), Gatersleben, and project funding ("SHAPE") from the German Federal Ministry of Education and Research (BMBF, Grant no. 031B0190A) to Dr. Nils Stein and Dr. Martin Mascher. We thank Dr. Erez Lieberman-Aiden and Dr. Olga Dudchenko for helpful discussions on the in situ Hi-C method. We thank Ines Walde and Manuela Knauft for excellent technical assistance during protocol development and TCC/Hi-C library construction.

References

1. Chekanova JA, Gregory BD, Reverdatto SV, Chen H, Kumar R, Hooker T, Yazaki J, Li P, Skiba N, Peng Q, Alonso J, Brukhin V, Grossniklaus U, Ecker JR, Belostotsky DA (2007) Genome-wide high-resolution mapping of exosome substrates reveals hidden features in the Arabidopsis transcriptome. Cell 131(7):1340–1353. https://doi.org/10.1016/j.cell.2007.10.056
2. Kapranov P, Cheng J, Dike S, Nix DA, Duttagupta R, Willingham AT, Stadler PF, Hertel J, Hackermüller J, Hofacker IL, Bell I, Cheung E, Drenkow J, Dumais E, Patel S, Helt G, Ganesh M, Ghosh S, Piccolboni A, Sementchenko V, Tammana H, Gingeras TR (2007) RNA maps reveal new RNA classes and a possible function for pervasive transcription. Science 316(5830):1484–1488. https://doi.org/10.1126/science.1138341
3. Jin J, Liu J, Wang H, Wong L, Chua N-H (2013) PLncDB: plant long non-coding RNA database. Bioinformatics 29(8):1068–1071. https://doi.org/10.1093/bioinformatics/btt107
4. Wang H, Chung PJ, Liu J, Jang I-C, Kean MJ, Xu J, Chua N-H (2014) Genome-wide identification of long noncoding natural antisense transcripts and their responses to light in Arabidopsis. Genome Res 24(3):444–453. https://doi.org/10.1101/gr.165555.113

5. Wang H, Chekanova JA (2017) Long noncoding RNAs in plants. Adv Exp Med Biol 1008:133–154. https://doi.org/10.1007/978-981-10-5203-3_5
6. Dekker J, Rippe K, Dekker M, Kleckner N (2002) Capturing chromosome conformation. Science 295(5558):1306–1311. https://doi.org/10.1126/science.1067799
7. Louwers M, Bader R, Haring M, van Driel R, de Laat W, Stam M (2009) Tissue and expression level specific chromatin looping at Maize b1 Epialleles. Plant Cell 21(3):832–842. https://doi.org/10.1105/tpc.108.064329
8. Simonis M, Klous P, Splinter E, Moshkin Y, Willemsen R, de Wit E (2006) Nuclear organization of active and inactive chromatin domains uncovered by chromosome conformation capture-on-chip (4C). Nat Genet 38:1348. https://doi.org/10.1038/ng1896
9. Dostie J, Richmond TA, Arnaout RA, Selzer RR, Lee WL, Honan TA (2006) Chromosome Conformation Capture Carbon Copy (5C): a massively parallel solution for mapping interactions between genomic elements. Genome Res 16:1299. https://doi.org/10.1101/gr.5571506
10. Lieberman-Aiden E, van Berkum NL, Williams L, Imakaev M, Ragoczy T, Telling A (2009) Comprehensive mapping of long-range interactions reveals folding principles of the human genome. Science 326:289. https://doi.org/10.1126/science.1181369
11. Kalhor R, Tjong H, Jayathilaka N, Alber F, Chen L (2011) Solid-phase chromosome conformation capture for structural characterization of genome architectures. Nat Biotechnol 30(1):90–98. https://doi.org/10.1038/nbt.2057
12. Rao SS, Huntley MH, Durand NC, Stamenova EK, Bochkov ID, Robinson JT, Sanborn AL, Machol I, Omer AD, Lander ES, Aiden EL (2014) A 3D map of the human genome at Kilobase resolution reveals principles of chromatin looping. Cell 159:1665. https://doi.org/10.1016/j.cell.2014.11.021
13. Nagano T, Várnai C, Schoenfelder S, Javierre BM, Wingett SW, Fraser P (2015) Comparison of Hi-C results using in-solution versus in-nucleus ligation. Genome Biol 16(1):175. https://doi.org/10.1186/s13059-015-0753-7
14. Mifsud B, Tavares-Cadete F, Young AN, Sugar R, Schoenfelder S, Ferreira L, Wingett SW, Andrews S, Grey W, Ewels PA, Herman B (2015) Mapping long-range promoter contacts in human cells with high-resolution capture Hi-C. Nat Genet 47(6):598. https://doi.org/10.1038/ng.3286
15. Nagano T, Lubling Y, Stevens TJ, Schoenfelder S, Yaffe E, Dean W, Laue ED, Tanay A, Fraser P (2013) Single-cell Hi-C reveals cell-to-cell variability in chromosome structure. Nature 502(7469):59. https://doi.org/10.1038/nature12593
16. Ma W, Ay F, Lee C, Gulsoy G, Deng X, Cook S, Hesson J, Cavanaugh C, Ware CB, Krumm A, Shendure J (2018) Using DNase Hi-C techniques to map global and local three-dimensional genome architecture at high resolution. Methods 142:59–73. https://doi.org/10.1101/184846
17. Fullwood MJ, Ruan Y (2009) ChIP-based methods for the identification of long-range chromatin interactions. J Cell Biochem 107(1):30–39. https://doi.org/10.1002/jcb.22116
18. Sexton T, Yaffe E, Kenigsberg E, Bantignies F, Leblanc B, Hoichman M, Parrinello H, Tanay A, Cavalli G (2012) Three-dimensional folding and functional organization principles of the Drosophila genome. Cell 148(3):458–472. https://doi.org/10.1016/j.cell.2012.01.010
19. Burton JN, Adey A, Patwardhan RP, Qiu R, Kitzman JO, Shendure J (2013) Chromosome-scale scaffolding of de novo genome assemblies based on chromatin interactions. Nat Biotechnol 31(12):1119–1125. https://doi.org/10.1038/nbt.2727
20. Mascher M, Gundlach H, Himmelbach A, Beier S, Twardziok SO, Wicker T, Radchuk V, Dockter C, Hedley PE, Russell J, Bayer M (2017) A chromosome conformation capture ordered sequence of the barley genome. Nature 544(7651):427. https://doi.org/10.1038/nature22043
21. Hovel I, Louwers M, Stam M (2012) 3C technologies in plants. Methods 58:204–211. https://doi.org/10.1016/j.ymeth.2012.06.010
22. Haring M, Offermann S, Danker T, Horst I, Peterhansel C, Stam M (2007) Chromatin immunoprecipitation: optimization, quantitative analysis and data normalization. Plant Methods 3:11. https://doi.org/10.1186/1746-4811-3-11
23. Green MR, Sambrook J (2012) Molecular cloning. A laboratory manual. Cold Spring Harbor Laboratory, New York, p 1834
24. Green MR, Sambrook J (2012) Molecular cloning. A laboratory manual. Cold Spring Harbor Laboratory, New York, p 1823
25. Green MR, Sambrook J (2012) Molecular cloning. A laboratory manual. Cold Spring Harbor Laboratory, New York, p 1815

26. Green MR, Sambrook J (2012) Molecular cloning. A laboratory manual. Cold Spring Harbor Laboratory, New York, p 1824
27. Mascher M, Richmond TA, Gerhardt DJ, Himmelbach A, Clissold L, Sampath D, Ayling S, Steuernagel B, Pfeifer M, D'Ascenzo M, Akhunov ED, Hedley PE, Gonzales AM, Morrell PL, Kilian B, Blattner FR, Scholz U, Mayer KF, Flavell AJ, Muehlbauer GJ, Waugh R, Jeddeloh JA, Stein N (2013) Barley whole exome capture: a tool for genomic research in the genus Hordeum and beyond. Plant J 76(3):494–505
28. Martin M (2011) Cutadapt removes adapter sequences from high-throughput sequencing reads. EMBnet.journal 17(1):10–12 ISSN 2226-6089
29. Li H (2013) Aligning sequence reads, clone sequences and assembly contigs with BWA-MEM. arXiv:1303.3997v1 [q-bio.GN]
30. Li H, Handsaker B, Wysoker A, Fennell T, Ruan J, Homer N, Marth G, Abecasis G, Durbin R (2009) The sequence alignment/map format and SAMtools. Bioinformatics 25 (16):2078–2079. https://doi.org/10.1093/bioinformatics/btp352
31. Quinlan A, Hall I (2010) BEDTools: a flexible suite of utilities for comparing genomic features. Bioinformatics 26(6):841–842. https://doi.org/10.1093/bioinformatics/btq033
32. Lajoie BR, Dekker J, Kaplan N (2015) The Hitchhiker's guide to Hi-C analysis: practical guidelines. Methods 72:65–75
33. Lieberman-Aiden E, Van Berkum NL, Williams L, Imakaev M, Ragoczy T, Telling A, Amit I, Lajoie BR, Sabo PJ, Dorschner MO (2009) Comprehensive mapping of long-range interactions reveals folding principles of the human genome. Science 326:289–293
34. Dixon JR, Selvaraj S, Yue F, Kim A, Li Y, Shen Y, Hu M, Liu JS, Ren B (2012) Topological domains in mammalian genomes identified by analysis of chromatin interactions. Nature 485:376
35. Harewood L, Kishore K, Eldridge MD, Wingett S, Pearson D, Schoenfelder S, Collins VP, Fraser P (2017) Hi-C as a tool for precise detection and characterisation of chromosomal rearrangements and copy number variation in human tumours. Genome Biol 18:125
36. Burton JN, Adey A, Patwardhan RP, Qiu R, Kitzman JO, Shendure J (2013) Chromosome-scale scaffolding of de novo genome assemblies based on chromatin interactions. Nat Biotechnol 31:1119
37. Kaplan N, Dekker J (2013) High-throughput genome scaffolding from in vivo DNA interaction frequency. Nat Biotechnol 31:1143
38. Burton JN, Liachko I, Dunham MJ, Shendure J (2014) Species-level deconvolution of metagenome assemblies with Hi-C-based contact probability maps. G3 (Bethesda) 4: 1339–1346
39. James G (1978) Inactivation of the protease inhibitor phenylmethylsulfonyl fluoride in buffers. Anal Biochem 86:574–579
40. Liu C (2017) In situ Hi-C library preparation for plants to study their three-dimensional chromatin interactions on a genome-wide scale. Methods Mol Biol 1629:155–166
41. Belaghzal H, Dekker J, Gibcus JH (2017) Hi-C 2.0: an optimized hi-C procedure for high-resolution genome-wide mapping of chromosome conformation. Methods 123:56–65. https://doi.org/10.1016/j.ymeth.2017.04.004
42. Putney S, Benkovic S, Schimmel P (1981) A DNA fragment with an alpha-phosphorothioate nucleotide at one end is asymmetrically blocked from digestion by exonuclease III and can be replicated in vivo. Proc Natl Acad Sci U S A 78:7350–7354

Index

A

Abiotic stresses
- drought 9, 151, 156, 158, 173–185, 198, 202, 382, 435
- salt stress 155, 156, 158, 163, 173–185
- salt tolerance 9, 175, 179, 181

Analysis tools
- BEDtools 38, 52, 58, 63, 227, 234, 365, 370, 375, 378, 466
- BioPerl 248
- biopython 52, 63, 211
- BLAST 63, 178, 179, 233, 267, 269–271
- BLAST2GO 270, 274
- blastx 232, 246, 253, 254, 261, 263, 269–271, 398–400, 425
- bowtie 38, 42, 229, 260, 369, 370
- bowtie2 .... 68, 78, 83, 226, 365, 379
- COME 434
- cuffcompare 191, 258, 421
- cuffdiff 191, 227, 229, 231, 235
- Cufflinks 52, 68, 81, 83, 141, 189, 191, 226, 227, 229, 230, 235, 258, 266, 421
- CummeRbund 227, 229, 235
- cutadapt 365, 374, 466
- DESeq2 .... 210, 211, 215, 216, 218, 266
- edgeR 210, 211, 215, 216, 218, 266
- FastQC 68, 75, 78, 79, 191, 226–228, 253, 267, 272
- FASTX-Toolkit 226, 227, 267, 268
- FEATnotator 38, 43, 44
- featureCounts 209, 210, 215, 218
- FEELnc 246, 248, 252
- Galaxy bioinformatics server 266
- getorf 198, 199, 201, 267, 271
- gffcompare 201, 203, 247, 249
- gffread 83, 199, 201, 258, 421
- HISAT2 .... 198, 247, 249
- MediaWiki 401
- MySQL 401, 432, 433
- Notepad 269
- ORF Finder 263
- Pandas 211
- psMimic 199, 202
- psRobot 198, 199, 202, 203
- RepeatMasker 400
- RMBLAST 400
- RNAmmer 254
- RNA-Star 266
- RSEM 68, 83
- RStudio 227, 235
- SAMtools 38, 68, 78, 141, 199, 200, 203, 227, 239, 247, 249, 365, 376, 466
- SRA Toolkit 76, 210, 218, 365
- sratools 83
- STAR 209, 210, 212, 213, 218, 219, 418, 427
- StringTie 198, 203, 209, 210, 212–214, 218, 219, 247, 249, 258, 418
- TextEdit 269
- TopHat 68, 78, 81, 189, 191, 226, 229, 258, 369, 370, 374, 379
- TopHat2 .... 140, 227, 229–231, 241, 266, 365
- trim_galore 140, 253
- Trimmomatic 38, 42, 78–80, 83, 253, 258, 260
- Trinity 81, 266–268, 273, 274
- Triplexator 198, 199, 202
- tRNAscan-SE 254
- Tuxedo suite 234
- *txCdsPredict* 260, 261
- WebLogo 38, 44

Antisense 5, 14, 19–21, 115, 121, 122, 124, 125, 128, 141, 152–154, 156, 158–161, 163, 188, 224, 232, 241, 424, 432

Argonaute (AGO)
- AGO1 .... 156, 159
- AGO4 .... 34, 35, 157, 386

5-Azacytidine-mediated RNA immunoprecipitation (Aza-IP) 18

B

Bimolecular fluorescence complementation (BiFC) 297, 298, 300

5′-BrU immunoprecipitation chase-deep sequencing analysis (BRIC-seq) 6, 10, 12

Julia A. Chekanova and Hsiao-Lin V. Wang (eds.), *Plant Long Non-Coding RNAs: Methods and Protocols*, Methods in Molecular Biology, vol. 1933, https://doi.org/10.1007/978-1-4939-9045-0,

BrU labeling and sequencing (BrU-seq) 6, 10, 12
BrU pulse-chase and sequencing
(BrUChase-Seq) 6, 10, 12

C

Cap analysis of gene expression (CAGE)
CAGE-seq 6, 8, 10
nano-CAGE 8
5′ Cap structure 9, 159
CAP-trapper 8
Capture hybridization analysis of RNA targets
(CHART) 14, 20
Capture oligonucleotides (C-oligos) 20
Cell lines
Hela cells 12, 15, 364
mouse ES cells 19
Cell-specific expression 132
Cell type-specific gene expression data 209, 212
Chemical modification 313, 314, 316, 317, 381
Chromatin 1, 33, 49, 67, 131, 160, 174, 187, 197, 287, 293, 297, 381, 397, 416, 442
Chromatin immunoprecipitation (ChIP) 10, 18, 91, 138, 145, 290, 291, 293, 364, 443
Chromatin interactions 2, 13, 14, 19–20, 188, 441–470
Chromatin isolation by RNA purification (ChIRP)
ChIRP-MS 14, 19
sequencing (ChIRP/ChIRP-Seq) 14, 19, 20, 136–137, 142–144
Chromatin modifications 1, 33, 67, 160, 161, 297
Chromosome conformation capture
capture Hi-C 442
CHIA-PET 443
CHIP-loop 443
chromosome conformation capture
(3C) 442, 469
chromosome conformation capture
carbon copy (5C) 442
chromosome conformation capture on
chip (4C) 442
Hi-C 442–444, 461, 463, 464, 466, 468–470
in situ Hi-C 13, 441–470
single cell Hi-C 442, 443
tethered chromosome conformation capture
(TCC) 442
Coding potential calculator (CPC) 4, 198, 199, 201, 202, 227, 231, 232, 241, 261, 267, 399, 400, 421, 434
Correlation analysis
co-expression analysis 198, 207–220, 225, 226, 234
correlation coefficients 234, 236, 237
expression correlation 236
negative correlation 225, 235
Pearson correlation coefficient (PCCs) 202, 217
positive co-expression 236
positively correlated co-expression 225
CRISPR technology
CRISPR 132, 192, 193
CRISPR/CAS9-mediated deletion/over
expression 239
CRISPR/Cas9-mediated genome editing
method 192
sgRNAs 193
Crosslinking
chemical cross-linking 21
UV cross-linking 18, 20, 21
Cross-linking immunoprecipitation (CLIP)
high-throughput sequencing cross-linking
immunoprecipitation (HITS-CLIP
/CLIP-seq) 14, 18
photoactivatable ribonucleotide-enhanced
cross-linking and immunoprecipitation
(PAR-CLIP) 18
Crosslinking, ligation and sequencing of
hybrids (CLASH) 15, 20, 21
Cryo-dissection
cryosectioning 13, 14, 53, 54
Cryo-EM 382
C-terminal domain (CTD) 35

D

Databases
CANTATAdb 1.0 .... 415, 427
CANTATAdb 2.0 .... 3, 415–428
Encyclopaedia of DNA Elements
(ENCODE) 397
ENSEMBL 415
Ensembl plants 75, 78, 218, 254, 369, 416, 417, 419, 421, 426, 427
EVLncRNAs 3, 4, 431–436
GreeNC 3, 4, 397–411, 415, 428, 432
Lnc2Cancer 432, 433
LncRANDisease 433
lncRInter 432
lncRNAdb 433
miRBase 404
NONCODE 4, 404, 415, 426, 428
Phytozome 45, 225, 254, 259, 369, 379, 398
plant long noncoding RNA database
(PLncDB) 4, 191, 198, 199, 202
Plant Natural Antisense Transcripts DataBase
(PlantNATsDB) 5, 432
plant ncRNA database (PNRD) 4, 432

PLNIncRbase 433
PLNlncRbase 428, 432, 433
RepBase 400, 404
Rfam 272, 399, 404, 421
Swiss-Prot 141, 267, 269, 399, 404, 424–426
The Arabidopsis Information Resource (TAIR) 4, 258, 259
Universal Protein Resource Knowledgebase (*UniProtKB*) database 261, 263, 267, 269
de Bruijn graphs 266
Decapping complex 159
Degradome-seq 9
Deoxy ribonuclease (DNase) 17, 19, 52, 59, 93, 95, 96, 132, 136–139, 144, 176, 181, 182, 281, 284, 292, 294, 295, 347, 350, 443, 446
Dicer-like (DCL) proteins
DCL2 .... 36, 159
dcl2/3/4 (*dcl2-5 dcl3-1 dcl4-2*) triple mutants
nrpd1-3 dcl2-5 dcl3-1 dcl4-2 (*pol IV dcl2/3/4*, quadruple mutant) 37
rdr2-1 dcl2-5 dcl3-1 dcl4-2 (*rdr2 dcl2/3/4*, quadruple mutant) 37
dcl2-5 37
dcl2-5 dcl3-1 37
dcl2-5 dcl4-2 37
dcl3 .... 34, 157
DCL3 .... 34–37, 157
dcl3-1 37
dcl3-1 dcl4-2 37
DCL3a 37
DCL4 .... 36, 159
dcl4-2 37
Dimethyl sulfide (DMS)
DMS modification 309, 333, 335
DNA methylation
CG 34
CHG 34
CHH 34, 157
Double-stranded RNA (dsRNA)
double-stranded RNA (dsRNA)-seq 364, 376

E

Enhancer
enhancer-associated RNAs 188
enhancer trap 16
Epigenetic 21, 157, 160, 164, 197, 381–383, 441
Epigenome 13
Exosome, the RNA exosome complex
RRP4 159
RRP40 159
RRP41 159
RRP42 159
RRP43 159
RRP45 159
RRP46 159
RRP6L1 159
RRP6L2
Expressed sequence tags (ESTs) 4, 72, 81, 202, 245

F

File formats
BED 214, 219, 364, 370, 378, 466
bigWig 366, 377, 378, 380
binary alignment/map (BAM) format 43, 78, 249, 370–377, 418, 466
FASTA 38, 58, 62, 75, 199, 214, 240, 247, 261, 374, 379, 401, 410, 421, 426
FASTQ 40, 42, 76, 77, 79, 135, 145, 227, 230, 370, 371, 374, 418, 427
general feature format (GFF, GFF3) 38, 44, 199, 246, 247, 253, 370, 416, 418, 427
general transfer format (GTF) 203, 209, 212–215, 218–220, 246–249, 416, 418, 421, 426, 427
sequence alignment/map (SAM) files 43
sequence read archive (SRA) 40, 74, 210, 212, 225, 365, 416, 422, 427
Flow cytometry 16
FLOWERING LOCUS
FLOWERING LOCUS C (FLC) 21, 68, 160, 161, 383, 398
FLOWERING LOCUS D (FLD) 161
Fluorescence-activated cell sorting (FACS) 13, 14, 16, 89, 100
Fluorescence in situ hybridization (FISH) 14, 16, 183, 184
Fluorescent in situ RNA sequencing (FISSEQ) 16
Folding energies 4, 398, 404
Functional analysis 9, 131–146, 344

G

GC content 193, 404
Generic run-on assays and global run-on sequencing assay (GRO-seq)
GRO-cap 11
5′ GRO-seq 11
Gene silencing
posttranscriptional gene silencing (PTGS) 159, 302
siRNA-mediated gene silencing 224
transcriptional gene silencing (TGS) 154, 159

Genome architectures 13, 187
Genome-wide contact matrix 443
Genome-wide mapping of uncapped and cleaved transcripts (GMUCT) 6, 9
Genomic imprinting 67, 187
Green fluorescent protein (GFP) 16, 89, 134, 139, 281, 283, 298

H

HEN1 .... 34, 35
HIDDEN TREASURE 1 (HID1) 162, 344
High-throughput era 1–23
High-throughput methods 3, 6, 21–23
High-throughput sequencing (HTS) 2, 3, 8, 11, 14–18, 22, 144, 198, 207, 389, 393
Hybridization-based approaches 3–5

I

Illumina NextSeq 134, 140
In situ hybridization (ISH) 3, 14, 16, 23, 50, 52–53, 99–128, 175, 183
Isolation of nuclei tagged in specific cell types (INTACT) 13, 14, 16, 132, 137–139, 350

L

Laser capture microdissection (LCM) 13, 15, 50, 53, 89–97, 100, 122
Laser microdissection (LM) 13–16
lncRNA identification 2, 3, 9, 12, 17, 49–63, 72–83, 197–203, 223–241, 245–254, 279–287, 398–400, 421, 424, 427
Long-range chromatin interactions 13, 441–470

M

Massively parallel signature sequencing (MPSS) 4, 6–8
Medium throughput RNA in situ hybridization 16, 99–128
Microarrays
  custom microarrays
    100 ATH 1.0R arrays 3
    Affymetrix ATH 1.0F arrays 3
    ATH lincRNAv1 array 5
  paraffin-embedded tissue microarrays 99–128
  tiling array (tiling microarray) 3, 6, 152
Micrococcal nuclease (MNase) 443
MicroRNAs (miRNAs)
  miR160 .... 36, 38
  miR168 156
  MIR168a 156
  miR173 156
  miR319a 163
  miR390 .... 92, 156
  miR393 156
  miR395 156
  miR398 155
  miR399 .... 156, 397
  miR400 155
  miR828 156
  miRbase 404
MiSeq 145, 313, 332

N

Nascent RNAs 6, 10–12
Nascent RNA sequencing 10–12
Nascent transcription 10
Nascent transcripts 10, 11, 34
Native elongating transcript sequencing (NET-seq)
  mammalian NET-seq (mNET-seq) 12
Non-protein-coding RNAs
  circular RNA (circRNAs) 163, 257
  competing endogenous RNAs (ceRNAs) 163
  housekeeping ncRNAs
    small nuclear RNA (snRNAs) 20, 43, 151, 153, 224
    small nucleolar RNAs (snoRNAs) 21, 43, 151, 153, 344
    transfer RNA (tRNA) 23, 43, 60, 151, 153, 224, 246, 252, 254, 263, 344, 374
  long noncoding RNAs (lncRNAs)
    functional characterization of lncRNAs 2, 132, 164
    functional study of lncRNAs 224, 297
    function of lncRNAs 2, 12–23, 50, 74, 131, 152–164, 173–185, 197–203, 279, 344, 364, 416, 428, 435
    intergenic lncRNAs 16, 162, 224
    intronic lncRNAs 224
    long intergenic noncoding RNAs (lincRNAs) 4, 5, 152, 154, 156, 161, 162, 187, 188, 192, 193, 197, 201, 202, 207–220, 424
    overexpression of the lncRNA 13, 132, 134, 146, 154
    viroids (sub-viral plant-pathogenic lncRNAs) 163
  metazoan lncRNAs
    Braveheart 385
    ciRS-7 163
    HOTAIR 19, 23, 385
    Malat1 20
    NEAT1 385
    P21-lncRNA 385
    roX2 20
    SRA-1 385
    Xist 17, 19, 20, 385

natural antisense transcripts (NATs)
cis-natural antisense RNAs (cis-NATs) 158
long noncoding natural antisense transcripts (lnc-NATs/lncNATs) 197, 201–203, 224, 241, 245
plant lncRNAs
APOLO 162
ASCO 161, 162, 344, 416
ASL 161, 164, 398
AtR18 155
COLDAIR 68, 160, 161, 383, 398
COLDWRAP 160, 161
COOLAIR 21, 68, 160, 161, 164, 383, 385, 398
ELENA1 .... 281, 298, 300
Enod40 .... 161, 162, 398, 434
intermediate-sized ncRNAs (im-ncRNAs) 153
IPS1 .... 156, 163, 397
LDMAR 398
MtENOD40 152
OsHID1 162
OsPI1 152
TPS11 152
ribosomal RNA (rRNAs)
5S rRNA 36, 41
45S rRNA 42, 43
Nonsense-mediated decay (NMD)
XRN2 159
XRN3 159
XRN4 159
Northern blotting analysis 3, 34, 433
Nuclear domain organization 441, 443
Nuclease protection assays (NPA) 3
Nuclei in specific cell types (INTACT) 100, 350
Nuclei purification 134–135

O

Open reading frame (ORF) 9, 58, 83, 141, 153, 162, 199, 214, 224, 232, 260, 261, 263, 271, 398, 404, 421, 424, 428
Organisms
Agrobacterium tumefaciens 141, 142, 178
Amborella trichopoda 417, 419, 422
Ananas comosus 416, 417, 419, 423
Arabidopsis lyrata 417, 419, 423
Arabidopsis thaliana (*see* Arabidopsis)
Chenopodium quinoa 416, 417, 419
Chlamydomonas reinhardtii 417, 419, 422
Chondrus crispus 417, 419, 422
Corchorus capsularis 417, 419, 422
Cucumis sativus 416, 417, 419, 422
Drosophila 20, 67
Fragaria vesca (*see* Strawberry)
Galdieria sulphuraria 417, 419, 423
Glycine max (*see* Soybean)
Hordeum vulgare (*see* Barley)
human 309, 369
Leersia perrieri 417, 419, 423
Magnoliophyta 257–263
Malus domestica 416, 417, 420, 422
Manihot esculenta 416, 417, 420, 422
Medicago truncatula 245, 398, 417, 420, 423, 434
mouse 369
Musa acuminata 417, 420, 422
Nicotiana tabacum 158
Oryza barthii 417, 420, 423
Oryza brachyantha 417, 420, 422
Oryza nivara 417, 420
Oryza punctata 417, 420, 423
Oryza rufipogon 417, 420, 423
Oryza sativa 37, 416, 417, 420, 423
Physcomitrella patens 37, 417, 420
Phytophthora infestans 188
Populus trichocarpa 417, 420, 422
Prunus persica 417, 420, 423
Saccharomyces cerevisiae 9, 12, 132
Schizosaccharomyces pombe 132
Selaginella moellendorffii 417, 420, 423
Setaria italica 418, 420, 422
Solanum lycopersicum 188, 418, 420, 422
Solanum tuberosum 418, 420, 422
Sorghum bicolor 418, 421, 423
Theobroma cacao 418, 421, 423
Trifolium pratense 418, 421, 423
Verticillium dahlia 188
Vitis vinifera 418, 421, 423
wheat 7
Zea mays (*see* Maize)

P

Paired-end analysis of transcription start sites (PEAT) 8
Parallel analysis of RNA-ends (PARE) 6, 9
Plant nuclear extracts 279–287
Plant tissues and organs
aleurone cell layer (AL) 54
basal endosperm transfer cell layer (BETL) 54
embryo sacs 69, 70, 72, 83
endosperm 13, 49–63, 132
florets 69, 70, 83
gametophyte 67–84
glumes 70, 83
lemma 83
ovule 69–72, 83

Plant tissues and organs (*cont.*)
palea 83
pollen 68–70, 72, 238
shoot apical meristem (SAM) 90, 92, 94, 95
starchy endosperm (SE) 54
Polyadenylations
poly-A 224
poly A+ RNA 192, 193
poly-A tail 224
poly(A) tail 6, 9, 159
Polycomb repressive complex 2 (PRC2) 160–162, 383, 398
Post translational histone modifications
H3K9ac 153
H3K4me3 153
H3K27me3 .... 160–162, 381
H3K36me 160
Precision nuclear run-on sequencing (PRO-seq)
PRO-cap 11
Programming languages
Perl 38, 75, 81, 248, 258, 271, 365, 366, 378, 466
Python 52, 58, 209, 211, 214, 217, 218, 267, 268, 365, 366, 369, 370, 374, 379, 421, 432, 433
R 38, 44, 198, 199, 209, 211, 216, 217, 220, 226, 227, 229, 235–237, 365, 369, 374, 421
UNIX shell 52, 209, 226, 267, 379, 466
Promoter
promoter-GFP fusions 16

R

Rapid amplification of cDNA ends (RACE) 188, 189, 191–193
Reactive oxygen species (ROS) 174, 178, 183
Regulation
cis-regulation 258
post-transcriptional regulation of gene expression 187
transcriptional regulation of gene expression 187
R-loop 161
RNA antisense purification (RAP)
RNA antisense purification followed by RNA-sequencing (RAP-RNA) 15, 19, 20
RNA binding proteins (RBPs) 17, 18, 134, 161, 162, 289, 290, 298, 343, 344, 363
RNA decoys 162, 163
RNA-dependent RNA polymerase
rdr2-1
rdr2-1 dcl2-5 dcl3-1 dcl4-2 (*rdr2 dcl2/3/4*, Quadruple mutant) 37
RDR6 .... 156, 159
RNA-dependent RNA polymerase 2 (RDR2) 34–36
RNA-directed DNA methylation (RdDM) 4, 9, 34, 35, 157, 386
RNA-DNA hybrids 17, 19, 162
RNA extraction 16, 50–55, 132, 135, 136, 138–139, 145, 189, 190, 193, 306, 309, 311, 314–318, 333, 335
RNA FISH 16, 19
RNA half-life 11
RNA immunoprecipitation
RIP smRNA-seq 17
RNA immunoprecipitation (RIP) 14, 389–394
RNA immunoprecipitation followed by sequencing (RIP-seq) 14, 17–18
RNA-induced silencing complex (RISC) 159
RNA in situ hybridization 16, 99–128, 175
RNA interactions
in vivo visualization of RNA-Protein interaction 297
protein interaction profiling (PIP)
protein interaction profile sequencing (PIP-seq) 15, 22, 343–360, 363–380
protein-lncRNA interaction 289–295
protein protected sites (PPSs) 346, 352, 366–368, 372, 374–375
RNA-protein interaction 2, 14, 15, 17–19, 279–287, 297, 343–360
RNA-RNA interactions 2, 15, 20–21, 152
RNA interference (RNAi)
microRNAs (miRNAs)
miR160 .... 36, 38
miR168 156
MIR168a 156
miR173 156
miR319a 163
miR390 .... 92, 156
miR393 156
miR395 156
miR398 155
miR399 .... 156, 163, 397
miR400 155
miR828 156
miRbase 199, 203, 399, 404
small interfering RNAs (siRNA)
siRNA-mediated gene silencing 224
small RNA 34, 45, 153
small RNA library 35, 42, 45
small RNA pathways 187
trans-acting siRNAs
TAS1 156
TAS1-siRNA 156
TAS2 156

TAS3 ... 156
ta-siRNAs ... 156
TAS3 ta-siRNA ... 156
RNA metabolism ... 152
RNA modifications
5-hydroxymethylcytosine (hm^{5}C) ... 389
5-methylcytosine (m^{5}C) ... 17, 389–394
inosine ... 389
m^{5}A-seq ... 18
m^{5}C-RIP-seq ... 14, 18, 389, 390
N^1-methyladenosine (m^{1}A) ... 389
N^6-methyladenosine (m^{6}A) ... 17, 389
pseudouridine ... 389
RNA bisulfite conversion with sequencing (bsRNA-seq) ... 18
RNA polymerases
RNA polymerase II (Pol II) ... 10, 34
RNA polymerase III ... 155, 257
RNA polymerases IV (Pol IV)
nrpd1-3 ... 37, 40
nrpd1-3 dcl2-5 dcl3-1 dcl4-2 (*pol IV dcl2/3/4,* Quadruple mutant) ... 37
nrpd2 ... 157
Pol IV/RDR2-dependent RNAs (P4R2 RNAs) ... 34, 36–41, 43, 45
RNA polymerases V (Pol V)
nrpd2 ... 157
NRPD2 ... 157
RNA processing machinery ... 187
RNA secondary structure
fragmentation sequencing (Frag-seq) ... 15, 23
parallel analysis of RNA structure (PARS) ... 15, 23
protein interaction profiling (PIP)
protein interaction profile sequencing (PIP-seq) ... 15, 22, 343–360, 363–380
RNA structure probing ... 23, 305–340
selective 2′-hydroxyl acylation by primer extension (SHAPE)
selective 2′-hydroxyl acylation by primer extension sequencing (SHAPE-seq) ... 15, 21, 22
shotgun secondary structure (3S) ... 385, 386
structure-seq
StructureFold2 ... 22, 307, 309, 332, 336
structure-seq2 ... 22, 305–340
structure-specific nucleases ... 346, 352
2-D structure ... 382
RNase III ... 34

S

Salicylic acid (SA) ... 155, 176, 182
Serial analysis of gene expression (SAGE) ... 4, 6–8
Single-hit kinetics (SHK) ... 305–307, 311, 313–316, 333–335, 339
Single-molecule, real-time (SMRT) sequencing technology
Oxford Nanopore ... 8, 9
Pacific Biosciences (PacBio) ... 8, 9
single-molecule-based sequencing technology ... 9
Small interfering RNAs (siRNA)
siRNA-mediated gene silencing ... 224
Spatiotemporal resolution ... 99
Splicing
alternative splicing ... 17, 155, 161, 162, 174, 416
Stem-loop RT-PCR ... 15, 90, 92, 96
Strand-specific RNA-sequencing
double-stranded RNA (dsRNA)-seq ... 345, 346, 351, 364, 376
single-stranded RNA (ssRNA)-seq ... 345, 346, 351, 364, 376

T

Tag-based methods ... 2–4, 7
Tandem repeats
trans-acting siRNAs
TAS1 ... 156
TAS1-siRNA ... 156
TAS2 ... 156
TAS3 ... 156
TAS3 ta-siRNA ... 156
ta-siRNAs ... 156
T-DNA, *see Agrobacterium tumefaciens*
3D-nuclei space ... 443
Tissue or cell-type specific analysis ... 2, 13–16
Tissue-specific ... 3, 13, 90, 100, 132, 152, 237, 238, 410
Transcript
quantification ... 266
visualization ... 83, 225
Transcriptional activation ... 67
Transcriptional interference ... 67
Transcript isoform sequencing (TIF-seq) ... 6, 9
Transcriptome
tissue-specific transcriptomics ... 100
transcriptome analysis ... 6, 50, 176, 181, 188–190
transcriptome assembly
de novo transcriptome assembly ... 266–268
reference-based transcriptome assembly ... 246, 266
Transposable elements (TEs) ... 34, 35, 57, 58, 62, 63, 157, 211, 214, 215, 312, 322, 326, 330, 331, 337
Trimolecular fluorescence complementation (TriFC) ... 297–302

V

Visualization tools
IGV 83, 203, 227, 238–240
JBrowse 36, 43, 410
UCSC genome browser 260, 369

X

X chromosome
inactivation 17
silencing 19, 67
Xist 19
X-ray crystallography 382

Printed by Printforce, the Netherlands